Annals of Mathematics Studies

Number 213

# The Einstein-Klein-Gordon Coupled System: Global Stability of the Minkowski Solution

Alexandru D. Ionescu
and Benoît Pausader

PRINCETON UNIVERSITY PRESS
PRINCETON AND OXFORD
2022

Published by Princeton University Press,
41 William Street, Princeton, New Jersey 08540
99 Banbury Road, Oxford OX2 6JX

press.princeton.edu



ISBN 978-0-691-23305-5
ISBN (pbk.) 978-0-691-23304-8
ISBN (e-book) 978-0-691-23303-1

British Library Cataloging-in-Publication Data is available

Editorial: Susannah Shoemaker, Diana Gillooly, and Kristen Hop
Production Editorial: Nathan Carr and Kathleen Cioffi
Production: Jacqueline Poirier
Publicity: Matthew Taylor and Charlotte Coyne
Copyeditor: Bhisham Bherwani

The publisher would like to acknowledge the authors of this volume for providing the camera-ready copy from which this book was printed.

This book has been composed in LaTeX

10 9 8 7 6 5 4 3 2 1

From A. Ionescu: To my daughters, Amelie and Catherine

From B. Pausader: To Katie

# Contents

# Chapter One

## Introduction

### 1.1 THE EINSTEIN-KLEIN-GORDON COUPLED SYSTEM

The Einstein field equations of General Relativity are a covariant geometric system that connect the Ricci tensor of a Lorentzian metric $\mathbf{g}$ on a manifold $M$ to the energy-momentum tensor of the matter fields in the spacetime, according to the equation

$$\mathbf{G}_{\alpha\beta} = 8\pi T_{\alpha\beta}. \tag{1.1.1}$$

Here $\mathbf{G}_{\alpha\beta} = \mathbf{R}_{\alpha\beta} - (1/2)\mathbf{R}\mathbf{g}_{\alpha\beta}$ is the Einstein tensor, where $\mathbf{R}_{\alpha\beta}$ is the Ricci tensor, $\mathbf{R}$ is the scalar curvature, and $T_{\alpha\beta}$ is the energy-momentum tensor of the matter in the spacetime.

In this monograph we are concerned with the Einstein-Klein-Gordon coupled system, which describes the coupled evolution of an unknown Lorentzian metric $\mathbf{g}$ and a massive scalar field $\psi$. In this case the associated energy momentum tensor $T_{\alpha\beta}$ is given by

$$T_{\alpha\beta} := \mathbf{D}_\alpha\psi\mathbf{D}_\beta\psi - \frac{1}{2}\mathbf{g}_{\alpha\beta}(\mathbf{D}_\mu\psi\mathbf{D}^\mu\psi + \psi^2), \tag{1.1.2}$$

where $\mathbf{D}$ denotes covariant derivatives.

Our goal is to prove definitive results on the global stability of the flat space among solutions of the Einstein-Klein-Gordon system. Our main theorems in this monograph include:

(1) A proof of global regularity (in wave coordinates) of solutions of the Einstein-Klein-Gordon coupled system, in the case of small, smooth, and localized perturbations of the stationary Minkowski solution $(\mathbf{g}, \psi) = (m, 0)$;

(2) Precise asymptotics of the metric components and the Klein-Gordon field as the time goes to infinity, including the construction of modified (nonlinear) scattering profiles and quantitative bounds for convergence;

(3) Classical estimates on the solutions at null and timelike infinity, such as bounds on the metric components, weak peeling estimates of the Riemann curvature tensor, ADM and Bondi energy identities and estimates, and asymptotic description of null and timelike geodesics.

The general plan is to work in a standard gauge (the classical wave coordinates) and transform the geometric Einstein-Klein-Gordon system into a coupled system of quasilinear wave and Klein-Gordon equations. We then analyze this system in a framework inspired by the recent advances in the global existence theory for quasilinear dispersive models, such as plasma models and water waves.

More precisely, we rely on a combination of energy estimates and Fourier analysis. At a very general level one should think that energy estimates are used, in combination with vector-fields, to control high regularity norms of the solutions. The Fourier analysis is used, mostly in connection with normal forms, analysis of resonant sets, and a special norm, to prove dispersion and decay in lower regularity norms.

The method we present here incorporates Fourier analysis in a critical way. Its main advantage over the classical physical space methods is the ability to identify clearly resonant and non-resonant nonlinear quadratic interactions. We can then use normal forms to dispose of the non-resonant interactions, and focus our attention on a small number of resonant quadratic interactions. This leads to very precise estimates.

In particular, some of our asymptotic results appear to be new even in the much-studied case of the Einstein-vacuum equations (corresponding to $\psi = 0$) mainly because we allow a large class of non-isotropic perturbations. Indeed, our assumptions on the metric on the initial slice are weak, essentially of the type

$$\mathbf{g}_{\alpha\beta} = m_{\alpha\beta} + \varepsilon_0 O(\langle x\rangle^{-1+}), \qquad \partial\mathbf{g}_{\alpha\beta} = \varepsilon_0 O(\langle x\rangle^{-2+}).$$

These assumptions are consistent with non-isotropic decay, in the sense that we do not assume that the metric has radial decay of the form $M/r$ up to lower order terms. Even with these weaker assumptions we are still able to derive suitable asymptotics of the spacetime, such as weak peeling estimates for the Riemann tensor, and construct a Bondi energy function.

### 1.1.1 Wave Coordinates and PDE Formulation of the Problem

The system of equations (1.1.1)–(1.1.2) is a geometric system, written in covariant form. To analyze it quantitatively and state our main theorems we need to fix a system of coordinates and reformulate our problem as a PDE problem.

We start by recalling some of the basic definitions and formulas of Lorentzian geometry. At this stage, all the formulas are completely analogous to the Riemannian case, hold in any dimension, and the computations can be performed in local coordinates. A standard reference is the book of Wald [73]. Assume $\mathbf{g}$ is a sufficiently smooth Lorentzian metric in a 4 dimensional open set $O$. We assume that we are working in a system of coordinates $x^0, x^1, x^2, x^3$ in $O$. We define the connection coefficients $\mathbf{\Gamma}$ and the covariant derivative $\mathbf{D}$ by

$$\mathbf{\Gamma}_{\mu\alpha\beta} := \mathbf{g}(\partial_\mu, \mathbf{D}_{\partial_\beta}\partial_\alpha) = \frac{1}{2}(\partial_\alpha \mathbf{g}_{\beta\mu} + \partial_\beta \mathbf{g}_{\alpha\mu} - \partial_\mu \mathbf{g}_{\alpha\beta}), \tag{1.1.3}$$

where $\partial_\mu := \partial_{x^\mu}$, $\mu \in \{0, 1, 2, 3\}$. Thus

$$\mathbf{D}_{\partial_\alpha}\partial_\beta = \mathbf{D}_{\partial_\beta}\partial_\alpha = \mathbf{\Gamma}^\nu{}_{\alpha\beta}\partial_\nu, \qquad \mathbf{\Gamma}^\nu{}_{\alpha\beta} := \mathbf{g}^{\mu\nu}\mathbf{\Gamma}_{\mu\alpha\beta}, \tag{1.1.4}$$

where $\mathbf{g}^{\alpha\beta}$ is the inverse of the matrix $\mathbf{g}_{\alpha\beta}$, i.e., $\mathbf{g}^{\alpha\beta}\mathbf{g}_{\mu\beta} = \delta^\alpha_\mu$. For $\mu, \nu \in$

$\{0,1,2,3\}$ let

$$\boldsymbol{\Gamma}_\mu := \mathbf{g}^{\alpha\beta}\boldsymbol{\Gamma}_{\mu\alpha\beta} = \mathbf{g}^{\alpha\beta}\partial_\alpha \mathbf{g}_{\beta\mu} - \frac{1}{2}\mathbf{g}^{\alpha\beta}\partial_\mu \mathbf{g}_{\alpha\beta}, \qquad \boldsymbol{\Gamma}^\nu := \mathbf{g}^{\mu\nu}\boldsymbol{\Gamma}_\mu. \tag{1.1.5}$$

We record also the useful general identity

$$\partial_\alpha \mathbf{g}^{\mu\nu} = -\mathbf{g}^{\mu\rho}\mathbf{g}^{\nu\lambda}\partial_\alpha \mathbf{g}_{\rho\lambda}, \tag{1.1.6}$$

and the Jacobi formula

$$\partial_\alpha(\log|\mathbf{g}|) = \mathbf{g}^{\mu\nu}\partial_\alpha \mathbf{g}_{\mu\nu}, \qquad \alpha \in \{0,1,2,3\}, \tag{1.1.7}$$

where $|\mathbf{g}|$ denotes the determinant of the matrix $\mathbf{g}_{\alpha\beta}$ in local coordinates.

Covariant derivatives can be calculated in local coordinates according to the general formula

$$\mathbf{D}_\alpha T_{\beta_1\dots\beta_n} = \partial_\alpha T_{\beta_1\dots\beta_n} - \sum_{j=1}^n \boldsymbol{\Gamma}^\nu{}_{\alpha\beta_j} T_{\beta_1\dots\nu\dots\beta_n}, \tag{1.1.8}$$

for any covariant tensor $T$. In particular, for any scalar function $f$

$$\square_{\mathbf{g}} f = \mathbf{g}^{\alpha\beta}\mathbf{D}_\alpha\mathbf{D}_\beta f = \widetilde{\square}_{\mathbf{g}} f - \boldsymbol{\Gamma}^\nu\partial_\nu f, \tag{1.1.9}$$

where $\widetilde{\square}_{\mathbf{g}} := \mathbf{g}^{\alpha\beta}\partial_\alpha\partial_\beta$ denotes the reduced wave operator.

The Riemann curvature tensor measures commutation of covariant derivatives according to the covariant formula

$$\mathbf{D}_\alpha\mathbf{D}_\beta\omega_\mu - \mathbf{D}_\beta\mathbf{D}_\alpha\omega_\mu = \mathbf{R}_{\alpha\beta\mu}{}^\nu\omega_\nu, \tag{1.1.10}$$

for any form $\omega$. The Riemann tensor $\mathbf{R}$ satisfies the symmetry properties

$$\begin{aligned} &\mathbf{R}_{\alpha\beta\mu\nu} = -\mathbf{R}_{\beta\alpha\mu\nu} = -\mathbf{R}_{\alpha\beta\nu\mu} = \mathbf{R}_{\mu\nu\alpha\beta}, \\ &\mathbf{R}_{\alpha\beta\mu\nu} + \mathbf{R}_{\beta\mu\alpha\nu} + \mathbf{R}_{\mu\alpha\beta\nu} = 0, \end{aligned} \tag{1.1.11}$$

and the covariant Bianchi identities

$$\mathbf{D}_\rho\mathbf{R}_{\alpha\beta\mu\nu} + \mathbf{D}_\alpha\mathbf{R}_{\beta\rho\mu\nu} + \mathbf{D}_\beta\mathbf{R}_{\rho\alpha\mu\nu} = 0. \tag{1.1.12}$$

Its components can be calculated in local coordinates in terms of the connection coefficients according to the formula

$$\mathbf{R}_{\alpha\beta\mu}{}^\rho = -\partial_\alpha\boldsymbol{\Gamma}^\rho{}_{\beta\mu} + \partial_\beta\boldsymbol{\Gamma}^\rho{}_{\alpha\mu} - \boldsymbol{\Gamma}^\rho{}_{\alpha\nu}\boldsymbol{\Gamma}^\nu{}_{\beta\mu} + \boldsymbol{\Gamma}^\rho{}_{\beta\nu}\boldsymbol{\Gamma}^\nu{}_{\alpha\mu}. \tag{1.1.13}$$

Therefore, the Ricci tensor $\mathbf{R}_{\alpha\mu} = g^{\beta\rho}\mathbf{R}_{\alpha\beta\mu\rho}$ is given by the formula

$$\mathbf{R}_{\alpha\mu} = -\partial_\alpha\boldsymbol{\Gamma}^\rho{}_{\rho\mu} + \partial_\rho\boldsymbol{\Gamma}^\rho{}_{\alpha\mu} - \boldsymbol{\Gamma}^\rho{}_{\nu\alpha}\boldsymbol{\Gamma}^\nu{}_{\rho\mu} + \boldsymbol{\Gamma}^\rho{}_{\rho\nu}\boldsymbol{\Gamma}^\nu{}_{\alpha\mu}.$$

Simple calculations using (1.1.3) and (1.1.5) show that the Ricci tensor is given by

$$2\mathbf{R}_{\alpha\mu} = -\widetilde{\square}_{\mathbf{g}}\mathbf{g}_{\alpha\mu} + \partial_\alpha\mathbf{\Gamma}_\mu + \partial_\mu\mathbf{\Gamma}_\alpha + F^{\geq 2}_{\alpha\mu}(g, \partial g), \tag{1.1.14}$$

where $F^{\geq 2}_{\alpha\beta}(g, \partial g)$ is a quadratic semilinear expression,

$$\begin{aligned} F^{\geq 2}_{\alpha\beta}(g, \partial g) = \frac{1}{2}\mathbf{g}^{\rho\mu}\mathbf{g}^{\nu\lambda}\{\partial_\nu\mathbf{g}_{\rho\mu}\partial_\beta\mathbf{g}_{\alpha\lambda} + \partial_\nu\mathbf{g}_{\rho\mu}\partial_\alpha\mathbf{g}_{\beta\lambda} - \partial_\nu\mathbf{g}_{\rho\mu}\partial_\lambda\mathbf{g}_{\alpha\beta}\} \\ + \mathbf{g}^{\rho\mu}\mathbf{g}^{\nu\lambda}\{ - \partial_\rho\mathbf{g}_{\mu\lambda}\partial_\alpha\mathbf{g}_{\beta\nu} - \partial_\rho\mathbf{g}_{\mu\lambda}\partial_\beta\mathbf{g}_{\alpha\nu} \\ + \partial_\rho\mathbf{g}_{\mu\lambda}\partial_\nu\mathbf{g}_{\alpha\beta} + \partial_\alpha\mathbf{g}_{\rho\lambda}\partial_\mu\mathbf{g}_{\beta\nu} + \partial_\beta\mathbf{g}_{\rho\lambda}\partial_\mu\mathbf{g}_{\alpha\nu}\} \\ - \frac{1}{2}\mathbf{g}^{\rho\mu}\mathbf{g}^{\nu\lambda}(\partial_\alpha\mathbf{g}_{\nu\mu} + \partial_\nu\mathbf{g}_{\alpha\mu} - \partial_\mu\mathbf{g}_{\alpha\nu})(\partial_\beta\mathbf{g}_{\rho\lambda} + \partial_\rho\mathbf{g}_{\beta\lambda} - \partial_\lambda\mathbf{g}_{\beta\rho}). \end{aligned} \tag{1.1.15}$$

We consider the Einstein field equations (1.1.1)–(1.1.2) for an unknown spacetime $(M, \mathbf{g})$; for simplicity, we drop the factor of $8\pi$ from the energy-momentum tensor. The covariant Bianchi identities $\mathbf{D}^\alpha\mathbf{G}_{\alpha\beta} = 0$ can be used to derive an evolution equation for the massive scalar field $\psi$. The equation is

$$\square_{\mathbf{g}}\psi - \psi = 0. \tag{1.1.16}$$

Therefore the main unknowns in the problem are the metric tensor $\mathbf{g}$ and the scalar field $\psi$, which satisfy the covariant coupled equations (1.1.1) and (1.1.16).

To construct solutions we need to fix a system of coordinates. In this paper we work in *wave coordinates*, which is the condition

$$\mathbf{\Gamma}^\alpha = -\square_{\mathbf{g}}x^\alpha \equiv 0 \quad \text{for} \quad \alpha \in \{0, 1, 2, 3\}. \tag{1.1.17}$$

Wave coordinates are known to be a good system of coordinates to prove global stability at least in the Einstein-vacuum equations due to the work of Lindblad-Rodnianski [63]. Our construction of global solutions of the Einstein-Klein-Gordon system is based on the following proposition, which can be proved by straightforward calculations.

**Proposition 1.1.** *Assume* $\mathbf{g}$ *is a Lorentzian metric in a* 4 *dimensional open set* $O$*, with induced covariant derivative* $\mathbf{D}$ *and Ricci curvature* $\mathbf{R}_{\alpha\beta}$*, and* $\psi : O \to \mathbb{R}$ *is a scalar. Let* $x^0, x^1, x^2, x^3$ *denote a system of coordinates in* $O$ *and let* $\mathbf{\Gamma}^\nu$ *be defined as in* (1.1.5).

*(i) Assume that* $(\mathbf{g}, \psi)$ *satisfies the Einstein-Klein-Gordon system*

$$\mathbf{R}_{\alpha\beta} - \mathbf{D}_\alpha\psi\mathbf{D}_\beta\psi - \frac{\psi^2}{2}\mathbf{g}_{\alpha\beta} = 0, \qquad \square_{\mathbf{g}}\psi - \psi = 0, \tag{1.1.18}$$

*in* $O$*. Assume also that* $\mathbf{\Gamma}^\mu \equiv 0$ *in* $O$*,* $\mu \in \{0, 1, 2, 3\}$ *(the harmonic gauge* *condition). Then*

$$\widetilde{\Box}_{\mathbf{g}}\mathbf{g}_{\alpha\beta}+2\partial_\alpha\psi\partial_\beta\psi+\psi^2\mathbf{g}_{\alpha\beta}-F^{\geq 2}_{\alpha\beta}(g,\partial g)=0,\\ \widetilde{\Box}_{\mathbf{g}}\psi-\psi=0, \tag{1.1.19}$$

*where the quadratic semilinear terms* $F^{\geq 2}_{\alpha\beta}(g,\partial g)$ *are defined in* (1.1.15) *and* $\widetilde{\Box}_{\mathbf{g}}:=\mathbf{g}^{\alpha\beta}\partial_\alpha\partial_\beta$ *denotes the reduced wave operator .*

*(ii) Conversely, assume that the equations* (1.1.19) *(the reduced Einstein-Klein-Gordon system) hold in* $O$*. Then*

$$\mathbf{R}_{\alpha\beta}-\partial_\alpha\psi\partial_\beta\psi-\frac{\psi^2}{2}\mathbf{g}_{\alpha\beta}-\frac{1}{2}(\partial_\alpha\mathbf{\Gamma}_\beta+\partial_\beta\mathbf{\Gamma}_\alpha)=0,\\ \Box_{\mathbf{g}}\psi-\psi+\mathbf{\Gamma}^\mu\partial_\mu\psi=0, \tag{1.1.20}$$

*and the functions* $\mathbf{\Gamma}_\beta=g_{\beta\nu}\mathbf{\Gamma}^\nu$ *satisfy the reduced wave equations*

$$\widetilde{\Box}_{\mathbf{g}}\mathbf{\Gamma}_\beta=2\mathbf{\Gamma}^\nu\partial_\nu\psi\partial_\beta\psi+\mathbf{g}^{\rho\alpha}[\mathbf{\Gamma}^\nu{}_{\rho\alpha}(\partial_\nu\mathbf{\Gamma}_\beta+\partial_\beta\mathbf{\Gamma}_\nu)\\ +\mathbf{\Gamma}^\nu{}_{\rho\beta}(\partial_\alpha\mathbf{\Gamma}_\nu+\partial_\nu\mathbf{\Gamma}_\alpha)]+\partial_\mu\mathbf{\Gamma}_\nu\partial_\beta\mathbf{g}^{\mu\nu}. \tag{1.1.21}$$

*In particular, the pair* $(\mathbf{g},\psi)$ *solves the Einstein-Klein-Gordon system* (1.1.18) *if* $\mathbf{\Gamma}_\mu\equiv 0$ *in* $O$*.*

Our basic strategy to construct global solutions of the Einstein-Klein-Gordon system is to use Proposition 1.1. We construct first the pair $(\mathbf{g},\psi)$ by solving the reduced Einstein-Klein-Gordon system (1.1.19) (regarded as a quasilinear Wave-Klein-Gordon system) in the domain $\mathbb{R}^3\times[0,\infty)$. In addition, we arrange that $\mathbf{\Gamma}_\mu,\partial_t\mathbf{\Gamma}_\mu$ vanish on the initial hypersurface, so they vanish in the entire open domain, as a consequence of the wave equations (1.1.21). Therefore the pair $(\mathbf{g},\psi)$ solves the Einstein-Klein-Gordon system as desired.

In other words, the problem is reduced to constructing global solutions of the quasilinear system (1.1.19) for initial data compatible with the wave coordinates condition.

## 1.2 THE GLOBAL REGULARITY THEOREM

To state our global regularity theorem we introduce first several spaces of functions on $\mathbb{R}^3$.

**Definition 1.2.** *For* $a\geq 0$ *let* $H^a$ *denote the usual Sobolev spaces of index* $a$ *on* $\mathbb{R}^3$*. We define the Banach spaces* $H^{a,b}_\Omega$*,* $a,b\in\mathbb{Z}_+$*, by the norms*

$$\|f\|_{H^{a,b}_\Omega}:=\sum_{|\alpha|\leq b}\|\Omega^\alpha f\|_{H^a}, \tag{1.2.1}$$

*where* $\Omega^\alpha = \Omega_{23}^{\alpha_1}\Omega_{31}^{\alpha_2}\Omega_{12}^{\alpha_3}$ *and* $\Omega_{jk} = x_j\partial_k - x_k\partial_j$ *are the rotation vector-fields of* $\mathbb{R}^3$. *We also define the weighted Sobolev spaces* $H^{a,b}_{S,wa}$ *and* $H^{a,b}_{S,kg}$ *by the norms*

$$\|f\|_{H^{a,b}_{S,wa}} := \sum_{|\beta'|\le|\beta|\le b} \|x^{\beta'}\partial^\beta f\|_{H^a}, \qquad \|f\|_{H^{a,b}_{S,kg}} := \sum_{|\beta|,|\beta'|\le b} \|x^{\beta'}\partial^\beta f\|_{H^a}, \tag{1.2.2}$$

*where* $x^{\beta'} = x_1^{\beta_1'}x_2^{\beta_2'}x_3^{\beta_3'}$ *and* $\partial^\beta := \partial_1^{\beta_1}\partial_2^{\beta_2}\partial_3^{\beta_3}$. *Notice that* $H^{a,b}_{S,kg} \hookrightarrow H^{a,b}_{S,wa} \hookrightarrow H^{a,b}_{\Omega} \hookrightarrow H^a$.

To implement the strategy described above and use Proposition 1.1 we need to prescribe suitable initial data. Let $\Sigma_0 = \{(x,t) \in \mathbb{R}^3 \times [0,\infty) : t = x^0 = 0\}$. We assume that $\overline{g}, k$ are given symmetric tensors on $\Sigma_0$, such that $\overline{g}$ is a Riemannian metric on $\Sigma_0$. We assume also that $\psi_0, \psi_1 : \Sigma_0 \to \mathbb{R}$ are given initial data for the scalar field $\psi$.

We start by prescribing the metric components on $\Sigma_0$,

$$\mathbf{g}_{ij} = \overline{g}_{ij}, \qquad \mathbf{g}_{0i} = 0, \qquad \mathbf{g}_{00} = -1.$$

The conditions $\mathbf{g}_{00} = -1$ and $\mathbf{g}_{0i} = 0$ hold only on the initial hypersurface and are not propagated by the flow. They are imposed mostly for convenience and do not play a significant role in the analysis. We also prescribe the time derivative of the metric tensor,

$$\partial_t \mathbf{g}_{ij} = -2k_{ij},$$

in such a way that $k$ is the second fundamental form of the surface $\Sigma_0$, $k(X,Y) = -\mathbf{g}(\mathbf{D}_X n, Y)$, where $n = \partial_0$ is the future-oriented unit normal vector-field on $\Sigma_0$. The conditions $\mathbf{\Gamma}_\alpha = 0$, $\alpha \in \{0,1,2,3\}$, can be used to determine the other components of the initial data for the pair $(\mathbf{g},\psi)$ on the hypersurface $\Sigma_0$, which are

$$\begin{aligned}
&\mathbf{g}_{ij} = \overline{g}_{ij}, \qquad \mathbf{g}_{0i} = \mathbf{g}_{i0} = 0, \qquad \mathbf{g}_{00} = -1,\\
&\partial_t \mathbf{g}_{ij} = -2k_{ij}, \qquad \partial_t \mathbf{g}_{00} = 2\overline{g}^{ij}k_{ij}, \qquad \partial_t \mathbf{g}_{n0} = \overline{g}^{ij}\partial_i\overline{g}_{jn} - \frac{1}{2}\overline{g}^{ij}\partial_n\overline{g}_{ij},\\
&\psi = \psi_0, \qquad \partial_t\psi = \psi_1.
\end{aligned} \tag{1.2.3}$$

The remaining restrictions $\partial_t \mathbf{\Gamma}_\alpha = 0$ lead to the constraint equations. In view of (1.1.20) the constraint equations are equivalent to the conditions $\mathbf{R}_{\alpha 0} - (1/2)\mathbf{R}\mathbf{g}_{\alpha 0} = T_{\alpha 0}$, $\alpha \in \{0,1,2,3\}$, where $T_{\alpha\beta}$ is as in (1.1.2). This leads to four constraint equations

$$\begin{aligned}
\overline{D}_n(\overline{g}^{ij}k_{ij}) - \overline{g}^{ij}\overline{D}_j k_{in} &= \psi_1 \overline{D}_n \psi_0, \qquad n \in \{1,2,3\},\\
\overline{R} + \overline{g}^{ij}\overline{g}^{mn}(k_{ij}k_{mn} - k_{im}k_{jn}) &= \psi_1^2 + \overline{g}^{ij}\overline{D}_i\psi_0\overline{D}_j\psi_0 + \psi_0^2,
\end{aligned} \tag{1.2.4}$$

where $\overline{D}$ denotes the covariant derivative induced by the metric $\overline{g}$ on $\Sigma_0$, and $\overline{R}$

is the scalar curvature of the metric $\overline{g}$ on $\Sigma_0$.

We are now ready to state our first main theorem, which concerns global regularity of the system (1.1.19) for small initial data $(\overline{g}_{ij}, k_{ij}, \psi_0, \psi_1)$.

**Theorem 1.3.** *Let $\Sigma_0 := \{(x,t) \in \mathbb{R}^4 : t = 0\}$ and assume that $(\overline{g}_{ij}, k_{ij}, \psi_0, \psi_1)$ is an initial data set on $\Sigma_0$, satisfying the constraint equations (1.2.4) and the smallness conditions*

$$\begin{aligned}\sum_{n=0}^{3}\sum_{i,j=1}^{3} \{\| \, |\nabla|^{1/2+\delta/4}(\overline{g}_{ij} - \delta_{ij})\|_{H^{N(n),n}_{S,wa}} + \| \, |\nabla|^{-1/2+\delta/4} k_{ij}\|_{H^{N(n),n}_{S,wa}}\} \\ + \sum_{n=0}^{3} \{\|\langle\nabla\rangle\psi_0\|_{H^{N(n),n}_{S,kg}} + \|\psi_1\|_{H^{N(n),n}_{S,kg}}\} \leq \varepsilon_0 \leq \overline{\varepsilon}.\end{aligned} \tag{1.2.5}$$

*Here $N_0 := 40$, $d := 10$, $\delta := 10^{-10}$, $N(0) := N_0 + 16d$, $N(n) = N_0 - nd$ for $n \geq 1$, $\overline{\varepsilon}$ is a small constant, and the operators $|\nabla|$ and $\langle\nabla\rangle$ are defined by the multipliers $|\xi|$ and $\langle\xi\rangle$.*

*(i) Then the reduced Einstein-Klein-Gordon system*

$$\begin{aligned}\widetilde{\Box}_{\mathbf{g}}\mathbf{g}_{\alpha\beta} + 2\partial_\alpha\psi\partial_\beta\psi + \psi^2\mathbf{g}_{\alpha\beta} - F^{\geq 2}_{\alpha\beta}(g, \partial g) = 0, \\ \widetilde{\Box}_{\mathbf{g}}\psi - \psi = 0,\end{aligned} \tag{1.2.6}$$

*admits a unique global solution $(\mathbf{g}, \psi)$ in $M := \{(x,t) \in \mathbb{R}^4 : t \geq 0\}$, with initial data $(\overline{g}_{ij}, k_{ij}, \psi_0, \psi_1)$ on $\Sigma_0$ (as described in (1.2.3)). Here $F^{\geq 2}_{\alpha\beta}(g, \partial g)$ are as in (1.1.15) and $\widetilde{\Box}_{\mathbf{g}} = \mathbf{g}^{\mu\nu}\partial_\mu\partial_\nu$. The solution satisfies the harmonic gauge conditions*

$$0 = \mathbf{\Gamma}_\mu = \mathbf{g}^{\alpha\beta}\partial_\alpha\mathbf{g}_{\beta\mu} - \frac{1}{2}\mathbf{g}^{\alpha\beta}\partial_\mu\mathbf{g}_{\alpha\beta}, \qquad \mu \in \{0,1,2,3\} \tag{1.2.7}$$

*in $M$. Moreover, the metric $\mathbf{g}$ stays close and converges to the Minkowski metric and $\psi$ stays small and converges to $0$ as $t \to \infty$ (in suitable norms).*

*(ii) In view of Proposition 1.1, the pair $(\mathbf{g}, \psi)$ is a global solution in $M$ of the Einstein-Klein-Gordon coupled system*

$$\mathbf{R}_{\alpha\beta} - \mathbf{D}_\alpha\psi\mathbf{D}_\beta\psi - \frac{\psi^2}{2}\mathbf{g}_{\alpha\beta} = 0, \qquad \Box_{\mathbf{g}}\psi - \psi = 0, \tag{1.2.8}$$

*with the prescribed initial data $(\overline{g}_{ij}, k_{ij}, \psi_0, \psi_1)$ on $\Sigma_0$. In our geometric context, globality means that all future directed timelike and null geodesics starting from points in $M$ extend forever with respect to their affine parametrization.*

The proof of Theorem 1.3 is based on a complex bootstrap argument, involving energy estimates, vector-fields, Fourier analysis, and nonlinear scattering. We summarize some of its main elements in subsection 1.3.1 below, and then provide a more extensive outline of its proof in section 2.2.

The global regularity conclusion of Theorem 1.3 is essentially a qualitative statement, which can only be proved by a precise quantitative analysis of the spacetime. In Chapter 7 we state and prove more precise theorems describing our spacetime. These theorems include global quantitative control and nonlinear scattering of the metric tensor and the Klein-Gordon field (Theorem 7.1), pointwise decay estimates in the physical space (Theorem 7.2 and Lemma 7.4), global control of timelike and null geodesics (Theorem 7.6), weak peeling estimates for the Riemann curvature tensor (Theorem 7.7 and Proposition 7.9), and ADM and Bondi energy formulas (Proposition 7.11, Proposition 7.13, Theorem 7.23, and Proposition 7.24). We will discuss some of these more precise conclusions in section 1.3 below.

In the rest of this section we discuss previous related work and motivate some of the assumptions on the initial data.

### 1.2.1 Global Stability Results in General Relativity

Global stability of physical solutions is an important topic in General Relativity. For example, the global nonlinear stability of the Minkowski spacetime among solutions of the Einstein-vacuum equation is a central theorem in the field, due to Christodoulou-Klainerman [12]. See also the more recent extensions of Klainerman-Nicolò [52], Lindblad-Rodnianski [62], Bieri and Zipser [6], Speck [72], and Hintz-Vasy [33].

More recently, small data global regularity theorems have also been proved for other coupled Einstein field equations. The Einstein-Klein-Gordon system (the same system we analyze here) was considered recently by LeFloch-Ma [58], who proved small data global regularity for restricted data, which agrees with a Schwarzschild solution with small mass outside a compact set. A similar result was announced by Wang [74].

Our main goals in this monograph are (1) to prove global nonlinear stability for general unrestricted small initial data, and (2) to develop the full asymptotic analysis of the spacetime. In particular, we answer the natural question, raised in the physics literature by Okawa-Cardoso-Pani [66], of whether the Minkowski solution is stable or unstable for small massive scalar field perturbations. A similar global regularity result for general small data was announced recently by LeFloch-Ma [59].

We also refer to the work by Fajman-Joudioux-Smulevici [19], Lindblad-Taylor [64], and, more recently, Bigorgne-Fajman-Joudioux-Smulevici-Thalleron [7] on the global stability of Einstein-Vlasov systems.

In a different direction, one can also raise the question of linear and nonlinear stability of other physical solutions of the Einstein equations. Stability of the Kerr family of solutions has been under intense study in recent years, first at the linearized level (see, for example, [13, 30] and the references therein) and more recently at the full nonlinear level (see [26, 32, 54]).

The stability of Kerr in the presence of a massive scalar field seems interesting as well. Solutions to the Klein-Gordon equation in Kerr can grow exponentially even from smooth initial data, as shown in [70], and this phenomenon was used by Chodosh and Shlapentokh-Rothman [10] to construct a curve of time-periodic solutions of the Einstein-Klein-Gordon system bifurcating from (empty) Kerr (see [31] for a prior numerical construction). Therefore a result on stability of Kerr similar to our main theorem could only be possible, if at all, in a stronger topology where this curve is not continuous (see also the discussion on the mini-bosons in subsection 1.2.5 below).

#### *1.2.1.1 Restricted initial data*

One can often simplify considerably the global analysis of wave and Klein-Gordon equations by considering initial data of compact support. The point is that the solutions have the finite speed of propagation, thus remain supported inside a light cone, and one can use the hyperbolic foliation method and its refinements (see [56] for a recent account) to analyze the evolution.

However, to implement this method one needs to first control the solution on an initial hyperboloid (the "initial data"), so the method is restricted to the case when one can establish such control. Due to the finite speed of propagation, this is possible for compactly supported data (for systems of wave or Klein-Gordon equations), or data that agrees with the Schwarzschild solution outside a compact set (in the case of the Einstein equations).

The use of "restricted initial data" coupled with the hyperbolic foliation method leads to significant simplifications of the global analysis, particularly at the level of proving decay. In the context of the Einstein equations these ideas have been used by many authors, such as Friedrich [21], Lindblad-Rodnianski [62], Fajman-Joudioux-Smulevici [19], Lindblad-Taylor [64], LeFloch-Ma [58], and Wang [74].

### 1.2.2 Simplified Wave-Klein-Gordon Models

Our system (1.2.6) is complicated, but one can gain intuition by looking at simpler models. For example, one can consider the simplified system

$$\begin{aligned} -\Box u &= A^{\alpha\beta}\partial_\alpha v\partial_\beta v + Dv^2,\\ (-\Box+1)v &= uB^{\alpha\beta}\partial_\alpha\partial_\beta v + Euv, \end{aligned} \tag{1.2.9}$$

where $u, v$ are real-valued functions, and $A^{\alpha\beta}$, $B^{\alpha\beta}$, $D$, and $E$ are real constants. This system was introduced by LeFloch-Ma [57] as a model for the full Einstein-Klein-Gordon system (1.2.6). Intuitively, the deviation of the Lorentzian metric $\mathbf{g}$ from the Minkowski metric is replaced by a scalar function $u$, and the massive scalar field $\psi$ is replaced by $v$. The system (1.2.9) has the same linear structure as the Einstein-Klein-Gordon system (1.2.6), but only keeps, schematically, quadratic interactions that involve the massive scalar field; for simplicity, all the quadratic interactions of the wave component with itself are neglected in this model.

Small data global regularity for the system (1.2.9) was proved by LeFloch-Ma [57] in the case of compactly supported initial data (the restricted data case), using the hyperbolic foliation method. For general small initial data, global regularity was proved by the authors [38].

Global regularity of Wave-Klein-Gordon coupled systems in 3 dimensions is a natural topic, motivated by physical models such as the Dirac-Klein-Gordon equations, and had been investigated earlier by Georgiev [22] and Katayama [44]. A similar system, the massive Maxwell-Klein-Gordon system, was analyzed recently by Klainerman-Wang-Yang [55], who also proved global regularity for general small initial data.

Coupled Wave-Klein-Gordon systems have also been considered in 2 dimensions, where the decay is slower and the global analysis requires nonlinearities with much more favorable structure (see, for example, Ifrim-Stingo [34] and the references therein).

### 1.2.3 Small Data Global Regularity Results

The system (1.2.6) can be easily transformed into a quasilinear coupled system of wave and Klein-Gordon equations. Indeed, let $m$ denote the Minkowski metric and write

$$\mathbf{g}_{\alpha\beta} = m_{\alpha\beta} + h_{\alpha\beta}, \qquad \mathbf{g}^{\alpha\beta} = m^{\alpha\beta} + g^{\alpha\beta}_{\geq 1}, \qquad \alpha, \beta \in \{0,1,2,3\}.$$

It follows from (1.2.6) that the metric components $h_{\alpha\beta}$ satisfy the nonlinear wave equations

$$(\partial_0^2 - \Delta)h_{\alpha\beta} = \mathcal{N}^h_{\alpha\beta} := \mathcal{KG}_{\alpha\beta} + g^{\mu\nu}_{\geq 1}\partial_\mu\partial_\nu h_{\alpha\beta} - F^{\geq 2}_{\alpha\beta}(g, \partial g) \tag{1.2.10}$$

where $F^{\geq 2}_{\alpha\beta}(g, \partial g)$ are the semilinear terms in (1.1.15) and $\mathcal{KG}_{\alpha\beta} := 2\partial_\alpha\psi\partial_\beta\psi + \psi^2(m_{\alpha\beta} + h_{\alpha\beta})$. Moreover, the field $\psi$ satisfies the quasilinear Klein-Gordon equation

$$(\partial_0^2 - \Delta + 1)\psi = \mathcal{N}^\psi := g^{\mu\nu}_{\geq 1}\partial_\mu\partial_\nu\psi. \tag{1.2.11}$$

Therefore Theorem 1.3 can be regarded as a small data global regularity result for a quasilinear evolution system. Several important techniques have been developed over the years in the study of such problems, starting with seminal contributions of John, Klainerman, Shatah, Simon, Christodoulou, Alinhac, and Delort [1, 2, 11, 12, 14, 15, 42, 43, 48, 49, 50, 51, 69, 71]. These include the vector-field method, normal forms, and the isolation of null structures.

In the case of Einstein equations and other hyperbolic systems, most global results have been proved mostly using the "physical space" framework, based on pointwise spacetime estimates. This is well adapted to geometric backgrounds with non-constant coefficients. The analysis is naturally carried out through weighted estimates and relies heavily on the presence of symmetries (vector-fields) that can be used to extract information about solutions. This is the main framework for many works on General Relativity, especially away from Minkowski space and in vacuum or with electromagnetic and massless scalar-fields, such as [6, 7, 12, 13, 19, 26, 52, 54, 58, 59, 62, 63, 64, 72].

#### *1.2.3.1 Fourier analysis and the Z-norm method*

In the last few years new ideas have emerged in the study of global solutions of quasilinear evolutions, inspired mainly by the advances in semilinear theory. The basic goal is to combine the classical energy and vector-fields methods with refined analysis of the Duhamel formula, using the Fourier transform. This starts by decomposing an unknown $U$ into a superposition of elementary waves

$$U(x,t) = \frac{1}{(2\pi)^d} \int_{\mathbb{R}^d} \widehat{V}(\xi,t) e^{i[\langle x,\xi\rangle - t\Lambda(\xi)]} d\xi, \tag{1.2.12}$$

for some appropriate dispersion relation $\Lambda$. The main objective is then to understand quantitatively properties of the "linear profile" $V$ during the evolution.

The main advantage of the Fourier transform method over physical space methods is the ability to identify clearly resonant and non-resonant nonlinear interactions, by decomposing the various waves as in (1.2.12) and examining their interactions. One can then dispose of the non-resonant interactions (using, for example, normal forms), and concentrate on a small number of resonant interactions. This is particularly important in low dimensions (like 1 or 2 dimension), when decay by itself cannot be enough to lead to global control of solutions.

In semilinear dispersive and hyperbolic equations Fourier analysis is a central tool that has led to major progress in the entire field. On the other hand, in the context of quasilinear evolutions, Fourier analysis has only been used more recently, starting essentially with the "method of spacetime resonances" of Germain-Masmoudi-Shatah [24, 25] and Gustafson-Nakanishi-Tsai [29]. The main difficulty in the quasilinear case is that the Duhamel formula cannot be used exclusively to study the evolution, due to derivative loss, and one has to rely also on energy estimates.

Our general philosophy, which we use in this monograph to prove Theorem 1.3, is to work both in the physical space, mainly to prove energy estimates (including vector-fields), and in the Fourier space, mainly to investigate resonances using the Duhamel formula and prove decay of the solutions in time. At the implementation level, the analysis in the Fourier space is based on a choice of a "$Z$-norm" to measure the size of the linear profiles dynamically in time. This choice is very important, and one should think of it as analogous to the choice of the "resolution norm" in the case of semilinear evolutions (the classical choices being Strichartz norms or $X^{s,b}$ norms). The key point is that the $Z$-norm has to complement well the information coming from energy estimates.

The $Z$-norm method, with different choices of the norm itself, depending on the problem, was used recently by the authors and their collaborators in several small data global regularity problems, for water waves and plasmas, such as [16, 17, 27, 28, 35, 36, 37, 39, 40, 41, 46]. It is particularly well suited to the study of systems with multiple characteristics, in which different components of the system evolve according to different dispersion relations and have different speeds of propagation, such as plasma models or the Einstein-Klein-Gordon system (1.2.10)–(1.2.11). The point is that such systems tend to have fewer joint symmetries, which complicates significantly the analysis in the physical space, but the Fourier analysis method is much less sensitive to the presence of symmetries.

### 1.2.4 Assumptions on the Initial Data

The precise form of the smallness assumptions (1.2.5) on the metric initial data $\overline{g}_{ij}$ and $k_{ij}$ is important. Indeed, in view of the positive mass theorem of Schoen-Yau [68], one expects the metric components $\overline{g}_{ij} - \delta_{ij}$ to decay no faster than $M/\langle x\rangle$ and the second fundamental form $k$ to decay no faster than $M/\langle x\rangle^2$, where $M \ll 1$ is the mass. Capturing this type of decay, using $L^2$-based norms, is precisely the role of the homogeneous multipliers $|\nabla|^{1/2+\delta/4}$ and $|\nabla|^{-1/2+\delta/4}$ in (1.2.5). Notice that these multipliers are sharp, up to the $\delta/4$ power.

Our assumptions on the metric are essentially of the type

$$g_{ij} = \delta_{ij} + \varepsilon_0 O(\langle x\rangle^{-1+\delta/4}), \qquad k_{ij} = \varepsilon_0 O(\langle x\rangle^{-2+\delta/4}) \tag{1.2.13}$$

at time $t = 0$. These are less restrictive than the assumptions used sometimes even in the vacuum case $\psi \equiv 0$—see, for example, [12, 52, 63]—in the sense that the initial data is not assumed to agree with the Schwarzschild initial data up to lower order terms. For maximal time foliations, our assumptions are, however, more restrictive than the ones in Bieri's work [6], but we are able to prove more precise asymptotic bounds on the metric and the Riemann curvature tensor; see section 1.3 below.

We remark also that our assumptions (1.2.5) allow for non-isotropic initial data, possibly with different "masses" in different directions. For the vacuum case, initial data of this type, satisfying the constraint equations, have been constructed recently by Carlotto-Schoen [9].

### 1.2.5 The Mini-bosons

A serious potential obstruction to small data global stability theorems is the presence of non-decaying "small" solutions, such as small solitons. A remarkable fact is that there are such small non-decaying solutions for the Einstein-Klein-Gordon system, namely the so-called mini-boson stars. These are time-periodic (therefore non-decaying) and spherically symmetric exact solutions of the Einstein-Klein-Gordon system. They were discovered numerically by physicists, such as Kaup [47], Friedberg-Lee-Pang [20] (see also [60]), and then constructed rigorously by Bizon-Wasserman [8].

These mini-bosons can be thought of as arbitrarily small (hence the name) in certain topologies, as explained in [8]. However, the mini-bosons (in particular the Klein-Gordon component) are not small in the stronger topology we use here, as described by (1.2.5), so we can thankfully avoid them in our analysis.

## 1.3 MAIN IDEAS AND FURTHER ASYMPTOTIC RESULTS

In this section we provide first a brief summary of some of the main ingredients in the proof of the global nonlinear stability result in Theorem 1.3. Then, in subsections 1.3.2–1.3.6 we present some of the additional theorems we prove in Chapter 7, concerning the global geometry of our spacetime.

### 1.3.1 Global Nonlinear Stability

The classical mechanism to establish small data global regularity for quasilinear dispersive and hyperbolic systems has two main components:

(1) Propagate control of energy functionals (high order Sobolev norms and vector-fields);

(2) Prove dispersion/decay of the solution over time.

These are our basic goals here as well, as we investigate solutions of the coupled Wave-Klein-Gordon system (1.2.10)–(1.2.11) in the variables $h_{\alpha\beta}$ and $\psi$. As expected, our analysis also involves vector-fields, corresponding to the natural symmetries of the linearized equations, namely the Lorentz vector-fields $\Gamma_a$ and the rotation vector-fields $\Omega_{ab}$,

$$\Gamma_a := x_a \partial_t + t\partial_a, \qquad \Omega_{ab} := x_a\partial_b - x_b\partial_a, \tag{1.3.1}$$

for $a, b \in \{1, 2, 3\}$. These vector-fields commute with both the wave operator and the Klein-Gordon operator in the flat Minkowski space. We note that the scaling vector-field $S = t\partial_t + x \cdot \nabla_x$ does not satisfy nice commutation properties with the linearized system (due to the Klein-Gordon field), so we cannot use it in our analysis.

The main objects we analyze in the proof of nonlinear stability are the normalized solutions $U^{\mathcal{L}h_{\alpha\beta}}$ and $U^{\mathcal{L}\psi}$ and the associated *linear profiles* $V^{\mathcal{L}h_{\alpha\beta}}$ and $V^{\mathcal{L}\psi}$, defined by

$$\begin{aligned} U^{\mathcal{L}h_{\alpha\beta}}(t) &:= \partial_t(\mathcal{L}h_{\alpha\beta})(t) - i\Lambda_{wa}(\mathcal{L}h_{\alpha\beta})(t), \qquad V^{\mathcal{L}h_{\alpha\beta}}(t) := e^{it\Lambda_{wa}} U^{\mathcal{L}h_{\alpha\beta}}(t), \\ U^{\mathcal{L}\psi}(t) &:= \partial_t(\mathcal{L}\psi)(t) - i\Lambda_{kg}(\mathcal{L}\psi)(t), \qquad V^{\mathcal{L}\psi}(t) := e^{it\Lambda_{kg}} U^{\mathcal{L}\psi}(t), \end{aligned} \tag{1.3.2}$$

where $\Lambda_{wa} = |\nabla|$, $\Lambda_{kg} = \langle\nabla\rangle = \sqrt{|\nabla|^2 + 1}$. Here $\mathcal{L}$ denotes differential operators obtained by applying up to three vector-fields $\Gamma_a$ or $\Omega_{ab}$, and these operators are applied to the metric components $h_{\alpha\beta}$ and the field $\psi$.

The complex-valued normalized solutions $U^{\mathcal{L}h_{\alpha\beta}}$ and $U^{\mathcal{L}\psi}$ capture both the time derivatives (as the real part) and the spatial derivatives (as the imaginary part) of the variables $h_{\alpha\beta}$ and $\psi$. The linear profiles $V^{\mathcal{L}h_{\alpha\beta}}$ and $V^{\mathcal{L}\psi}$, which are constructed by going forward in time along the nonlinear evolution, and then going backwards in time along the linear flow, capture the cumulative effect of the nonlinearity over time.

Our proof of global stability relies on controlling simultaneously three types of norms, as part of a bootstrap argument:

(1) High order energy norms, involving Sobolev derivatives and the vector-fields $\Gamma_a$ and $\Omega_{ab}$, with slow growth in time;

(2) Matching weighted estimates on the profiles $V^{\mathcal{L}h_{\alpha\beta}}$ and $V^{\mathcal{L}\psi}$ in Sobolev spaces, again with slow growth in time;

(3) Sharp uniform in time estimates on the Klein-Gordon profile $V^{\psi}$ and on some parts of the metric profiles $V^{h_{\alpha\beta}}$, in a suitable $Z$-norm to be defined.

We discuss these estimates in more detail in the rest of this subsection.

### 1.3.1.1 *Energy estimates and weighted estimates on the profiles*

The main energy estimates we prove as part of our bootstrap argument are

$$\big\|(\langle t\rangle|\nabla|_{\leq 1})^{\delta/4}|\nabla|^{-1/2}U^{\mathcal{L}h_{\alpha\beta}}(t)\big\|_{H^{n(\mathcal{L})}} + \big\|U^{\mathcal{L}\psi}(t)\big\|_{H^{n(\mathcal{L})}} \lesssim \varepsilon_0\langle t\rangle^{H(\mathcal{L})\delta}, \quad (1.3.3)$$

for a suitable hierarchy of parameters $n(\mathcal{L})$ and $H(\mathcal{L})$ that depend on the differential operator $\mathcal{L}$. We remark that the energy estimates we prove for the metric variables $U^{\mathcal{L}h_{\alpha\beta}}$ also contain significant information at low frequencies, due to the operators $|\nabla|^{-1/2}$ and $|\nabla|_{\leq 1}$, which are connected to the natural $|x|^{-1+}$ decay of the metric components $h_{\alpha\beta}$. The nonlinear propagation of the low-frequency energy bounds is, in fact, the more subtle part of the argument.

The second component of our bootstrap argument consists of compatible weighted estimates on the profiles $V^{\mathcal{L}h_{\alpha\beta}}$ and $V^{\mathcal{L}\psi}$, of the form

$$\begin{aligned} &2^{k/2}(2^{k^-}\langle t\rangle)^{\delta/4}\|P_k(x_l V^{\mathcal{L}h_{\alpha\beta}})(t)\|_{L^2} \\ &\quad + 2^{k^+}\|P_k(x_l V^{\mathcal{L}\psi})(t)\|_{L^2} \lesssim \varepsilon_0\langle t\rangle^{H'(\mathcal{L})\delta}2^{-n'(\mathcal{L})k^+}, \end{aligned} \quad (1.3.4)$$

for any $k \in \mathbb{Z}$, $l \in \{1,2,3\}$, and differential operator $\mathcal{L}$ containing at most two vector-fields $\Gamma_a$ or $\Omega_{ab}$. Here $P_k$ denote Littlewood-Paley projections to frequencies $\approx 2^k$ and $x^+ = \max(x,0)$ and $x^- = \min(x,0)$ for any $x \in \mathbb{R}$.

The energy estimates (1.3.3) and the weighted estimates (1.3.4) are compatible, at the level of the important parameters $H(\mathcal{L})$, $n(\mathcal{L})$, $H'(\mathcal{L})$, and $n'(\mathcal{L})$ that measure the slow growth in time and the Sobolev smoothness of the various components.

The weighted estimates (1.3.4) imply almost optimal pointwise decay estimates on the metric components and the Klein-Gordon field, with improved decay at low and high frequencies, due to Lemma 3.9. We emphasize, however, that weighted estimates on linear profiles are a lot stronger than pointwise decay estimates on solutions, and serve many other purposes. For example, space localization of the linear profiles allows us to decompose the main variables both in frequency and space, which leads to precise control in nonlinear estimates.

#### *1.3.1.2 Weak null structure and decomposition of the metric tensor*

The proof of the global stability theorem is involved, mainly because the nonlinearities $\mathcal{N}^h_{\alpha\beta}$ and $\mathcal{N}^\psi$ have complicated structure, both at the semilinear level (for $\mathcal{N}^h_{\alpha\beta}$) and at the quasilinear level.

In particular, it is well known that the semilinear terms $F^{\geq 2}_{\alpha\beta}(g, \partial g)$ do not have the classical null structure. They have, however, a remarkable weak null structure in harmonic coordinates, which is still suitable for global analysis as discovered by Lindblad-Rodnianski [62]. To identify and use this weak null structure we need to decompose the tensor $h_{\alpha\beta}$.

The standard way to decompose the metric tensor in General Relativity is based on null frames (see, for instance, [12] or [62]). Here we use a different decomposition of the metric tensor, reminiscent of the div-curl decomposition of vector-fields in fluid models, which is connected to the classical work of Arnowitt-Deser-Misner [3] on the Hamiltonian formulation of General Relativity. For us, this decomposition has the advantage of being more compatible with the Fourier transform and the vector-fields $\Omega_{ab}$ and $\Gamma_a$.

More precisely, let $R_j = |\nabla|^{-1}\partial_j$, $j \in \{1, 2, 3\}$, denote the Riesz transforms on $\mathbb{R}^3$, and let

$$\begin{aligned} &F := (1/2)[h_{00} + R_j R_k h_{jk}], \qquad \underline{F} := (1/2)[h_{00} - R_j R_k h_{jk}],\\ &\rho := R_j h_{0j}, \qquad \omega_j := \in_{jkl} R_k h_{0l},\\ &\Omega_j := \in_{jkl} R_k R_m h_{lm}, \qquad \vartheta_{jk} := \in_{jmp} \in_{knq} R_m R_n h_{pq}. \end{aligned} \tag{1.3.5}$$

Geometrically, the variables $F + \underline{F}$, $\rho$, and $\omega$ are linked to the lapse and the shift vector, $F - \underline{F}$ and $\Omega$ are gauge components associated to spatial coordinates, while $\vartheta$ corresponds to the (linearized) coordinate-free component of the spatial metric (see Proposition 7.14). The metric tensor $h$ can be recovered linearly from the components $F, \underline{F}, \rho, \omega_j, \Omega_j, \vartheta_{jk}$.

Our analysis shows that the components $F, \omega_j, \Omega_j, \vartheta_{jk}$ satisfy good wave equations, with all the quadratic semilinear terms having suitable null structure. On the other hand, the components $\underline{F}$ and $\rho$ (which are related elliptically due to the harmonic gauge conditions) satisfy wave equations with some quadratic semilinear terms with no null structure. However, these non-null quadratic semilinear terms have the redeeming feature that they can be expressed only in terms of the good components $\vartheta_{jk}$.

This algebraic structure suggests that we should aim to prove that the good components $F, \omega_j, \Omega_j, \vartheta_{jk}$ do not grow during the evolution, in suitable norms to be made precise. On the other hand, the components $\underline{F}, \rho$, as well as all the components $\mathcal{L}h_{\alpha\beta}$ and $\mathcal{L}\psi$ which contain some weighted vector-fields $\Omega_{ab}$ or $\Gamma_a$, should be allowed to grow in time slowly, at suitable rates to be determined. We note that our vector-fields are adapted to the Minkowski geometry, containing the coordinate functions $x_a$ and $t$, not to the true geometry of the spacetime; thus it is expected that they can only be useful only up to $\langle t\rangle^{0+}$ losses. At a qualitative level, this is precisely what our final conclusions are.

#### *1.3.1.3 Uniform bounds and the Z-norm*

To prove uniform control on the good metric components $F, \omega_j, \Omega_j, \vartheta_{jk}$ and the field $\psi$ we use what we call *the Z-norm method*: we define the special norms

$$\|f\|_{Z_{wa}} := \sup_{k\in\mathbb{Z}} 2^{N_0 k^+} 2^{k^-(1+\kappa)} \|\widehat{P_k f}\|_{L^\infty},$$
$$\|f\|_{Z_{kg}} := \sup_{k\in\mathbb{Z}} 2^{N_0 k^+} 2^{k^-(1/2-\kappa)} \|\widehat{P_k f}\|_{L^\infty}, \tag{1.3.6}$$

where $N_0 = 40$ and $\kappa = 10^{-3}$. The last component of our bootstrap construction involves uniform bounds of the form

$$\|V^F(t)\|_{Z_{wa}} + \|V^{\omega_a}(t)\|_{Z_{wa}} + \|V^{\vartheta_{ab}}(t)\|_{Z_{wa}} + \|V^\psi(t)\|_{Z_{kg}} \lesssim \varepsilon_0, \tag{1.3.7}$$

for any $t \in [0,\infty)$ and $a, b \in \{1,2,3\}$, where the profiles $V^G$ are defined in as (1.3.2),

$$U^G(t) := \partial_t G(t) - i\Lambda_{wa} G(t), \qquad V^G(t) := e^{it\Lambda_{wa}} U^G(t), \tag{1.3.8}$$

for $G \in \{F, \omega_a, \vartheta_{ab}\}$. The main point of the estimates (1.3.7) is the uniformity in time, in particular allowing us to prove sharp $\varepsilon_0\langle t\rangle^{-1}$ pointwise decay on some components of the metric tensor.

The $Z$-norms defined in (1.3.6) measure the $L^\infty$ norm of solutions in the Fourier space, with weights that are particularly important at low frequencies. They cannot be propagated using energy estimates, since they are not $L^2$-based norms. We use instead the Duhamel formula, in the Fourier space, which leads to derivative loss. Because of this the $Z$-norm bounds (1.3.7) are weaker than the energy bounds (1.3.3) at very high frequencies. One should think of the $Z$-norm bounds as effective at middle frequencies, say $\langle t\rangle^{-1/2} \lesssim 2^k \lesssim \langle t\rangle^{1/2}$.

### 1.3.2 Nonlinear Scattering

The global dynamics of solutions is complicated mainly because they do not scatter linearly as $t \to \infty$. This is due to the low frequencies of the metric tensor in the quasilinear terms $g^{\mu\nu}_{\geq 1}\partial_\mu\partial_\nu h_{\alpha\beta}$ and $g^{\mu\nu}_{\geq 1}\partial_\mu\partial_\nu\psi$, which create a long-range perturbation.

To understand the asymptotic behavior of our spacetime we need to renormalize the profiles. More precisely, we define the wave phase correction (related to optical functions)

$$
\begin{aligned}
\Theta_{wa}(\xi,t) := \int_0^t \Big\{ & h_{00}^{low}(s\xi/\Lambda_{wa}(\xi),s)\frac{\Lambda_{wa}(\xi)}{2} \\
& + h_{0j}^{low}(s\xi/\Lambda_{wa}(\xi),s)\xi_j + h_{jk}^{low}(s\xi/\Lambda_{wa}(\xi),s)\frac{\xi_j\xi_k}{2\Lambda_{wa}(\xi)}\Big\}\, ds
\end{aligned}
\tag{1.3.9}
$$

and the Klein-Gordon phase correction

$$
\begin{aligned}
\Theta_{kg}(\xi,t) := \int_0^t \Big\{ & h_{00}^{low}(s\xi/\Lambda_{kg}(\xi),s)\frac{\Lambda_{kg}(\xi)}{2} \\
& + h_{0j}^{low}(s\xi/\Lambda_{kg}(\xi),s)\xi_j + h_{jk}^{low}(s\xi/\Lambda_{kg}(\xi),s)\frac{\xi_j\xi_k}{2\Lambda_{kg}(\xi)}\Big\}\, ds,
\end{aligned}
\tag{1.3.10}
$$

where $h_{\alpha\beta}^{low}$ are low frequency components of the metric tensor,

$$
\widehat{h_{\alpha\beta}^{low}}(\rho,s) := \varphi_{\leq 0}(\langle s\rangle^{p_0}\rho)\widehat{h_{\alpha\beta}}(\rho,s), \qquad p_0 := 0.68. \tag{1.3.11}
$$

The choice of $p_0$, slightly bigger than $2/3$, is important in the proof to justify the correction. Geometrically, the two phase corrections $\Theta_{wa}$ and $\Theta_{kg}$ are obtained by integrating suitable low frequency components of the metric tensor along the characteristics of the wave and the Klein-Gordon linear flows.

The nonlinear profiles are obtained by multiplication in the Fourier space,

$$
\widehat{V_*^G}(\xi,t) := e^{-i\Theta_{wa}(\xi,t)}\widehat{V^G}(\xi,t), \qquad \widehat{V_*^\psi}(\xi,t) := e^{-i\Theta_{kg}(\xi,t)}\widehat{V^\psi}(\xi,t), \tag{1.3.12}
$$

for $G \in \{F, \omega_a, \vartheta_{ab}\}$. Notice that $\|V_*^G\|_{Z_{wa}} = \|V^G\|_{Z_{wa}}$ and $\|V_*^\psi\|_{Z_{kg}} = \|V^\psi\|_{Z_{kg}}$, since the phases $\Theta_{wa}$ and $\Theta_{kg}$ are real-valued. The point of this construction is that the new nonlinear profiles $V_*^F$, $V_*^{\omega_a}$, $V_*^{\vartheta_{ab}}$, and $V_*^\psi$ converge as the time goes to infinity, i.e.,

$$
\begin{aligned}
\|V_*^F(t) - V_\infty^F\|_{Z_{wa}} + \|V_*^{\omega_a}(t) - V_\infty^{\omega_a}\|_{Z_{wa}} + \|V_*^{\vartheta_{ab}}(t) - V_\infty^{\vartheta_{ab}}\|_{Z_{wa}} &\lesssim \varepsilon_0\langle t\rangle^{-\delta/2}, \\
\|V_*^\psi(t) - V_\infty^\psi\|_{Z_{kg}} &\lesssim \varepsilon_0\langle t\rangle^{-\delta/2},
\end{aligned}
\tag{1.3.13}
$$

where $V_\infty^F, V_\infty^{\omega_a}, V_\infty^{\vartheta_{ab}} \in Z_{wa}$ and $V_\infty^\psi \in Z_{kg}$ are the *nonlinear scattering data*. These functions, in particular the components $V_\infty^{\vartheta_{ab}}$ and $V_\infty^\psi$, are important in the asymptotic analysis of our spacetime. Chapter 5 is mainly concerned with the proofs of the bounds (1.3.13).

### 1.3.3 Asymptotic Bounds and Causal Geodesics

Our core bootstrap argument relies on controlling the solution both in the physical space and in the Fourier space, as summarized above. However, after closing the main bootstrap argument, we can derive classical bounds on the solutions in the physical space, without explicit use of the Fourier transform.

We start with decay estimates in the physical space. Let

$$L := \partial_t + \partial_r, \qquad \underline{L} := \partial_t - \partial_r, \tag{1.3.14}$$

where $r := |x|$ and $\partial_r := |x|^{-1}x^j\partial_j$. Let

$$\mathcal{T} := \{L, r^{-1}\Omega_{12}, r^{-1}\Omega_{23}, r^{-1}\Omega_{31}\} \tag{1.3.15}$$

denote the set of "good" vector-fields, tangential to the (Minkowski) light cones.

In Theorem 7.2 we prove that the metric components satisfy the bounds

$$|h(x,t)| + \langle t+r\rangle|\partial_V h(x,t)| + \langle t-r\rangle|\partial_{\underline{L}} h(x,t)| \lesssim \varepsilon_0\langle t+r\rangle^{2\delta'-1}, \tag{1.3.16}$$

in the manifold $M := \{(x,t) \in \mathbb{R}^3 \times [0,\infty)\}$, where $r = |x|$, $V \in \mathcal{T}$, $h \in \{h_{\alpha\beta}\}$, $\partial_W := W^\alpha\partial_\alpha$, and $\delta' = 2000\delta$. The scalar field decays faster but with no derivative improvement,

$$\begin{aligned} |\psi(x,t)| + |\partial_0\psi(x,t)| &\lesssim \varepsilon_0\langle t+r\rangle^{\delta'/2-1}\langle r\rangle^{-1/2}, \\ |\partial_b\psi(x,t)| &\lesssim \varepsilon_0\langle t+r\rangle^{\delta'/2-3/2}, \qquad b \in \{1,2,3\}. \end{aligned} \tag{1.3.17}$$

Also, in Lemma 7.4 we show that the second order derivatives to the metric satisfy the bounds

$$\begin{aligned} \langle r\rangle^2|\partial_{V_1}\partial_{V_2}h(x,t)| &+ \langle t-r\rangle^2|\partial_{\underline{L}}^2 h(x,t)| \\ &+ \langle t-r\rangle\langle r\rangle|\partial_{\underline{L}}\partial_{V_1}h(x,t)| \lesssim \varepsilon_0\langle r\rangle^{3\delta'-1}, \end{aligned} \tag{1.3.18}$$

in the region $M' := \{(x,t) \in M : t \geq 1, |x| \geq 2^{-8}t\}$, where $V_1, V_2 \in \mathcal{T}$ are good vector-fields.

The pointwise bounds (1.3.16)–(1.3.18) are as expected, including the small $\delta'$ losses that are due to our weak assumptions (1.2.13) on the initial data. These bounds follow mainly from the profile bounds (1.3.4) and linear estimates.

As an application, we can describe precisely the future-directed causal geodesics in our spacetime $M$. Indeed, in Theorem 7.6 we show that if $p = (p^0, p^1, p^2, p^3)$ is a point in $M$ and $v = v^\alpha\partial_\alpha$ is a null or timelike vector at $p$, normalized with $v^0 = 1$, then there is a unique affinely parametrized global geodesic curve $\gamma : [0,\infty) \to M$ with

$$\gamma(0) = p = (p^0, p^1, p^2, p^3), \qquad \dot\gamma(0) = v = (v^0, v^1, v^2, v^3).$$

Moreover, the geodesic curve $\gamma$ becomes asymptotically parallel to a geodesic line of the Minkowski space, i.e., there is a vector $v_\infty = (v^0_\infty, v^1_\infty, v^2_\infty, v^3_\infty)$ such that, for any $s \in [0,\infty)$,

$$|\dot\gamma(s) - v_\infty| \lesssim \varepsilon_0(1+s)^{-1+6\delta'} \quad \text{and} \quad |\gamma(s) - v_\infty s - p| \lesssim \varepsilon_0(1+s)^{6\delta'}.$$

### 1.3.4 Weak Peeling Estimates

These are classical estimates on asymptotically flat spacetimes, which assert, essentially, that certain components of the Riemann curvature tensor have improved decay compared to the general estimate $|\mathbf{R}| \lesssim \varepsilon_0 \langle t+r\rangle^{-1+}\langle t-r\rangle^{-2}$. The rate of decay is mainly determined by the *signature* of the component.

More precisely, we use the Minkowski frames $(L, \underline{L}, e_a)$, where $L, \underline{L}$ are as in (1.3.14) and $e_a \in \mathcal{T}_h := \{r^{-1}\Omega_{12}, r^{-1}\Omega_{23}, r^{-1}\Omega_{31}\}$, and assign signature $+1$ to the vector-field $L$, $-1$ to the vector-field $\underline{L}$, and $0$ to the horizontal vector-fields in $\mathcal{T}_h$. With $e_1, e_2, e_3, e_4 \in \mathcal{T}_h$, we define $\mathrm{Sig}(a)$ as the set of components of the Riemann tensor of total signature $a$, so

$$
\begin{aligned}
\mathrm{Sig}(-2) &:= \{\mathbf{R}(\underline{L}, e_1, \underline{L}, e_2)\},\\
\mathrm{Sig}(2) &:= \{\mathbf{R}(L, e_1, L, e_2)\},\\
\mathrm{Sig}(-1) &:= \{\mathbf{R}(\underline{L}, e_1, e_2, e_3), \mathbf{R}(\underline{L}, L, \underline{L}, e_1)\},\\
\mathrm{Sig}(1) &:= \{\mathbf{R}(L, e_1, e_2, e_3), \mathbf{R}(L, \underline{L}, L, e_1)\},\\
\mathrm{Sig}(0) &:= \{\mathbf{R}(e_1, e_2, e_3, e_4), \mathbf{R}(L, \underline{L}, e_1, e_2), \mathbf{R}(L, e_1, \underline{L}, e_2), \mathbf{R}(L, \underline{L}, L, \underline{L})\}.
\end{aligned}
\tag{1.3.19}
$$

These components capture the entire curvature tensor, due to the symmetries (1.1.11).

In Theorem 7.7 we prove that if $\Psi_{(a)} \in \mathrm{Sig}(a)$, $a \in \{-2, -1, 1, 2\}$, then

$$
\begin{aligned}
|\Psi_{(-2)}(x,t)| &\lesssim \varepsilon_0 \langle r\rangle^{7\delta'-1}\langle t-r\rangle^{-2},\\
|\Psi_{(-1)}(x,t)| &\lesssim \varepsilon_0 \langle r\rangle^{7\delta'-2}\langle t-r\rangle^{-1},\\
|\Psi_{(2)}(x,t)| + |\Psi_{(1)}(x,t)| + |\Psi_{(0)}(x,t)| &\lesssim \varepsilon_0 \langle r\rangle^{7\delta'-3},
\end{aligned}
\tag{1.3.20}
$$

in the region $M' = \{(x,t) \in M : t \geq 1 \text{ and } |x| \geq 2^{-8}t\}$. This holds in all cases except if $\Psi_{(0)}$ is of the form $\mathbf{R}(L, e_1, \underline{L}, e_2) \in \mathrm{Sig}(0)$, in which case we can only prove the weaker bounds

$$
|\mathbf{R}(L, e_1, \underline{L}, e_2)(x,t)| \lesssim \varepsilon_0 \langle r\rangle^{7\delta'-2}\langle t-r\rangle^{-1}. \tag{1.3.21}
$$

Notice that we define our decomposition in terms of the Minkowski null pair $(L, \underline{L})$ instead of more canonical null frames (or tetrads) adapted to the metric $\mathbf{g}$ (see, for example, [12], [52], [53]). This is not important however, since the weak peeling estimates are invariant under natural changes of the frame of the form $(L, \underline{L}, e_a) \to (L', \underline{L}', e'_a)$, satisfying

$$
|(L-L')(x,t)| + |(\underline{L}-\underline{L}')(x,t)| + |(e_a - e'_a)(x,t)| \lesssim r^{-1+2\delta'} \qquad \text{in } M'.
$$

As we show in Proposition 7.9, one can in fact restore the full $\varepsilon_0\langle r\rangle^{7\delta'-3}$ decay of the component $\mathbf{R}(L', e'_1, \underline{L}', e'_2)$, provided that $L'$ is almost null, i.e., $|\mathbf{g}(L', L')(x,t)| \lesssim \langle r\rangle^{-2+4\delta'}$ in $M'$.

The almost cubic decay we prove in (1.3.20)–(1.3.21) seems optimal in our problem, for two reasons. First, the Ricci components themselves involve squares of the massive field, and cannot decay better than $\langle r\rangle^{-3+}$ in $M'$. Moreover, the almost cubic decay is also formally consistent with the weak peeling estimates of Klainerman-Nicolò [53, Theorem 1.2 (b)] in the setting of our more general metrics (formally, one would take $\gamma = -1/2-$ and $\delta = 2+$ with the notation in [53], to match our decay assumptions (1.2.13) on the initial data; this range of parameters is not allowed, however, in [53] as $\delta$ is assumed to be $< 3/2$).

### 1.3.5 The ADM Energy and the Linear Momentum

The ADM energy (or the ADM mass) measures the total deviation of our space-time from the Minkowski solution. It is calculated according to the standard formula (see, for example, [4])

$$E_{ADM}(t) := \frac{1}{16\pi} \lim_{R\to\infty} \int_{S_{R,t}} (\partial_j \mathbf{g}_{nj} - \partial_n \mathbf{g}_{jj}) \frac{x^n}{|x|}\, dx, \tag{1.3.22}$$

where the integration is over large (Euclidean) spheres $S_{R,t} \subseteq \Sigma_t = \{(x,t) : x \in \mathbb{R}^3\}$ of radius $R$. In our case we show in Proposition 7.11 that the energy $E_{ADM}(t) = E_{ADM}$ is well defined and constant in time. Moreover, it is non-negative and can be expressed in terms of the scattering profiles $V_\infty^\psi$ and $V_\infty^{\vartheta_{mn}}$ (see (1.3.13)) according to the formula

$$E_{ADM} = \frac{1}{16\pi} \|V_\infty^\psi\|_{L^2}^2 + \frac{1}{64\pi} \sum_{m,n\in\{1,2,3\}} \|V_\infty^{\vartheta_{mn}}\|_{L^2}^2. \tag{1.3.23}$$

We can also prove conservation of one other natural quantity, namely the linear momentum. Let $N$ denote the future unit normal vector-field to the hypersurface $\Sigma_t$, let $\overline{g}_{ab} = \mathbf{g}_{ab}$ denote the induced (Riemannian) metric on $\Sigma_t$, and define the second fundamental form

$$k_{ab} := -\mathbf{g}(\mathbf{D}_{\partial_a} N, \partial_b) = \mathbf{g}(N, \mathbf{D}_{\partial_a}\partial_b) = N^\alpha \mathbf{\Gamma}_{\alpha ab}, \qquad a,b \in \{1,2,3\}.$$

Then we define the linear momentum $\mathbf{p}_a$, $a \in \{1,2,3\}$,

$$\mathbf{p}_a(t) := \frac{1}{8\pi} \lim_{R\to\infty} \int_{S_{R,t}} \pi_{ab} \frac{x^b}{|x|}\, dx, \qquad \pi_{ab} := k_{ab} - (\mathrm{tr} k)\overline{g}_{ab},$$

In Proposition 7.13 we prove that the functions $\mathbf{p}_a$ are well defined and constant in time. Moreover, we show that $\sum_{a\in\{1,2,3\}} \mathbf{p}_a^2 \le E_{ADM}^2$, so the ADM mass $M_{ADM} := \big(E_{ADM}^2 - \sum_{a\in\{1,2,3\}} \mathbf{p}_a^2\big)^{1/2} \ge 0$ is well defined.

We remark that the momentum $\mathbf{p}_a$ vanishes in the case of metrics $\mathbf{g}$ that agree with the Schwarzschild metric (including time derivatives) up to lower order terms. In particular, it vanishes in the case of metrics considered in earlier work on the stability for the Einstein vacuum equations, such as [12, 52, 62]. However, in our non-isotropic case the linear momentum does not necessarily vanish, and the quantities $\mathbf{p}_a$ defined above are natural conserved quantities of the evolution.

### 1.3.6 The Bondi Energy

To define a Bondi energy we have to be more careful. We would like to compute integrals over large spheres as in (1.3.22), and then take the limit along outgoing null cones towards null infinity. But the limit exists only if we account properly for the geometry of the problem.

First we need to understand the bending of the light cones caused by the long-range effect of the nonlinearity (i.e., the modified scattering). For this we construct (in Lemma 7.19) an *almost optical function* $u : M' \to \mathbb{R}$, satisfying the properties

$$u(x,t) = |x| - t + u^{cor}(x,t), \qquad \mathbf{g}^{\alpha\beta}\partial_\alpha u \partial_\beta u = O(\varepsilon_0 \langle r\rangle^{-2+6\delta'}). \tag{1.3.24}$$

In addition, the correction $u^{cor} = O(\varepsilon_0\langle r\rangle^{3\delta'})$ is close to $\Theta_{wa}/|x|$ (see (1.3.9)) near the light cone,

$$\Big| u^{cor}(x,t) - \frac{\Theta_{wa}(x,t)}{|x|}\Big| \lesssim \varepsilon_0 \langle r\rangle^{-1+3\delta'}(\langle r\rangle^{0.68} + \langle t - |x|\rangle), \tag{1.3.25}$$

if $(x,t) \in M'$, $\big|t - |x|\big| \leq t/10$. Notice that we work with an approximate optical condition $\mathbf{g}^{\alpha\beta}\partial_\alpha u \partial_\beta u = O(\varepsilon_0\langle r\rangle^{-2+6\delta'})$ instead of the classical optical condition $\mathbf{g}^{\alpha\beta}\partial_\alpha u \partial_\beta u = 0$. This is mostly for convenience, since the weaker condition is still good enough for our analysis and almost optical functions are much easier to construct than exact optical functions.

For any $t \geq 1$ we define the hypersurface $\Sigma_t := \{(x,t) \in M : x \in \mathbb{R}^3\}$, and let $\overline{g}_{jk} = \mathbf{g}_{jk}$ denote the induced (Riemannian) metric on $\Sigma_t$. With $u$ as above, we define the modified spheres $S^u_{R,t} := \{x \in \Sigma_t : u(x,t) = R\}$ and let $\mathbf{n}_j := \partial_j u(\overline{g}^{ab}\partial_a u \partial_b u)^{-1/2}$ denote the unit vector-field normal to the spheres $S^u_{R,t}$. For $R \in \mathbb{R}$ and $t$ large (say $t \geq 2|R| + 10$) we define

$$E_{Bondi}(R) := \frac{1}{16\pi}\lim_{t\to\infty}\int_{S^u_{R,t}} \overline{g}^{ab}(\partial_a h_{jb} - \partial_j h_{ab})\mathbf{n}^j \, d\sigma, \tag{1.3.26}$$

where $d\sigma = d\sigma(\overline{g})$ is the surface measure induced by the metric $\overline{g}$. Notice that this definition is a more geometric version of the definition (1.3.22), in the sense that the integration is with respect to the metric $\overline{g}$. Geometrically, we fix $R$ and integrate on surfaces $S^u_{R,t}$ that live on the "light cone" $\{u(x,t) = R\}$

In Theorem 7.23 we prove our main result: the limit in (1.3.26) exists, and $E_{Bondi} : \mathbb{R} \to \mathbb{R}$ is a well-defined increasing and continuous function on $\mathbb{R}$, which increases from the Klein-Gordon energy $E_{KG}$ to the ADM energy $E_{ADM}$, i.e.,

$$\lim_{R\to-\infty} E_{Bondi}(R) = E_{KG} := \frac{1}{16\pi}\|V_\infty^\psi\|_{L^2}^2, \qquad \lim_{R\to\infty} E_{Bondi}(R) = E_{ADM}. \tag{1.3.27}$$

The definition (1.3.26) of the Bondi energy is consistent with the general heuristics in [73, Chapter 11] and with the definition in [67, Section 4.3.4]. It also has expected properties, like monotonicity, continuity, and satisfies the limits (1.3.27).

However, it is not clear to us if this definition is identical to the definition used by Klainerman-Nicolò [52, Chapter 8.5], starting from the Hawking mass. In fact, at the level of generality of our metrics (1.2.13), it not even clear that one can prove sharp $r^{-3}$ pointwise decay on some of the signature 0 components of the curvature tensor, which is one of the ingredients of the argument in [52].

We notice that the Klein-Gordon energy $E_{KG}$ is part of $E_{Bondi}(R)$, for all $R \in \mathbb{R}$. This is consistent with the geometric intuition, since the matter travels at speeds lower than the speed of light and accumulates at timelike infinity, not at null infinity. We can further measure its radiation by taking limits along timelike cones. Indeed, for $\alpha \in (0,1)$ let

$$E_{i^+}(\alpha) := \frac{1}{16\pi}\lim_{t\to\infty}\int_{S_{\alpha t,t}} (\partial_j h_{nj} - \partial_n h_{jj})\frac{x^n}{|x|}\,dx, \tag{1.3.28}$$

where the integration is over the Euclidean spheres $S_{\alpha t,t} \subseteq \Sigma_t$ of radius $\alpha t$. In Proposition 7.24 we prove that the limit in (1.3.28) exists, and $E_{i^+} : (0,1) \to \mathbb{R}$ is a well-defined continuous and increasing function, satisfying

$$\lim_{\alpha\to 0} E_{i^+}(\alpha) = 0, \qquad \lim_{\alpha\to 1} E_{i^+}(\alpha) = E_{KG}. \tag{1.3.29}$$

### 1.3.7 Organization

The rest of this monograph is organized as follows:

In Chapter 2 we introduce our main notations and definitions and state precisely our main bootstrap Proposition 2.3. This proposition is the key quantitative result leading to global nonlinear stability, and its proof covers Chapters 3, 4, 5, and 6. Then we provide a detailed outline of the proof of this proposition, describing at a conceptual level the entire construction and the main ingredients of the proof.

In Chapter 3 we prove several important lemmas that are being used in the rest of the analysis, such as Lemmas 3.4 and 3.6 on the structure and bounds on quadratic resonances, Lemma 3.9 concerning linear estimates for wave and Klein-Gordon evolutions, and Lemmas 3.10–3.12 concerning bilinear estimates. Finally, we use these lemmas and the bootstrap hypothesis to prove linear estimates on the solutions and the profiles, such as Lemmas 3.15 (localized $L^2$ bounds) and Lemma 3.16 (pointwise decay).

In Chapter 4 we analyze our main nonlinearities $\mathcal{LN}^h_{\alpha\beta}$ and $\mathcal{LN}^\psi$ at a fixed time $t$. The main results in this chapter are Proposition 4.7 (localized $L^2$, $L^\infty$, and weighted $L^2$ bounds on these nonlinearities), Lemmas 4.19–4.20 (identification of the energy disposable nonlinear components), and Proposition 4.22 (decomposition of the main nonlinearities).

In Chapter 5 we prove the main bootstrap bounds (2.1.50) on the energy functionals. We start from the decomposition in Proposition 4.22, perform energy estimates, and prove bounds on all the resulting spacetime integrals. The main spacetime bounds are stated in Proposition 5.2, and are proved in the rest of the chapter, using normal forms, null structures, angular decompositions, and paradifferential calculus in some of the harder cases.

In Chapter 6 we first prove the main bootstrap bounds (2.1.51) (weighted estimates on profiles) in Proposition 6.2, as a consequence of the improved energy estimates and the nonlinear bounds in Proposition 4.7. Then we prove the main bootstrap bounds (2.1.52) (the $Z$-norm estimates). This proof has several steps, such as the renormalization procedures in (6.2.4)–(6.2.6) and (6.3.3)–(6.3.4), and the estimates (6.2.14) and (6.3.16) showing boundedness and convergence of the nonlinear profiles in suitable norms in the Fourier space.

In Chapter 7 we prove a full, quantitative version of our main global regularity result (Theorem 7.1) as well as all the other consequences on the asymptotic structure of our spacetimes, as described in detail in subsections 1.3.3–1.3.6 above.

### 1.3.8 Acknowledgements

The first author was supported in part by NSF grant DMS-1600028 and NSF-FRG grant DMS-1463753. The second author was supported in part by NSF grant DMS-1362940 and by a Sloan Research fellowship.

# Chapter Two

# The Main Construction and Outline of the Proof

## 2.1 SETUP AND THE MAIN BOOTSTRAP PROPOSITION

In this section we introduce most of our notations and definitions and state our main bootstrap proposition.

### 2.1.1 The Nonlinearities $\mathcal{N}^h_{\alpha\beta}$ and $\mathcal{N}^\psi$

Let $m$ denote the Minkowski metric and write

$$\mathbf{g}_{\alpha\beta} = m_{\alpha\beta} + h_{\alpha\beta}, \qquad \mathbf{g}^{\alpha\beta} = m^{\alpha\beta} + g^{\alpha\beta}_{\geq 1}. \tag{2.1.1}$$

We start by rewriting our system as a Wave-Klein-Gordon coupled system:

**Proposition 2.1.** *Assume $(g, \psi)$ in a solution in $\mathbb{R}^3 \times [0, T]$ of the reduced Einstein-Klein-Gordon system in harmonic gauge (1.2.6)–(1.2.7). For $\alpha, \beta \in \{0, 1, 2, 3\}$ we have*

$$(\partial_0^2 - \Delta)h_{\alpha\beta} = \mathcal{N}^h_{\alpha\beta} := \mathcal{KG}_{\alpha\beta} + g^{\mu\nu}_{\geq 1}\partial_\mu\partial_\nu h_{\alpha\beta} - F^{\geq 2}_{\alpha\beta}(g, \partial g), \tag{2.1.2}$$

*where*

$$\mathcal{KG}_{\alpha\beta} := 2\partial_\alpha\psi\partial_\beta\psi + \psi^2(m_{\alpha\beta} + h_{\alpha\beta}). \tag{2.1.3}$$

*Moreover,*

$$(\partial_0^2 - \Delta + 1)\psi = \mathcal{N}^\psi := g^{\mu\nu}_{\geq 1}\partial_\mu\partial_\nu\psi. \tag{2.1.4}$$

*In addition, the nonlinearities $F^{\geq 2}_{\alpha\beta}(g, \partial g)$ admit the decompositions*

$$F^{\geq 2}_{\alpha\beta}(g, \partial g) = Q_{\alpha\beta} + P_{\alpha\beta}, \tag{2.1.5}$$

*where*

$$\begin{aligned} Q_{\alpha\beta} := &\, \mathbf{g}^{\rho\rho'}\mathbf{g}^{\lambda\lambda'}(\partial_\alpha h_{\rho'\lambda'}\partial_\rho h_{\beta\lambda} - \partial_\rho h_{\rho'\lambda'}\partial_\alpha h_{\beta\lambda}) \\ &+ \mathbf{g}^{\rho\rho'}\mathbf{g}^{\lambda\lambda'}(\partial_\beta h_{\rho'\lambda'}\partial_\rho h_{\alpha\lambda} - \partial_\rho h_{\rho'\lambda'}\partial_\beta h_{\alpha\lambda}) \\ &+ (1/2)\mathbf{g}^{\rho\rho'}\mathbf{g}^{\lambda\lambda'}(\partial_{\lambda'} h_{\rho\rho'}\partial_\beta h_{\alpha\lambda} - \partial_\beta h_{\rho\rho'}\partial_{\lambda'} h_{\alpha\lambda}) \\ &+ (1/2)\mathbf{g}^{\rho\rho'}\mathbf{g}^{\lambda\lambda'}(\partial_{\lambda'} h_{\rho\rho'}\partial_\alpha h_{\beta\lambda} - \partial_\alpha h_{\rho\rho'}\partial_{\lambda'} h_{\beta\lambda}) \\ &- \mathbf{g}^{\rho\rho'}\mathbf{g}^{\lambda\lambda'}(\partial_\lambda h_{\alpha\rho'}\partial_\rho h_{\beta\lambda'} - \partial_\rho h_{\alpha\rho'}\partial_\lambda h_{\beta\lambda'}) + \mathbf{g}^{\rho\rho'}\mathbf{g}^{\lambda\lambda'}\partial_{\rho'} h_{\alpha\lambda'}\partial_\rho h_{\beta\lambda} \end{aligned} \tag{2.1.6}$$

*and*

$$P_{\alpha\beta} := -\frac{1}{2}\mathbf{g}^{\rho\rho'}\mathbf{g}^{\lambda\lambda'}\partial_\alpha h_{\rho'\lambda'}\partial_\beta h_{\rho\lambda} + \frac{1}{4}\mathbf{g}^{\rho\rho'}\mathbf{g}^{\lambda\lambda'}\partial_\alpha h_{\rho\rho'}\partial_\beta h_{\lambda\lambda'}. \tag{2.1.7}$$

*Proof.* The identities (2.1.2) and (2.1.4) follow directly from the system (1.2.6). The identities (2.1.5), which allow us to extract the null components of the semilinear nonlinearities, follow by explicit calculations from the identities (1.1.15); see, for example, [58, Lemma 4.1] and notice that $\partial_\rho h_{\mu\nu} = \partial_\rho g_{\mu\nu}$. □

We often need to extract the linear part of the matrix $g^{\alpha\beta}_{\geq 1}$. So we write $g^{\alpha\beta}_{\geq 1} = g^{\alpha\beta}_1 + g^{\alpha\beta}_{\geq 2}$ and use the identity

$$\begin{aligned}\delta^\alpha_\beta &= \mathbf{g}^{\alpha\rho}\mathbf{g}_{\beta\rho} = (m^{\alpha\rho} + g^{\alpha\rho}_1 + g^{\alpha\rho}_{\geq 2})(m_{\beta\rho} + h_{\beta\rho}) \\ &= \delta^\alpha_\beta + (m^{\alpha\rho}h_{\beta\rho} + g^{\alpha\rho}_1 m_{\beta\rho}) + (g^{\alpha\rho}_1 h_{\beta\rho} + g^{\alpha\rho}_{\geq 2}m_{\beta\rho} + g^{\alpha\rho}_{\geq 2}h_{\beta\rho}).\end{aligned}$$

Therefore, we can define $g^{\alpha\beta}_1$ and $g^{\alpha\beta}_{\geq 2}$ by

$$\begin{aligned} &g^{00}_1 = -h_{00}, \qquad g^{0j}_1 = g^{j0}_1 = h_{0j}, \qquad g^{jk}_1 = -h_{jk}, \\ &g^{\alpha\rho}_{\geq 2}m_{\beta\rho} + g^{\alpha\rho}_{\geq 2}h_{\beta\rho} + g^{\alpha\rho}_1 h_{\beta\rho} = 0. \end{aligned} \tag{2.1.8}$$

### 2.1.1.1 *The quadratic nonlinearities*

We identify now the quadratic components of the nonlinearities $\mathcal{N}^h_{\alpha\beta}$ and $\mathcal{N}^\psi$, which play a key role in the nonlinear evolution. We start from the identities (2.1.2) and rewrite them in the form

$$\mathcal{N}^h_{\alpha\beta} = g^{00}_{\geq 1}\partial_0^2 h_{\alpha\beta} + \sum_{(\mu,\nu)\neq(0,0)} g^{\mu\nu}_{\geq 1}\partial_\mu\partial_\nu h_{\alpha\beta} + \mathcal{KG}_{\alpha\beta} - F^{\geq 2}_{\alpha\beta}(g, \partial g).$$

We would like to eliminate the terms that contain two time derivatives, in order to express the nonlinearities in terms of the normalized solutions $U^{h_{\alpha\beta}}$ and $U^\psi$ (defined in (2.1.32) below). Indeed, since $\partial_0^2 h_{\alpha\beta} = \Delta h_{\alpha\beta} + \mathcal{N}^h_{\alpha\beta}$, we have

$$\mathcal{N}^h_{\alpha\beta} = (1 - g^{00}_{\geq 1})^{-1}\Big[\sum_{(\mu,\nu)\neq(0,0)} g^{\mu\nu}_{\geq 1}\partial_\mu\partial_\nu h_{\alpha\beta} + g^{00}_{\geq 1}\Delta h_{\alpha\beta} + \mathcal{KG}_{\alpha\beta} - F^{\geq 2}_{\alpha\beta}(g, \partial g)\Big]. \tag{2.1.9}$$

Recall the decomposition of the metric components,

$$\mathbf{g}^{\alpha\beta} = m^{\alpha\beta} + g^{\alpha\beta}_{\geq 1} = m^{\alpha\beta} + g^{\alpha\beta}_1 + g^{\alpha\beta}_{\geq 2}, \tag{2.1.10}$$

into the Minkowski metric, a linearized metric, and a quadratic metric. Using this decomposition we extract the quadratic components of the nonlinearity

$$\mathcal{N}^{h,2}_{\alpha\beta} := \mathcal{KG}^2_{\alpha\beta} + \mathcal{Q}^2_{\alpha\beta} + \mathcal{S}^2_{\alpha\beta}, \tag{2.1.11}$$

where $\mathcal{KG}^2_{\alpha\beta}$ are semilinear quadratic terms that involve the Klein-Gordon field,

$$\mathcal{KG}^2_{\alpha\beta} := 2\partial_\alpha\psi\partial_\beta\psi + \psi^2 m_{\alpha\beta}, \tag{2.1.12}$$

$\mathcal{Q}^2_{\alpha\beta}$ are quasilinear quadratic terms,

$$\begin{aligned}\mathcal{Q}^2_{\alpha\beta} &:= \sum_{(\mu,\nu)\neq(0,0)} g_1^{\mu\nu}\partial_\mu\partial_\nu h_{\alpha\beta} + g_1^{00}\Delta h_{\alpha\beta} \\ &= -h_{00}\Delta h_{\alpha\beta} + 2h_{0j}\partial_0\partial_j h_{\alpha\beta} - h_{jk}\partial_j\partial_k h_{\alpha\beta},\end{aligned} \tag{2.1.13}$$

(compare with the formulas (2.1.9)), and

$$\mathcal{S}^2_{\alpha\beta} := -(Q^2_{\alpha\beta} + P^2_{\alpha\beta})$$

are semilinear quadratic terms that involve the metric components, where

$$\begin{aligned}Q^2_{\alpha\beta} := {} & m^{\rho\rho'}m^{\lambda\lambda'}(\partial_\alpha h_{\rho'\lambda'}\partial_\rho h_{\beta\lambda} - \partial_\rho h_{\rho'\lambda'}\partial_\alpha h_{\beta\lambda}) \\ & + m^{\rho\rho'}m^{\lambda\lambda'}(\partial_\beta h_{\rho'\lambda'}\partial_\rho h_{\alpha\lambda} - \partial_\rho h_{\rho'\lambda'}\partial_\beta h_{\alpha\lambda}) \\ & + (1/2)m^{\rho\rho'}m^{\lambda\lambda'}(\partial_{\lambda'} h_{\rho\rho'}\partial_\beta h_{\alpha\lambda} - \partial_\beta h_{\rho\rho'}\partial_{\lambda'} h_{\alpha\lambda}) \\ & + (1/2)m^{\rho\rho'}m^{\lambda\lambda'}(\partial_{\lambda'} h_{\rho\rho'}\partial_\alpha h_{\beta\lambda} - \partial_\alpha h_{\rho\rho'}\partial_{\lambda'} h_{\beta\lambda}) \\ & - m^{\rho\rho'}m^{\lambda\lambda'}(\partial_\lambda h_{\alpha\rho'}\partial_\rho h_{\beta\lambda'} - \partial_\rho h_{\alpha\rho'}\partial_\lambda h_{\beta\lambda'} - \partial_{\rho'} h_{\alpha\lambda'}\partial_\rho h_{\beta\lambda})\end{aligned} \tag{2.1.14}$$

and

$$P^2_{\alpha\beta} := -\frac{1}{2}m^{\rho\rho'}m^{\lambda\lambda'}\partial_\alpha h_{\rho'\lambda'}\partial_\beta h_{\rho\lambda} + \frac{1}{4}m^{\rho\rho'}m^{\lambda\lambda'}\partial_\alpha h_{\rho\rho'}\partial_\beta h_{\lambda\lambda'}. \tag{2.1.15}$$

Compare with the formulas (2.1.6)–(2.1.7). Let $\mathcal{N}^{h,\geq 3}_{\alpha\beta} := \mathcal{N}^h_{\alpha\beta} - \mathcal{N}^{h,2}_{\alpha\beta}$ denote the cubic and higher order components of $\mathcal{N}^h_{\alpha\beta}$.

Similarly, Klein-Gordon nonlinearities defined in (2.1.4) can be written as

$$\mathcal{N}^\psi = \sum_{(\mu,\nu)\neq(0,0)} g^{\mu\nu}_{\geq 1}\partial_\mu\partial_\nu\psi + g^{00}_{\geq 1}\partial_0^2\psi = \sum_{(\mu,\nu)\neq(0,0)} g^{\mu\nu}_{\geq 1}\partial_\mu\partial_\nu\psi + g^{00}_{\geq 1}(\Delta\psi - \psi + \mathcal{N}^\psi).$$

Therefore

$$\mathcal{N}^\psi = (1 - g^{00}_{\geq 1})^{-1}\Big[\sum_{(\mu,\nu)\neq(0,0)} g^{\mu\nu}_{\geq 1}\partial_\mu\partial_\nu\psi + g^{00}_{\geq 1}(\Delta\psi - \psi)\Big]. \tag{2.1.16}$$

Using also the identities (2.1.8) we extract the quadratic component of $\mathcal{N}^\psi$,

$$\begin{aligned}\mathcal{N}^{\psi,2} &:= \sum_{(\mu,\nu)\neq(0,0)} g_1^{\mu\nu}\partial_\mu\partial_\nu\psi + g_1^{00}(\Delta\psi - \psi) \\ &= -h_{00}(\Delta\psi - \psi) + 2h_{0j}\partial_0\partial_j\psi - h_{jk}\partial_j\partial_k\psi,\end{aligned} \tag{2.1.17}$$

and let $\mathcal{N}^{\psi,\geq 3} := \mathcal{N}^{\psi} - \mathcal{N}^{\psi,2}$ denote its cubic and higher order component.

### 2.1.2 The Fourier Transform and Frequency Projections

We will use extensively the Fourier transform and the Fourier inversion formula on $\mathbb{R}^3$,

$$\widehat{f}(\xi) = \mathcal{F}(f)(\xi) := \int_{\mathbb{R}^3} f(x)e^{-ix\cdot\xi}\,dx, \qquad f(x) = \frac{1}{(2\pi)^3}\int_{\mathbb{R}^3} \widehat{f}(\xi)e^{-ix\cdot\xi}\,d\xi, \tag{2.1.18}$$

defined for suitable functions $f : \mathbb{R}^3 \to \mathbb{C}$. We fix $\varphi : \mathbb{R} \to [0,1]$ an even smooth function supported in $[-8/5, 8/5]$ and equal to 1 in $[-5/4, 5/4]$. For simplicity of notation, we also let $\varphi : \mathbb{R}^3 \to [0,1]$ denote the corresponding radial function on $\mathbb{R}^3$. For any $k \in \mathbb{Z}$ and $I \subseteq \mathbb{R}$ let

$$\varphi_k(x) := \varphi(|x|/2^k) - \varphi(|x|/2^{k-1}), \qquad \varphi_I := \sum_{m\in I\cap\mathbb{Z}} \varphi_m.$$

For any $B \in \mathbb{R}$ let

$$\varphi_{\leq B} := \varphi_{(-\infty,B]}, \quad \varphi_{\geq B} := \varphi_{[B,\infty)}, \quad \varphi_{<B} := \varphi_{(-\infty,B)}, \quad \varphi_{>B} := \varphi_{(B,\infty)}.$$

For any $a < b \in \mathbb{Z}$ and $j \in [a,b] \cap \mathbb{Z}$ let

$$\varphi_j^{[a,b]} := \begin{cases} \varphi_j & \text{if } a < j < b, \\ \varphi_{\leq a} & \text{if } j = a, \\ \varphi_{\geq b} & \text{if } j = b. \end{cases} \tag{2.1.19}$$

Let $P_k$, $k \in \mathbb{Z}$, (respectively $P_I$, $I \subseteq \mathbb{R}$) denote the operators on $\mathbb{R}^3$ defined by the Fourier multipliers $\xi \to \varphi_k(\xi)$ (respectively $\xi \to \varphi_I(\xi)$). For simplicity of notation let $P'_k = P_{[k-2,k+2]}$.

For any $x \in \mathbb{Z}$ let $x^+ = \max(x,0)$ and $x^- := \min(x,0)$. Let

$$\mathcal{J} := \{(k,j) \in \mathbb{Z} \times \mathbb{Z}_+ : k + j \geq 0\}.$$

For any $(k,j) \in \mathcal{J}$ let

$$\widetilde{\varphi}_j^{(k)}(x) := \begin{cases} \varphi_{\leq -k}(x) & \text{if } k + j = 0 \text{ and } k \leq 0, \\ \varphi_{\leq 0}(x) & \text{if } j = 0 \text{ and } k \geq 0, \\ \varphi_j(x) & \text{if } k + j \geq 1 \text{ and } j \geq 1, \end{cases}$$

and notice that, for any $k \in \mathbb{Z}$ fixed, $\sum_{j\geq -\min(k,0)} \widetilde{\varphi}_j^{(k)} = 1$.

For $(k,j) \in \mathcal{J}$ let $Q_{j,k}$ denote the operator

$$(Q_{j,k}f)(x) := \widetilde{\varphi}_j^{(k)}(x) \cdot P_k f(x). \tag{2.1.20}$$

In view of the uncertainty principle the operators $Q_{j,k}$ are relevant only when $2^j 2^k \gtrsim 1$, which explains the definitions above.

We will often estimate bilinear interactions, like products of two functions. For $k \in \mathbb{Z}$ let

$$\mathcal{X}_k := \{(k_1, k_2) \in \mathbb{Z}^2 : |\max(k, k_1, k_2) - \mathrm{med}(k, k_1, k_2)| \leq 4\}. \tag{2.1.21}$$

Notice that $P_k(P_{k_1} f \cdot P_{k_2} g) \equiv 0$ if $(k_1, k_2) \notin \mathcal{X}_k$.

### 2.1.3 Vector-fields

Recall the vector-fields $\Gamma_j$ and $\Omega_{jk}$ defined in (1.3.1),

$$\Gamma_j := x_j \partial_t + t\partial_j, \qquad \Omega_{jk} := x_j \partial_k - x_k \partial_j,$$

for $j, k \in \{1, 2, 3\}$. These vector-fields satisfy simple commutation relations, which can be written schematically in the form

$$\begin{aligned} &[\partial, \partial] = 0, \qquad [\partial, \Omega] = \partial, \qquad [\partial, \Gamma] = \partial, \\ &[\Omega, \Omega] = \Omega, \qquad [\Omega, \Gamma] = \Gamma, \qquad [\Gamma, \Gamma] = \Omega, \end{aligned} \tag{2.1.22}$$

where $\partial$ denotes generic coordinate vector-fields, $\Omega$ denotes generic rotation vector-fields, and $\Gamma$ denotes generic Lorentz vector-fields. For $\alpha = (\alpha_1, \alpha_2, \alpha_3) \in (\mathbb{Z}_+)^3$ we define

$$\partial^\alpha := \partial_1^{\alpha_1} \partial_2^{\alpha_2} \partial_3^{\alpha_3}, \qquad \Omega^\alpha := \Omega_{23}^{\alpha_1} \Omega_{31}^{\alpha_2} \Omega_{12}^{\alpha_3}, \qquad \Gamma^\alpha := \Gamma_1^{\alpha_1} \Gamma_2^{\alpha_2} \Gamma_3^{\alpha_3}. \tag{2.1.23}$$

For any $n, q \in \mathbb{Z}_+$ we define $\mathcal{V}_n^q$ as the set of differential operators of the form

$$\mathcal{V}_n^q := \{\mathcal{L} = \Gamma^a \Omega^b : |a| + |b| \leq n,\, q(\mathcal{L}) := |a| \leq q\}. \tag{2.1.24}$$

Here $q(\mathcal{L})$ denotes the number of vector-fields transversal to the surfaces $\Sigma_a := \{(x, t) \in \mathbb{R}^3 \times \mathbb{R} : t = a\}$. We remark that in our proof we distinguish between the Lorentz vector-fields $\Gamma$ (which lead to slightly faster growth rates; see the definition (2.1.49)) and the rotational vector-fields $\Omega$. Notice that, for any $\alpha \in \{0, 1, 2, 3\}$ and any $\mathcal{L}_1 \in \mathcal{V}_{n_1}^{q_1}$, $\mathcal{L}_2 \in \mathcal{V}_{n_2}^{q_2}$, we have

$$\begin{aligned} \mathcal{L}_1 \mathcal{L}_2 &= \text{ sum of operators in } \mathcal{V}_{n_1+n_2}^{q_1+q_2}, \\ [\partial_\alpha, \mathcal{L}_1] &= \text{ sum of operators of the form } \partial_\beta \mathcal{L}', \beta \in \{0, 1, 2, 3\},\, \mathcal{L}' \in \mathcal{V}_{n_1-1}^{q_1}. \end{aligned} \tag{2.1.25}$$

### 2.1.4 Decomposition of the Metric Tensor

Let $R_j = |\nabla|^{-1}\partial_j$, $j \in \{1, 2, 3\}$, denote the Riesz transforms on $\mathbb{R}^3$, and notice that $\delta_{jk} R_j R_k = -I$. To identify null structures we use a double Hodge decomposition for the metric tensor, which is connected to the work of Arnowitt-

Deser-Misner [3] on the Hamiltonian formulation of General Relativity.

Recall the variables $F, \underline{F}, \rho, \omega_j, \Omega_j, \vartheta_{jk}$ defined in (1.3.5). To include vector-fields we need to expand this definition. More precisely, given a symmetric covariant 2-tensor $H_{\alpha\beta}$ we define

$$\begin{aligned} F = F[H] &:= (1/2)[H_{00} + R_j R_k H_{jk}], \\ \underline{F} = \underline{F}[H] &:= (1/2)[H_{00} - R_j R_k H_{jk}], \\ \rho = \rho[H] &:= R_j H_{0j}, \\ \omega_j = \omega_j[H] &:= \in_{jkl} R_k H_{0l}, \\ \Omega_j = \Omega_j[H] &:= \in_{jkl} R_k R_m H_{lm}, \\ \vartheta_{jk} = \vartheta_{jk}[H] &:= \in_{jmp} \in_{knq} R_m R_n H_{pq}. \end{aligned} \tag{2.1.26}$$

Notice that $\omega[H]$ and $\Omega[H]$ are divergence-free vector-fields,

$$R_j \omega_j[H] = 0, \qquad R_j \Omega_j[H] = 0, \tag{2.1.27}$$

and $\vartheta[H]$ is a symmetric and divergence-free tensor-field,

$$\vartheta_{jk}[H] = \vartheta_{kj}[H], \qquad R_j \vartheta_{jk}[H] = 0, \qquad R_k \vartheta_{jk}[H] = 0. \tag{2.1.28}$$

This provides an orthogonal decomposition of $h_{\alpha\beta}$ in terms of $\{F, \underline{F}, \rho, \omega, \Omega, \vartheta\}$ (see Lemma 7.16). Using the general formula $\in_{mnk} \in_{pqk} = \delta_{mp}\delta_{nq} - \delta_{mq}\delta_{np}$ we notice that we can recover the tensor $H$ according to the identities

$$\begin{aligned} H_{00} &= F[H] + \underline{F}[H], \\ H_{0j} &= -R_j \rho[H] + \in_{jkl} R_k \omega_l[H], \\ H_{jk} &= R_j R_k (F[H] - \underline{F}[H]) - (\in_{klm} R_j + \in_{jlm} R_k) R_l \Omega_m[H] \\ &\quad + \in_{jpm} \in_{kqn} R_p R_q \vartheta_{mn}[H]. \end{aligned} \tag{2.1.29}$$

We often apply this decomposition to the tensor $H_{\alpha\beta} = \mathcal{L} h_{\alpha\beta}$, $\mathcal{L} \in \mathcal{V}_3^3$. As a general rule, here and in other places, we first apply all the vector-fields to the components $h_{\alpha\beta}$, and then take the Riesz transforms. So we define the variables

$$G^{\mathcal{L}} := G[\mathcal{L}h], \qquad G \in \{F, \underline{F}, \rho, \omega_j, \Omega_j, \vartheta_{jk}\}. \tag{2.1.30}$$

For simplicity of notation, let $G = G^{Id}$, $G \in \{F, \underline{F}, \rho, \omega_j, \Omega_j, \vartheta_{jk}\}$, corresponding to the identity operator $\mathcal{L} = Id$. As a consequence of the harmonic gauge condition, we will show in Lemma 4.15 that the main dynamical variables are $F, \underline{F}, \omega_j$ and the traceless part of $\vartheta_{jk}$, while the variables $\rho$ and $\Omega_j$, can be expressed elliptically in terms of these variables, up to quadratic remainders. This is important in identifying suitable null structures of the nonlinearities $\mathcal{N}^h_{\alpha\beta}$ in section 4.3.

This $3+1$ formalism, where the spacetime is foliated by time slices $\Sigma_t$, is commonly used in general relativity, and is connected to the Hamiltonian formulation of Einstein's equations (see e.g. [3, 65]). In this context, one can think of the unknowns being evolved as the spatial metric $\overline{g}$ and the second fundamental form $k$. The foliation defines two kinematic objects: the *lapse* $N$ and *shift* vector-field $N_j$, which in our setting determine $F+\underline{F}$ and $\rho$, $\omega$ through the Hodge decomposition on $\Sigma_t$:

$$\begin{aligned} N &= (-g^{00})^{-1/2} = 1 - h_{00}/2 + O(h^2) = -(F+\underline{F})/2 + O(h^2),\\ N_j &= h_{0j} = -\partial_j(|\nabla|^{-1}\rho) + \in_{jkl}\partial_k(|\nabla|^{-1}\omega)_l. \end{aligned} \tag{2.1.31}$$

Two other gauge components $(F-\underline{F})$ and $\Omega$ are associated to choices of coordinates on $\Sigma_t$ through the Lie-derivative of the Euclidean metric $\delta_{Eucl}$:

$$\begin{aligned} &\overline{g}_{jk} = g_{jk} = (\mathcal{L}_X\delta_{Eucl})_{jk} + \in_{jpm}\in_{kqn} R_pR_q\vartheta_{mn},\\ &X_a := \frac{1}{2}\partial_a|\nabla|^{-2}(F-\underline{F}) - \in_{abc}\partial_b(|\nabla|^{-2}\Omega)_c, \quad (\mathcal{L}_X\delta_{Eucl})_{jk} = \partial_jX_k + \partial_kX_j. \end{aligned}$$

Finally, the last component $\vartheta$ is (to first order) coordinate-independent on $\Sigma_t$. Proposition 7.14 shows that it can be understood as the expression of the Riemannian metric on $\Sigma_t$ *in spatially harmonic coordinates*. This quantity plays a central role in our analysis.

### 2.1.5 Linear Profiles and the $Z$-norms

We define the *normalized solutions* $U^{h_{\alpha\beta}}$, $U^F$, $U^{\underline{F}}$, $U^\rho$, $U^{\omega_a}$, $U^{\Omega_a}$, $U^{\vartheta_{ab}}$, $U^\psi$ and their associated *linear profiles* $V^{h_{\alpha\beta}}$, $V^F$, $V^{\underline{F}}$, $V^\rho$, $V^{\omega_a}$, $V^{\Omega_a}$, $V^{\vartheta_{ab}}$, $V^\psi$, $\alpha,\beta\in\{0,1,2,3\}$, $a,b\in\{1,2,3\}$, by

$$\begin{aligned} U^G(t) &:= \partial_tG(t) - i\Lambda_{wa}G(t), &\quad V^G(t) &:= e^{it\Lambda_{wa}}U^G(t),\\ U^\psi(t) &:= \partial_t\psi(t) - i\Lambda_{kg}\psi(t), &\quad V^\psi(t) &:= e^{it\Lambda_{kg}}U^\psi(t), \end{aligned} \tag{2.1.32}$$

where $G\in\{h_{\alpha\beta}, F, \underline{F}, \rho, \omega_a, \Omega_a, \vartheta_{ab}\}$ and

$$\Lambda_{wa} := |\nabla|, \qquad \Lambda_{kg} := \langle\nabla\rangle = \sqrt{|\nabla|^2+1}. \tag{2.1.33}$$

More generally, for $\mathcal{L}\in\mathcal{V}_3^3$ (see definition (2.1.24)) we define the *weighted normalized solutions* $U^*$ and the *weighted linear profiles* $V^*$ by the formulas

$$\begin{aligned} U^{\mathcal{L}h_{\alpha\beta}}(t) &:= (\partial_t - i\Lambda_{wa})(\mathcal{L}h_{\alpha\beta})(t), &\quad V^{\mathcal{L}h_{\alpha\beta}}(t) &:= e^{it\Lambda_{wa}}U^{\mathcal{L}h_{\alpha\beta}}(t),\\ U^{G^{\mathcal{L}}}(t) &:= (\partial_t - i\Lambda_{wa})(G^{\mathcal{L}})(t), &\quad V^{G^{\mathcal{L}}}(t) &:= e^{it\Lambda_{wa}}U^{G^{\mathcal{L}}}(t),\\ U^{\mathcal{L}\psi}(t) &:= (\partial_t - i\Lambda_{kg})(\mathcal{L}\psi)(t), &\quad V^{\mathcal{L}\psi}(t) &:= e^{it\Lambda_{kg}}U^{\mathcal{L}\psi}(t), \end{aligned} \tag{2.1.34}$$

for $G^{\mathcal{L}}\in\{F^{\mathcal{L}}, \underline{F}^{\mathcal{L}}, \rho^{\mathcal{L}}, \omega_a^{\mathcal{L}}, \Omega_a^{\mathcal{L}}, \vartheta_{ab}^{\mathcal{L}}\}$. Finally, let

$$U^{*,+} := U^*, \qquad U^{*,-} := \overline{U^*}, \qquad V^{*,+} := V^*, \qquad V^{*,-} := \overline{V^*}. \tag{2.1.35}$$

for any $* \in \{F^{\mathcal{L}}, \underline{F}^{\mathcal{L}}, \rho^{\mathcal{L}}, \omega_a^{\mathcal{L}}, \Omega_a^{\mathcal{L}}, \vartheta_{ab}^{\mathcal{L}}, \mathcal{L}h_{\alpha\beta}, \mathcal{L}\psi\}$.

The functions $F^{\mathcal{L}}, \underline{F}^{\mathcal{L}}, \rho^{\mathcal{L}}, \omega_a^{\mathcal{L}}, \Omega_a^{\mathcal{L}}, \vartheta_{ab}^{\mathcal{L}}, \mathcal{L}h_{\alpha\beta}, \mathcal{L}\psi$ can be recovered linearly from the normalized variables $U^{F^{\mathcal{L}}}$, $U^{\underline{F}^{\mathcal{L}}}$, $U^{\rho^{\mathcal{L}}}$, $U^{\omega_a^{\mathcal{L}}}$, $U^{\Omega_a^{\mathcal{L}}}$, $U^{\vartheta_{ab}^{\mathcal{L}}}$, $U^{\mathcal{L}h_{\alpha\beta}}$, $U^{\mathcal{L}\psi}$ by the formulas

$$\begin{aligned} \partial_0 G &= (U^G + \overline{U^G})/2, \qquad & \Lambda_{wa} G &= i(U^G - \overline{U^G})/2, \\ \partial_0 \mathcal{L}\psi &= (U^{\mathcal{L}\psi} + \overline{U^{\mathcal{L}\psi}})/2, \qquad & \Lambda_{kg} \mathcal{L}\psi &= i(U^{\mathcal{L}\psi} - \overline{U^{\mathcal{L}\psi}})/2, \end{aligned} \tag{2.1.36}$$

where $G \in \{F^{\mathcal{L}}, \underline{F}^{\mathcal{L}}, \rho^{\mathcal{L}}, \omega_a^{\mathcal{L}}, \Omega_a^{\mathcal{L}}, \vartheta_{ab}^{\mathcal{L}}, \mathcal{L}h_{\alpha\beta}\}$.

The identities (2.1.2) and (2.1.4) show that

$$(\partial_t^2 + \Lambda_{wa}^2)(\mathcal{L}h_{\alpha\beta}) = \mathcal{L}\mathcal{N}_{\alpha\beta}^h, \qquad (\partial_t^2 + \Lambda_{kg}^2)(\mathcal{L}\psi) = \mathcal{L}\mathcal{N}^{\psi}, \tag{2.1.37}$$

for any $\alpha, \beta \in \{0,1,2,3\}$ and $\mathcal{L} \in \mathcal{V}_3^3$. Using the definitions (2.1.34), we have

$$(\partial_t + i\Lambda_{wa}) U^{\mathcal{L}h_{\alpha\beta}} = \mathcal{L}\mathcal{N}_{\alpha\beta}^h, \qquad (\partial_t + i\Lambda_{kg}) U^{\mathcal{L}\psi} = \mathcal{L}\mathcal{N}^{\psi}. \tag{2.1.38}$$

In terms of the linear profiles, these basic identities become

$$\partial_t V^{\mathcal{L}h_{\alpha\beta}}(t) = e^{it\Lambda_{wa}} \mathcal{L}\mathcal{N}_{\alpha\beta}^h(t), \qquad \partial_t V^{\mathcal{L}\psi}(t) = e^{it\Lambda_{kg}} \mathcal{L}\mathcal{N}^{\psi}(t), \tag{2.1.39}$$

for any $t \in [0,T]$, $\alpha, \beta \in \{0,1,2,3\}$, and $\mathcal{L} \in \mathcal{V}_3^3$.

Let

$$\mathcal{P} := \{(wa,+), (wa,-), (kg,+), (kg,-)\}. \tag{2.1.40}$$

Let $\Lambda_{wa,+}(\xi) = \Lambda_{wa}(\xi) = |\xi|$, $\Lambda_{wa,-}(\xi) = -\Lambda_{wa,+}(\xi)$, $\Lambda_{kg,+}(\xi) = \Lambda_{kg}(\xi) = \sqrt{|\xi|^2 + 1}$, $\Lambda_{kg,-}(\xi) = -\Lambda_{kg,+}(\xi)$. For any $\sigma, \mu, \nu \in \mathcal{P}$ we define the *quadratic phase function*

$$\Phi_{\sigma\mu\nu} : \mathbb{R}^3 \times \mathbb{R}^3 \to \mathbb{R}, \qquad \Phi_{\sigma\mu\nu}(\xi, \eta) := \Lambda_\sigma(\xi) - \Lambda_\mu(\xi - \eta) - \Lambda_\nu(\eta). \tag{2.1.41}$$

In our analysis we will need a few parameters:

$$N_0 := 40, \quad d := 10, \quad \kappa := 10^{-3}, \quad \delta := 10^{-10}, \quad \delta' := 2000\delta, \quad \gamma := \delta/4. \tag{2.1.42}$$

We define also the numbers $N(n)$ (which measure the number of Sobolev derivatives under control at the level of $n$ vector-fields),

$$N(0) := N_0 + 16d, \qquad N(n) := N_0 - dn \ \text{ for } n \in \{1,2,3\}. \tag{2.1.43}$$

Let $|\xi|_{\leq 1}$ denote a smooth increasing radial function on $\mathbb{R}^3$ equal to $|\xi|$ if $|\xi| \leq 1/2$ and equal to 1 if $|\xi| \geq 2$. Let $|\nabla|_{\leq 1}^{\gamma}$ denote the associated operator defined by the multiplier $\xi \to |\xi|_{\leq 1}^{\gamma}$.

We are now ready to define the main $Z$-norms.

**Definition 2.2.** *For any* $x \in \mathbb{R}$ *let* $x^+ = \max(x,0)$ *and* $x^- = \min(x,0)$. *We* *define the spaces* $Z_{wa}$ *and* $Z_{kg}$ *by the norms*

$$\|f\|_{Z_{wa}} := \sup_{k\in\mathbb{Z}} 2^{N_0 k^+} 2^{k^-(1+\kappa)} \|\widehat{P_k f}\|_{L^\infty} \tag{2.1.44}$$

*and*

$$\|f\|_{Z_{kg}} := \sup_{k\in\mathbb{Z}} 2^{N_0 k^+} 2^{k^-(1/2-\kappa)} \|\widehat{P_k f}\|_{L^\infty}. \tag{2.1.45}$$

### 2.1.6 The Main Bootstrap Proposition

Our main result is the following proposition:

**Proposition 2.3.** *Assume that* $(g, \psi)$ *is a solution of the system* (1.2.6)–(1.2.7) *on the time interval* $[0, T]$, $T \geq 1$, *with an initial data set* $(\overline{g}_{ij}, k_{ij}, \psi_0, \psi_1)$ *that satisfies the smallness conditions* (1.2.5) *and the constraint equations* (1.2.4).

*Define* $U^G, U^{\mathcal{L}h_{\alpha\beta}}, U^{\mathcal{L}\psi}$ *as in* (2.1.32)–(2.1.34) *and recall* (2.1.24). *Assume that, for any* $t \in [0, T]$, *the solution satisfies the bootstrap hypothesis*

$$\begin{aligned}\sup_{q\leq n\leq 3,\, \mathcal{L}\in\mathcal{V}_n^q} \langle t\rangle^{-H(q,n)\delta} \{ &\|(\langle t\rangle |\nabla|_{\leq 1})^\gamma |\nabla|^{-1/2} U^{\mathcal{L}h_{\alpha\beta}}(t)\|_{H^{N(n)}} \\ &+ \|U^{\mathcal{L}\psi}(t)\|_{H^{N(n)}} \} \leq \varepsilon_1,\end{aligned} \tag{2.1.46}$$

$$\begin{aligned}&\sup_{q\leq n\leq 2,\, \mathcal{L}\in\mathcal{V}_n^q}\ \sup_{k\in\mathbb{Z},\, l\in\{1,2,3\}} 2^{N(n+1)k^+} \langle t\rangle^{-H(q+1,n+1)\delta} \\ &\{2^{k/2}(2^{k^-}\langle t\rangle)^\gamma \|P_k(x_l V^{\mathcal{L}h_{\alpha\beta}})(t)\|_{L^2} + 2^{k^+}\|P_k(x_l V^{\mathcal{L}\psi})(t)\|_{L^2}\} \leq \varepsilon_1,\end{aligned} \tag{2.1.47}$$

*and*

$$\begin{aligned}&\|V^F(t)\|_{Z_{wa}} + \|V^{\omega_a}(t)\|_{Z_{wa}} + \|V^{\vartheta_{ab}}(t)\|_{Z_{wa}} \\ &\quad + \langle t\rangle^{-\delta}\|V^{h_{\alpha\beta}}(t)\|_{Z_{wa}} + \|V^{\psi}(t)\|_{Z_{kg}} \leq \varepsilon_1,\end{aligned} \tag{2.1.48}$$

*for any* $\alpha, \beta \in \{0, 1, 2, 3\}$ *and* $a, b \in \{1, 2, 3\}$. *Here* $\langle t\rangle := \sqrt{1+t^2}$,

$$H(q,n) := \begin{cases} 1 & \text{if } q = 0 \text{ and } n = 0, \\ 60(n-1)+20 & \text{if } q = 0 \text{ and } n \geq 1, \\ 200(n-1)+30 & \text{if } q = 1 \text{ and } n \geq 1, \\ 100(q+1)(n-1) & \text{if } q \geq 2, \end{cases} \tag{2.1.49}$$

*and* $\varepsilon_1 := \varepsilon_0^{2/3}$. *Then, for any* $t \in [0, T]$, $\alpha, \beta \in \{0, 1, 2, 3\}$, *and* $a, b \in \{1, 2, 3\}$, *one has the improved bounds*

$$\sup_{q\leq n\leq 3,\,\mathcal{L}\in\mathcal{V}_n^q}\langle t\rangle^{-H(q,n)\delta}\{\|(\langle t\rangle|\nabla|_{\leq 1})^{\gamma}|\nabla|^{-1/2}U^{\mathcal{L}h_{\alpha\beta}}(t)\|_{H^{N(n)}} + \|U^{\mathcal{L}\psi}(t)\|_{H^{N(n)}}\}\lesssim\varepsilon_0, \tag{2.1.50}$$

$$\sup_{q\leq n\leq 2,\,\mathcal{L}\in\mathcal{V}_n^q}\ \sup_{k\in\mathbb{Z},\,l\in\{1,2,3\}}2^{N(n+1)k^+}\langle t\rangle^{-H(q+1,n+1)\delta}\{2^{k/2}(2^{k^-}\langle t\rangle)^{\gamma}\|P_k(x_lV^{\mathcal{L}h_{\alpha\beta}})(t)\|_{L^2}+2^{k^+}\|P_k(x_lV^{\mathcal{L}\psi})(t)\|_{L^2}\}\lesssim\varepsilon_0, \tag{2.1.51}$$

*and*

$$\|V^F(t)\|_{Z_{wa}}+\|V^{\omega_a}(t)\|_{Z_{wa}}+\|V^{\vartheta_{ab}}(t)\|_{Z_{wa}} + \langle t\rangle^{-\delta}\|V^{h_{\alpha\beta}}(t)\|_{Z_{wa}}+\|V^{\psi}(t)\|_{Z_{kg}}\lesssim\varepsilon_0. \tag{2.1.52}$$

We will show in section 7.1 below that the smallness assumptions (1.2.5) on the initial data imply the bounds (2.1.50)–(2.1.52) at time $t = 0$, and for all $t \in [0, 2]$. Then we will show that Proposition 2.3 implies our main conclusions, in the quantitative form of Theorem 7.1, and use it to derive some additional asymptotic information about the solutions (in Chapter 7).

Most of the work in this monograph, Chapters 3, 4, 5, and 6, is concerned with the proof of Proposition 2.3. As summarized in section 1.3.1, our goal is to control simultaneously three types of norms: (i) energy norms involving up to three vector-fields $\Gamma_a$ and $\Omega_{ab}$, (ii) weighted norms on the linear profiles $V^{\mathcal{L}h_{\alpha\beta}}$ and $V^{\mathcal{L}\psi}$, and (iii) the $Z$-norms on the undifferentiated profiles.

*Remark 2.4.* The function $H$ defined in (2.1.49) is important, as it establishes a hierarchy of growth of the various energy norms. At the conceptual level this is needed because we define the weighted vector-fields $\Gamma_a, \Omega_{ab}$ in terms of the Minkowski coordinate functions $x_a$ and $t$, and thus we expect (at least logarithmic) losses as we apply these vector-fields.

At the technical level, the function $H$ satisfies superlinear inequalities like

$$H(q_1, n_1) + H(q_2, n_2) \leq H(q_1 + q_2, n_1 + n_2) - 40, \tag{2.1.53}$$

if $n_1, n_2 \geq 1$ and $n_1 + n_2 \leq 3$, and more refined versions. Such inequalities are helpful to estimate nonlinear interactions when the vector-fields split among the different components.

We notice also that we treat differently the two types of weighted vector-fields $\Gamma_a$ and $\Omega_{ab}$, in the sense that the application of the non-tangential vector-fields $\Gamma_a$ leads to more loss in terms of time growth than the application of the tangential vector-fields $\Omega_{ab}$ (for example, $H(0,1) = 20 < H(1,1) = 30$). This is a subtle technical point to keep in mind, connected to a more general difficulty of estimating the effect of non-tangential vector-fields.

## 2.2 OUTLINE OF THE PROOF

The proof of Proposition 2.3 is involved and covers Chapters 3, 4, 5, and 6 of this monograph. In this section we provide an expanded outline of this proof.

### 2.2.1 Chapter 3: Preliminary Estimates

In this chapter we start by proving a few lemmas, such as estimates on multilinear operators, lower bounds on the phases of bilinear interactions, paradifferential inequalities, linear and bilinear estimates on solutions of wave and Klein-Gordon operators, and interpolation inequalities. Then we use the bootstrap assumptions (2.1.46)–(2.1.48) to prove several linear estimates concerning the main variables $U^{\mathcal{L}h_{\alpha\beta}}$ and $U^{\mathcal{L}\psi}$ and the associated profiles $V^{\mathcal{L}h_{\alpha\beta}}$ and $V^{\mathcal{L}\psi}$.

We provide now some details on some of the main results of this chapter.

#### 2.2.1.1 *Normal forms and the bilinear phases* $\Phi_{\sigma\mu\nu}$

Our goal is to set up the application of normal forms. Indeed, in many of the estimates in Chapters 5 and 6 we need to control integrals of the form

$$J(\xi) = \int_{\mathbb{R}}\int_{\mathbb{R}^3} q(s)e^{is\Phi_{\sigma\mu\nu}(\xi,\eta)}\widehat{V^{\mu}}(\xi-\eta,s)\widehat{V^{\nu}}(\eta,s)m(\xi-\eta,\eta)\,d\eta ds, \qquad (2.2.1)$$

where $\sigma,\mu,\nu \in \{(wa,+),(wa,-),(kg,+),(kg,-)\}$, and $\Phi_{\sigma\mu\nu}(\xi,\eta) = \Lambda_\sigma(\xi) - \Lambda_\mu(\xi-\eta) - \Lambda_\nu(\eta)$ as in (2.1.40)-(2.1.41). Here $V^\mu, V^\nu$ are associated profiles, i.e., $V^\rho \in \{V^{\mathcal{L}h_{\alpha\beta},\pm}\}$ if $\rho = (wa,\pm)$ and $V^\rho \in \{V^{\mathcal{L}\psi,\pm}\}$ if $\rho = (kg,\pm)$. Integrals of this type arise, for example, when using the Duhamel formula starting from the identities (2.1.39), with various multipliers $m$. Slightly different integrals arise in energy estimates—see Proposition 5.2—or could include the phase corrections $\Theta_{kg}(\xi,s)$ and $\Theta_{wa}(\xi,s)$, as in sections 6.2 and 6.3.

The basic idea to estimate such integrals is to integrate by parts in time, so

$$\begin{aligned} J(\xi) &= i\int_{\mathbb{R}}\int_{\mathbb{R}^3} \frac{m(\xi-\eta,\eta)}{\Phi_{\sigma\mu\nu}(\xi,\eta)}e^{is\Phi_{\sigma\mu\nu}(\xi,\eta)}\cdot \partial_s q(s)\widehat{V^{\mu}}(\xi-\eta,s)\widehat{V^{\nu}}(\eta,s)\,d\eta ds \\ &+ i\int_{\mathbb{R}}\int_{\mathbb{R}^3} \frac{m(\xi-\eta,\eta)}{\Phi_{\sigma\mu\nu}(\xi,\eta)}e^{is\Phi_{\sigma\mu\nu}(\xi,\eta)}\cdot q(s)(\partial_s\widehat{V^{\mu}})(\xi-\eta,s)\widehat{V^{\nu}}(\eta,s)\,d\eta ds \\ &+ i\int_{\mathbb{R}}\int_{\mathbb{R}^3} \frac{m(\xi-\eta,\eta)}{\Phi_{\sigma\mu\nu}(\xi,\eta)}e^{is\Phi_{\sigma\mu\nu}(\xi,\eta)}\cdot q(s)\widehat{V^{\mu}}(\xi-\eta,s)(\partial_s\widehat{V^{\nu}})(\eta,s)\,d\eta ds. \end{aligned} \qquad (2.2.2)$$

The point of this procedure is to gain integrability. Indeed, if $q$ is localized to an interval of times $s \approx 2^m \gg 1$ then the integrands in the three integrals in (2.2.2) all gain a factor of almost $2^{-m}$ compared to the integrand in (2.2.1), upon application of the formulas (2.1.39).

The main obstruction is the potential presence of "small denominators" in the integrals in (2.2.2), coming from the resonant frequencies $(\xi, \eta)$ where $\Phi_{\sigma\mu\nu}(\xi, \eta) = 0$. In our problem we have two types of phases: mixed Wave-Klein-Gordon phases of the form

$$\Lambda_{wa,\iota}(\xi) - \Lambda_{kg,\iota_1}(\xi - \eta) - \Lambda_{kg,\iota_2}(\eta) \quad \text{or} \quad \Lambda_{kg,\iota}(\xi) - \Lambda_{kg,\iota_1}(\xi - \eta) - \Lambda_{wa,\iota_2}(\eta), \tag{2.2.3}$$

containing two Klein-Gordon dispersions and one wave dispersion, or pure wave phases of the form

$$\Lambda_{wa,\iota}(\xi) - \Lambda_{wa,\iota_1}(\xi - \eta) - \Lambda_{wa,\iota_2}(\eta). \tag{2.2.4}$$

Thus, to apply normal forms we need to understand the multipliers $(\xi, \eta) \to \Phi_{\sigma\mu\nu}(\xi, \eta)^{-1}$. We show that the mixed Wave-Klein-Gordon phases can only vanish when the frequency of the wave component is 0, and in fact satisfy the quantitative bounds

$$\begin{aligned} \big|\Lambda_{wa,\iota}(\xi) - \Lambda_{kg,\iota_1}(\xi - \eta) - \Lambda_{kg,\iota_2}(\eta)\big|^{-1} &\lesssim (1 + |\xi|^2 + |\xi - \eta|^2 + |\eta|^2)/|\xi|, \\ \big|\Lambda_{kg,\iota}(\xi) - \Lambda_{kg,\iota_1}(\xi - \eta) - \Lambda_{wa,\iota_2}(\eta)\big|^{-1} &\lesssim (1 + |\xi|^2 + |\xi - \eta|^2 + |\eta|^2)/|\eta|, \end{aligned} \tag{2.2.5}$$

for any $\xi, \eta \in \mathbb{R}^3$. On the other hand, the pure wave phases can vanish on much larger sets, corresponding to parallel and anti-parallel interactions, in the quantitative form

$$\big|\Lambda_{wa,\iota}(\xi) - \Lambda_{wa,\iota_1}(\xi - \eta) - \Lambda_{wa,\iota_2}(\eta)\big|^{-1} \lesssim \frac{|\eta|^{-1} + |\xi - \eta|^{-1}}{|\Xi_{\iota_1\iota_2}(\xi - \eta, \eta)|^2}, \tag{2.2.6}$$

where $\Xi_{\iota_1\iota_2}(v, w)$ denotes the angle between the vectors $v$ and $w$,

$$\Xi_{\iota_1\iota_2}(v, w) = \iota_1 v/|v| - \iota_2 w/|w|.$$

In fact, in order to bound multilinear integrals such as those in (2.2.2), in Lemmas 3.4 and 3.6 we prove stronger estimates on the multipliers $\Phi_{\sigma\mu\nu}(\xi, \eta)^{-1}$, involving suitable frequency localizations and the $L^1$ norms of the inverse Fourier transforms of the localized multipliers. These estimates are consistent with the pointwise bounds (2.2.5) and (2.2.6), and are also compatible with multilinear estimates such as those in Lemma 3.2.

### 2.2.1.2 *Linear estimates*

Estimates on solutions of linear wave and Klein-Gordon equations are the main building blocks of our nonlinear analysis. To prove efficient estimates we decompose functions $f : \mathbb{R}^3 \to \mathbb{C}$ both in frequency and space,

$$f = \sum_{k \in \mathbb{Z}} P_k f = \sum_{(k,j) \in \mathcal{J}} P'_k Q_{j,k} f,$$

where the operators $Q_{j,k}$ are defined in (2.1.20) and $P'_k = P_{[k-2,k+2]}$. In our case the functions $f$ we decompose are the linear profiles $V^{\mathcal{L}h_{\alpha\beta}}$ and $V^{\mathcal{L}\psi}$.

Our linear estimates are stated and proved in Lemma 3.9. For example, we prove general dispersive estimates of the form

$$\|e^{-it\Lambda_{wa}} f_{j,k}\|_{L^\infty} \lesssim 2^{3k/2} \min(1, 2^j \langle t\rangle^{-1}) \|Q_{j,k} f\|_{L^2},$$

where $f_{j,k} = P'_k Q_{j,k} f$, $|t| \geq 1$, and $j \geq \max(-k, 0)$. We also prove more specialized dispersive estimates, like

$$\|e^{-it\Lambda_{wa}} f_{\leq j,k}\|_{L^\infty} \lesssim 2^{2k} \langle t\rangle^{-1} \|\widehat{Q_{\leq j,k} f}\|_{L^\infty} \qquad \text{if } 2^j \lesssim \langle t\rangle^{1/2} 2^{-k/2},$$

which give sharp decay of the linear solutions in terms of the $L^\infty$ norms of the profiles in the Fourier space, thus connecting to our $Z$-norms. For the Klein-Gordon components we prove similar bounds, such as

$$\|e^{-it\Lambda_{kg}} f_{j,k}\|_{L^\infty} \lesssim \min\left\{2^{3k/2}, 2^{3k^+} \langle t\rangle^{-3/2} 2^{3j/2}\right\} \|Q_{j,k} f\|_{L^2},$$

if $|t| \geq 1$ and $j \geq \max(-k, 0)$, as well as the stronger bounds

$$\|e^{-it\Lambda_{kg}} f_{\leq j,k}\|_{L^\infty} \lesssim 2^{5k^+} \langle t\rangle^{-3/2} \|\widehat{Q_{\leq j,k} f}\|_{L^\infty} \qquad \text{if } 2^j \lesssim \langle t\rangle^{1/2}.$$

Notice that these bounds are expressed in terms of three basic parameters: the frequency parameter $k$, the spatial location parameter $j$, and the length of the time of evolution $|t|$.

Our linear estimates also include super-localized dispersive bounds such as (3.2.16)–(3.2.18), which are useful in High $\times$ High $\to$ Low estimates, as well as non-dispersive bounds such as

$$\|\widehat{f_{j,k}}\|_{L^\infty} \lesssim \min\left\{2^{3j/2} \|Q_{j,k} f\|_{L^2}, 2^{j/2-k} 2^{\delta^3(j+k)} \|Q_{j,k} f\|_{H^{0,1}_\Omega}\right\}$$

and

$$\|\widehat{f_{j,k}}(r\theta)\|_{L^2(r^2 dr) L^p_\theta} \lesssim_p \|Q_{j,k} f\|_{H^{0,1}_\Omega}, \qquad p \in [2, \infty),$$

which provide additional control on the profiles.

#### *2.2.1.3 Bilinear estimates*

The linear estimates in Lemma 3.9 are sufficient to control most nonlinear interactions, at least after suitable decompositions in frequency and space. In a small number of cases, however, these estimates are not sufficient, and we need to prove bilinear estimates directly. Our bilinear estimates are stated and proved in Lemma 3.10–3.12. For example, we prove sharp bilinear estimates such as

$$\begin{aligned} &\|P_k I_m[e^{-it\Lambda_{wa}} f_{\leq j_1, k_1}, P_{k_2} g]\|_{L^2} \\ &\lesssim 2^{\min(k,k_2)/2} \langle t\rangle^{-1} 2^{3k_1/2} \|\widehat{Q_{\leq j_1, k_1} f}\|_{L^\infty} \|P_{k_2} g\|_{L^2}, \end{aligned} \tag{2.2.7}$$

if $|t| \geq 1$ and $2^{j_1} \leq \langle t\rangle^{1/2} 2^{-k_1/2} + 2^{-\min(k,k_1,k_2)}$, where $f, g \in L^2(\mathbb{R}^3)$ and $I_m$ denotes a suitable paraproduct. The proof of (2.2.7) relies on a $TT^*$ argument, thus going beyond the conclusions one could prove by using linear estimates on each component.

#### 2.2.1.4 *Bounds on the functions $U^{\mathcal{L}h_{\alpha\beta}}$, $U^{\mathcal{L}\psi}$, $V^{\mathcal{L}h_{\alpha\beta}}$, and $V^{\mathcal{L}\psi}$*

We combine the bootstrap assumptions (2.1.46)–(2.1.48) and linear estimates to derive bounds on our solutions. For example, in section 3.3 we prove global $L^2$ bounds on the profiles $V^{\mathcal{L}h}$ and $V^{\mathcal{L}\psi}$ of the form

$$\|(\langle t\rangle|\nabla|_{\leq 1})^{\gamma}|\nabla|^{-1/2} V^{\mathcal{L}h}(t)\|_{H^{N(n)}} + \|V^{\mathcal{L}\psi}(t)\|_{H^{N(n)}} \lesssim \varepsilon_1 \langle t\rangle^{H(q,n)\delta},$$

for any $\mathcal{L} \in \mathcal{V}_n^q$ and $h \in \{h_{\alpha\beta}\}$, as well as localized $L^2$ bounds like

$$\begin{aligned} 2^{k/2}(2^{k^-}\langle t\rangle)^{\gamma} 2^j \|Q_{j,k} V^{\mathcal{L}h}(t)\|_{L^2} &+ 2^{k^+} 2^j \|Q_{j,k} V^{\mathcal{L}\psi}(t)\|_{L^2} \\ &\lesssim \varepsilon_1 \langle t\rangle^{H(q+1,n+1)\delta} 2^{-N(n+1)k^+}, \end{aligned}$$

for any $(k,j) \in \mathcal{J}$ and $\mathcal{L} \in \mathcal{V}_n^q$, $n \leq 2$. We prove also general pointwise decay bounds on the normalized solutions $U^{\mathcal{L}h}$ and $U^{\mathcal{L}\psi}$ of the form

$$\|P_k U^{\mathcal{L}h}(t)\|_{L^\infty} \lesssim \varepsilon_1 \langle t\rangle^{-1+H(q+1,n+1)\delta} 2^{k^-} 2^{-N(n+1)k^+ + 2k^+} \min\{1, \langle t\rangle 2^{k^-}\}^{1-\delta}$$

and

$$\|P_k U^{\mathcal{L}\psi}(t)\|_{L^\infty} \lesssim \varepsilon_1 \langle t\rangle^{-1+H(q+1,n+1)\delta} 2^{k^-/2} 2^{-N(n+1)k^+ + 2k^+} \min\{1, 2^{2k^-}\langle t\rangle\},$$

for any $k \in \mathbb{Z}$, $t \in [0,T]$, and $\mathcal{L} \in \mathcal{V}_n^q$, $n \leq 2$.

The $Z$-norm assumptions (2.1.48) lead to estimates on the $L^\infty$ norms of the unweighted profiles $\widehat{V^h}$ and $\widehat{V^\psi}$ corresponding to $\mathcal{L} = Id$, as in (3.3.7). They also lead to sharp pointwise decay bounds on certain parts of the unweighted normalized solutions $U^h$ and $U^\psi$, with no $\langle t\rangle^{C\delta}$ loss, as discussed in Lemma 3.16 (ii). All these estimates are stated and proved in section 3.3 and are used extensively in the nonlinear analysis.

### 2.2.2 Chapter 4: the Nonlinearities $\mathcal{N}_{\alpha\beta}^h$ and $\mathcal{N}^\psi$

The goal in this chapter is to investigate the structure of the nonlinearities $\mathcal{N}_{\alpha\beta}^h$ and $\mathcal{N}^\psi$. We decompose these nonlinearities into quadratic and cubic and higher order components, as in subsection 2.1.1, and then decompose these components dyadically in the Fourier space along every time-slice $\Sigma_t = \{(x,t) : x \in \mathbb{R}^3\}$, using the Littlewood-Paley projections $P_k$.

Ideally, we would like to prove that the nonlinearities satisfy bounds like

$$\begin{aligned}\sum_{\alpha,\beta\in\{0,1,2,3\}}\|\mathcal{L}\mathcal{N}^h_{\alpha\beta}(t)\|\lesssim\varepsilon_1\langle t\rangle^{-1}\sum_{\alpha,\beta\in\{0,1,2,3\}}\|\nabla_{x,t}\mathcal{L}h_{\alpha\beta}(t)\|,\\ \|\mathcal{L}\mathcal{N}^\psi(t)\|\lesssim\varepsilon_1\langle t\rangle^{-1}\|(\langle\nabla\rangle_x,\partial_t)\mathcal{L}\psi(t)\|,\end{aligned}\tag{2.2.8}$$

in suitable norms, as if $\mathcal{N}^h$ and $\mathcal{N}^\psi$ were of the size of $\epsilon_1\langle t\rangle^{-1}(|\nabla_x|h,\partial_t h)$ and $\epsilon_1\langle t\rangle^{-1}(\langle\nabla_x\rangle h,\partial_t h)$ respectively. Such optimal bounds are, of course, not true, as we have both derivative loss, due to the quasilinear nature of the system, and loss of decay in time, due to the slower decay of the metric components $h_{\alpha\beta}$. But the bounds (2.2.8) can still serve as a guideline for the type of control we are looking for, and we are able to prove such bounds up to small losses, both at the level of time decay and at the level of smoothness.

We summarize now the main conclusions of this chapter.

#### 2.2.2.1 *Weighted bounds on the nonlinearities $\mathcal{N}^h_{\alpha\beta}$ and $\mathcal{N}^\psi$*

As part of our analysis, we prove two weighted bounds on the full nonlinearities $\mathcal{N}^h_{\alpha\beta}$ and $\mathcal{N}^\psi$,

$$\begin{aligned}2^{-k/2}(2^{k^-}\langle t\rangle)^\gamma\|P_k(x_l\mathcal{L}\mathcal{N}^h_{\alpha\beta})(t)\|_{L^2}\lesssim\varepsilon_1^2\langle t\rangle^{H(q+1,n+1)\delta}2^{-N(n+1)k^+},\\ \|P_k(x_l\mathcal{L}\mathcal{N}^\psi)(t)\|_{L^2}\lesssim\varepsilon_1^2\langle t\rangle^{H(q+1,n+1)\delta}2^{-N(n+1)k^+},\end{aligned}\tag{2.2.9}$$

for any $k\in\mathbb{Z}$, $t\in[0,T]$, $l\in\{1,2,3\}$, and operators $\mathcal{L}\in\mathcal{V}^q_n$, $n\leq 2$. These bounds, which follow from the stronger bounds in Proposition 4.7, play a crucial role in our bootstrap scheme, as they allow us to prove the main bootstrap estimates (2.1.51) in section 6.1.

The polynomial factors $x_l$ in (2.2.9) correspond to $\partial_{\xi_l}$ derivatives in the Fourier space. These derivatives can hit the various factors of the nonlinear terms, including the oscillatory factors $e^{\pm it\Lambda_*(\xi)}$ and the frequency cutoffs. Therefore, at the analytical level, the polynomial factors $x_l$ are essentially equivalent to multiplication by $\langle t\rangle+2^{-k}$, and the weighted bounds (2.2.9) essentially follow from the $L^2$ bounds

$$\begin{aligned}2^{-k/2}(2^{k^-}\langle t\rangle)^\gamma\|P_k\mathcal{L}\mathcal{N}^h_{\alpha\beta}(t)\|_{L^2}\lesssim\varepsilon_1^2\langle t\rangle^{H(q+1,n+1)\delta}2^{-N(n+1)k^+}\min\{2^k,\langle t\rangle^{-1}\},\\ \|P_k\mathcal{L}\mathcal{N}^\psi(t)\|_{L^2}\lesssim\varepsilon_1^2\langle t\rangle^{H(q+1,n+1)\delta}2^{-N(n+1)k^+}\min\{2^k,\langle t\rangle^{-1}\},\end{aligned}\tag{2.2.10}$$

for any $k\in\mathbb{Z}$, $t\in[0,T]$, and operators $\mathcal{L}\in\mathcal{V}^q_n$, $n\leq 2$. These $L^2$ bounds are consequences of the stronger bounds (4.2.3)–(4.2.4) in Proposition 4.7, and are consistent with the heuristics that ideal bounds like (2.2.8) hold up to small loss of time decay and derivative loss.

### 2.2.2.2 *Energy disposable nonlinear terms*

Some of the nonlinear terms can be treated perturbatively in energy estimates, as they have sufficient time decay and do not lose derivatives. More precisely, wave-disposable remainders of order $(q, n)$ are defined as functions $L : \mathbb{R}^3 \times [0, T] \to \mathbb{C}$ that satisfy the bounds

$$\big\||\nabla|^{-1/2} L(t)\big\|_{H^{N(n)}} \lesssim \varepsilon_1^2 \langle t\rangle^{-1+H(q,n)\delta-\delta/2}$$

for any $t \in [0, T]$. Similarly, KG-disposable remainders of order $(q, n)$ are functions $L : \mathbb{R}^3 \times [0, T] \to \mathbb{C}$ that satisfy the bounds

$$\big\|L(t)\big\|_{H^{N(n)}} \lesssim \varepsilon_1^2 \langle t\rangle^{-1+H(q,n)\delta-\delta/2}.$$

Our analysis shows that most cubic and higher order nonlinear terms and all of the commutator terms arising from commuting Sobolev derivatives and the vector-fields $\Omega$ and $\Gamma$ are energy disposable. See Lemmas 4.19 and 4.20 for precise statements.

As a consequence, for the purpose of proving energy estimates we can concentrate on the quadratic components of the nonlinearities $\mathcal{N}^h_{\alpha\beta}$ and $\mathcal{N}^\psi$ (as expressed in terms of the variables $h_{\alpha\beta}$ and $\psi$) and on the quasilinear terms of the form $g_{\geq 1}\partial^2 h$ and $g_{\geq 1}\partial^2\psi$.

### 2.2.2.3 *Elliptic consequences of the harmonic gauge condition*

The harmonic gauge conditions $\mathbf{g}^{\alpha\beta}\partial_\alpha h_{\beta\mu} = (1/2)\mathbf{g}^{\alpha\beta}\partial_\mu h_{\alpha\beta}$ can be used to derive approximate identities for some of the metric components. Indeed, in terms of the functions $F^{\mathcal{L}}, \underline{F}^{\mathcal{L}}, \rho^{\mathcal{L}}, \omega_j^{\mathcal{L}}, \Omega_j^{\mathcal{L}}, \vartheta_{jk}^{\mathcal{L}}$ defined in section 2.1.4, we have the approximate identities

$$|\nabla|\rho^{\mathcal{L}} \sim \partial_0 \underline{F}^{\mathcal{L}}, \qquad |\nabla|\Omega_j^{\mathcal{L}} \sim \partial_0\omega_j^{\mathcal{L}}, \qquad \delta_{jk}\vartheta_{jk}^{\mathcal{L}} \sim 0, \tag{2.2.11}$$

up to suitable quadratic errors (which only lead to energy disposable terms). Precise identities of this type are derived in Lemma 4.15 and play an important role in identifying the null structures of the metric nonlinearities $\mathcal{N}^h_{\alpha\beta}$.

### 2.2.2.4 *Null structures and decompositions of the nonlinearities $\mathcal{N}^h_{\alpha\beta}$ and $\mathcal{N}^\psi$*

The presence of null structures is an important feature of many nonlinear hyperbolic equations. In our case we define the class of "semilinear null forms of order $(q, n)$" as the set of finite sums of paraproducts $I^{null}_{n_{\iota_1\iota_2}}[U^{\mathcal{L}_1 h_1,\iota_1}, U^{\mathcal{L}_2 h_2,\iota_2}]$ of the form

$$\begin{aligned}&\mathcal{F}\{I^{null}_{n_{\iota_1\iota_2}}[U^{\mathcal{L}_1 h_1,\iota_1}, U^{\mathcal{L}_2 h_2,\iota_2}]\}(\xi)\\ &= \frac{1}{8\pi^3}\int_{\mathbb{R}^3} n_{\iota_1\iota_2}(\xi-\eta,\eta)\widehat{U^{\mathcal{L}_1 h_1,\iota_1}}(\xi-\eta)\widehat{U^{\mathcal{L}_2 h_2,\iota_2}}(\eta)\,d\eta,\end{aligned} \tag{2.2.12}$$

where $\iota_1, \iota_2 \in \{+,-\}$, $h_1, h_2 \in \{h_{\alpha\beta}\}$, $\mathcal{L}_1 \in \mathcal{V}^{q_1}_{n_1}$, $\mathcal{L}_2 \in \mathcal{V}^{q_2}_{n_2}$, $(q_1, n_1) + (q_2, n_2) \leq (q, n)$. The multipliers $n_{\iota_1\iota_2}$ belong to the set of null multipliers

$$\begin{aligned}\mathcal{M}^{null}_{\iota_1\iota_2} = \{n : (\mathbb{R}^3 \setminus 0)^2 \to \mathbb{C} : \, & n(x,y) = (\iota_1 x_i/|x| - \iota_2 y_i/|y|)m(x,y) \\ & \text{for some } m \in \mathcal{M} \text{ and } i \in \{1,2,3\}\}.\end{aligned} \tag{2.2.13}$$

Here $\mathcal{M}$ denotes a general class of acceptable multipliers, as defined in (3.2.41). Our class of semilinear null forms contains the classical null forms

$$\partial_\alpha h_1 \partial_\beta h_2 - \partial_\beta h_1 \partial_\alpha h_2 \qquad \text{and} \qquad m^{\alpha\beta}\partial_\alpha h_1 \partial_\beta h_2,$$

but is more flexible, i.e., we allow Fourier multipliers and bilinear paraproducts.

We are now finally ready to decompose the nonlinearities $\mathcal{L}\mathcal{N}^h_{\alpha\beta}$ and $\mathcal{L}\mathcal{N}^\psi$. More precisely, in Proposition 4.22 we show that if $\alpha, \beta \in \{0,1,2,3\}$ and $\mathcal{L} \in \mathcal{V}^q_n$, $n \leq 3$, then

$$\begin{aligned}\mathcal{L}\mathcal{N}^h_{\alpha\beta} = & \sum_{\mu,\nu\in\{0,1,2,3\}} \widetilde{g}^{\mu\nu}_{\geq 1}\partial_\mu\partial_\nu(\mathcal{L}h_{\alpha\beta}) + \mathcal{Q}^{\mathcal{L}}_{wa}(h_{\alpha\beta}) \\ & + \mathcal{S}^{\mathcal{L},1}_{\alpha\beta} + \mathcal{S}^{\mathcal{L},2}_{\alpha\beta} + \mathcal{KG}^{\mathcal{L}}_{\alpha\beta} + \mathcal{R}^{\mathcal{L}h}_{\alpha\beta}\end{aligned} \tag{2.2.14}$$

and

$$\mathcal{L}\mathcal{N}^\psi = \sum_{\mu,\nu\in\{0,1,2,3\}} \widetilde{g}^{\mu\nu}_{\geq 1}\partial_\mu\partial_\nu(\mathcal{L}\psi) + h_{00}\mathcal{L}\psi + \mathcal{Q}^{\mathcal{L}}_{kg}(\psi) + \mathcal{R}^{\mathcal{L}\psi}, \tag{2.2.15}$$

where the remainders $\mathcal{R}^{\mathcal{L}h}_{\alpha\beta}$ and $\mathcal{R}^{\mathcal{L}\psi}$ are wave-disposable and KG-disposable of order $(q,n)$. The reduced metric components $\widetilde{g}^{\mu\nu}_{\geq 1}$ are defined by

$$\widetilde{g}^{00}_{\geq 1} := 0, \qquad \widetilde{g}^{0j}_{\geq 1} := (1 - g^{00}_{\geq 1})^{-1} g^{0j}_{\geq 1}, \qquad \widetilde{g}^{jk}_{\geq 1} := (1 - g^{00}_{\geq 1})^{-1}\big[g^{jk}_{\geq 1} + g^{00}_{\geq 1}\delta^{jk}\big].$$

The origin of most of the terms in (2.2.14)–(2.2.15) can be traced back by examining the formulas (2.1.9) and (2.1.16). The terms $\widetilde{g}^{\mu\nu}_{\geq 1}\partial_\mu\partial_\nu(\mathcal{L}h_{\alpha\beta})$ and $\widetilde{g}^{\mu\nu}_{\geq 1}\partial_\mu\partial_\nu(\mathcal{L}\psi)$ come from the main quasilinear terms in the nonlinearities $\mathcal{N}^h_{\alpha\beta}$ and $\mathcal{N}^\psi$, when the entire differential operator $\mathcal{L}$ hits the functions $h_{\alpha\beta}$ and $\psi$ respectively (recall that commutators between derivatives and the operators $\mathcal{L}$ are energy disposable).

The quadratic terms $\mathcal{Q}^{\mathcal{L}}_{wa}(h_{\alpha\beta})$ (which have a certain type of quasilinear null structure) and $h_{00}\mathcal{L}\psi + \mathcal{Q}^{\mathcal{L}}_{kg}(\psi)$ also come from the main quasilinear terms in the nonlinearities $\mathcal{N}^h_{\alpha\beta}$ and $\mathcal{N}^\psi$, after distributing the operator $\mathcal{L}$ between the factors and noticing that all cubic and higher order terms are energy disposable remainders. Similarly, the semilinear quadratic terms $\mathcal{KG}^{\mathcal{L}}_{\alpha\beta}$ come from the Klein-Gordon terms $\mathcal{KG}_{\alpha\beta}$ in (2.1.9), after distributing the operator $\mathcal{L}$ and removing energy disposable remainders.

Finally, the semilinear quadratic terms $\mathcal{S}^{\mathcal{L},1}_{\alpha\beta}$ and $\mathcal{S}^{\mathcal{L},2}_{\alpha\beta}$ can be written as

$$\begin{aligned}\mathcal{S}^{\mathcal{L},1}_{\alpha\beta} &= \sum_{h_1,h_2\in\{h_{\mu\nu}\}}\sum_{\iota_1,\iota_2\in\{+,-\}}\sum_{\mathcal{L}_1+\mathcal{L}_2=\mathcal{L}} I^{null}_{\mathfrak{n}_{\iota_1\iota_2}}[U^{\mathcal{L}_1h_1,\iota_1},U^{\mathcal{L}_2h_2,\iota_2}],\\ \mathcal{S}^{\mathcal{L},2}_{\alpha\beta} &= \sum_{\mathcal{L}_1+\mathcal{L}_2=\mathcal{L}}(1/2)c_{\mathcal{L}_1,\mathcal{L}_2}R_pR_q\partial_\alpha\vartheta^{\mathcal{L}_1}_{mn}\cdot R_pR_q\partial_\beta\vartheta^{\mathcal{L}_2}_{mn},\end{aligned} \tag{2.2.16}$$

for suitable null multipliers $\mathfrak{n}_{\iota_1\iota_2}\in\mathcal{M}^{null}_{\iota_1\iota_2}$, and suitable constants $c_{\mathcal{L}_1,\mathcal{L}_2}$. The point of this decomposition is that semilinear Wave × Wave interactions are either null, as in $\mathcal{S}^{\mathcal{L},1}_{\alpha\beta}$, or involve only the good components $\vartheta$ of the metric tensor. This is a key weak null structure property of the Einstein equations in harmonic gauge, as identified by Linblad-Rodnianski [63]. The formulas (2.2.16) are derived using the identities (2.1.14), (2.1.15), and (2.1.29), and the approximate identities (2.2.11).

### 2.2.3 Chapter 5: Improved Energy Estimates

In this chapter we prove the main energy estimates (2.1.50). We define the operators $P^n_{wa}:=\langle\nabla\rangle^{N(n)}|\nabla|^{-1/2}|\nabla|^{\gamma}_{\leq 1}$ and $P^n_{kg}:=\langle\nabla\rangle^{N(n)}$, apply the operators $P^n_{wa}\mathcal{L}$ and $P^n_{kg}\mathcal{L}$ to the equations (2.1.2) and (2.1.4) respectively, construct suitable energy functionals, and perform energy estimates. The main issue is to estimate the spacetime cubic (and higher order) bulk terms generated by the nonlinearities.

After eliminating the contributions of the energy disposable remainders, we are left with eight main spacetime integrals that need to be estimated. The required bounds are stated in Proposition 5.2. These integrals involve either Wave × Wave × Wave interactions (coming from the metric nonlinearities) or Wave × KG × KG interactions (coming from both the metric and the Klein-Gordon nonlinearities). Our main tool to estimate these integrals is the method of normal forms (integration by parts in time).

We discuss below some of the main bounds we prove.

#### *2.2.3.1 Semilinear wave interactions*

There are two types of semilinear Wave × Wave × Wave spacetime contributions, coming from the terms $\mathcal{S}^{\mathcal{L},1}_{\alpha\beta}$ and $\mathcal{S}^{\mathcal{L},2}_{\alpha\beta}$ in (2.2.14). In particular, we need to prove that if $\mathfrak{n}_{\iota_1\iota_2}\in\mathcal{M}^{null}_{\iota_1\iota_2}$ then

$$\begin{aligned}\Big|\int_{J_m}\int_{\mathbb{R}^3} q_m(s)\mathcal{F}\{P^n_{wa}I_{\mathfrak{n}_{\iota_1\iota_2}}[U^{\mathcal{L}_1h_1,\iota_1},U^{\mathcal{L}_2h_2,\iota_2}]\}(\xi,s)&\\ \times\overline{P^n_{wa}\widehat{U^{\mathcal{L}h,\iota}}(\xi,s)}\,d\xi ds\Big| &\lesssim \varepsilon_1^3 2^{-\delta m},\end{aligned} \tag{2.2.17}$$

where $\iota_1, \iota_2, \iota_3 \in \{+,-\}$, $h_1, h_2, h \in \{h_{\alpha\beta}\}$, $\mathcal{L} \in \mathcal{V}^q_n$, $\mathcal{L}_1 \in \mathcal{V}^{q_1}_{n_1}$, $\mathcal{L}_1 \in \mathcal{V}^{q_1}_{n_1}$, $(q_1, n_1) + (q_2, n_2) \leq (q, n)$. We also prove that if $\mathfrak{m} \in \mathcal{M}$ and $\vartheta \in \{\vartheta_{mn}\}$ then

$$\begin{aligned}\Big|\int_{J_m}\int_{\mathbb{R}^3} q_m(s)\mathcal{F}\{P^n_{wa} I_{\mathfrak{m}}[U^{\vartheta^{\mathcal{L}_1},\iota_1}, U^{\vartheta^{\mathcal{L}_2},\iota_2}]\}(\xi, s) \cdot \overline{P^n_{wa}\widehat{U^{\mathcal{L}h,\iota}}(\xi, s)}\, d\xi ds\Big| \\ \lesssim \varepsilon_1^3 2^{2H(q,n)\delta m - 2\gamma m}.\end{aligned} \tag{2.2.18}$$

The cutoff function $q_m$ in (2.2.17)–(2.2.18) restricts to times $s \approx 2^m \gg 1$.

To illustrate the main ideas we discuss only the simpler bounds (2.2.17), in the main case $\mathcal{L}_1 = Id$ and $\mathcal{L}_2 = \mathcal{L}$. After dyadic decompositions in frequency it suffices to prove that

$$\begin{aligned}\sum_{k,k_1,k_2 \in \mathbb{Z}} 2^{2N(n)k^+ - k}\Big|\int_{J_m} q_m(s) \\ \times \mathcal{G}_{\mathfrak{n}_{\iota_1\iota_2}}\left[P_{k_1}U^{h_1,\iota_1}(s), P_{k_2}U^{\mathcal{L}h_2,\iota_2}(s), P_k U^{\mathcal{L}h,\iota}(s)\right] ds\Big| \lesssim \varepsilon_1^3 2^{-\delta m},\end{aligned} \tag{2.2.19}$$

where, for suitable multipliers $\mathfrak{m}$, the trilinear operators $\mathcal{G}_{\mathfrak{m}}$ are defined by

$$\mathcal{G}_{\mathfrak{m}}[f, g, h] = \int_{\mathbb{R}^3\times\mathbb{R}^3} \mathfrak{m}(\xi - \eta, \eta)\widehat{f}(\xi - \eta)\widehat{g}(\eta)\overline{\widehat{h}(\xi)}\, d\xi d\eta.$$

To prove (2.2.19) we fix $m$ (the parameter connected to the size of the time) and analyze the contributions of the various frequency parameters $(k, k_1, k_2)$.

In some cases, for example, if $\min(k, k_1, k_2) \leq -m - \delta' m$ or if $\max(k, k_1, k_2) \geq \delta' m$, we can simply use $L^2$ or $L^\infty$ estimates on every factor in the trilinear expressions, and bound the corresponding contributions.

In the more complicated cases, for example, if $k, k_1, k_2 \in [-\delta' m, \delta' m]$, we need to exploit the null structure of the multiplier. For this we insert angular cutoffs of the form $\varphi_{\leq q_0}(\Xi_{\iota_1\iota_2}(\xi-\eta, \eta))$ and $\varphi_{>q_0}(\Xi_{\iota_1\iota_2}(\xi-\eta, \eta))$ in the definition of the trilinear operator $\mathcal{G}_{\mathfrak{m}}$, where $q_0 = -8\delta' m$. The point is that the multiplier $\mathfrak{n}_{\iota_1\iota_2}$ is small in the region where the angle $|\Xi_{\iota_1\iota_2}(\xi - \eta, \eta)|$ is small, due to the null structure assumption. On the other hand, if the angle $|\Xi_{\iota_1\iota_2}(\xi - \eta, \eta)|$ is large then we are in the non-resonant region (due to (2.2.6)) and we can use normal forms, as described in subsection 2.2.1.1 to gain time integrability.

There is one more issue that comes up in the analysis of semilinear cubic terms, in the case when $2^{k_1} \approx 1$ and $2^k \approx 2^{k_2} \gg 1$. We can still insert the angular decomposition and try to use normal forms to bound the non-resonant contributions, but this leads to a loss of derivative. This is a well-known issue that comes up when using normal forms in the context of energy estimates in quasilinear problems, and we deal with it as in some of our earlier work [16, 17] using paradifferential calculus and a second symmetrization.

### 2.2.3.2 *Low frequencies and the quasilinear terms*

The most difficult contributions come from the quasilinear terms $\widetilde{g}^{\mu\nu}_{\geq 1}\partial_\mu\partial_\nu h_{\alpha\beta}$ and $\widetilde{g}^{\mu\nu}_{\geq 1}\partial_\mu\partial_\nu\psi$, when the entire operator $\mathcal{L}$ hits the undifferentiated metric components $\widetilde{g}^{\mu\nu}_{\geq 1}$, i.e., the terms

$$\mathcal{L}(\widetilde{g}^{\mu\nu}_{\geq 1}) \cdot \partial_\mu\partial_\nu h_{\alpha\beta} \qquad \text{and} \qquad \mathcal{L}(\widetilde{g}^{\mu\nu}_{\geq 1}) \cdot \partial_\mu\partial_\nu\psi. \tag{2.2.20}$$

These terms have their own quasilinear null structure, which helps with the analysis of medium frequencies, but the difficulty is to bound the contributions of very low frequencies of the factor $\mathcal{L}(\widetilde{g}^{\mu\nu}_{\geq 1})$, which does not have a spatial derivative. When $\mathcal{L} = Id$, we can symmetrize the system and obtain improved estimates, but no algebraic symmetrization is possible when $\mathcal{L} \neq Id$.

After dyadic decompositions in time and frequency, we need to prove bounds of the form

$$\begin{aligned}\sum_{k,k_1,k_2\in\mathbb{Z}} & 2^{2N(n)k^+ - k + k_1 - k_2} 2^{2\gamma(m+k^-)} \Big| \int_{J_m} q_m(s) \\ & \times \mathcal{G}_{\mathfrak{q}_{\iota_1\iota_2}}\big[P_{k_1}U^{h_1,\iota_1}(s), P_{k_2}U^{\mathcal{L}h_2,\iota_2}(s), P_k U^{\mathcal{L}h,\iota}(s)\big]\, ds\Big| \lesssim \varepsilon_1^3 2^{2H(q,n)\delta m},\end{aligned} \tag{2.2.21}$$

where $\mathfrak{q}_{\iota_1\iota_2}$ is a null multiplier and $\mathcal{L} \in \mathcal{V}^q_n$, $n \geq 1$. This looks similar to the semilinear bounds (2.2.19), but with an additional factor of $2^{k_1-k_2}$ in the sum in the left-hand side.

We can still exploit the null structure of the multipliers and use normal forms (with paradifferential calculus again to set up a second symmetrization) to deal with the contribution of frequencies that are not too small. But a different idea is needed to deal with the contribution of triples $(k_1, k_2, k)$ for which $2^{k_1} \approx 2^k \approx 1$ and $2^{k_2} \approx 2^{-m}$, due to the large additional factor $2^{k_1-k_2}$. In this case we "trivialize" one vector-field, in the sense that we notice that a vector-field essentially contributes a factor of about $2^m 2^{k_2}$ at such low frequencies, in the context of multilinear estimates (the precise statement is in Lemma 4.13). Then we can close the desired estimates (2.2.21) provided that the hierarchy function $H(q, n)$ allows for sufficiently large gaps between the growth rates of different differential operators.

This is the most difficult case of the energy analysis. Some of the main choices we make in the proof, like the precise choice of the function $H(q, n)$ in (2.1.49) that determines the hierarchy of energy growth, are motivated by the energy analysis of the terms in (2.2.21).

### 2.2.3.3 *Wave-Klein-Gordon undifferentiated interactions*

At the conceptual level, the new difficulty in the analysis of interactions of the metric tensor and the massive field is the presence of terms that do not contain derivatives. There are two such quadratic interactions,

$$\text{terms like } \psi^2 \text{ in } \mathcal{N}^h_{\alpha\beta} \qquad \text{and} \qquad \text{terms like } h\psi \text{ in } \mathcal{N}^\psi, \tag{2.2.22}$$

as one can see from the formulas (2.1.12) and (2.1.17). These undifferentiated terms do not have null structure, of course, but the situation is worse mainly because the low frequencies of the Klein-Gordon field $\psi$ disperse slowly and lead to slow decay. The Klein-Gordon $Z$-norm defined in (2.1.45) is designed as a very strong norm at low frequencies mainly to be able to capture these terms.

As before, our main tool to analyze these interactions is the method of normal forms. The point is that the resonant structure of these interactions is determined by the Wave-Klein-Gordon quadratic phases $\Phi(\xi,\eta) = \Lambda_{kg}(\xi) \pm \Lambda_{kg}(\eta) \pm \Lambda_{wa}(\xi+\eta)$. As we have seen earlier, in (2.2.4)–(2.2.5), these phases are resonant only when the wave component has frequency 0. This means that the normal forms lead to stronger bounds, which are able to compensate for the fact that the factors themselves are undifferentiated at low frequencies.

### 2.2.4 Chapter 6: Improved Profile Bounds

In this chapter we prove the main profile bounds (2.1.51) and (2.1.52), thus completing the proof of the bootstrap. We accomplish this in three main steps.

#### *2.2.4.1 The weighted bounds* (2.1.51)

These bounds follow from the main energy estimates (2.1.50), already proved in Chapter 5, the main weighted estimates (2.2.10) on the nonlinearities $\mathcal{LN}^h_{\alpha\beta}$ and $\mathcal{LN}^\psi$, proved in Proposition 4.7, and the general identities

$$e^{-it\Lambda_\mu(\xi)}\partial_{\xi_l}[\Lambda_\mu(\xi)\widehat{V}(\xi,t)] = \widehat{\Gamma_l U}(\xi,t) - i(\partial_{\xi_l}\widehat{\mathcal{N}})(\xi,t),$$

which hold if $\mu \in \{wa, kg\}$, $(\partial_t + i\Lambda_\mu)U = \mathcal{N}$ and $V(t) = e^{it\Lambda_\mu}U(t)$ on $\mathbb{R}^3 \times [0,T]$. These identities, which are proved in Lemma 6.1, show explicitly that application of Lorentz vector-fields on linear solutions of wave and Klein-Gordon equations is connected to differentiation of the associated profiles in the Fourier space.

#### *2.2.4.2 The $Z$-norm bounds on the Klein-Gordon profile*

We would like to use the formula $\partial_t V^\psi(t) = e^{it\Lambda_{kg}}\mathcal{N}^\psi(t)$ in (2.1.39) and integrate in time to prove the desired uniform bounds $\|V^\psi(t)\|_{Z_{kg}} \lesssim \varepsilon_0$. However, this does not work directly since the profile $V^\psi(t)$ itself does not converge as $t \to \infty$, due to a long-range effect.

Our solution is to renormalize the profile $V^\psi$, and this can be done efficiently in the Fourier space. We start from the quadratic nonlinearity $\mathcal{N}^{\psi,2}$ in (2.1.17) and use the identities (2.1.36) and (2.1.32) to rewrite it in the Fourier space in the form

$$
\begin{aligned}
&e^{it\Lambda_{kg}(\xi)}\widehat{\mathcal{N}^{\psi,2}}(\xi,t)\\
&=\frac{1}{8\pi^3}\sum_{\pm}\int_{\mathbb{R}^3} ie^{it\Lambda_{kg}(\xi)}e^{\mp it\Lambda_{kg}(\xi-\eta)}\widehat{V^{\psi,\pm}}(\xi-\eta,t)\mathfrak{q}_{kg,\pm}(\xi-\eta,\eta,t)\,d\eta,
\end{aligned}
\tag{2.2.23}
$$

where

$$
\mathfrak{q}_{kg,\pm}(\rho,\eta,t):=\pm\widehat{h_{00}}(\eta,t)\frac{\Lambda_{kg}(\rho)}{2}+\widehat{h_{0j}}(\eta,t)\rho_j\pm\widehat{h_{jk}}(\eta,t)\frac{\rho_j\rho_k}{2\Lambda_{kg}(\rho)}. \tag{2.2.24}
$$

We would like to eliminate the resonant bilinear interaction between $h_{\alpha\beta}$ and $V^{\psi,+}$ in (2.2.23) corresponding to $|\eta|\ll 1$. To identify the main term we approximate, heuristically,

$$
\begin{aligned}
&\frac{1}{(2\pi)^3}\int_{|\eta|\ll\langle t\rangle^{-1/2}} ie^{it\Lambda_{kg}(\xi)}e^{-it\Lambda_{kg}(\xi-\eta)}\widehat{V^{\psi,+}}(\xi-\eta,t)\mathfrak{q}_{kg,+}(\xi-\eta,\eta,t)\,d\eta\\
&\approx i\frac{\widehat{V^{\psi,+}}(\xi,t)}{(2\pi)^3}\int_{|\eta|\ll\langle t\rangle^{-1/2}} e^{it\eta\cdot\nabla\Lambda_{kg}(\xi)}\\
&\qquad\times\Big\{\widehat{h_{00}}(\eta,t)\frac{\Lambda_{kg}(\xi)}{2}+\widehat{h_{0j}}(\eta,t)\xi_j+\widehat{h_{jk}}(\eta,t)\frac{\xi_j\xi_k}{2\Lambda_{kg}(\xi)}\Big\}\,d\eta\\
&\approx i\widehat{V^{\psi}}(\xi,t)\Big\{h_{00}^{low}\Big(\frac{t\xi}{\Lambda_{kg}(\xi)},t\Big)\frac{\Lambda_{kg}(\xi)}{2}\\
&\qquad+h_{0j}^{low}\Big(\frac{t\xi}{\Lambda_{kg}(\xi)},t\Big)\xi_j+h_{jk}^{low}\Big(\frac{t\xi}{\Lambda_{kg}(\xi)},t\Big)\frac{\xi_j\xi_k}{2\Lambda_{kg}(\xi)}\Big\},
\end{aligned}
\tag{2.2.25}
$$

where $h_{\alpha\beta}^{low}$ are suitable low-frequency components of $h_{\alpha\beta}$.

To eliminate this term, we define the nonlinear (modified) Klein-Gordon profile $V_*^{\psi}$ by

$$
\widehat{V_*^{\psi}}(\xi,t):=e^{-i\Theta_{kg}(\xi,t)}\widehat{V^{\psi}}(\xi,t), \tag{2.2.26}
$$

where the Klein-Gordon phase correction $\Theta_{kg}$ is defined by

$$
\begin{aligned}
\Theta_{kg}(\xi,t)=\int_0^t&\Big\{h_{00}^{low}(s\xi/\Lambda_{kg}(\xi),s)\frac{\Lambda_{kg}(\xi)}{2}\\
&+h_{0j}^{low}(s\xi/\Lambda_{kg}(\xi),s)\xi_j+h_{jk}^{low}(s\xi/\Lambda_{kg}(\xi),s)\frac{\xi_j\xi_k}{2\Lambda_{kg}(\xi)}\Big\}\,ds,
\end{aligned}
\tag{2.2.27}
$$

and the low frequency components $h_{\alpha\beta}^{low}$ are given by

$$
\widehat{h_{\alpha\beta}^{low}}(\rho,s):=\varphi_{\le 0}(\langle s\rangle^{p_0}\rho)\widehat{h_{\alpha\beta}}(\rho,s),\qquad p_0:=0.68. \tag{2.2.28}
$$

The choice of $p_0$, slightly larger than $2/3$, is important in the proof, to justify the validity of the approximation in (2.2.25). The phase correction $\Theta_{kg}$ is defined such that $(\partial_t\Theta_{kg})(\xi,t)$ matches the factor in the last line of (2.2.25), and has a simple geometric interpretation: it is obtained by integrating low frequency components of the metric tensor along the characteristics of the Klein-Gordon linear flow given by $\nabla\Lambda_{kg}(\xi)=\xi/\Lambda_{kg}(\xi)$.

Since $\Theta_{kg}$ is real-valued we have $|\widehat{V_*^\psi}(\xi,t)|=|\widehat{V^\psi}(\xi,t)|$ for any $(\xi,t)\in\mathbb{R}^3\times[0,T]$, therefore $\|V_*^\psi(t)\|_{Z_{kg}}=\|V^\psi(t)\|_{Z_{kg}}$. The modified profile $V_*^\psi$ satisfies the better equation

$$\partial_t\widehat{V_*^\psi}(\xi,t)=e^{-i\Theta_{kg}(\xi,t)}\{\partial_t\widehat{V^\psi}(\xi,t)-i\widehat{V^\psi}(\xi,t)\dot{\Theta}_{kg}(\xi,t)\}, \tag{2.2.29}$$

which allows us to replace the resonant long-range interaction between $V^{\psi,+}$ and the metric tensor by a perturbative term of the form

$$\begin{aligned}\mathcal{R}_2^\psi(\xi,t)=\frac{e^{-i\Theta_{kg}(\xi,t)}}{(2\pi)^3}\int_{\mathbb{R}^3}&i\{e^{it(\Lambda_{kg}(\xi)-\Lambda_{kg}(\xi-\eta))}\widehat{V^\psi}(\xi-\eta,t)\\ &\times\mathfrak{q}_{kg,+}^{low}(\xi-\eta,\eta,t)-e^{it(\xi\cdot\eta)/\Lambda_{kg}(\xi)}\widehat{V^\psi}(\xi,t)\mathfrak{q}_{kg,+}^{low}(\xi,\eta,t)\}\,d\eta.\end{aligned} \tag{2.2.30}$$

We would like to prove that the nonlinear profile $V_*^\psi(t)$ is uniformly bounded and converges in the $Z_{kg}$ norm as $t\to\infty$. For this it suffices to prove that

$$\|\varphi_k(\xi)\{\widehat{V_*^\psi}(\xi,t_2)-\widehat{V_*^\psi}(\xi,t_1)\}\|_{L^\infty_\xi}\lesssim\varepsilon_0 2^{-\delta m/2}2^{-k^-/2+\kappa k^-}2^{-N_0k^+} \tag{2.2.31}$$

for any $k\in\mathbb{Z}$, $m\geq 1$, and $t_1,t_2\in[2^m-2,2^{m+1}]\cap[0,T]$.

This is proved in subsection 6.2.2, in several steps. First we dispose of very low or very high frequency parameters $k$, in which case the energy norms already provide suitable control. In the intermediate range $k\in[-\kappa m,\delta' m]$, $m\geq 100$, we integrate the identity (2.2.29) between times $t_1$ and $t_2$, use the formula $\partial_t\widehat{V^\psi}(\xi,t)=e^{it\Lambda_{kg}(\xi)}\widehat{\mathcal{N}^\psi}(\xi,t)$, and expand the nonlinearity $\widehat{\mathcal{N}^\psi}$. This produces four terms $\mathcal{R}_a^\psi$, $a\in\{1,2,3,4\}$, one of them being the integral in (2.2.30).

To prove (2.2.31) we need to bound spacetime oscillatory integrals, involving the Wave-Klein-Gordon phases $\Lambda_{kg}(\xi)-\Lambda_{kg,\iota_1}(\xi-\eta)-\Lambda_{wa,\iota_2}(\eta)$, which are similar to the integral in (2.2.1). The point is that the relevant interactions are all non-resonant, due to the bounds (2.2.5) and the renormalization procedure that removed the low frequencies of the wave component. We can therefore integrate by parts in time, as in (2.2.2), and gain sufficient decay to prove the desired bounds (2.2.31).

#### *2.2.4.3 The $Z$-norm bounds on the metric profiles*

These bounds are similar to the $Z$-norm bounds for the Klein-Gordon profile, at least at the conceptual level. For $G\in\{h_{\alpha\beta},F,\omega_a,\vartheta_{ab}\}$ we define the renormalized nonlinear profiles $V_*^G$ by

$$\widehat{V_*^G}(\xi,t) := e^{-i\Theta_{wa}(\xi,t)}\widehat{V^G}(\xi,t), \tag{2.2.32}$$

where the wave phase correction $\Theta_{wa}$ is defined by

$$\begin{aligned}\Theta_{wa}(\xi,t) = \int_0^t \Big\{ & h_{00}^{low}(s\xi/\Lambda_{wa}(\xi),s)\frac{\Lambda_{wa}(\xi)}{2}\\ &+ h_{0j}^{low}(s\xi/\Lambda_{wa}(\xi),s)\xi_j + h_{jk}^{low}(s\xi/\Lambda_{wa}(\xi),s)\frac{\xi_j\xi_k}{2\Lambda_{wa}(\xi)}\Big\}\,ds,\end{aligned} \tag{2.2.33}$$

and the low frequency components $h_{\alpha\beta}^{low}$ are as in (2.2.28). The phase correction $\Theta_{wa}$ is designed to eliminate the long-range effect of the quasilinear component $\mathcal{Q}^2_{\alpha\beta}$ defined in (2.1.13). As in the Klein-Gordon case, it is obtained by integrating low frequency components of the metric tensor along the characteristics of the wave linear flow. It is also connected to the construction of optical functions, as we discuss later in section 7.3

To prove the bootstrap bounds (2.1.52) it suffices to show that

$$\begin{aligned}\|\varphi_k(\xi)\{\widehat{V_*^{h_{\alpha\beta}}}(\xi,t_2)-\widehat{V_*^{h_{\alpha\beta}}}(\xi,t_1)\}\|_{L^\infty_\xi} &\lesssim \varepsilon_0 2^{\delta m}2^{-k^- -\kappa k^-}2^{-N_0k^+},\\ \|\varphi_k(\xi)\{\widehat{V_*^{H}}(\xi,t_2)-\widehat{V_*^{H}}(\xi,t_1)\}\|_{L^\infty_\xi} &\lesssim \varepsilon_0 2^{-\delta m/2}2^{-k^- -\kappa k^-}2^{-N_0k^+},\end{aligned} \tag{2.2.34}$$

for any $H\in\{F,\omega_a,\vartheta_{ab}\}$, $k\in\mathbb{Z}$, $m\geq 1$, and $t_1,t_2\in[2^m-2,2^{m+1}]\cap[0,T]$. We may also assume that $k$ is in the intermediate range $k\in[-\kappa m/4,\delta' m]$, since in the other cases the energy estimates are already stronger. The bounds (2.2.34) are conceptually similar to the bounds (2.2.31), but more involved at the technical level.

Starting from the formula $\partial_t V^{h_{\alpha\beta}} = e^{it\Lambda_{wa}}\mathcal{N}^h_{\alpha\beta}$ in (2.1.39) and expanding $\mathcal{N}^h_{\alpha\beta} = \mathcal{K}\mathcal{G}^2_{\alpha\beta}+\mathcal{Q}^2_{\alpha\beta}+\mathcal{S}^2_{\alpha\beta}+\mathcal{N}^{h,\geq 3}_{\alpha\beta}$, we can express $\partial_t\widehat{V_*^{h_{\alpha\beta}}}$ as a sum of six terms $\mathcal{R}_a^{h_{\alpha\beta}}$, $a\in\{1,\dots,6\}$. Most of these terms lead to non-resonant or cubic contributions that can be bounded using normal forms as in the Klein-Gordon case. However, the term $\mathcal{S}^2_{\alpha\beta}$ leads to a more difficult contribution,

$$\mathcal{R}_5^{h_{\alpha\beta}}(\xi,t) = e^{-i\Theta_{wa}(\xi,t)}e^{it\Lambda_{wa}(\xi)}\widehat{\mathcal{S}^2_{\alpha\beta}}(\xi,t).$$

The contribution of the term $\mathcal{R}_5^{h_{\alpha\beta}}$ is analyzed in subsections 6.3.3 and 6.3.5. It requires not only angular decompositions and normal forms, but also careful Fourier analysis of the resulting cubic nonlinearities, to show that the three interacting factors are not coherent (in the spirit of the spacetime resonances method). Finally, we also need to use again the weak null structure of the semilinear term $\mathcal{S}^2_{\alpha\beta}$, in order to be able to distinguish between the good and the bad components of the metric, and prove the bounds (2.2.34).

# Chapter Three

# Preliminary Estimates

## 3.1 SOME LEMMAS

We are now ready to start the proof of the main bootstrap proposition 2.3. In this section we prove several results that are used in the rest of the monograph.

### 3.1.1 General Lemmas

We start with a lemma that is used often in integration by parts arguments.

**Lemma 3.1.** *Assume that* $0 < \epsilon \leq 1/\epsilon \leq K$, $N \geq 1$ *is an integer, and* $f, g \in C^{N+1}(\mathbb{R}^3)$. *Then*

$$\Big| \int_{\mathbb{R}^3} e^{iKf} g\, dx \Big| \lesssim_N (K\epsilon)^{-N} \big[ \sum_{|\alpha| \leq N} \epsilon^{|\alpha|} \|D_x^\alpha g\|_{L^1} \big], \tag{3.1.1}$$

*provided that* $f$ *is real-valued,*

$$|\nabla_x f| \geq \mathbf{1}_{\mathrm{supp}\, g}, \quad \text{and} \quad \|D_x^\alpha f \cdot \mathbf{1}_{\mathrm{supp}\, g}\|_{L^\infty} \lesssim_N \epsilon^{1-|\alpha|},\ 2 \leq |\alpha| \leq N+1. \tag{3.1.2}$$

*Proof.* We localize first to balls of size $\approx \epsilon$. Using the assumptions in (3.1.2) we may assume that inside each small ball, one of the directional derivatives of $f$ is bounded away from 0, say $|\partial_1 f| \gtrsim_N 1$. Then we integrate by parts $N$ times in $x_1$, gaining a factor of $K$ and losing a factor of $1/\epsilon$ at every step, and the desired bounds (3.1.1) follow. □

To bound multilinear operators, we often use the following simple lemma.

**Lemma 3.2.** *(i) Assume that* $l \geq 2$, $f_1, \ldots, f_l, f_{l+1} \in L^2(\mathbb{R}^3)$, *and* $M : (\mathbb{R}^3)^l \to \mathbb{C}$ *is a continuous compactly supported function. Then*

$$\Big| \int_{(\mathbb{R}^3)^l} M(\xi_1, \ldots, \xi_l) \cdot \widehat{f_1}(\xi_1) \cdot \ldots \cdot \widehat{f_l}(\xi_l) \cdot \widehat{f_{l+1}}(-\xi_1 - \ldots - \xi_l)\, d\xi_1 \ldots d\xi_l \Big|$$
$$\lesssim \big\|\mathcal{F}^{-1} M\big\|_{L^1((\mathbb{R}^3)^l)} \|f_1\|_{L^{p_1}} \cdot \ldots \cdot \|f_{l+1}\|_{L^{p_{l+1}}}, \tag{3.1.3}$$

*for any exponents* $p_1, \ldots, p_{l+1} \in [1, \infty]$ *satisfying* $1/p_1 + \ldots + 1/p_{l+1} = 1$.

*(ii) As a consequence, if $q, p_2, p_3 \in [1, \infty]$ satisfy $1/p_2 + 1/p_3 = 1/q$ then*

$$\Big\| \mathcal{F}_\xi^{-1}\Big\{ \int_{\mathbb{R}^3} M(\xi,\eta)\widehat{f}(\eta)\widehat{g}(-\xi-\eta)\,d\eta \Big\}\Big\|_{L^q} \lesssim \big\|\mathcal{F}^{-1}M\big\|_{L^1}\|f\|_{L^{p_2}}\|g\|_{L^{p_3}}. \quad (3.1.4)$$

*Proof.* Let $K := \mathcal{F}^{-1}M$. The integral in the left-hand side of (3.1.3) is equal to

$$C\int_{(\mathbb{R}^3)^{l+1}} K(y_1,\ldots,y_l)\cdot f_1(x-y_1)\cdot\ldots\cdot f_l(x-y_l)\,dxdy_1\ldots dy_l,$$

as a consequence of the Fourier inversion formula. The desired bounds (3.1.3) follow using the Hölder inequality in the variable $x$ and the $L^1$ integrability of the kernel $K$.

The bounds (3.1.4) can be proved in a similar way, using duality. □

We will use also a Hardy-type estimate.

**Lemma 3.3.** *(i) For $f \in L^2(\mathbb{R}^3)$ and $k \in \mathbb{Z}$ let*

$$\begin{aligned} A_k &:= \|P_k f\|_{L^2} + \sum_{l=1}^{3} \|\varphi_k(\xi)(\partial_{\xi_l}\widehat{f})(\xi)\|_{L^2_\xi}, \\ B_k &:= \Big[\sum_{j\geq \max(-k,0)} 2^{2j}\|Q_{j,k}f\|_{L^2}^2\Big]^{1/2}. \end{aligned} \quad (3.1.5)$$

*Then, for any $k \in \mathbb{Z}$,*

$$A_k \lesssim \sum_{|k'-k|\leq 4} B_{k'} \quad (3.1.6)$$

*and*

$$B_k \lesssim \begin{cases} \sum_{|k'-k|\leq 4} A_{k'} & \text{if } k \geq 0, \\ \sum_{k'\in\mathbb{Z}} A_{k'} 2^{-|k-k'|/2}\min(1, 2^{k'-k}) & \text{if } k \leq 0. \end{cases} \quad (3.1.7)$$

*(ii) If $m \in \mathcal{M}_0$—see (3.2.40)—then, for any $(k,j) \in \mathcal{J}$,*

$$\|Q_{j,k}\{\mathcal{F}^{-1}(m\widehat{f})\}\|_{L^2} \lesssim \sum_{j'\geq\max(-k,0)} \|Q_{j',k}f\|_{L^2} 2^{-4|j-j'|}. \quad (3.1.8)$$

*Proof.* (i) The bounds (3.1.6)–(3.1.7) were proved in [38, Lemma 3.5]. For convenience we reproduce the proofs here. Clearly, by almost orthogonality,

$$\begin{aligned} B_k &\approx 2^{\max(-k,0)}\|P_k f\|_{L^2} + \||x|\cdot P_k f\|_{L^2} \\ &\approx 2^{\max(-k,0)}\|P_k f\|_{L^2} + \sum_{l=1}^{3}\|\partial_{\xi_l}(\varphi_k(\xi)\widehat{f}(\xi))\|_{L^2_\xi}. \end{aligned} \quad (3.1.9)$$

The bound (3.1.6) follows. The bound in (3.1.7) also follows when $k \geq 0$. On the other hand, if $k \leq 0$ then it suffices to prove that

$$2^{-k}\|P_k f\|_{L^2} \lesssim \sum_{k'\in\mathbb{Z}} A_{k'} 2^{-|k-k'|/2} \min(1, 2^{k'-k}). \tag{3.1.10}$$

For this we let $f_l := x_l f$, $l \in \{1,2,3\}$, so

$$f = \frac{1}{|x|^2+1} f + \sum_{l=1}^{3} \frac{x_l}{|x|^2+1} f_l$$

and, for any $k' \in \mathbb{Z}$,

$$\|P_{k'} f\|_{L^2} + \sum_{l=1}^{3} \|P_{k'} f_l\|_{L^2} \lesssim A_{k'}.$$

Since $|\mathcal{F}\{(x^2+1)^{-1}\}(\xi)| \lesssim |\xi|^{-2}$ and $|\mathcal{F}\{x_l(x^2+1)^{-1}\}(\xi)| \lesssim |\xi|^{-2}$ for $l \in \{1,2,3\}$, for (3.1.10) it suffices to prove that

$$2^{-k}\|\varphi_k(\xi)(g * K)(\xi)\|_{L^2} \lesssim \sum_{k'\in\mathbb{Z}} A_{k'} 2^{-|k-k'|/2} \min(1, 2^{k'-k}), \tag{3.1.11}$$

provided that $\|\varphi_{k'} \cdot g\|_{L^2} \lesssim A_{k'}$ and $K(\eta) = |\eta|^{-2}$. With $g_{k'} = \varphi_{k'} \cdot g$ we estimate

$$\|\varphi_k(\xi)(g_{k'} * K)(\xi)\|_{L^2} \lesssim \|g_{k'}\|_{L^2} \|K \cdot \varphi_{\leq k+10}\|_{L^1} \lesssim 2^k \|g_{k'}\|_{L^2} \quad \text{if } |k-k'| \leq 6;$$

$$\begin{aligned}\|\varphi_k(\xi)(g_{k'} * K)(\xi)\|_{L^2} &\lesssim 2^{3k/2} \|g_{k'}\|_{L^2} \|K \cdot \varphi_{[k'-4,k'+4]}\|_{L^2} \\ &\lesssim 2^{3k/2} 2^{-k'/2} \|g_{k'}\|_{L^2} \quad \text{if } k' \geq k+6;\end{aligned}$$

$$\begin{aligned}\|\varphi_k(\xi)(g_{k'} * K)(\xi)\|_{L^2} &\lesssim \|g_{k'}\|_{L^1} \|K \cdot \varphi_{[k-4,k+4]}\|_{L^2} \\ &\lesssim 2^{3k'/2} 2^{-k/2} \|g_{k'}\|_{L^2} \quad \text{if } k' \leq k-6.\end{aligned}$$

The desired bound (3.1.11) follows, which completes the proof of the lemma. □

### 3.1.2 The Phases $\Phi_{\sigma\mu\nu}$

Our normal form analysis relies on precise bounds on the phases $\Phi_{\sigma\mu\nu}$ defined in (2.1.41). We summarize the results we need in this subsection.

We consider first Wave × KG → KG and KG × KG → Wave interactions. These interactions are weakly elliptic, in the sense that the corresponding phases $\Phi_{\sigma\mu\nu}$ do not vanish, except when the wave frequency vanishes. More precisely, we have the following quantitative estimates.

**Lemma 3.4.** *(i) Assume that $\Phi_{\sigma\mu\nu}$ is as in (2.1.41). If $|\xi|, |\xi-\eta|, |\eta| \in [0, b]$,*

$1 \leq b$, *then*

$$
\begin{aligned}
&|\Phi_{\sigma\mu\nu}(\xi,\eta)| \geq |\xi|/(4b^2) && \text{if } (\sigma,\mu,\nu) = ((wa,\iota),(kg,\iota_1),(kg,\iota_2)),\\
&|\Phi_{\sigma\mu\nu}(\xi,\eta)| \geq |\eta|/(4b^2) && \text{if } (\sigma,\mu,\nu) = ((kg,\iota),(kg,\iota_1),(wa,\iota_2)).
\end{aligned}
\tag{3.1.12}
$$

*(ii) Assume that* $k, k_1, k_2 \in \mathbb{Z}$ *and* $n$ *satisfies* $\|\mathcal{F}^{-1}n\|_{L^1(\mathbb{R}^3\times\mathbb{R}^3)} \leq 1$. *Let* $\overline{k} = \max(k, k_1, k_2)$. *If* $(\sigma,\mu,\nu) = ((wa,\iota),(kg,\iota_1),(kg,\iota_2))$ *then*

$$
\big\|\mathcal{F}^{-1}\{\Phi_{\sigma\mu\nu}(\xi,\eta)^{-1}n(\xi,\eta)\cdot\varphi_k(\xi)\varphi_{k_1}(\xi-\eta)\varphi_{k_2}(\eta)\}\big\|_{L^1(\mathbb{R}^3\times\mathbb{R}^3)} \lesssim 2^{-k}2^{4\overline{k}^+}.
\tag{3.1.13}
$$

*Moreover, if* $(\sigma,\mu,\nu) = ((kg,\iota),(kg,\iota_1),(wa,\iota_2))$ *then*

$$
\big\|\mathcal{F}^{-1}\{\Phi_{\sigma\mu\nu}(\xi,\eta)^{-1}n(\xi,\eta)\cdot\varphi_k(\xi)\varphi_{k_1}(\xi-\eta)\varphi_{k_2}(\eta)\}\big\|_{L^1(\mathbb{R}^3\times\mathbb{R}^3)} \lesssim 2^{-k_2}2^{4\overline{k}^+}.
\tag{3.1.14}
$$

*Proof.* The conclusions were all proved in Lemma 3.3 in [38]. For convenience we reproduce the proof here.

(i) The bounds follow from the elementary inequalities

$$
\begin{aligned}
&\sqrt{1+x^2}+\sqrt{1+y^2}-(x+y) \geq 1/(2b),\\
&x+\sqrt{1+y^2}-\sqrt{1+(x+y)^2} \geq x/(4b^2),
\end{aligned}
\tag{3.1.15}
$$

which hold if $x, y, x+y \in [0,b]$. The second inequality can be proved by setting $F(x) := x+\sqrt{1+y^2}-\sqrt{1+(x+y)^2}$ and noticing that $F'(x) \geq 1/(4b^2)$ as long as $y, x+y \in [0,b]$.

(ii) By symmetry, it suffices to prove (3.1.13). Also, since

$$
\|\mathcal{F}^{-1}(fg)\|_{L^1} \lesssim \|\mathcal{F}^{-1}f\|_{L^1}\|\mathcal{F}^{-1}g\|_{L^1},
\tag{3.1.16}
$$

without loss of generality we may assume that $n \equiv 1$ and $\iota = +$. Let

$$
m(v,\eta) := 2^{-k}\Phi_{\sigma\mu\nu}(2^k v,\eta)^{-1} = \frac{1}{|v| - 2^{-k}\Lambda_{kg,\iota_1}(\eta-2^k v) - 2^{-k}\Lambda_{kg,\iota_2}(\eta)}.
\tag{3.1.17}
$$

For (3.1.13) it suffices to prove that

$$
\big\|\mathcal{F}^{-1}\{m(v,\eta)\cdot\varphi_0(v)\varphi_{k_1}(\eta-2^k v)\varphi_{k_2}(\eta)\}\big\|_{L^1(\mathbb{R}^3\times\mathbb{R}^3)} \lesssim 2^{4\overline{k}^+}.
\tag{3.1.18}
$$

We consider two cases, depending on the signs $\iota_1$ and $\iota_2$.

**Case 1.** $\iota_1 \neq \iota_2$. By symmetry we may assume that $\iota_2 = -, \iota_1 = +$, so

$$\begin{aligned} m(v,\eta) &= \frac{1}{|v| - 2^{-k}\sqrt{1+|\eta - 2^k v|^2} + 2^{-k}\sqrt{1+|\eta|^2}} \\ &= \frac{2^k|v| + \sqrt{1+|\eta|^2} + \sqrt{1+|\eta-2^k v|^2}}{2(|v|\sqrt{1+|\eta|^2} + v\cdot\eta)} \\ &= \frac{\left[2^k|v| + \sqrt{1+|\eta|^2} + \sqrt{1+|\eta - 2^k v|^2}\right]\left[|v|\sqrt{1+|\eta|^2} - v\cdot\eta\right]}{2[|v|^2 + |v|^2|\eta|^2 - (v\cdot\eta)^2]}. \end{aligned} \tag{3.1.19}$$

The first identity follows by algebraic simplifications, after multiplying both the numerator and the denominator by $|v| + 2^{-k}\sqrt{1+|\eta - 2^k v|^2} + 2^{-k}\sqrt{1+|\eta|^2}$. The second identity follows by multiplying both the numerator and the denominator by $|v|\sqrt{1+|\eta|^2} - v\cdot\eta$. The numerator in the formula above is a sum of simple products and its contribution is a factor of $2^{2\overline{k}^+}$. In view of the general bound (3.1.16), for (3.1.18) it suffices to prove that, for $l \geq 0$,

$$\left\| \int_{\mathbb{R}^3\times\mathbb{R}^3} e^{ix\cdot v} e^{iy\cdot\eta} \frac{1}{|v|^2 + |v|^2|\eta|^2 - (v\cdot\eta)^2} \varphi_0(v)\varphi_{\leq l}(\eta)\, dv d\eta \right\|_{L^1_{x,y}} \lesssim 2^{2l}. \tag{3.1.20}$$

We insert thin angular cutoffs in $v$. Due to rotation invariance it suffices to prove that

$$\left\| \int_{\mathbb{R}^3\times\mathbb{R}^3} e^{ix\cdot v} e^{iy\cdot\eta} \frac{\varphi_{\leq -l-10}(v_2)\varphi_{\leq -l-10}(v_3)}{|v|^2 + |v|^2|\eta|^2 - (v\cdot\eta)^2} \varphi_0(v)\varphi_{\leq l}(\eta)\, dv d\eta \right\|_{L^1_{x,y}} \lesssim 1.$$

We make the changes of variables $v_1 \leftrightarrow w_1, v_2 \leftrightarrow 2^{-l}w_2, v_3 \leftrightarrow 2^{-l}w_3$, $\eta_1 \leftrightarrow 2^l\rho_1, \eta_2 \leftrightarrow \rho_2, \eta_3 \leftrightarrow \rho_3$. After rescaling the spatial variables appropriately, it suffices to prove that

$$\begin{aligned} \Big\| \int_{\mathbb{R}^3\times\mathbb{R}^3} & e^{ix\cdot w} e^{iy\cdot\rho} m'(w,\rho)\varphi_{[-4,4]}(w_1)\varphi_{\leq -10}(w_2)\varphi_{\leq -10}(w_3) \\ & \times \varphi_{\leq 4}(\rho_1)\varphi_{\leq l+4}(\rho_2)\varphi_{\leq l+4}(\rho_3)\, dw d\rho \Big\|_{L^1_{x,y}} \lesssim 1, \end{aligned} \tag{3.1.21}$$

where

$$\begin{aligned} m'(w,\rho) := \big\{ & w_1^2(1+\rho_2^2+\rho_3^2) + \rho_1^2(w_2^2+w_3^2) \\ & + 2^{-2l}(w_2^2 + w_3^2 + (w_2\rho_3 - w_3\rho_2)^2) - 2\rho_1 w_1(w_2\rho_2 + w_3\rho_3)\big\}^{-1}. \end{aligned}$$

It is easy to see that $|m'(w,\rho)| \approx (1+|\rho|^2)^{-1}$ and $|D_w^\alpha D_\rho^\beta m'(w,\rho)| \lesssim (1+|\rho|^2)^{-1-|\beta|/2}$ in the support of the integral, for all multi-indices $\alpha$ and $\beta$ with $|\alpha| \leq 4$, $|\beta| \leq 4$. The bound (3.1.21) follows by a standard integration by parts argument, which completes the proof of (3.1.18).

**Case 2.** $\iota_1 = \iota_2$. If $\iota_1 = \iota_2 = +$ then we write, as in (3.1.19),

$$m(v,\eta) = \frac{1}{|v| - 2^{-k}\sqrt{1+|\eta-2^k v|^2} - 2^{-k}\sqrt{1+|\eta|^2}}$$
$$= \frac{-\big[2^k|v| - \sqrt{1+|\eta|^2} + \sqrt{1+|\eta-2^k v|^2}\big]\big[|v|\sqrt{1+|\eta|^2} + v\cdot\eta\big]}{2[|v|^2 + |v|^2|\eta|^2 - (v\cdot\eta)^2]}.$$

On the other hand, if $\iota_1 = \iota_2 = -$ then we write, as in (3.1.19),

$$m(v,\eta) = \frac{1}{|v| + 2^{-k}\sqrt{1+|\eta-2^k v|^2} + 2^{-k}\sqrt{1+|\eta|^2}}$$
$$= \frac{\big[2^k|v| + \sqrt{1+|\eta|^2} - \sqrt{1+|\eta-2^k v|^2}\big]\big[|v|\sqrt{1+|\eta|^2} - v\cdot\eta\big]}{2[|v|^2 + |v|^2|\eta|^2 - (v\cdot\eta)^2]}.$$

The desired conclusion follows in both cases using (3.1.20) and (3.1.16). Since $\|\mathcal{F}^{-1}\{\varphi_0(v)(2^k|v| \pm \sqrt{1+|\eta|^2} \mp \sqrt{1+|\eta-2^k v|^2})\}\|_{L^1(\mathbb{R}^3\times\mathbb{R}^3)} \lesssim 2^k$, we get in fact a stronger bound when $\sigma = (wa,\iota)$ and $\mu = \nu \in \{(kg,+),(kg,-)\}$,

$$\big\|\mathcal{F}^{-1}\{\Phi_{\sigma\mu\nu}(\xi,\eta)^{-1}n(\xi,\eta)\cdot\varphi_k(\xi)\varphi_{k_1}(\xi-\eta)\varphi_{k_2}(\eta)\}\big\|_{L^1(\mathbb{R}^3\times\mathbb{R}^3)} \lesssim 2^{3\overline{k}^+}, \tag{3.1.22}$$

as desired. □

We will also consider Wave × Wave → Wave interactions. In this case the corresponding bilinear phases $\Phi_{\sigma\mu\nu}$ can vanish on large sets, when the frequencies are parallel, and the strength of these interactions depends significantly on the angle between the frequencies of the inputs. To measure this angle, for $\iota_1, \iota_2 \in \{+,-\}$ we define the functions

$$\Xi_{\iota_1\iota_2} : (\mathbb{R}^3\setminus\{0\})^2 \to B_2, \qquad \Xi_{\iota_1\iota_2}(\theta,\eta) := \iota_1\frac{\theta}{|\theta|} - \iota_2\frac{\eta}{|\eta|}, \tag{3.1.23}$$

where $B_R := \{x \in \mathbb{R}^3 : |x| \le R\}$. Let $\Xi_{\iota_1\iota_2,k}(\theta,\eta) := \iota_1\theta_k/|\theta| - \iota_2\eta_k/|\eta|$, $k \in \{1,2,3\}$. We will often use the following elementary lemma:

**Lemma 3.5.** *(i) By convention, let* $++ = -- = +$ *and* $+- = -+ = -$. *Then*

$$|\Xi_{\iota_1\iota_2}(\theta,\eta)| = \sqrt{\frac{2(|\theta||\eta| - \iota_1\iota_2\theta\cdot\eta)}{|\theta||\eta|}}. \tag{3.1.24}$$

*(ii) We define the functions* $\widetilde{\Xi} : (\mathbb{R}^3\setminus\{0\})^2 \to [0,2]$,

$$\widetilde{\Xi}(\theta,\eta) := \Big|\frac{\theta}{|\theta|} - \frac{\eta}{|\eta|}\Big|\Big|\frac{\theta}{|\theta|} + \frac{\eta}{|\eta|}\Big| = 2\sqrt{1 - \frac{|\theta\cdot\eta|^2}{|\theta|^2|\eta|^2}}. \tag{3.1.25}$$

*Then, for any $\theta, \eta \in \mathbb{R}^3 \setminus \{0\}$ we have*

$$\min\{|\Xi_+(\theta,\eta)|, |\Xi_-(\theta,\eta)|\} \le \widetilde{\Xi}(\theta,\eta) \le 2\min\{|\Xi_+(\theta,\eta)|, |\Xi_-(\theta,\eta)|\}. \quad (3.1.26)$$

*Moreover, for any $x, y, z \in \mathbb{R}^3 \setminus \{0\}$ we have*

$$\widetilde{\Xi}(x,z) \le 2[\widetilde{\Xi}(x,y) + \widetilde{\Xi}(y,z)]. \quad (3.1.27)$$

*In addition, if $y + z \neq 0$ then*

$$\widetilde{\Xi}(x, y+z)|y+z| \le \widetilde{\Xi}(x,y)|y| + \widetilde{\Xi}(x,z)|z|. \quad (3.1.28)$$

*Proof.* The identities (3.1.24) follow directly from definitions. The inequalities (3.1.26) follow from definitions as well, once we notice that $\widetilde{\Xi}(\theta,\eta) = |\Xi_+(\theta,\eta)| \cdot |\Xi_-(\theta,\eta)|$ and $|\Xi_+(\theta,\eta)| + |\Xi_-(\theta,\eta)| \ge 2$. For (3.1.27) we notice that

$$\begin{aligned}&\min\{|\Xi_+(x,z)|, |\Xi_-(x,z)|\}\\ &\quad\le \min\{|\Xi_+(x,y)|, |\Xi_-(x,y)|\} + \min\{|\Xi_+(y,z)|, |\Xi_-(y,z)|\},\end{aligned}$$

and then use (3.1.26). Finally, to prove (3.1.28) we may assume that $x = (1,0,0)$, $y = (y_1, y')$, $z = (z_1, z')$, and estimate, using just (3.1.25),

$$\widetilde{\Xi}(x,y+z)|y+z| = 2|y'+z'| \le 2|y'| + 2|z'| = \widetilde{\Xi}(x,y)|y| + \widetilde{\Xi}(x,z)|z|,$$

which gives the desired conclusion. □

We consider now trilinear expressions localized with respect to angular separation, as well as expressions resulting from normal form transformations.

**Lemma 3.6.** *(i) Assume $\chi_1 : \mathbb{R}^3 \to [0,1]$ is a smooth function supported in the ball $B_2$, $\iota_1, \iota_2 \in \{+,-\}$, $b \le 2$, $f, f_1, f_2 \in L^2(\mathbb{R}^3)$, and $k, k_1, k_2 \in \mathbb{Z}$. Let*

$$\begin{aligned} L^b_{k,k_1,k_2} := \int_{\mathbb{R}^3\times\mathbb{R}^3} & m(\xi-\eta,\eta)\chi_1(2^{-b}\Xi_{\iota_1\iota_2}(\xi-\eta,\eta)) \\ & \times \widehat{P_{k_1}f_1}(\xi-\eta)\widehat{P_{k_2}f_2}(\eta)\widehat{P_k f}(\xi)\,d\xi d\eta, \end{aligned} \quad (3.1.29)$$

*where $m$ is a symbol satisfying $\|\mathcal{F}^{-1}(m)\|_{L^1(\mathbb{R}^6)} \le 1$. Then*

$$|L^b_{k,k_1,k_2}| \lesssim \min\{2^{-b}, 2^{k_1-k}+1\}\|P_{k_1}f_1\|_{L^\infty}\|P_{k_2}f_2\|_{L^2}\|P_k f\|_{L^2}. \quad (3.1.30)$$

*(ii) Let $\chi_2 : \mathbb{R}^3 \to [0,1]$ be a smooth function supported in $B_2 \setminus B_{1/2}$, $(\sigma,\mu,\nu) = ((wa,\iota),(wa,\iota_1),(wa,\iota_2))$, $\iota,\iota_1,\iota_2 \in \{+,-\}$. Then for any $k, k_1, k_2 \in \mathbb{Z}$,*

$$\begin{aligned}&|\Phi_{\sigma\mu\nu}(\xi,\eta)^{-1}\cdot\varphi_k(\xi)\varphi_{k_1}(\xi-\eta)\varphi_{k_2}(\eta)\chi_2(2^{-b}\Xi_{\iota_1\iota_2}(\xi-\eta,\eta))| \\ &\qquad\lesssim 2^{-2b}2^{-\min(k_1,k_2)},\end{aligned} \quad (3.1.31)$$

*where* $\Phi_{\sigma\mu\nu}(\xi,\eta)=\Lambda_\sigma(\xi)-\Lambda_\mu(\xi-\eta)-\Lambda_\nu(\eta)$ *as in* (2.1.41). *In addition, if*

$$\begin{aligned} M^b_{k,k_1,k_2} := \int_{\mathbb{R}^3\times\mathbb{R}^3} m(\xi-\eta,\eta)\frac{\chi_2(2^{-b}\Xi_{\iota_1\iota_2}(\xi-\eta,\eta))}{\Phi_{\sigma\mu\nu}(\xi,\eta)} \\ \times\widehat{P_{k_1}f_1}(\xi-\eta)\widehat{P_{k_2}f_2}(\eta)\widehat{P_kf}(\xi)\,d\xi d\eta, \end{aligned} \tag{3.1.32}$$

*and* $\|\mathcal{F}^{-1}(m)\|_{L^1(\mathbb{R}^6)}\le 1$, *then*

$$\begin{aligned} |M^b_{k,k_1,k_2}| \lesssim 2^{-2b}2^{-\min(k_1,k_2)}\min\{2^{-b},2^{k_1-k}+1\} \\ \times\|P_{k_1}f_1\|_{L^\infty}\|P_{k_2}f_2\|_{L^2}\|P_kf\|_{L^2}. \end{aligned} \tag{3.1.33}$$

*(iii) If* $\chi_3:\mathbb{R}\to[0,1]$ *is a smooth function supported in* $[-2,2]\setminus[-1/2,1/2]$ *and* $\Phi_{\sigma,\mu,\nu}$ *is as above then, for any* $k,k_1,k_2\in\mathbb{Z}$,

$$\begin{aligned} \sup_{\xi\in\mathbb{R}^3}\Big\|\int_{\mathbb{R}^3} e^{iy\cdot\eta}\frac{\chi_3(2^{-2b}|\Xi_{\iota_1\iota_2}(\xi-\eta,\eta)|^2)}{\Phi_{\sigma\mu\nu}(\xi,\eta)}\varphi_k(\xi)\varphi_{k_1}(\xi-\eta)\varphi_{k_2}(\eta)\,d\eta\Big\|_{L^1_y} \\ \lesssim 2^{-2b}2^{-\min(k_1,k_2)}. \end{aligned} \tag{3.1.34}$$

*Remark 3.7.* All estimates would follow from the "natural" localization bounds

$$\big\|\mathcal{F}^{-1}\{\chi_1(2^{-b}\Xi_{\iota_1\iota_2}(\xi-\eta,\eta))\varphi_0(\xi-\eta)\varphi_0(\eta)\}\big\|_{L^1(\mathbb{R}^6)}\lesssim 1 \tag{3.1.35}$$

and the identities (3.1.48). We are not able to prove these bounds, but we prove the weaker bounds (3.1.36) and (3.1.44), which still allow us to derive the conclusions of the lemma.

*Proof.* **Step 1.** We show first that, for any $k_1,k_2\in\mathbb{Z}$ and $\iota_1,\iota_2\in\{+,-\}$,

$$\big\|\mathcal{F}^{-1}\{\chi_1(2^{-b}\Xi_{\iota_1\iota_2}(\xi-\eta,\eta))\varphi_{k_1}(\xi-\eta)\varphi_{k_2}(\eta)\}\big\|_{L^1(\mathbb{R}^6)}\lesssim 2^{-b}. \tag{3.1.36}$$

After changes of variables we may assume that $\iota_1=\iota_2=+$, $k_1=k_2=0$. We have to prove that $\big\|F_b\big\|_{L^1}\lesssim 2^{-b}$ for any $b\le 2$, where

$$F_b(x,y):=\int_{\mathbb{R}^6} e^{ix\cdot\theta}e^{iy\cdot\eta}\chi_1(2^{-b}\Xi_{++}(\theta,\eta))\varphi_0(\theta)\varphi_0(\eta)\,d\theta d\eta. \tag{3.1.37}$$

In fact, we will prove the stronger pointwise bounds

$$\begin{aligned} |F_b(x,y)| \lesssim \frac{1}{\big[1+2^{2b}|x|^2+2^{2b}|y|^2\big]^8} \\ \times\frac{1}{\big[1+\min\{|x|,|y|\}\widetilde{\Xi}(x,y)\big]^8}\frac{2^{2b}}{(1+|x|^2+|y|^2)^{1/2}}, \end{aligned} \tag{3.1.38}$$

for any $x,y\in\mathbb{R}^3$, which would imply the desired conclusion $\big\|F_b\big\|_{L^1}\lesssim 2^{-b}$.

In proving (3.1.38), we may assume that $|y| \le |x|$, $x = (x_1, 0, 0)$, $x_1 \ge 0$, $y = (y_1, y_2, y_3)$, $|y_2| \ge |y_3|$. The desired bounds (3.1.38) are then equivalent to

$$|F_b(x, y)| \lesssim \frac{1}{[1 + 2^{2b}x_1^2]^8} \frac{1}{[1 + |y_2|]^8} \frac{2^{2b}}{(1 + x_1^2)^{1/2}}, \qquad (3.1.39)$$

where $y = (y_1, y')$. Notice that this follows easily if $x_1 \lesssim 1$, since the $\theta, \eta$ integration is taken over a set of volume $\approx 2^{2b}$.

We will integrate by parts in $\theta$ and $\eta$ using the operators

$$2^{2b}\Delta_\theta, \quad 2^{2b}\Delta_\eta, \quad L_\theta := \theta_j \partial_{\theta_j}, \quad L_\eta := \eta_j \partial_{\eta_j}, \quad S_{ij} := \theta_i \partial_{\theta_j} + \eta_i \partial_{\eta_j}. \quad (3.1.40)$$

The main point is that the vector-fields $L_\theta, L_\eta, S_{ij}$ act well on $\Xi_{++}(\theta, \eta)$, i.e.,

$$\begin{aligned} &(L_\theta \Xi_{++,k})(\theta, \eta) = 0, \qquad (L_\eta \Xi_{++,k})(\theta, \eta) = 0, \\ &(S_{ij}\Xi_{++,k})(\theta, \eta) = \delta_{jk}\big[\theta_i/|\theta| - \eta_i/|\eta|\big] - \big[\theta_i\theta_j\theta_k/|\theta|^3 - \eta_i\eta_j\eta_k/|\eta|^3\big]. \end{aligned} \qquad (3.1.41)$$

In particular, if $\mathcal{O}$ is any combination of the operators $L_\theta, L_\eta, S_{ij}, 2^{2b}\Delta_\theta$ then $\mathcal{O}\{\chi_1(2^{-b}\Xi_{++}(\theta, \eta))\varphi_0(\theta)\varphi_0(\eta)\}$ is a function of the form $\widetilde{\chi}[2^{-b}\Xi_{++}(\theta, \eta), \theta, \eta]$, where $\widetilde{\chi} = \widetilde{\chi}_{\mathcal{O}}$ is a smooth function supported in the set $B_2 \times (B_2 \setminus B_{1/2})^2$.

We notice now that

$$S_{j1}\{e^{ix\cdot\theta}e^{iy\eta}\} = i(\theta_j x_1 + \eta_j y_1)\{e^{ix\cdot\theta}e^{iy\eta}\}.$$

Therefore, if $|y_1| \le 2^{-20}x_1$ then we can integrate by parts using only the vector-fields $S_{j1}$ and gain a factor of $x_1$ at every iteration. It follows that $|F_b(x, y)| \lesssim 2^{2b}(1 + x_1^2)^{-20}$, which is better than the desired bounds (3.1.39).

On the other hand, if $|y_1| \ge 2^{-20}x_1$ then we first integrate by parts using $2^{2b}\Delta_\theta$ and $S_{j2}$, $j \in \{1, 2, 3\}$, to gain the factors $(1 + 2^{2b}x_1^2)^{-10}$ and $(1 + |y_2|)^{-10}$ in (3.1.39). Letting

$$F_b'(x, y) := \int_{\mathbb{R}^6} e^{ix\cdot\theta}e^{iy\cdot\eta}\widetilde{\chi}[2^{-b}\Xi_{++}(\theta, \eta), \theta, \eta]\, d\theta d\eta, \qquad (3.1.42)$$

where $\widetilde{\chi}$ is a smooth function supported in the set $B_2 \times (B_2 \setminus B_{1/2})^2$, it remains to prove that

$$|F_b'(x, y)| \lesssim 2^{2b}(1 + |x_1|)^{-1}. \qquad (3.1.43)$$

For this we use the scaling vector-field $L_\theta$. For integers $n \ge 0$ we define $F'_{b;n}$ by inserting cutoff functions of the form $\varphi_n^{[0,\infty)}(x \cdot \theta)$ in the integral in (3.1.42). Clearly, $|F'_{b;0}(x, y)| \lesssim 2^{2b}(1 + |x_1|)^{-1}$. We integrate by parts twice using the scaling vector-field $L_\theta$ to show that $|F'_{b;n}(x, y)| \lesssim 2^{-n}2^{2b}(1 + |x_1|)^{-1}$, and the desired bounds (3.1.43) follow.

**Step 2.** With $\chi_1$ as in (i), we show now that

$$\|\mathcal{F}^{-1}\{\chi_1(2^{-b}\Xi_{\iota_1\iota_2}(\xi-\eta,\eta))\chi'(2^{-b}(\eta/|\eta|-e))\varphi_{k_1}(\xi-\eta)\varphi_{k_2}(\eta)\}\|_{L^1(\mathbb{R}^6)}\lesssim 1, \tag{3.1.44}$$

for any smooth function $\chi':\mathbb{R}^3\to[0,1]$ supported in the ball $B_2$ and any vector $e\in\mathbb{S}^2$. Indeed, to prove this we can rescale to $k_1=k_2=0$. Then we define

$$G_b(x,y):=\int_{\mathbb{R}^6}e^{ix\cdot\theta}e^{iy\cdot\eta}\chi_1(2^{-b}\Xi_{++}(\theta,\eta))\chi'(2^{-b}(\eta/|\eta|-e))\varphi_0(\theta)\varphi_0(\eta)\,d\theta d\eta, \tag{3.1.45}$$

and notice that it suffices to prove the pointwise bounds

$$|G_b(x,y)|\lesssim\frac{1}{\big[1+2^{2b}|x|^2+2^{2b}|y|^2\big]^8}\frac{2^{4b}}{(1+|x\cdot e|^2+|y\cdot e|^2)^8}, \tag{3.1.46}$$

for any $x,y\in\mathbb{R}^3$. These bounds follow by integration by parts as before, using (3.1.41) and the operators $2^{2b}\Delta_\theta$, $2^{2b}\Delta_\eta$, $L_\theta$, $L_\eta$, and $S_{ij}$ defined in (3.1.40).

**Step 3.** We prove now the bounds (3.1.30). The estimates with the factor $2^{-b}$ follow from (3.1.36) and (3.1.3). To prove the estimates with the factor $2^{k_1-k}+1$ we need to introduce an angular decomposition. Given $q\in\mathbb{Z}$, $q\le 2$, we fix a $2^q$-net $\mathfrak{N}_q$ on $\mathbb{S}^2$ and define

$$\begin{aligned}\varphi_{k;q,e}(\xi)&:=\varphi_k(\xi)\frac{\varphi_{\le 0}(2^{-2q}|\xi/|\xi|-e|^2)}{\sum_{e'\in\mathfrak{N}_q}\varphi_{\le 0}(2^{-2q}|\xi/|\xi|-e'|^2)},\\ P_{k;q,e}f&:=\mathcal{F}^{-1}\{\varphi_{k;q,e}\widehat{f}\},\end{aligned} \tag{3.1.47}$$

for any $k\in\mathbb{Z}$ and $e\in\mathfrak{N}_q$. We insert the partition of unity $\{\varphi_{k_2;b,e}(\eta)\}_{e\in\mathfrak{N}_b}$ in the integrals in (3.1.29). Notice also that if $|\Xi_{\iota_1\iota_2}(\xi-\eta,\eta)|\lesssim 2^b$ and $|\eta/|\eta|-e|\lesssim 2^b$ then $\widetilde{\Xi}(\xi,e)\lesssim 2^b(2^{k_1-k}+1)$ in the support of the integral, as a consequence of (3.1.27)–(3.1.28). Thus

$$L^b_{k,k_1,k_2}=\sum_{e\in\mathfrak{N}_b}L^{b,e}_{k,k_1,k_2},$$

$$\begin{aligned}L^{b,e}_{k,k_1,k_2}:=\int_{\mathbb{R}^3\times\mathbb{R}^3}&m(\xi-\eta,\eta)\chi_1(2^{-b}\Xi_{\iota_1\iota_2}(\xi-\eta,\eta))\\ &\times\widehat{P_{k_1}f_1}(\xi-\eta)\widehat{P_{k_2;b,e}f_2}(\eta)\widehat{P_kA_{b',e}f}(\xi)\,d\xi d\eta,\end{aligned}$$

where $b':=b+\max\{k_1-k,0\}+C$, and $\widehat{A_{b',e}f}(\xi)=\widehat{f}(\xi)\varphi_{\le 0}(2^{-b'}\widetilde{\Xi}(\xi,e))$. Thus

$$\begin{aligned}|L^b_{k,k_1,k_2}|^2&\lesssim\|P_{k_1}f_1\|^2_{L^\infty}\Big\{\sum_{e\in\mathfrak{N}_b}\|P_{k_2;b,e}f_2\|^2_{L^2}\Big\}\Big\{\sum_{e\in\mathfrak{N}_b}\|P_kA_{b',e}f\|^2_{L^2}\Big\}\\ &\lesssim\|P_{k_1}f_1\|^2_{L^\infty}\|P_{k_2}f_2\|^2_{L^2}\cdot 2^{2(b'-b)}\|P_kf\|^2_{L^2},\end{aligned}$$

using (3.1.44), (3.1.3), and orthogonality. The desired bounds (3.1.30) follow.

**Step 4.** We prove now the bounds in (ii). With $\theta = \xi - \eta$ we write

$$\begin{aligned}\frac{1}{\Phi_{\sigma\mu\nu}(\xi,\eta)} &= \frac{1}{\iota|\xi| - \iota_1|\xi-\eta| - \iota_2|\eta|} = \frac{\iota|\theta+\eta| + \iota_1|\theta| + \iota_2|\eta|}{2(\theta\cdot\eta - \iota_1\iota_2|\theta||\eta|)} \\ &= \frac{\iota|\theta+\eta| + \iota_1|\theta| + \iota_2|\eta|}{-\iota_1\iota_2|\theta||\eta||\Xi_{\iota_1\iota_2}(\theta,\eta)|^2}.\end{aligned} \tag{3.1.48}$$

The bounds (3.1.31) and (3.1.33) follow, using also (3.1.30).

**Step 5.** Since $\|\mathcal{F}^{-1}(m\cdot m')\|_{L^1} \lesssim \|\mathcal{F}^{-1}m\|_{L^1}\|\mathcal{F}^{-1}m'\|_{L^1}$, for (3.1.34) it suffices to prove that, for any $\xi \in \mathbb{R}^3$,

$$\Big\|\varphi_k(\xi)\int_{\mathbb{R}^3} e^{iy\cdot\eta}\chi_3(2^{-2b}|\Xi_{\iota_1\iota_2}(\xi-\eta,\eta)|^2)\cdot\varphi_{k_1}(\xi-\eta)\varphi_{k_2}(\eta)\,d\eta\Big\|_{L^1_y} \lesssim 1. \tag{3.1.49}$$

By symmetry and rotation, we may assume that $k_2 \le k_1$ and $\xi = (\xi_1, 0, 0)$, $\xi_1 > 0$. The bounds (3.1.49) follow by standard integration by parts if $b \ge -20$, since the function

$$H_{b,\xi}(\eta) := 2^{-2b}|\Xi_{\iota_1\iota_2}(\xi-\eta,\eta)|^2$$

satisfies differential bounds of the form $|D^\alpha_\eta H_{b;\xi}(\eta)| \lesssim_{|\alpha|} 2^{-|\alpha|k_2}$ in the support of the integral, for all multi-indices $\alpha \in \mathbb{Z}^3_+$.

When $b \le -20$ we have to be slightly more careful. Notice that if $b \le -20$ then the function $\chi_3(2^{-2b}|\Xi_{\iota_1\iota_2}(\xi-\eta,\eta)|^2)\cdot\varphi_k(\xi)\varphi_{k_1}(\xi-\eta)\varphi_{k_2}(\eta)$ is nontrivial only if $b \le k - \max(k_1,k_2) + 10$. This can be verified easily by considering the two cases $\iota_1 = \iota_2$ and $\iota_1 = -\iota_2$, and examining the definitions. Notice also that

$$\begin{aligned}H_{b;\xi}(\eta) &= 2^{-2b+1}\frac{|\xi-\eta||\eta| - \iota_1\iota_2(\xi-\eta)\cdot\eta}{|\xi-\eta||\eta|} \\ &= \frac{2^{-2b+1}[|\xi-\eta|^2|\eta|^2 - ((\xi-\eta)\cdot\eta)^2]}{|\xi-\eta||\eta|[|\xi-\eta||\eta| + \iota_1\iota_2(\xi-\eta)\cdot\eta]} \\ &= \frac{2^{-2b+1}\xi_1^2(\eta_2^2+\eta_3^2)}{|\xi-\eta||\eta|[|\xi-\eta||\eta| + \iota_1\iota_2(\xi-\eta)\cdot\eta]},\end{aligned} \tag{3.1.50}$$

using (3.1.24). Notice that the denominator of fraction in the right-hand side above is $\approx 2^{2k_1+2k_2}$ in the support of the integral. Thus the $\eta$ integral in (3.1.49) is supported in the set $\mathcal{R}_{b;\xi} := \big\{|\eta| \approx 2^{k_2},\ |\xi-\eta| \approx 2^{k_1},\ \sqrt{\eta_2^2+\eta_3^2} \approx 2^{b+k_1+k_2-k}\big\}$ (according to the remark above, we may assume that $2^{b+k_1+k_2-k} \lesssim 2^{\min(k_1,k_2)} = 2^{k_2}$). Moreover, it is easy to see that

$$|\partial^\alpha_{\eta_1}H_{b;\xi}(\eta)| \lesssim 2^{-k_2|\alpha|}, \qquad |\partial^\alpha_{\eta_l}H_{b;\xi}(\eta)| \lesssim 2^{-|\alpha|(b+k_1+k_2-k)},$$

for $\eta \in \mathcal{R}_{b;\xi}$, $l \in \{2,3\}$, and $\alpha \in [0,10]$. The bounds (3.1.49) follow by integration by parts in $\eta$. □

### 3.1.3 Elements of Paradifferential Calculus

In quasilinear equations the application of normal forms to prove energy estimates leads to loss of derivative. As in some of our earlier work [16, 17], we deal with this issue using paradifferential calculus.

In this subsection we summarize some basic tools of paradifferential calculus. We recall first the definition of paradifferential operators (Weyl quantization): given a symbol $a = a(x, \zeta) : \mathbb{R}^3 \times \mathbb{R}^3 \to \mathbb{C}$, we define the operator $T_a$ by

$$\mathcal{F}\{T_a f\}(\xi) = \frac{1}{8\pi^3} \int_{\mathbb{R}^3} \chi_0\Big(\frac{|\xi - \eta|}{|\xi + \eta|}\Big) \widetilde{a}(\xi - \eta, (\xi + \eta)/2) \widehat{f}(\eta) d\eta, \tag{3.1.51}$$

where $\widetilde{a}$ denotes the Fourier transform of $a$ in the first variable and $\chi_0 = \varphi_{\leq -40}$.

We will use a simple norm to estimate symbols: we define

$$\begin{aligned} \|a\|_{\mathcal{L}^q_l} &:= \sup_{\zeta \in \mathbb{R}^3} (1 + |\zeta|^2)^{-l/2} \| \, |a|(., \zeta)\|_{L^q_x}, \\ |a|(x, \zeta) &:= \sum_{|\beta| \leq 20,\, |\alpha| \leq 3} |\zeta|^{|\beta|} |(D^\beta_\zeta D^\alpha_x a)(x, \zeta)|, \end{aligned} \tag{3.1.52}$$

for $q \in [1, \infty]$ and $l \in \mathbb{R}$. The index $l$ is called the *order* of the symbol, and it measures the contribution of the symbol in terms of derivatives on $f$. Notice that we have the simple product rule

$$\|ab\|_{\mathcal{L}^p_{l_1 + l_2}} \lesssim \|a\|_{\mathcal{L}^q_{l_1}} \|b\|_{\mathcal{L}^r_{l_2}}, \qquad 1/p = 1/q + 1/r. \tag{3.1.53}$$

Our main result in this subsection is the following lemma.

**Lemma 3.8.** *(i) If $1/p = 1/q + 1/r$ and $k \in \mathbb{Z}$, and $l \in [-10, 10]$ then*

$$\|P_k T_a f\|_{L^p} \lesssim 2^{lk_+} \|a\|_{\mathcal{L}^q_l} \|P_{[k-2,k+2]} f\|_{L^r}. \tag{3.1.54}$$

*(ii) For any symbols $a, b$ let $E(a, b) := T_a T_b - T_{ab}$. If $1/p = 1/q_1 + 1/q_2 + 1/r$, $k \in \mathbb{Z}$, and $l_1, l_2 \in [-4, 4]$ then*

$$2^{k_+} \|P_k E(a, b) f\|_{L^p} \lesssim (2^{l_1 k_+} \|a\|_{\mathcal{L}^{q_1}_{l_1}})(2^{l_2 k_+} \|b\|_{\mathcal{L}^{q_2}_{l_2}}) \cdot \|P_{[k-4,k+4]} f\|_{L^r}. \tag{3.1.55}$$

*(iii) If $\|a\|_{\mathcal{L}^\infty_0} < \infty$ is real-valued then $T_a$ is a bounded self-adjoint operator on $L^2$. Moreover, we have*

$$\overline{T_a f} = T_{a'} \overline{f}, \qquad \text{where} \qquad a'(y, \zeta) := \overline{a(y, -\zeta)}. \tag{3.1.56}$$

*Proof.* The conclusions were all proved in Appendix A in [17]. For convenience we provide complete proofs here as well.

(i) Using the definition (3.1.51) we write

$$\langle P_kT_af,g\rangle = C\int_{\mathbb{R}^6}\overline{g}(x)f(y)I(x,y)dxdy,$$

$$I(x,y) := \int_{\mathbb{R}^9} a(z,(\xi+\eta)/2)e^{i\xi\cdot(x-z)}e^{i\eta\cdot(z-y)}\chi_0\Big(\frac{|\xi-\eta|}{|\xi+\eta|}\Big)\varphi_k(\xi)\,d\eta d\xi dz$$
$$= \int_{\mathbb{R}^9} a(z,\xi+\theta/2)e^{i\theta\cdot(z-y)}e^{i\xi\cdot(x-y)}\chi_0\Big(\frac{|\theta|}{|2\xi+\theta|}\Big)\varphi_k(\xi)\varphi_{\le k-10}(\theta)\,d\xi d\theta dz. \tag{3.1.57}$$

We observe that

$$(1+2^{2k}|x-y|^2)^2I(x,y) = \int_{\mathbb{R}^9}\frac{a(z,\xi+\theta/2)}{(1+2^{2k}|z-y|^2)^2}\chi_0\Big(\frac{|\theta|}{|2\xi+\theta|}\Big)\varphi_k(\xi)\varphi_{\le k-10}(\theta)$$
$$\times\Big[(1-2^{2k}\Delta_\theta)^2(1-2^{2k}\Delta_\xi)^2\{e^{i\theta\cdot(z-y)}e^{i\xi\cdot(x-y)}\}\Big]\,d\xi d\theta dz.$$

By integration by parts in $\xi$ and $\theta$ it follows that

$$(1+2^{2k}|x-y|^2)^2|I(x,y)| \lesssim \int_{\mathbb{R}^9}\frac{|a|(z,\xi+\theta/2)\varphi_{[k-2,k+2]}(\xi)\varphi_{\le k-8}(\theta)}{(1+2^{2k}|z-y|^2)^2}\,d\xi d\theta dz,$$

where $|a|$ is defined as in (3.1.52). Notice that $2^{3k}\|(1+2^{2k}|y|^2)^{-2}\|_{L^1(\mathbb{R}^3)} \lesssim 1$. The bounds (3.1.54) follow using (3.1.57).

(ii) The point is the gain of one derivative in the left-hand side for the operator $T_aT_b - T_{ab}$. In view of part (i) we may assume that $k \ge 0$ and replace the symbols $a$ and $b$ with $P_{\le k-50}a$ and $P_{\le k-50}b$ respectively. As before,

$$\langle P_kE(a,b)f,g\rangle = C\int_{\mathbb{R}^6}\overline{g}(x)(P_{[k-4,k+4]}f)(y)J(x,y)dxdy, \tag{3.1.58}$$

where

$$J(x,y) := \int_{\mathbb{R}^9}\varphi_k(\xi)\varphi_{\le k-50}(\xi-\rho)\varphi_{\le k-50}(\rho-\eta)e^{ix\cdot\xi}e^{-iy\cdot\eta}$$
$$\times\Big[\widetilde{a}(\xi-\rho,\frac{\xi+\rho}{2})\widetilde{b}(\rho-\eta,\frac{\rho+\eta}{2}) - \widetilde{a}(\xi-\rho,\frac{\xi+\eta}{2})\widetilde{b}(\rho-\eta,\frac{\xi+\eta}{2})\Big]\,d\eta d\rho d\xi.$$

We decompose

$$\widetilde{a}(\xi-\rho,\frac{\xi+\rho}{2})\widetilde{b}(\rho-\eta,\frac{\rho+\eta}{2}) - \widetilde{a}(\xi-\rho,\frac{\xi+\eta}{2})\widetilde{b}(\rho-\eta,\frac{\xi+\eta}{2})$$
$$= m_1(\xi,\eta,\rho) + m_2(\xi,\eta,\rho),$$

where, for any $\xi, \eta, \rho \in \mathbb{R}^3$,

$$\begin{aligned}m_1(\xi,\eta,\rho) &:= \widetilde{a}(\xi-\rho, \frac{\xi+\rho}{2})\widetilde{b}(\rho-\eta, \frac{\rho+\eta}{2}) - \widetilde{a}(\xi-\rho, \frac{\xi+\rho}{2})\widetilde{b}(\rho-\eta, \frac{\xi+\eta}{2})\\ &= \widetilde{a}(\xi-\rho, \frac{\xi+\rho}{2}) \int_0^1 \frac{(\rho-\xi)_j}{2}(\partial_{\zeta_j}\widetilde{b})(\rho-\eta, \frac{\xi+\eta+s(\rho-\xi)}{2})\, ds\end{aligned} \tag{3.1.59}$$

and

$$\begin{aligned}m_2(\xi,\eta,\rho) &:= \widetilde{a}(\xi-\rho, \frac{\xi+\rho}{2})\widetilde{b}(\rho-\eta, \frac{\xi+\eta}{2}) - \widetilde{a}(\xi-\rho, \frac{\xi+\eta}{2})\widetilde{b}(\rho-\eta, \frac{\xi+\eta}{2})\\ &= \widetilde{b}(\rho-\eta, \frac{\xi+\eta}{2}) \int_0^1 \frac{(\rho-\eta)_j}{2}(\partial_{\zeta_j}\widetilde{a})(\xi-\rho, \frac{\xi+\eta+s(\rho-\eta)}{2})\, ds.\end{aligned} \tag{3.1.60}$$

Then we decompose $J = J_1 + J_2$ where

$$J_n(x,y) := \int_{\mathbb{R}^9} \varphi_k(\xi)\varphi_{\le k-50}(\xi-\rho)\varphi_{\le k-50}(\rho-\eta)e^{ix\cdot\xi}e^{-iy\cdot\eta}m_n(\xi,\eta,\rho)\, d\eta d\rho d\xi.$$

As before we would like to estimate $|J_n(x,y)|$. We rewrite

$$\begin{aligned}m_1(\xi,\eta,\rho) = C\int_0^1\int_{\mathbb{R}^6} &(\partial_{z_j}a)(z, \frac{\xi+\rho}{2})(\partial_{\zeta_j}b)(w, \frac{\xi+\eta+s(\rho-\xi)}{2})\\ &\times e^{-iz\cdot(\xi-\rho)}e^{-iw\cdot(\rho-\eta)}\, dzdwds.\end{aligned}$$

Therefore

$$\begin{aligned}J_1(x,y) = C\int_0^1\int_{\mathbb{R}^6}\int_{\mathbb{R}^9} &\varphi_k(\xi)\varphi_{\le k-50}(\xi-\rho)\varphi_{\le k-50}(\rho-\eta)\\ &\times(\partial_{z_j}a)(z, \frac{\xi+\rho}{2})(\partial_{\zeta_j}b)(w, \frac{\xi+\eta+s(\rho-\xi)}{2})\frac{[1-2^{2k}\Delta_\xi]^2e^{i(x-z)\cdot\xi}}{[1+2^{2k}|x-z|^2]^2}\\ &\times\frac{[1-2^{2k}\Delta_\eta]^2e^{i(w-y)\cdot\eta}}{[1+2^{2k}|w-y|^2]^2}\frac{[1-2^{2k}\Delta_\rho]^2e^{i(z-w)\cdot\rho}}{[1+2^{2k}|z-w|^2]^2}\, d\eta d\rho d\xi dzdwds.\end{aligned}$$

We integrate by parts in $\xi, \eta, \rho$ and use the definition (3.1.52) to bound

$$\begin{aligned}2^k|J_1(x,y)| \lesssim \int_0^1\int_{\mathbb{R}^6}\int_{\mathbb{R}^9} &|\varphi_{[k-4,k+4]}(\xi)\varphi_{\le k-40}(\xi-\rho)\varphi_{\le k-40}(\rho-\eta)|\\ &\times\frac{|a|(z, \frac{\xi+\rho}{2})|b|(w, \frac{\xi+\eta+s(\rho-\xi)}{2})}{[1+2^{2k}|x-z|^2]^2[1+2^{2k}|w-y|^2]^2[1+2^{2k}|z-w|^2]^2}\, d\eta d\rho d\xi dzdwds.\end{aligned}$$

Notice that $\|(1+2^{2k}|y|^2)^{-2}\|_{L^1(\mathbb{R}^3)} \lesssim 2^{-3k}$. Using the Hölder inequality we have

$$\int_{\mathbb{R}^6} 2^k |J_1(x,y)||g(x)(P_{[k-4,k+4]}f)(y)|\,dxdy$$
$$\lesssim \|P_{[k-4,k+4]}f\|_{L^r}\|g\|_{L^{p'}}(2^{l_1k_+}\|a\|_{\mathcal{L}^{q_1}_{l_1}})(2^{l_2k_+}\|b\|_{\mathcal{L}^{q_2}_{l_2}})$$

if $f \in L^r$, $g \in L^{p'}$, and $1/p' + 1/q_1 + 1/q_2 + 1/r = 1$.

The contribution of the kernel $J_2$ defined by the multiplier $m_2$ in (3.1.60) can be estimated in a similar way, and the desired bounds (3.1.55) follow.

(iii) The operators $T_a$ associated to symbols $a \in \mathcal{L}^\infty_0$ are bounded on $L^2$ due to (3.1.54). Self-adjointness and the identities (3.1.56) follow easily from the definition (3.1.51). □

We remark that the operators $E(a,b) = T_aT_b - T_{ab}$ gain one derivative, compared to the individual operators $T_aT_b$ and $T_{ab}$, as shown in (3.1.55). One could gain two derivatives by subtracting also the contribution of the Poisson bracket of the symbols $a$ and $b$, defined by

$$\{a,b\} := \nabla_x a \nabla_\zeta b - \nabla_\zeta a \nabla_x b,$$

but we do not need a refinement of this type in our applications.

## 3.2 LINEAR AND BILINEAR ESTIMATES

Localized linear and bilinear estimates, localized in both frequency and space, are the main building blocks to prove nonlinear estimates. In this section we state and prove several such estimates that are used in the nonlinear analysis.

### 3.2.1 Linear Estimates

We start with our main linear estimates, which are localized in both frequency and space. In fact, in some estimates we need to localize in the Fourier space to rotational invariant sets that are thinner than dyadic. For this we fix a smooth function $\chi : \mathbb{R} \to [0,1]$ supported in $[-2,2]$ with the property that $\sum_{n\in\mathbb{Z}} \chi(x-n) = 1$ for all $x \in \mathbb{R}$. Then we define the operators $\mathcal{C}_{n,l}$, $n \geq 4$, $l \in \mathbb{Z}$, by

$$\widehat{\mathcal{C}_{n,l}g}(\xi) := \chi(|\xi|2^{-l} - n)\widehat{g}(\xi). \tag{3.2.1}$$

We prove now several linear dispersive estimates.

**Lemma 3.9.** *For any* $f \in L^2(\mathbb{R}^3)$ *and* $(k,j) \in \mathcal{J}$ *let*

$$f_{j,k} := P'_k Q_{j,k} f, \qquad Q_{\leq j,k} f := \sum_{j' \in [\max(-k,0),j]} Q_{j',k} f, \qquad f_{\leq j,k} := P'_k Q_{\leq j,k} f, \tag{3.2.2}$$

*where* $P'_k = P_{[k-2,k+2]}$. *For simplicity of notation, let*

$$f^*_{j,k} := Q_{j,k} f, \qquad f^*_{\leq j,k} := Q_{\leq j,k} f. \tag{3.2.3}$$

*(i) Then, for any* $\alpha \in (\mathbb{Z}_+)^3$,

$$\|D^\alpha_\xi \widehat{f_{j,k}}\|_{L^2} \lesssim 2^{|\alpha| j} \|\widehat{f^*_{j,k}}\|_{L^2}, \qquad \|D^\alpha_\xi \widehat{f_{j,k}}\|_{L^\infty} \lesssim 2^{|\alpha| j} \|\widehat{f^*_{j,k}}\|_{L^\infty}. \tag{3.2.4}$$

*Moreover, we have*

$$\|\widehat{f_{j,k}}\|_{L^\infty} \lesssim \min\{2^{3j/2} \|f^*_{j,k}\|_{L^2}, 2^{j/2-k} 2^{\delta^3 (j+k)} \|f^*_{j,k}\|_{H^{0,1}_\Omega}\}, \tag{3.2.5}$$

$$\|\widehat{f_{j,k}}(r\theta)\|_{L^2(r^2 dr) L^\infty_\theta} \lesssim 2^{j+k} \|f^*_{j,k}\|_{L^2}, \tag{3.2.6}$$

$$\|\widehat{f_{j,k}}(r\theta)\|_{L^2(r^2 dr) L^p_\theta} \lesssim_p \|f^*_{j,k}\|_{H^{0,1}_\Omega}, \qquad p \in [2,\infty), \tag{3.2.7}$$

*and*

$$\|\widehat{f^*_{j,k}} - \widehat{f_{j,k}}\|_{L^\infty} \lesssim 2^{3j/2} 2^{-4(j+k)} \|P_k f\|_{L^2}. \tag{3.2.8}$$

*(ii) For any* $t \in \mathbb{R}$, $(k,j) \in \mathcal{J}$, *and* $f \in L^2(\mathbb{R}^3)$ *we have*

$$\|e^{-it\Lambda_{wa}} f_{j,k}\|_{L^\infty} \lesssim 2^{3k/2} \min(1, 2^j \langle t \rangle^{-1}) \|f^*_{j,k}\|_{L^2}. \tag{3.2.9}$$

*In addition, if* $|t| \geq 1$ *and* $j \geq \max(-k,0)$, *then we have the stronger bounds*

$$\|\varphi_{[-80,80]}(\langle t \rangle^{-1} x)(e^{-it\Lambda_{wa}} f_{j,k})(x)\|_{L^\infty_x} \lesssim \langle t \rangle^{-1} 2^{k/2} (1 + \langle t \rangle 2^k)^{\delta^3} \|f^*_{j,k}\|_{H^{0,1}_\Omega}; \tag{3.2.10}$$

$$\|e^{-it\Lambda_{wa}} f_{j,k}\|_{L^\infty} \lesssim \langle t \rangle^{-1} 2^{k/2} (1 + \langle t \rangle 2^k)^{\delta^3} \|f^*_{j,k}\|_{H^{0,1}_\Omega} \qquad \textit{if } 2^j \leq 2^{-10} \langle t \rangle; \tag{3.2.11}$$

$$\|e^{-it\Lambda_{wa}} f_{\leq j,k}\|_{L^\infty} \lesssim 2^{2k} \langle t \rangle^{-1} \|\widehat{f^*_{\leq j,k}}\|_{L^\infty} \qquad \textit{if } 2^j \lesssim \langle t \rangle^{1/2} 2^{-k/2}. \tag{3.2.12}$$

*(iii) For any* $t \in \mathbb{R}$, $(k,j) \in \mathcal{J}$, *and* $f \in L^2(\mathbb{R}^3)$ *we have*

$$\|e^{-it\Lambda_{kg}} f_{j,k}\|_{L^\infty} \lesssim \min\{2^{3k/2}, 2^{3k^+} \langle t \rangle^{-3/2} 2^{3j/2}\} \|f^*_{j,k}\|_{L^2}. \tag{3.2.13}$$

*Moreover, if* $|t| \geq 1$ *and* $j \geq \max(-k, 0)$, *then we have the stronger bounds*

$$\|e^{-it\Lambda_{kg}} f_{j,k}\|_{L^\infty} \lesssim 2^{5k^+} \langle t \rangle^{-3/2} 2^{j/2-k^-} (1 + \langle t \rangle 2^{2k^-})^{\delta^3} \|f^*_{j,k}\|_{H^{0,1}_\Omega} \quad \text{if } 2^j \leq 2^{k^- -20} \langle t \rangle; \tag{3.2.14}$$

$$\|e^{-it\Lambda_{kg}} f_{\leq j,k}\|_{L^\infty} \lesssim 2^{5k^+} \langle t \rangle^{-3/2} \|\widehat{f^*_{\leq j,k}}\|_{L^\infty} \qquad \text{if } 2^j \lesssim \langle t \rangle^{1/2}. \tag{3.2.15}$$

*(iv) The bounds* (3.2.11), (3.2.12), (3.2.14) *can be improved by using the super-localization operators* $C_{n,l}$ *defined in* (3.2.1). *Indeed, assume that* $|t| \geq 1$, $j \geq \max(-k, 0)$, *and* $l \leq k - 6$. *Then*

$$\Big\{ \sum_{n \geq 4} \|e^{-it\Lambda_{wa}} C_{n,l} f_{j,k}\|^2_{L^\infty} \Big\}^{1/2} \lesssim \langle t \rangle^{-1} 2^{l/2} (1 + \langle t \rangle 2^k)^{\delta^3} \|f^*_{j,k}\|_{H^{0,1}_\Omega} \tag{3.2.16}$$

*provided that* $2^j + 2^{-l} \lesssim \langle t \rangle (1 + \langle t \rangle 2^k)^{-\delta^3}$. *Moreover, if* $2^j + 2^{-l} \lesssim \langle t \rangle^{1/2} 2^{-k/2}$ *then*

$$\sup_{n \geq 4} \|e^{-it\Lambda_{wa}} C_{n,l} f_{\leq j,k}\|_{L^\infty} \lesssim 2^k 2^l \langle t \rangle^{-1} \|\widehat{f^*_{\leq j,k}}\|_{L^\infty}. \tag{3.2.17}$$

*Finally, if* $2^j + 2^{-l} \lesssim \langle t \rangle 2^{k^-} (1 + \langle t \rangle 2^{2k^-})^{-\delta^3}$ *then*

$$\Big\{ \sum_{n \geq 4} \|e^{-it\Lambda_{kg}} C_{n,l} f_{j,k}\|^2_{L^\infty} \Big\}^{1/2} \lesssim 2^{5k^+} \langle t \rangle^{-1} 2^{l/2} 2^{-k^-} (1 + \langle t \rangle 2^{2k^-})^{\delta^3} \|f^*_{j,k}\|_{H^{0,1}_\Omega}. \tag{3.2.18}$$

*Proof.* The conclusions were all proved in Lemma 3.4 in [38]. For convenience we reproduce the proof here.

(i) The bound (3.2.4) follows from definitions, since every $\xi$ derivative corresponds to multiplication by $x$ in the physical space. Similarly,

$$\|\widehat{f_{j,k}}\|_{L^\infty} \lesssim \|\widehat{f^*_{j,k}} * \widehat{\varphi_{\leq j+4}}\|_{L^\infty} \lesssim 2^{3j/2} \|\widehat{f^*_{j,k}}\|_{L^2},$$

which gives the first inequality in (3.2.5). A similar argument also gives (3.2.8).

Using the Sobolev embedding along the spheres $\mathbb{S}^2$, for any $g \in H^{0,1}_\Omega$ we have

$$\|\widehat{g}(r\theta)\|_{L^2(r^2dr)L^p_\theta} \lesssim_p \sum_{m_1+m_2+m_3 \leq 1} \|\Omega_{23}^{m_1} \Omega_{31}^{m_2} \Omega_{12}^{m_3} \widehat{g}\|_{L^2} \lesssim_p \|\widehat{g}\|_{H^{0,1}_\Omega}, \tag{3.2.19}$$

for any $p \in [2, \infty)$. This gives (3.2.7). Moreover, for $\xi \in \mathbb{R}^3$ with $|\xi| \approx 2^k$ we estimate

$$\begin{aligned}|\widehat{f_{j,k}}(\xi)| &\lesssim \int_{\mathbb{R}^3} |\widehat{f_{j,k}^*}(r\theta)||\widehat{\varphi_{\le j+4}}(\xi - r\theta)|r^2 dr d\theta \\ &\lesssim \|\widehat{f_{j,k}^*}(r\theta)\|_{L^2(r^2dr)L^p_\theta} \|2^{3j}(1+2^j|\xi - r\theta|)^{-8}\|_{L^2(r^2dr)L^{p'}_\theta} \\ &\lesssim_p \|\widehat{f_{j,k}^*}\|_{H^{0,1}_\Omega} \cdot 2^{3j}2^{-j/2}2^k 2^{-2(j+k)/p'}.\end{aligned}$$

The second bound in (3.2.5) follows. The proof of (3.2.6) is similar.

We prove the remaining bounds (3.2.9)–(3.2.18) in several steps.

**Step 1: proof of** (3.2.16) **and** (3.2.17). Let

$$f_{j,k;n} := \mathcal{C}_{n,l} f_{j,k}, \qquad \widehat{g_{j,k;n}}(\xi) := \widehat{f_{j,k}^*}(\xi)\varphi_{\le 4}(2^{-l}|\xi| - n). \tag{3.2.20}$$

By orthogonality,

$$\Big\{\sum_{n\ge 4} \|g_{j,k;n}\|^2_{H^{0,1}_\Omega}\Big\}^{1/2} \lesssim \|f^*_{j,k}\|_{H^{0,1}_\Omega}.$$

For (3.2.16) it suffices to prove that, for any $n \ge 4$ and $x \in \mathbb{R}^3$,

$$\begin{aligned}\Big|\int_{\mathbb{R}^3} e^{-it|\xi|}e^{ix\cdot\xi}\widehat{g_{j,k;n}}(\xi)\varphi_{[k-2,k+2]}(\xi)\chi(|\xi|2^{-l} - n)\, d\xi\Big| \\ \lesssim \langle t\rangle^{-1}2^{l/2}(\langle t\rangle 2^k)^{\delta^3}\|g_{j,k;n}\|_{H^{0,1}_\Omega}.\end{aligned} \tag{3.2.21}$$

This follows easily if $2^k\langle t\rangle \lesssim 1$. Recall that $2^j + 2^{-l} \lesssim \langle t\rangle(1+\langle t\rangle 2^k)^{-\delta^3}$ and $k \ge l+6$. The bounds (3.2.21) also follow directly from Lemma 3.1 (integration by parts in $\xi$) if $|x| \notin [2^{-40}\langle t\rangle, 2^{40}\langle t\rangle]$.

It remains to prove (3.2.21) when

$$2^k\langle t\rangle \ge 2^{50}, \qquad |x| \in [2^{-50}\langle t\rangle, 2^{50}\langle t\rangle]. \tag{3.2.22}$$

By rotation invariance we may assume $x = (x_1, 0, 0)$. Then we decompose $e^{-it\Lambda_{wa}} f_{j,k;n}(x) = \sum_{b,c\ge 0} J_{b,c}$, where

$$\begin{aligned}J_{b,c} &:= C\int_{\mathbb{R}^3} \widehat{g_{j,k;n}}(\xi)\varphi_{[k-2,k+2]}(\xi)\chi(|\xi|2^{-l} - n)e^{ix_1\xi_1 - it|\xi|}\psi_{b,c}(\xi)\, d\xi, \\ \psi_{b,c}(\xi) &:= \varphi_b^{[0,\infty)}(\xi_2/2^\lambda)\varphi_c^{[0,\infty)}(\xi_3/2^\lambda), \qquad 2^\lambda := \langle t\rangle^{-1/2}2^{k/2}.\end{aligned} \tag{3.2.23}$$

We estimate first $|J_{0,0}|$. For any $p \in [2,\infty)$, using also (3.2.19) we have

$$\begin{aligned}|J_{0,0}| &\lesssim \|\widehat{g_{j,k;n}}(r\theta)\|_{L^2(r^2dr)L^p_\theta}(2^{\lambda - k})^{2/p'} \cdot 2^k 2^{l/2} \\ &\lesssim_p \|g_{j,k;n}\|_{H^{0,1}_\Omega} \cdot \langle t\rangle^{-1}2^{l/2}(\langle t\rangle 2^k)^{1/p}.\end{aligned} \tag{3.2.24}$$

This is consistent with the desired bound (3.2.21), by taking $p$ large enough.

To estimate $|J_{b,c}|$ when $(b,c) \neq (0,0)$ we may assume without loss of generality that $b \geq c$. It suffices to show that if $b \geq \max(c,1)$ then

$$|J_{b,c}| \lesssim \langle t\rangle^{-1} 2^{l/2} (\langle t\rangle 2^k)^{\delta^3/8} \|g_{j,k;n}\|_{H^{0,1}_\Omega}. \tag{3.2.25}$$

We integrate by parts in the integral in (3.2.23), up to three times, using the rotation vector-field $\Omega_{12} = \xi_1\partial_{\xi_2} - \xi_2\partial_{\xi_1}$. Since $\Omega_{12}\{x_1\xi_1 - t|\xi|\} = -\xi_2 x_1$, every integration by parts gains a factor of $|t|2^{\lambda+b} \approx \langle t\rangle^{1/2}2^{k/2+b}$ and loses a factor $\lesssim \langle t\rangle^{1/2}2^{k/2}$. If $\Omega_{12}$ hits the function $\widehat{g_{j,k;n}}$ then we stop integrating by parts and bound the integral by estimating $\Omega_{12}\widehat{g_{j,k;n}}$ in $L^2$. As in (3.2.24), we have

$$\begin{aligned}|J_{b,c}| \lesssim\ & \|\widehat{g_{j,k;n}}(r\theta)\|_{L^2(r^2dr)L^p_\theta} (2^{\lambda-k})^{2/p'} 2^k 2^{l/2} 2^{-b}\\ & + \|\Omega_{12}\widehat{g_{j,k;n}}\|_{L^2} (2^{\lambda+b} 2^{l/2}) (\langle t\rangle^{1/2} 2^{k/2+b})^{-1},\end{aligned}$$

which gives the desired bound (3.2.25). This completes the proof of the main bounds (3.2.21).

The proof of (3.2.17) is easier. We define $f_{\leq j,k;n} := C_{n,l} f_{\leq j,k}$. For (3.2.17) it suffices to prove that, for any $n \geq 4$ and $x \in \mathbb{R}^3$,

$$\Big| \int_{\mathbb{R}^3} e^{-it|\xi|} e^{ix\cdot\xi} \widehat{f^*_{\leq j,k}}(\xi) \varphi_{[k-2,k+2]}(\xi) \chi(|\xi|2^{-l} - n)\, d\xi \Big| \lesssim 2^k 2^l \langle t\rangle^{-1} \|\widehat{f^*_{\leq j,k}}\|_{L^\infty}. \tag{3.2.26}$$

As before, we may assume $x = (x_1, 0, 0)$ and decompose $e^{-it\Lambda_{wa}} f_{\leq j,k;n}(x) = \sum_{b,c\geq 0} J'_{b,c}$, where

$$\begin{aligned} J'_{b,c} &:= C \int_{\mathbb{R}^3} \widehat{f^*_{\leq j,k}}(\xi) \varphi_{[k-2,k+2]}(\xi) \chi(|\xi|2^{-l} - n) e^{ix_1\xi_1 - it|\xi|} \psi_{b,c}(\xi)\, d\xi,\\ \psi_{b,c}(\xi) &:= \varphi_b^{[0,\infty)}(\xi_2/2^\lambda) \varphi_c^{[0,\infty)}(\xi_3/2^\lambda), \qquad 2^\lambda := \langle t\rangle^{-1/2} 2^{k/2}. \end{aligned} \tag{3.2.27}$$

Using polar coordinates, it is easy to see that $|J'_{0,0}| \lesssim 2^k 2^l \langle t\rangle^{-1} \|\widehat{f^*_{\leq j,k}}\|_{L^\infty}$. Then we integrate by parts in $\xi_2$ or $\xi_3$ (using the assumption $2^j + 2^{-l} \lesssim \langle t\rangle 2^\lambda 2^{-k}$) to show that

$$|J'_{b,c}| \lesssim 2^{-\max(b,c)} 2^k 2^l \langle t\rangle^{-1} \|\widehat{f^*_{\leq j,k}}\|_{L^\infty}$$

for any $b, c \geq 0$. The desired conclusion (3.2.26) follows.

**Step 2: proof of** (3.2.9) **and** (3.2.10)**.** We start with (3.2.10). By rotation invariance we may assume $x = (x_1, 0, 0)$, $|x_1| \approx \langle t\rangle$. We may also assume that $2^k\langle t\rangle \geq 2^{40}$. As before we decompose $e^{-it\Lambda_{wa}} f_{j,k}(x) = \sum_{b,c\geq 0} J''_{b,c}$, where

$$\begin{aligned} J''_{b,c} &:= \int_{\mathbb{R}^3} \widehat{f_{j,k}}(\xi) \varphi_{[k-4,k+4]}(\xi) e^{ix_1\xi_1 - it|\xi|} \psi_{b,c}(\xi)\, d\xi,\\ \psi_{b,c}(\xi) &:= \varphi_b^{[0,\infty)}(\xi_2/2^\lambda) \varphi_c^{[0,\infty)}(\xi_3/2^\lambda), \qquad 2^\lambda := \langle t\rangle^{-1/2} 2^{k/2}. \end{aligned} \tag{3.2.28}$$

This is similar to the decomposition (3.2.23) with $l = k - 6$, once we notice that super-localization is not important if $2^l \approx 2^k$. As in (3.2.24)–(3.2.25), we have

$$|J''_{0,0}| \lesssim_p \|f_{j,k}\|_{H^{0,1}_\Omega} \cdot \langle t\rangle^{-1} 2^{k/2} (\langle t\rangle 2^k)^{\delta^3/8},$$

and, if $b \geq \max(c, 1)$,

$$|J''_{b,c}| \lesssim \langle t\rangle^{-1} 2^{k/2} (\langle t\rangle 2^k)^{\delta^3/8} \|f^*_{j,k}\|_{H^{0,1}_\Omega}.$$

The proof of this second bound uses integration by parts with the rotation vector-field $\Omega_{12} = \xi_1 \partial_{\xi_2} - \xi_2 \partial_{\xi_1}$, and relies on the assumption $|x_1| \approx \langle t\rangle$. The desired conclusion (3.2.10) follows from these two bounds.

The bounds (3.2.9) follow by the same argument, using the decomposition (3.2.28), but using (3.2.6) instead of (3.2.7) in the estimate of $|J_{0,0}|$. Also, we integrate by parts in $\xi_2$ or $\xi_3$ to bound $|J_{b,c}|$ when $2^{\lambda+\max(b,c)} \gtrsim 2^{j+k}\langle t\rangle^{-1}$.

**Step 3: proof of** (3.2.11) **and** (3.2.12)**.** The bounds (3.2.12) follow directly from (3.2.17) by taking $2^l \approx 2^k$. To prove (3.2.11) we may assume that $x = (x_1, 0, 0)$ and $\langle t\rangle 2^k \geq 2^{40}$. If $|x_1| \in [2^{-10}|t|, 2^{10}|t|]$ then the desired bounds follow from (3.2.10). On the other hand, if $|x_1| \leq 2^{-10}|t|$ or $|x_1| \geq 2^{10}|t|$ then we write

$$[e^{-it\Lambda_{wa}} f_{j,k}](x) = C \int_{\mathbb{R}^3\times\mathbb{R}^3} f^*_{j,k}(y) e^{-iy\cdot\xi} e^{ix_1\xi_1} e^{-it|\xi|} \varphi_{[k-2,k+2]}(\xi)\, d\xi dy. \tag{3.2.29}$$

Here we use the fact that $2^j \leq \langle t\rangle 2^{-10}$ and integrate by parts in $\xi$ sufficiently many times (using Lemma 3.1) to see that

$$|e^{-it\Lambda_{wa}} f_{j,k}(x)| \lesssim (\langle t\rangle 2^k)^{-4} 2^{3k} 2^{3j/2} \|f^*_{j,k}\|_{L^2} \lesssim (\langle t\rangle 2^k)^{-4} 2^{3k} \langle t\rangle^{3/2} \|f^*_{j,k}\|_{L^2},$$

which is better than what we need.

**Step 4: proof of** (3.2.18)**.** This is similar to the proof of (3.2.16). It suffices to show that for any $n \geq 4$ and $x \in \mathbb{R}^3$,

$$\begin{aligned}\Big|\int_{\mathbb{R}^3} & e^{-it\langle\xi\rangle} e^{ix\cdot\xi} \widehat{g_{j,k;n}}(\xi) \varphi_{[k-2,k+2]}(\xi) \chi(|\xi| 2^{-l} - n)\, d\xi\Big| \\ &\lesssim 2^{5k^+} \langle t\rangle^{-1} 2^{l/2} 2^{-k^-} (1 + \langle t\rangle 2^{2k^-})^{\delta^3} \|g_{j,k;n}\|_{H^{0,1}_\Omega}.\end{aligned} \tag{3.2.30}$$

This follows easily if $2^{2k^-}\langle t\rangle \lesssim 1$. Recall that $2^j + 2^{-l} \lesssim \langle t\rangle 2^{k^-}(1 + \langle t\rangle 2^{2k^-})^{-\delta^3}$ and $k \geq l + 6$. The bounds (3.2.30) also follow directly from Lemma 3.1 if $|x| \notin [2^{-40} 2^{k^-}\langle t\rangle, 2^{40} 2^{k^-}\langle t\rangle]$.

It remains to prove (3.2.30) when

$$2^{2k^-}\langle t\rangle \geq 2^{50}, \qquad |x| \in [2^{-50} 2^{k^-}\langle t\rangle, 2^{50} 2^{k^-}\langle t\rangle]. \tag{3.2.31}$$

We may assume $x = (x_1, 0, 0)$ and decompose $e^{-it\Lambda_{kg}} f_{j,k;n}(x) = \sum_{b,c\geq 0} J'''_{b,c}$,

where

$$\begin{aligned} J'''_{b,c} &:= C\int_{\mathbb{R}^3} \widehat{g_{j,k;n}}(\xi)\varphi_{[k-2,k+2]}(\xi)\chi(|\xi|2^{-l}-n)e^{ix_1\xi_1-it\langle\xi\rangle}\psi'_{b,c}(\xi)\,d\xi, \\ \psi'_{b,c}(\xi) &:= \varphi_b^{[0,\infty)}(\xi_2/2^{\lambda'})\varphi_c^{[0,\infty)}(\xi_3/2^{\lambda'}), \qquad 2^{\lambda'} := \langle t\rangle^{-1/2}2^{k^+}. \end{aligned} \tag{3.2.32}$$

As in the proof of (3.2.16), we estimate first $|J'''_{0,0}|$, using (3.2.19). Thus, for any $p \in [2,\infty)$,

$$\begin{aligned} |J'''_{0,0}| &\lesssim \|\widehat{g_{j,k;n}}(r\theta)\|_{L^2(r^2dr)L^p_\theta}(2^{\lambda'-k})^{2/p'}2^k2^{l/2} \\ &\lesssim_p \|g_{j,k;n}\|_{H^{0,1}_\Omega}\cdot 2^{5k^+}2^{-k^-}\langle t\rangle^{-1}2^{l/2}(\langle t\rangle 2^{2k^-})^{1/p}. \end{aligned}$$

Moreover, if $b \geq \max(c,1)$ then we show that

$$|J'''_{b,c}| \lesssim 2^{5k^+}\langle t\rangle^{-1}2^{l/2}2^{-k^-}(\langle t\rangle 2^{2k^-})^{\delta^3/8}\|g_{j,k;n}\|_{H^{0,1}_\Omega}. \tag{3.2.33}$$

These two bounds clearly suffice to prove (3.2.30).

To prove (3.2.33) we integrate by parts in the integral in (3.2.32), up to three times, using the rotation vector-field $\Omega_{12} = \xi_1\partial_{\xi_2} - \xi_2\partial_{\xi_1}$. Since $\Omega_{12}\{x_1\xi_1 - t\langle\xi\rangle\} = -\xi_2 x_1$, every integration by parts gains a factor of $2^{k^-}|t|2^{\lambda'+b} \approx \langle t\rangle^{1/2}2^{k+b}$ (see (3.2.31)) and loses a factor $\lesssim \langle t\rangle^{1/2}2^k$. If $\Omega_{12}$ hits the function $\widehat{g_{j,k;n}}$ then we stop integrating by parts and bound the integral by estimating $\Omega_{12}\widehat{g_{j,k;n}}$ in $L^2$. As before it follows that

$$\begin{aligned} |J'''_{b,c}| &\lesssim \|\widehat{g_{j,k;n}}(r\theta)\|_{L^2(r^2dr)L^p_\theta}(2^{\lambda'-k})^{2/p'}2^k2^{l/2}2^{-b} \\ &\quad + \|\Omega_{12}\widehat{g_{j,k;n}}\|_{L^2}(2^{\lambda'+b}2^{l/2})(\langle t\rangle^{1/2}2^{k+b})^{-1}, \end{aligned}$$

which gives the desired bound (3.2.33). This completes the proof of the main bounds (3.2.30).

**Step 5: proof of** (3.2.13)–(3.2.15). Clearly $\|e^{-it\Lambda_{kg}}f_{j,k}\|_{L^\infty} \lesssim \|\widehat{f_{j,k}}\|_{L^1} \lesssim 2^{3k/2}\|f_{j,k}\|_{L^2}$. Moreover, the standard dispersive bounds

$$\|e^{-it\Lambda_{kg}}P_{\leq k}\|_{L^1\to L^\infty} \lesssim (1+|t|)^{-3/2}2^{3k^+}$$

can then be used to prove (3.2.13), i.e.,

$$\|e^{-it\Lambda_{kg}}f_{j,k}\|_{L^\infty} \lesssim (1+|t|)^{-3/2}2^{3k^+}\|f^*_{j,k}\|_{L^1} \lesssim (1+|t|)^{-3/2}2^{3k^+}2^{3j/2}\|f^*_{j,k}\|_{L^2}.$$

To prove (3.2.14) we consider first the harder case $2^j \geq \langle t\rangle^{1/2}$. By rotation invariance we may assume $x = (x_1,0,0)$, $x_1/t > 0$. We may also assume that $2^{j+k} \geq 2^{3k^++10}$ (otherwise the desired conclusion follows from (3.2.13)) and

$\langle t\rangle 2^{-3k^+} \gg 1$. If $|x_1| \leq 2^{-100}|t|2^{k^-}$ or $|x_1| \geq 2^{100}|t|2^{k^-}$ then we write

$$[e^{-it\Lambda_{kg}} f_{j,k}](x) = C \int_{\mathbb{R}^3\times\mathbb{R}^3} f^*_{j,k}(y) e^{-iy\cdot\xi} e^{ix_1\xi_1} e^{-it\sqrt{|\xi|^2+1}} \varphi_{[k-2,k+2]}(\xi)\, d\xi dy. \tag{3.2.34}$$

We integrate by parts in $\xi$ sufficiently many times (using Lemma 3.1 and recalling that $|y| \leq 2^{j+1} \leq 2^{k^- -19}\langle t\rangle$) to see that

$$|e^{-it\Lambda_{kg}} f_{j,k}(x)| \lesssim (\langle t\rangle 2^{2k^-})^{-4} 2^{3k} 2^{3j/2} \|f^*_{j,k}\|_{L^2}.$$

This is better than what we need.

It remains to consider the main case $|x_1| \approx |t|2^{k^-}$. Let $\rho \in (0,\infty)$ denote the unique number with the property that $t\rho/\sqrt{\rho^2+1} = x_1$, such that $(\rho,0,0)$ is the stationary point of the phase $\xi \to x_1\xi_1 - t\sqrt{|\xi|^2+1}$ and $\rho \gtrsim 2^{k^-}$. Using integration by parts (Lemma 3.1), we may assume that $\xi_1, \xi_2, \xi_3$ are restricted to $|\xi_2|, |\xi_3| \leq 2^{k-10}$ and $\xi_1 \in [2^{k-10}, 2^{k+10}]$ (for the other contributions we use (3.2.34) and get stronger bounds as before). Then we let

$$\begin{aligned} J_{a,b,c} &:= \int_{\mathbb{R}^3} \widehat{f_{j,k}}(\xi) \varphi_{[k-4,k+4]}(\xi) \mathbf{1}_+(\xi_1) \varphi_{\leq k-9}(\xi_2) \varphi_{\leq k-9}(\xi_3) \\ &\qquad\qquad \times e^{ix_1\xi_1 - it\sqrt{|\xi|^2+1}} \psi_{a,b,c}(\xi)\, d\xi, \\ \psi_{a,b,c}(\xi) &:= \varphi_a^{[0,\infty)}((\xi_1-\rho)/2^{\lambda_1}) \varphi_b^{[0,\infty)}(\xi_2/2^{\lambda_2}) \varphi_c^{[0,\infty)}(\xi_3/2^{\lambda_2}), \end{aligned} \tag{3.2.35}$$

where, for some sufficiently large constant $C$,

$$2^{\lambda_1} := 2^j \langle t\rangle^{-1} 2^{3k^+ + C} (\langle t\rangle 2^{2k^-})^{\delta^3/4}, \qquad 2^{\lambda_2} := \langle t\rangle^{-1/2} 2^{k^+}. \tag{3.2.36}$$

Compared to the earlier decompositions, such as (7.1.66), we insert an additional decomposition in the variable $\xi_1$ around the stationary point $(\rho,0,0)$.

Recall that $2^j \geq \langle t\rangle^{1/2}$. We estimate first $|J_{0,0,0}|$, using (3.2.7), for any $p \in [2,\infty)$,

$$\begin{aligned} |J_{0,0,0}| &\lesssim_p \|\widehat{f_{j,k}}(r\theta)\|_{L^2(r^2dr)L^p_\theta} (2^{\lambda_2-k})^{2/p'} 2^k 2^{\lambda_1/2} \\ &\lesssim_p \|f_{j,k}\|_{H^{0,1}_\Omega} \langle t\rangle^{-3/2} 2^{j/2} 2^{-k^-} 2^{4k^+} (\langle t\rangle 2^{2k^-})^{1/p+\delta^3/8}. \end{aligned} \tag{3.2.37}$$

This is consistent with the desired bound (3.2.14), by taking $p$ large enough.

To estimate $|J_{a,b,c}|$ when $(a,b,c) \neq (0,0,0)$ we may assume without loss of generality that $b \geq c$. If $2^{\lambda_2+b} \geq 2^j\langle t\rangle^{-1} 2^{k^+} (\langle t\rangle 2^{2k^-})^{\delta^3/40}$ then we integrate by parts in $\xi_2$ many times, using Lemma 3.1, to show that

$$|J_{a,b,c}| \lesssim \|f_{j,k}\|_{L^2} (\langle t\rangle 2^{2k^-})^{-4} 2^{3k/2},$$

which is better than what we need. This also holds, using integration by parts in $\xi_1$, if $2^{\lambda_2+b} \leq 2^j \langle t\rangle^{-1} 2^{k^+}(\langle t\rangle 2^{2k^-})^{\delta^3/40}$ and $a \geq 1$. It remains to prove that

$$|J_{0,b,c}| \lesssim 2^{5k^+}\langle t\rangle^{-3/2} 2^{j/2-k^-}(\langle t\rangle 2^{2k^-})^{\delta^3/2} \|f^*_{j,k}\|_{H^{0,1}_\Omega} \tag{3.2.38}$$

provided that

$$b \geq \max(c,1) \qquad \text{and} \qquad 2^{\lambda_2+b} \leq 2^j \langle t\rangle^{-1} 2^{k^+}(\langle t\rangle 2^{2k^-})^{\delta^3/40}. \tag{3.2.39}$$

To prove (3.2.38) we integrate by parts in (3.2.35), up to three times, using the rotation vector-field $\Omega_{12} = \xi_1\partial_{\xi_2} - \xi_2\partial_{\xi_1}$. Since $\Omega_{12}\{x_1\xi_1 - t\sqrt{|\xi|^2+1}\} = -\xi_2 x_1$, every integration by parts gains a factor of $|t|2^{k^-}2^{\lambda_2+b} \approx \langle t\rangle^{1/2}2^{k+b}$ and loses a factor $\lesssim \langle t\rangle^{1/2}2^k$. If $\Omega_{12}$ hits the function $\widehat{f_{j,k}}$ then we stop integrating by parts and bound the integral by estimating $\Omega_{12}\widehat{f_{j,k}}$ in $L^2$. As in (3.2.37) it follows that

$$\begin{aligned}|J_{0,b,c}| \lesssim_p\ & \|\widehat{f_{j,k}}(r\theta)\|_{L^2(r^2dr)L^p_\theta}(2^{\lambda_2-k})^{2/p'}2^k 2^{\lambda_1/2}2^{-b} \\ & + \|\Omega_{12}\widehat{f_{j,k}}\|_{L^2} 2^{\lambda_2}2^{\lambda_1/2}(\langle t\rangle^{1/2}2^k)^{-1},\end{aligned}$$

which gives the desired bound (3.2.38). This completes the proof of (3.2.14) when $2^j \geq \langle t\rangle^{1/2}$.

The bound (3.2.15) follows by a similar argument. We decompose the integral dyadically around the critical point $(\rho,0,0)$, as in (3.2.35), with $2^{\lambda_1} = \langle t\rangle^{-1/2}2^{3k^++C}$ and $2^{\lambda_2} = \langle t\rangle^{-1/2}2^{k^+}$, and integrate by parts in $\xi_1$, $\xi_2$, or $\xi_3$.

The bound (3.2.14) when $2^j \leq \langle t\rangle^{1/2}$ follows from (3.2.15) and (3.2.5). □

### 3.2.2 Multipliers and Bilinear Operators

We define two classes of multipliers $\mathcal{M}_0$ and $\mathcal{M}$ by

$$\begin{aligned}\mathcal{M}_0 := \{m : \mathbb{R}^3 \to \mathbb{C} : |x|^{|\alpha|}\,|D^\alpha_x m(x)| \lesssim_{|\alpha|} 1 \\ \text{for any } \alpha \in \mathbb{Z}^3_+ \text{ and } x \in \mathbb{R}^3 \setminus \{0\}\},\end{aligned} \tag{3.2.40}$$

and

$$\begin{aligned}&\mathcal{M} := \{m : \mathbb{R}^6 \to \mathbb{C} :\ m(x,y) = m_1(x,y)m'(x+y),\ m' \in \mathcal{M}_0, \\ &|x|^{|\alpha|}|y|^\beta\,|D^\alpha_x D^\beta_y m_1(x,y)| \lesssim_{|\alpha|,|\beta|} 1 \text{ for any } \alpha,\beta \in \mathbb{Z}^3_+ \text{ and } x,y \in \mathbb{R}^3 \setminus \{0\}\}.\end{aligned} \tag{3.2.41}$$

In most of our applications the multipliers in $\mathcal{M}$ will be of the form $m_1(x)m_2(y)$, where $m_1, m_2 \in \mathcal{M}_0$. We will also need to allow sums of such multipliers in order to be able to define the important classes of null multipliers $\mathcal{M}^{null}_\pm \subseteq \mathcal{M}$; see Definition 4.21.

In some of our constructions, in particular connected to the application of normal forms and associated angular cutoffs, the class of multipliers $\mathcal{M}$ is too restrictive. To treat such situations we define a more general class of multipliers

$$\mathcal{M}^* := \{m \in L^\infty(\mathbb{R}^6) : \|\mathcal{F}^{-1}\{m \cdot \varphi_{k_1}(x)\varphi_{k_2}(y)\varphi_k(x+y)\}\|_{L^1(\mathbb{R}^6)} \lesssim 1 \\ \text{for any } k_1, k_2, k \in \mathbb{Z}\}. \tag{3.2.42}$$

Given a bounded multiplier $m$ let $I = I_m$ denote the bilinear operator

$$\widehat{I[f,g]}(\xi) = \widehat{I_m[f,g]}(\xi) := \frac{1}{8\pi^3}\int_{\mathbb{R}^3} m(\xi-\eta,\eta)\widehat{f}(\xi-\eta)\widehat{g}(\eta)\,d\eta. \tag{3.2.43}$$

We will often use the simple $L^2$ bounds

$$\|P_k I_m[P_{k_1}f, P_{k_2}g]\|_{L^2} \lesssim 2^{3\min\{k,k_1,k_2\}/2}\|P_{k_1}f\|_{L^2}\|P_{k_2}g\|_{L^2} \tag{3.2.44}$$

for any multiplier $m$ satisfying $\|m\|_{L^\infty} \le 1$, $f, g \in L^2(\mathbb{R}^3)$, and $k, k_1, k_2 \in \mathbb{Z}$.

### 3.2.3 Bilinear Estimates

Linear estimates are insufficient to bound some of the quadratic terms in our nonlinearities. In this subsection we use $TT^*$ arguments to prove several additional bilinear estimates involving solutions of wave and Klein-Gordon equations:

**Lemma 3.10.** *Assume $k, k_2 \in \mathbb{Z}$, $(k_1, j_1) \in \mathcal{J}$, $t \in \mathbb{R}$, and $f, g \in L^2(\mathbb{R}^3)$. Define $f_{j_1,k_1}$, $f_{\le j_1,k_1}$, $f^*_{j,k}$, $f^*_{\le j,k}$ as in (3.2.2)-(3.2.3). Assume that $m \in \mathcal{M}^*$ and $I_m$ is a bilinear operator as in (3.2.43). If $|t| \ge 1$ and*

$$2^{j_1} \le \langle t\rangle^{1/2}2^{-k_1/2} + 2^{-\min(k,k_1,k_2)} \tag{3.2.45}$$

*then*

$$\|P_k I_m[e^{-it\Lambda_{wa}}f_{\le j_1,k_1}, P_{k_2}g]\|_{L^2} \lesssim 2^{\min(k,k_2)/2}\langle t\rangle^{-1}2^{3k_1/2}\|\widehat{f^*_{\le j_1,k_1}}\|_{L^\infty}\|P_{k_2}g\|_{L^2}. \tag{3.2.46}$$

*Proof.* By duality we may assume that $k \le k_2$, and the point is to gain both factors $\langle t\rangle^{-1}$ and $2^{k/2}$ in the right-hand side of (3.2.46). We write

$$m(\xi-\eta,\eta)\varphi_{[k-4,k+4]}(\xi)\varphi_{[k_1-4,k_1+4]}(\xi-\eta)\varphi_{[k_2-4,k_2+4]}(\eta) \\ = C\int_{\mathbb{R}^6} K(x,y)e^{-ix\cdot\xi}e^{-i(y-x)\cdot\eta}\,dxdy,$$

for some kernel $K$ satisfying $\|K\|_{L^1} \le 1$. Combining the factors $e^{-ix\cdot\xi}$ and $e^{-i(y-x)\cdot\eta}$ with the $L^2$ functions, we may also assume $m \equiv 1$ and write, for simplicity, $I = I_m$.

We estimate first

$$\begin{aligned}\|P_kI[e^{-it\Lambda_{wa}}f_{\le j_1,k_1},P_{k_2}g]\|_{L^2}&\lesssim 2^{3k/2}\|f_{\le j_1,k_1}\|_{L^2}\|P_{k_2}g\|_{L^2}\\&\lesssim 2^{3k/2}2^{3k_1/2}\|\widehat{f_{\le j_1,k_1}}\|_{L^\infty}\|P_{k_2}g\|_{L^2}.\end{aligned}$$

This suffices if $\langle t\rangle 2^k\lesssim 1$. On the other hand, if $\langle t\rangle 2^k\ge 2^{40}$ and $\langle t\rangle 2^{k_1}\le 2^{40}$ then we estimate

$$\|P_kI[e^{-it\Lambda_{wa}}f_{\le j_1,k_1},P_{k_2}g]\|_{L^2}\lesssim 2^{3k_1}\|\widehat{f_{\le j_1,k_1}}\|_{L^\infty}\|P_{k_2}g\|_{L^2},$$

which suffices. If $\langle t\rangle 2^k\ge 2^{40}$, $\langle t\rangle 2^{k_1}\ge 2^{40}$, and $k_1\le k+10$ then $2^{-k}\le 2^{-k_1+10}\le\langle t\rangle^{1/2}2^{-k_1/2}$. Therefore $2^{j_1}\le\langle t\rangle^{1/2}2^{-k_1/2+1}$ and (3.2.12) gives

$$\begin{aligned}\|P_kI[e^{-it\Lambda_{wa}}f_{\le j_1,k_1},P_{k_2}g]\|_{L^2}&\lesssim\|e^{-it\Lambda_{wa}}f_{\le j_1,k_1}\|_{L^\infty}\|P_{k_2}g\|_{L^2}\\&\lesssim 2^{2k_1}\langle t\rangle^{-1}\|\widehat{f^*_{\le j_1,k_1}}\|_{L^\infty}\|P_{k_2}g\|_{L^2},\end{aligned}$$

which suffices. It remains to prove (3.2.46) when

$$k\le k_1-10,\qquad\langle t\rangle 2^k\ge 2^{40},\qquad 2^{j_1}\le 2^{-k}+\langle t\rangle^{1/2}2^{-k_1/2}.\tag{3.2.47}$$

**Case 1.** Assume first that (3.2.47) holds and, in addition,

$$2^{-k}\ge\langle t\rangle^{1/2}2^{-k_1/2}.\tag{3.2.48}$$

In particular, $2^{j_1}\le 2^{-k+1}$ and $k\le 1$. We pass to the Fourier space and write

$$\|P_kI[e^{-it\Lambda_{wa}}f_{\le j_1,k_1},P_{k_2}g]\|_{L^2}^2=C\int_{\mathbb{R}^3\times\mathbb{R}^3}\widehat{P_{k_2}g}(\eta)\overline{\widehat{P_{k_2}g}(\rho)}L(\eta,\rho)\,d\eta d\rho$$

where

$$L(\eta,\rho):=\int_{\mathbb{R}^3}\varphi_k^2(\xi)e^{-it\Lambda_{wa}(\xi-\eta)}\widehat{f_{\le j_1,k_1}}(\xi-\eta)e^{it\Lambda_{wa}(\xi-\rho)}\overline{\widehat{f_{\le j_1,k_1}}(\xi-\rho)}\,d\xi.\tag{3.2.49}$$

Using Schur's lemma, for (3.2.46) it suffices to prove that

$$\begin{aligned}&\sup_{\rho\in\mathbb{R}^3}\int_{\mathbb{R}^3}\varphi_{[k_1-4,k_1+4]}(\eta)\varphi_{[k_1-4,k_1+4]}(\rho)|L(\eta,\rho)|\,d\eta\lesssim 2^k\langle t\rangle^{-2}2^{3k_1}\|\widehat{f^*_{\le j_1,k_1}}\|_{L^\infty}^2,\\&\sup_{\eta\in\mathbb{R}^3}\int_{\mathbb{R}^3}\varphi_{[k_1-4,k_1+4]}(\eta)\varphi_{[k_1-2,k_1+4]}(\rho)|L(\eta,\rho)|\,d\rho\lesssim 2^k\langle t\rangle^{-2}2^{3k_1}\|\widehat{f^*_{\le j_1,k_1}}\|_{L^\infty}^2.\end{aligned}\tag{3.2.50}$$

Since $L(\rho,\eta)=\overline{L(\eta,\rho)}$, it suffices to prove the first bound in (3.2.50).

We would like to integrate by parts in $\xi$ in the integral definition of the kernel

$L$. Let $n_0$ denote the smallest integer satisfying $2^{n_0} \geq (2^k \langle t \rangle)^{-1}$ and let

$$L_n(\eta,\rho) := \int_{\mathbb{R}^3} \varphi_n^{[n_0,\infty)}(\Lambda'_{wa}(\xi-\eta) - \Lambda'_{wa}(\xi-\rho))\varphi_k^2(\xi) \times e^{-it(\Lambda_{wa}(\xi-\eta)-\Lambda_{wa}(\xi-\rho))}\widehat{f_{\leq j_1,k_1}}(\xi-\eta)\overline{\widehat{f_{\leq j_1,k_1}}(\xi-\rho)}\, d\xi. \tag{3.2.51}$$

We may assume that $\|\widehat{f^*_{\leq j_1,k_1}}\|_{L^\infty} = 1$. Assume $n = n_0 + p$, $p \geq 0$. If $p \geq 1$ then we integrate by parts in $\xi$, using Lemma 3.1 with $K \approx \langle t \rangle 2^n$ and $\epsilon \approx 2^k$ (recall that $2^{j_1} \lesssim \epsilon^{-1}$ and $2^{-n-k_1} \lesssim 2^{-k_1}\langle t \rangle 2^k \lesssim \epsilon^{-1}$, due to (3.2.48)). It follows that, for all $p \geq 0$,

$$|L_n(\eta,\rho)| \lesssim 2^{-4p} \int_{\mathbb{R}^3} \varphi_{\leq n+4}(\Lambda'_{wa}(\xi-\eta) - \Lambda'_{wa}(\xi-\rho))\varphi_{[k-4,k+4]}(\xi) \times \varphi_{[k_1-4,k_1+4]}(\xi-\eta)\varphi_{[k_1-4,k_1+4]}(\xi-\rho)\, d\xi.$$

Therefore, after changes of variables,

$$\int_{\mathbb{R}^3} \varphi_{[k_1-4,k_1+4]}(\eta)|L_n(\eta,\rho)|\, d\eta \lesssim 2^{-4p} \int_{\mathbb{R}^3\times\mathbb{R}^3} \varphi_{\leq n+4}(\Lambda'_{wa}(y) - \Lambda'_{wa}(x))\varphi_{[k-4,k+4]}(\rho+x) \times \varphi_{[k_1-4,k_1+4]}(\rho+x-y)\varphi_{[k_1-4,k_1+4]}(y)\varphi_{[k_1-4,k_1+4]}(x)\, dxdy,$$

for any $\rho \in \mathbb{R}^3$ with $|\rho| \in [2^{k_1-6}, 2^{k_1+6}]$. Since $\Lambda'_{wa}(z) = z/|z|$, the integration in $y$ in the expression above is essentially in a rectangle of sides smaller than $C2^n2^{k_1} \times C2^n2^{k_1} \times C2^{k_1}$. Thus

$$\int_{\mathbb{R}^3} \varphi_{[k_1-4,k_1+4]}(\eta)|L_n(\eta,\rho)|\, d\eta \lesssim 2^{-4p}2^{2n}2^{3k_1}2^{3k} \lesssim 2^{-2p}\langle t \rangle^{-2}2^{3k_1}2^k.$$

The desired conclusion (3.2.50) follows.

**Case 2.** Assume now that (3.2.47) holds and, in addition,

$$2^{-k} \leq \langle t \rangle^{1/2} 2^{-k_1/2}. \tag{3.2.52}$$

We fix a smooth function $\chi : \mathbb{R} \to [0,1]$ supported in $[-2,2]$ with the property that $\sum_{n\in\mathbb{Z}} \chi(x-n) = 1$ for all $x \in \mathbb{R}$. Then we decompose

$$f_{\leq j_1,k_1} = \sum_n f_{\leq j_1,k_1;n}, \qquad \widehat{f_{\leq j_1,k_1;n}}(\xi) := \widehat{f_{\leq j_1,k_1}}(\xi)\chi(2^{-k}|\xi| - n).$$

Let $\widehat{g_{k_2;n}}(\xi) := \widehat{P_{k_2}g}(\xi)\varphi_{\leq 4}(2^{-k}|\xi| - n)$. Clearly

$$P_kI[e^{-it\Lambda_{wa}}f_{\leq j_1,k_1}, P_{k_2}g] = \sum_n P_kI[e^{-it\Lambda_{wa}}f_{\leq j_1,k_1;n}, g_{k_2;n}].$$

The sum in $n$ has at most $C2^{k_1-k}$ nontrivial terms, so, by orthogonality,

$$\begin{aligned}\|P_kI[e^{-it\Lambda_{wa}}f_{\leq j_1,k_1},P_{k_2}g]\|_{L^2} &\lesssim \sum_n \|I[e^{-it\Lambda_{wa}}f_{\leq j_1,k_1;n},g_{k_2;n}]\|_{L^2}\\ &\lesssim 2^{(k_1-k)/2}\Big\{\sum_n \|e^{-it\Lambda_{wa}}f_{\leq j_1,k_1;n}\|^2_{L^\infty}\|g_{k_2;n}\|^2_{L^2}\Big\}^{1/2}\\ &\lesssim 2^{(k_1-k)/2}\|P_{k_2}g\|_{L^2}\sup_n \|e^{-it\Lambda_{wa}}f_{\leq j_1,k_1;n}\|_{L^\infty}.\end{aligned}$$

For (3.2.46) it suffices to prove that, for any $n\approx 2^{k_1-k}$,

$$\|e^{-it\Lambda_{wa}}f_{\leq j_1,k_1;n}\|_{L^\infty}\lesssim 2^k2^{k_1}\langle t\rangle^{-1}\|\widehat{f^*_{\leq j_1,k_1}}\|_{L^\infty},$$

which follows from (3.2.17). □

We also have some variants using only rotational derivatives:

**Lemma 3.11.** *Assume $k,k_1,k_2\in\mathbb{Z}$, $(k_1,j_1),(k_2,j_2)\in\mathcal{J}$, $t\in\mathbb{R}$, $|t|\geq 1$, and $f,g\in L^2(\mathbb{R}^3)$. Define $f_{j_1,k_1},f^*_{j_1,k_1},g_{j_2,k_2},g^*_{j_2,k_2}$ as in (3.2.2)–(3.2.3).*

*(i) If $m\in\mathcal{M}^*$, $I_m$ is the associated bilinear operator as in (3.2.43), and*

$$2^{j_1}\lesssim\langle t\rangle(1+2^{k_1}\langle t\rangle)^{-\delta/20}+2^{-\min(k,k_1,k_2)} \tag{3.2.53}$$

*then*

$$\begin{aligned}&\|P_kI_m[e^{-it\Lambda_{wa}}f_{j_1,k_1},P_{k_2}g]\|_{L^2}\\ &\quad\lesssim 2^{\min(k,k_2)/2}\langle t\rangle^{-1}(1+\langle t\rangle 2^{k_1})^{\delta/20}\|f^*_{j_1,k_1}\|_{H^{0,1}_\Omega}\|P_{k_2}g\|_{L^2}.\end{aligned} \tag{3.2.54}$$

*(ii) If $m\in\mathcal{M}$ and $I_m$ is the associated bilinear operator as in (3.2.41)–(3.2.43), $\iota_2\in\{+,-\}$,*

$$2^{k_1},2^{k_2}\in[\langle t\rangle^{-1+\delta/2},\langle t\rangle^{2/\delta}]\qquad \textit{and}\qquad 2^{j_2}\leq\langle t\rangle^{1-\delta/2}, \tag{3.2.55}$$

*then, with $g^+_{j_2,k_2}:=g_{j_2,k_2}$ and $g^-_{j_2,k_2}:=\overline{g_{j_2,k_2}}$,*

$$\|I_m[e^{-it\Lambda_{wa}}f_{j_1,k_1},e^{-it\Lambda_{wa,\iota_2}}g^{\iota_2}_{j_2,k_2}]\|_{L^2}\lesssim 2^{k_1/2}\langle t\rangle^{-1+\delta/2}\|g^*_{j_2,k_2}\|_{L^2}\|f^*_{j_1,k_1}\|_{H^{0,1}_\Omega}. \tag{3.2.56}$$

*Proof.* (i) As before, by duality we may assume that $k\leq k_2$ and the point is to gain both factors $\langle t\rangle^{-1}$ and $2^{k/2}$. We may assume that $m\equiv 1$ and write $I=I_m$.

We estimate first, using just the Cauchy-Schwarz inequality,

$$\|P_kI[e^{-it\Lambda_{wa}}f_{j_1,k_1},P_{k_2}g]\|_{L^2}\lesssim 2^{3\min(k,k_1)/2}\|f_{j_1,k_1}\|_{L^2}\|P_{k_2}g\|_{L^2}.$$

This suffices to prove (3.2.54) if $2^{\min(k,k_1)}\lesssim\langle t\rangle^{-1}(1+\langle t\rangle 2^{k_1})^{\delta/20}$. On the other hand, if $2^{\min(k,k_1)}\gg\langle t\rangle^{-1}(1+\langle t\rangle 2^{k_1})^{\delta/20}$ and $k_1\leq k+20$ then we use (3.2.11) to estimate

$$\|P_kI[e^{-it\Lambda_{wa}}f_{j_1,k_1},P_{k_2}g]\|_{L^2} \lesssim \|e^{-it\Lambda_{wa}}f_{j_1,k_1}\|_{L^\infty}\|P_{k_2}g]\|_{L^2}$$
$$\lesssim \langle t\rangle^{-1}2^{k_1/2}(1+\langle t\rangle 2^{k_1})^{\delta/20}\|f^*_{j_1,k_1}\|_{H^{0,1}_\Omega}\|P_{k_2}g\|_{L^2},$$

which suffices. It remains to prove (3.2.54) when

$$k \le k_1-20, \qquad \langle t\rangle 2^k \ge (1+\langle t\rangle 2^{k_1})^{\delta/20}, \qquad 2^{j_1} \lesssim \langle t\rangle(1+2^{k_1}\langle t\rangle)^{-\delta/20}. \quad (3.2.57)$$

We decompose

$$f_{j_1,k_1} = \sum_{n\ge 4} f_{j_1,k_1;n}, \qquad f_{j_1,k_1;n} := \mathcal{C}_{n,k}f_{j_1,k_1}.$$

Let $\widehat{g_{k_2;n}}(\xi) := \widehat{P_{k_2}g}(\xi)\varphi_{\le 4}(2^{-k}|\xi|-n)$. Clearly

$$P_kI[e^{-it\Lambda_{wa}}f_{j_1,k_1},P_{k_2}g] = \sum_{n\ge 4} P_kI[e^{-it\Lambda_{wa}}f_{j_1,k_1;n},g_{k_2;n}].$$

Therefore

$$\|P_kI[e^{-it\Lambda_{wa}}f_{j_1,k_1},P_{k_2}g]\|_{L^2} \lesssim \sum_{n\ge 4}\|e^{-it\Lambda_{wa}}f_{j_1,k_1;n}\|_{L^\infty}\big\|g_{k_2;n}\big\|_{L^2}$$
$$\lesssim \Big\{\sum_{n\ge 4}\|e^{-it\Lambda_{wa}}f_{j_1,k_1;n}\|^2_{L^\infty}\Big\}^{1/2}\Big\{\sum_{n\ge 4}\|g_{k_2;n}\|^2_{L^2}\Big\}^{1/2}$$
$$\lesssim \langle t\rangle^{-1}2^{k/2}(1+\langle t\rangle 2^{k_1})^{\delta/20}\|f^*_{j_1,k_1}\|_{H^{0,1}_\Omega}\cdot\|P_{k_2}g\|_{L^2},$$

where we used (3.2.16) (see the restrictions (3.2.57)) and orthogonality in the last inequality. This completes the proof of (3.2.54).

(ii) We decompose $e^{-it\Lambda_{wa}}f_{j_1,k_1} = P_{[k_1-4,k_1+4]}H^1_{k_1} + P_{[k_1-4,k_1+4]}H^2_{k_1}$, where

$$H^1_{k_1}(x) := \varphi_{[-40,40]}(x/\langle t\rangle)e^{-it\Lambda_{wa}}f_{j_1,k_1}(x),$$
$$H^2_{k_1}(x) := (1-\varphi_{[-40,40]}(x/\langle t\rangle))e^{-it\Lambda_{wa}}f_{j_1,k_1}(x),$$

for $x\in\mathbb{R}^3$. In view of (3.2.10), we have

$$\|H^1_{k_1}\|_{L^\infty} \lesssim 2^{k_1/2}\langle t\rangle^{-1+\delta/2}\|f^*_{j_1,k_1}\|_{H^{0,1}_\Omega},$$

so the contribution of $H^1_{k_1}$ is bounded as claimed.

On the other hand, we claim that the contribution of $H^2_{k_1}$ is negligible,

$$\|I_m[P_{[k_1-4,k_1+4]}H^2_{k_1},e^{-it\Lambda_{wa,\iota_2}}g^{\iota_2}_{j_2,k_2}]\|_{L^2} \lesssim \langle t\rangle^{-2}\|g^*_{j_2,k_2}\|_{L^2}\|f^*_{j_1,k_1}\|_{L^2}. \quad (3.2.58)$$

Indeed, the definitions (3.2.41) and (3.2.43) show that

$$|I_m[P_{l_1}F, P_{l_2}G](x)| \lesssim_N (|F| * K_{l_1}^N)(x) \cdot (|G| * K_{l_2}^N)(x),$$

for any $l_1, l_2 \in \mathbb{Z}$ and $N \geq 10$, where $K_l^N(y) := 2^{3l}(1 + 2^{l^-}|y|)^{-N}$. Moreover,

$$\|\varphi_{[-30,30]}(x/\langle t\rangle) \cdot (|H_{k_1}^2| * K_{l_1}^N)(x)\|_{L^\infty} \lesssim \langle t\rangle^{-8/\delta}\|f_{j_1,k_1}^*\|_{L^2},$$

for $l_1 \in [k_1 - 4, k_1 + 4]$, in view of the support restriction on $H_{k_1}^2$ and the assumption $2^{k_1^-}\langle t\rangle \gtrsim \langle t\rangle^{\delta/2}$. Also, using Lemma 3.1 and the assumption $2^{j_2} \leq \langle t\rangle^{1-\delta/2}$, we have

$$\|(1 - \varphi_{[-30,30]}(x/\langle t\rangle)) \cdot (|e^{-it\Lambda_{wa}}g_{j_2,k_2}| * K_{l_2}^N)(x)\|_{L^2} \lesssim \langle t\rangle^{-8/\delta}\|g_{j_2,k_2}^*\|_{L^2},$$

for $l_2 \in [k_2 - 4, k_2 + 4]$. The desired estimates (3.2.58) follow from the last three bounds. □

We also need some bilinear estimates involving the Klein-Gordon flow:

**Lemma 3.12.** *Assume $k, k_1, k_2 \in \mathbb{Z}$, $(k_1, j_1) \in \mathcal{J}$, $t \in \mathbb{R}$, $|t| \geq 1$, and $f, g \in L^2(\mathbb{R}^3)$. Define $f_{j_1,k_1}, f_{j_1,k_1}^*$ as in (3.2.2)–(3.2.3). If $\|\mathcal{F}^{-1}m\|_{L^1} \leq 1$, $I_m$ is the bilinear operator as in (3.2.43),*

$$k \leq k_1^- - 10 \qquad \text{and} \qquad 2^{j_1} \lesssim \langle t\rangle 2^{k_1^-}(1 + 2^{2k_1^-}\langle t\rangle)^{-\delta/20}, \tag{3.2.59}$$

*then*

$$\begin{aligned}&\|P_k I_m[e^{-it\Lambda_{kg}}f_{j_1,k_1}, P_{k_2}g]\|_{L^2}\\ &\quad\lesssim 2^{k/2}\langle t\rangle^{-1}2^{5k_1^+}2^{-k_1^-}(1 + \langle t\rangle 2^{2k_1^-})^{\delta/20}\|f_{j_1,k_1}^*\|_{H_\Omega^{0,1}}\|P_{k_2}g\|_{L^2}.\end{aligned} \tag{3.2.60}$$

*Proof.* This is similar to the proof of Lemma 3.11 (i). We may assume $m \equiv 1$ and write $I = I_m$. We estimate first, using just the Cauchy-Schwarz inequality,

$$\|P_k I[e^{-it\Lambda_{kg}}f_{j_1,k_1}, P_{k_2}g]\|_{L^2} \lesssim 2^{3k/2}\|f_{j_1,k_1}\|_{L^2}\|P_{k_2}g\|_{L^2},$$

which suffices if $2^{k+k_1^-} \lesssim \langle t\rangle^{-1}(1 + \langle t\rangle 2^{2k_1^-})^{\delta/20}$. On the other hand, if $2^{k+k_1^-} \gg \langle t\rangle^{-1}(1 + \langle t\rangle 2^{2k_1^-})^{\delta/20}$ and $k \leq k_1^- - 10$ then $2^{2k_1^-} \gg \langle t\rangle^{-1}$ and we decompose

$$f_{j_1,k_1} = \sum_{n\geq 4} f_{j_1,k_1;n}, \qquad f_{j_1,k_1;n} := \mathcal{C}_{n,k}f_{j_1,k_1}(\xi).$$

Let $\widehat{g_{k_2;n}}(\xi) := \widehat{P_{k_2}g}(\xi)\varphi_{\leq 4}(2^{-k}|\xi| - n)$. Clearly

$$P_k I[e^{-it\Lambda_{kg}}f_{j_1,k_1}, P_{k_2}g] = \sum_{n\geq 4} P_k I[e^{-it\Lambda_{wa}}f_{j_1,k_1;n}, g_{k_2;n}].$$

Therefore, as in the proof of Lemma 3.11 (i),

$$\begin{aligned}\|P_kI[e^{-it\Lambda_{kg}}f_{j_1,k_1},P_{k_2}g]\|_{L^2} &\lesssim \sum_{n\geq 4}\|e^{-it\Lambda_{kg}}f_{j_1,k_1;n}\|_{L^\infty}\|g_{k_2;n}\|_{L^2}\\ &\lesssim \Big\{\sum_{n\geq 4}\|e^{-it\Lambda_{kg}}f_{j_1,k_1;n}\|_{L^\infty}^2\Big\}^{1/2}\Big\{\sum_{n\geq 4}\|g_{k_2;n}\|_{L^2}^2\Big\}^{1/2}\\ &\lesssim \langle t\rangle^{-1}2^{k/2}2^{5k_1^+}2^{-k_1^-}(1+\langle t\rangle 2^{2k_1^-})^{\delta/20}\|f_{j_1,k_1}^*\|_{H_\Omega^{0,1}}\|P_{k_2}g\|_{L^2},\end{aligned}$$

using (3.2.18) (see (3.2.59) and recall that $2^{k+k_1^-}\gg\langle t\rangle^{-1}(1+\langle t\rangle 2^{2k_1^-})^{\delta/20}$) and orthogonality in the last inequality. This completes the proof of (3.2.60). □

### 3.2.4 Interpolation Inequalities

Finally, we need some interpolation bounds involving $L^p$ spaces and rotation vector-fields.

**Lemma 3.13.** *(i) Assume that $f\in H_\Omega^{0,1}$, $k\in\mathbb{Z}$, and $0\leq A_0\lesssim A_1\lesssim B$. If*

$$\|Q_{j,k}f\|_{L^2}\leq A_0,\qquad \|Q_{j,k}f\|_{H_\Omega^{0,1}}\leq A_1,\qquad 2^{j+k}\|Q_{j,k}f\|_{H_\Omega^{0,1}}\leq B \tag{3.2.61}$$

*for all $j\geq -k^-$, then*

$$\|\widehat{P_kf}\|_{L^\infty}\lesssim 2^{-3k/2}A_0^{(1-\delta)/4}B^{(3+\delta)/4} \tag{3.2.62}$$

*and*

$$\|\widehat{P_kf}\|_{L^\infty}\lesssim 2^{-3k/2}A_1^{(1-\delta)/2}B^{(1+\delta)/2}. \tag{3.2.63}$$

*(ii) If $f\in H_\Omega^{0,2}$, $\Omega\in\{\Omega_{23},\Omega_{31},\Omega_{12}\}$, and $k\in\mathbb{Z}$ then*

$$\|P_k\Omega f\|_{L^4}\lesssim\|P_kf\|_{L^\infty}^{1/2}\|P_kf\|_{H_\Omega^{0,2}}^{1/2}. \tag{3.2.64}$$

*Similarly, if $f\in H_\Omega^{0,3}$ and $\Omega^2\in\{\Omega_{23}^{a_1}\Omega_{31}^{a_2}\Omega_{12}^{a_3}: a_1+a_2+a_3=2\}$ then*

$$\begin{aligned}\|P_k\Omega f\|_{L^6}&\lesssim\|P_kf\|_{L^\infty}^{2/3}\|P_kf\|_{H_\Omega^{0,3}}^{1/3},\\ \|P_k\Omega^2 f\|_{L^3}&\lesssim\|P_kf\|_{L^\infty}^{1/3}\|P_kf\|_{H_\Omega^{0,3}}^{2/3}.\end{aligned} \tag{3.2.65}$$

*Finally, we have the $L^2$ interpolation estimates*

$$\|P_k\Omega f\|_{L^2}\lesssim\|P_kf\|_{L^2}^{1/2}\|P_kf\|_{H_\Omega^{0,2}}^{1/2} \tag{3.2.66}$$

*and*

$$\|P_k\Omega f\|_{L^2} \lesssim \|P_k f\|_{L^2}^{2/3}\|P_k f\|_{H^{0,3}_\Omega}^{1/3},$$
$$\|P_k\Omega^2 f\|_{L^2} \lesssim \|P_k f\|_{L^2}^{1/3}\|P_k f\|_{H^{0,3}_\Omega}^{2/3}. \tag{3.2.67}$$

*Proof.* (i) The bounds follow from (3.2.5): with $f_{j,k} = P_{k'}Q_{j,k}f$ we have

$$\begin{aligned}\|\widehat{f_{j,k}}\|_{L^\infty} &\lesssim \min\{2^{3j/2}\|Q_{j,k}f\|_{L^2}, 2^{j/2-k}2^{\delta^3(j+k)}\|Q_{j,k}f\|_{H^{0,1}_\Omega}\}\\ &\lesssim 2^{-3k/2}\min\{2^{3(j+k)/2}A_0, 2^{-(j+k)/2}2^{\delta^3(j+k)}B\}.\end{aligned}$$

The desired bounds (3.2.62) follow by summing over $j$ and considering the two cases $2^{j+k} \leq (B/A_0)^{1/2}$ and $2^{j+k} \geq (B/A_0)^{1/2}$. Similarly,

$$\begin{aligned}\|\widehat{f_{j,k}}\|_{L^\infty} &\lesssim 2^{j/2-k}2^{\delta^3(j+k)}\|Q_{j,k}f\|_{H^{0,1}_\Omega}\\ &\lesssim 2^{-3k/2}2^{(j+k)(1/2+\delta^3)}\min\{A_1, B2^{-(j+k)}\}.\end{aligned}$$

The desired bounds (3.2.63) follow again by summing over $j$.

(ii) For (3.2.64) we let $g := P_k f$ and use integration by parts to write

$$\|\Omega g\|_{L^4}^4 = \int_{\mathbb{R}^3} \Omega g\,\Omega g\,\overline{\Omega g}\,\overline{\Omega g}\,dx = -\int_{\mathbb{R}^3} g\cdot\Omega\{\Omega g\,\overline{\Omega g}\,\overline{\Omega g}\}\,dx.$$

Therefore we can estimate

$$\|\Omega g\|_{L^4}^4 \lesssim \int_{\mathbb{R}^3} |g|\,|\Omega g|^2\,|\Omega^2 g|\,dx \lesssim \|g\|_{L^\infty}\|\Omega^2 g\|_{L^2}\|\Omega g\|_{L^4}^2,$$

which gives (3.2.64).

Similarly, to prove (3.2.65) we estimate as above

$$\|\Omega g\|_{L^6}^6 \lesssim \int_{\mathbb{R}^3} |g|\,|\Omega g|^4\,|\Omega^2 g|\,dx \lesssim \|g\|_{L^\infty}\|\Omega^2 g\|_{L^3}\|\Omega g\|_{L^6}^4 \tag{3.2.68}$$

and

$$\begin{aligned}\|\Omega^2 g\|_{L^3}^3 = \left|\int_{\mathbb{R}^3} \Omega^2 g\,\overline{\Omega^2 g}(\Omega^2 g\overline{\Omega^2 g})^{1/2}\,dx\right| &\lesssim \int_{\mathbb{R}^3} |\Omega g|\,|\Omega^2 g|\,|\Omega^3 g|\,dx\\ &\lesssim \|g\|_{H^{0,3}_\Omega}\|\Omega^2 g\|_{L^3}\|\Omega g\|_{L^6},\end{aligned}$$

where $\Omega^3 g$ denotes vector-fields of the form $\Omega_{23}^{a_1}\Omega_{31}^{a_2}\Omega_{12}^{a_3}$ with $a_1 + a_2 + a_3 = 3$. Therefore, using also (3.2.68) and simplifying,

$$\|\Omega^2 g\|_{L^3}^2 \lesssim \|g\|_{H^{0,3}_\Omega}\|\Omega g\|_{L^6} \lesssim \|g\|_{H^{0,3}_\Omega}\big(\|g\|_{L^\infty}\|\Omega^2 g\|_{L^3}\big)^{1/2},$$

which gives the bounds in the second line of (3.2.65). The bounds in the first line now follow from (3.2.68).

The $L^2$ bounds (3.2.66) and (3.2.67) follow by similar arguments. □

We need also a bilinear interpolation lemma:

**Lemma 3.14.** *Assume $k, k_1, k_2 \in \mathbb{Z}$, $k + 4 \leq K := \min(k_1, k_2)$, $t \in \mathbb{R}$, and $f, g \in L^2(\mathbb{R}^3)$. Assume $J \geq \max(-K, 0)$ satisfies*

$$2^J \leq \langle t \rangle^{1/2} 2^{-K/2-2} + 2^{-k}, \tag{3.2.69}$$

*and define $f_{\leq J,k_1}$ and $g_{\leq J,k_2}$ as in (3.2.2). Then*

$$\|P_k I[e^{-it\Lambda_{wa}} f_{\leq J,k_1}, e^{-it\Lambda_{wa}} g_{\leq J,k_2}]\|_{L^2} \lesssim 2^{k/2} 2^{3K/2} \langle t \rangle^{-1} \|\widehat{P_{k_1} f}\|_{L^a} \|\widehat{P_{k_2} g}\|_{L^b}, \tag{3.2.70}$$

*for all exponents $a, b \in [2, \infty]$ satisfying $1/a + 1/b = 1/2$.*

*Proof.* The conclusion follows directly from Lemma 3.10 (which corresponds to the cases $(a, b) = (\infty, 2)$ and $(a, b) = (2, \infty)$) and bilinear interpolation. □

## 3.3 ANALYSIS OF THE LINEAR PROFILES

In this section we use the main bootstrap assumptions (2.1.46)–(2.1.48) to derive many linear bounds on the profiles $V^*$ and the normalized solutions $U^*$.

For $t \in [0, T]$, $(j, k) \in \mathcal{J}$, $J \geq \max(-k, 0)$, and

$$X \in \{F, \underline{F}, \rho, \omega_a, \Omega_a, \vartheta_{ab}, \mathcal{L}h_{\alpha\beta}, \mathcal{L}\psi : \mathcal{L} \in \mathcal{V}_3^3\},$$

we define the profiles $V^{X,\pm}$ as in (2.1.35). If $\mathcal{L} \in \mathcal{V}_2^2$ we define also the *space-localized profiles*

$$\begin{aligned} &V_{j,k}^{X,\pm}(t) := P'_k Q_{j,k} V^{X,\pm}(t), \\ &V_{\leq J,k}^{X,\pm}(t) := \sum_{j \leq J} V_{j,k}^{X,\pm}(t), \qquad V_{>J,k}^{X,\pm}(t) := \sum_{j>J} V_{j,k}^{X,\pm}(t) \end{aligned} \tag{3.3.1}$$

and the associated localized solutions

$$\begin{aligned} &U_{j,k}^{X,\pm}(t) := e^{-it\Lambda_{\mu,\pm}} V_{j,k}^{X,\pm}(t), \\ &U_{\leq J,k}^{X,\pm}(t) := \sum_{j \leq J} U_{j,k}^{X,\pm}(t), \qquad U_{>J,k}^{X,\pm}(t) := \sum_{j>J} U_{j,k}^{X,\pm}(t), \end{aligned} \tag{3.3.2}$$

where $\mu = kg$ if $X = \mathcal{L}\psi$ and $\mu = wa$ otherwise. For simplicity of notation, we sometimes let $V_*^X := V_*^{X,+}$ and $U_*^X := U_*^{X,+}$, and notice that $V_*^{X,-} = \overline{V_*^X}$ and $U_*^{X,-} = \overline{U_*^X}$.

Given two pairs $(q,n),(q',n') \in \mathbb{Z}_+^2$ we write $(q,n) \leq (q',n')$ if $q \leq q'$ and $n \leq n'$. In the lemmas below we let $h$ denote generic components of the linearized metric, i.e., $h \in \{h_{\alpha\beta} : \alpha, \beta \in \{0,1,2,3\}\}$.

**Lemma 3.15.** *Assume that $(g,\psi)$ is a solution of the system (1.2.6)–(1.2.7) on the interval $[0,T]$, $T \geq 1$, satisfying the bootstrap hypothesis (2.1.46)–(2.1.48).*

*(i) For any $t \in [0,T]$ and $\mathcal{L} \in \mathcal{V}_n^q$, $n \leq 3$, we have*

$$\|(\langle t\rangle|\nabla|_{\leq 1})^{\gamma}|\nabla|^{-1/2}V^{\mathcal{L}h}(t)\|_{H^{N(n)}} + \|V^{\mathcal{L}\psi}(t)\|_{H^{N(n)}} \lesssim \varepsilon_1 \langle t\rangle^{H(q,n)\delta}. \tag{3.3.3}$$

*In addition, if $(k,j) \in \mathcal{J}$ and $\mathcal{L} \in \mathcal{V}_n^q$, $n \leq 2$, then*

$$2^{k/2}(2^{k^-}\langle t\rangle)^{\gamma}2^j\|Q_{j,k}V^{\mathcal{L}h}(t)\|_{L^2} + 2^{k^+}2^j\|Q_{j,k}V^{\mathcal{L}\psi}(t)\|_{L^2} \lesssim \varepsilon_1 Y(k,t;q,n) \tag{3.3.4}$$

*and*

$$2^{k^+}\|P_kV^{\mathcal{L}\psi}(t)\|_{L^2} + \|P_kV^{\mathcal{L}\psi}(t)\|_{H^{0,1}_\Omega} \lesssim \varepsilon_1 2^{k^-}Y(k,t;q,n), \tag{3.3.5}$$

*where, for $q,n \in \{0,1,2\}$,*

$$Y(k,t;q,n) := \langle t\rangle^{H(q+1,n+1)\delta}2^{-N(n+1)k^+}. \tag{3.3.6}$$

*(ii) For any $H \in \{F, \omega_a, \vartheta_{ab} : a,b \in \{1,2,3\}\}$ and $k \in \mathbb{Z}$*

$$\begin{aligned}\|\widehat{P_kV^H}(t)\|_{L^\infty} + \langle t\rangle^{-\delta}\|\widehat{P_kV^h}(t)\|_{L^\infty} &\lesssim \varepsilon_1 2^{-k^- - \kappa k^-}2^{-N_0k^+}\\ \|\widehat{P_kV^\psi}(t)\|_{L^\infty} &\lesssim \varepsilon_1 2^{-k^-/2+\kappa k^-}2^{-N_0k^+}.\end{aligned} \tag{3.3.7}$$

*Moreover, if $(k,j) \in \mathcal{J}$ and $H' \in \{F, \underline{F}, \rho, \omega_a, \Omega_a, \vartheta_{ab}\}$ then*

$$\begin{aligned}\|P_kV^{H'}(t)\|_{H^{0,a}_\Omega} &\lesssim \varepsilon_1\langle t\rangle^{H(0,a)\delta}2^{-N(a)k^+}2^{k/2}(\langle t\rangle 2^{k^-})^{-\gamma}, \qquad \textit{if } a \leq 3,\\ 2^j\|Q_{j,k}V^{H'}(t)\|_{H^{0,a}_\Omega} &\lesssim \varepsilon_1\langle t\rangle^{H(1,a+1)\delta}2^{-N(a+1)k^+}2^{-k/2}(\langle t\rangle 2^{k^-})^{-\gamma}, \qquad \textit{if } a \leq 2.\end{aligned} \tag{3.3.8}$$

*Proof.* The bounds (3.3.3) follow directly from (2.1.46), and the bounds (3.3.7) follow from (2.1.48) and Definition 2.2.

It follows from (2.1.47) that

$$\begin{aligned}2^{k/2}(2^{k^-}\langle t\rangle)^{\gamma}\|\varphi_k(\xi)(\partial_{\xi_l}\widehat{V^{\mathcal{L}h}})(\xi,t)\|_{L^2_\xi} + 2^{k^+}\|\varphi_k(\xi)(\partial_{\xi_l}\widehat{V^{\mathcal{L}\psi}})(\xi,t)\|_{L^2_\xi}&\\ \lesssim \varepsilon_1 Y(k,t;q,n),&\end{aligned} \tag{3.3.9}$$

for any $k \in \mathbb{Z}$, $l \in \{1,2,3\}$, and $\mathcal{L} \in \mathcal{V}_n^q$, $(q,n) \leq (2,2)$. The bounds in (3.3.4) follow using also (3.3.3) and Lemma 3.3 (i). The bounds (3.3.5) follow from

(3.3.9) and the estimates

$$\|P_k V^{\mathcal{L}\psi}(t)\|_{H^{0,1}_\Omega} \lesssim 2^k \|\varphi_k(\xi)\nabla_\xi(\widehat{V^{\mathcal{L}\psi}})(\xi,t)\|_{L^2_\xi} \lesssim \varepsilon_1 2^{k^-} Y(k,t;q,n).$$

To prove (3.3.8) we notice first that if $\Omega \in \{\Omega_{23}, \Omega_{31}, \Omega_{12}\}$ then

$$\Omega Q_{j,k} V^X = Q_{j,k}\Omega V^X = Q_{j,k} V^{\Omega X} \quad \text{and} \quad \Omega P_k V^X = P_k \Omega V^X = P_k V^{\Omega X}, \tag{3.3.10}$$

for suitable profiles $X$. Recall that the functions $H'$ are defined by taking Riesz transforms of the metric components $h_{\alpha\beta}$ (see (2.1.26)). The bounds in the first line of (3.3.8) follow from (3.3.3), while the bounds in the second line follow from (3.3.4) and Lemma 3.3 (ii). □

We prove now several pointwise decay bounds on the normalized solutions.

**Lemma 3.16.** *(i) For any $k \in \mathbb{Z}$, $t \in [0,T]$, and $\mathcal{L} \in \mathcal{V}^q_n$, $n \leq 2$, we have*

$$\begin{aligned} &\sum_{j \geq -k^-} \|U^{\mathcal{L}h}_{j,k}(t)\|_{L^\infty} \\ &\lesssim \varepsilon_1 \langle t\rangle^{-1+H(q+1,n+1)\delta} 2^{k^-} 2^{-N(n+1)k^+ + 2k^+} \min\{1, \langle t\rangle 2^{k^-}\}^{1-\delta}, \end{aligned} \tag{3.3.11}$$

*where $h \in \{h_{\alpha\beta} : \alpha, \beta \in \{0,1,2,3\}\}$ as before. In addition, if $2^{k^-}\langle t\rangle \geq 2^{20}$ then*

$$\sum_{2^j \in [2^{-k^-}, 2^{-20}\langle t\rangle]} \|U^{\mathcal{L}h}_{j,k}(t)\|_{L^\infty} \lesssim \varepsilon_1 \langle t\rangle^{-1+H(q,n+1)\delta} 2^{k^-} 2^{-N(n+1)k^+ + 2k^+}. \tag{3.3.12}$$

*Moreover,*

$$\begin{aligned} &\sum_{j \geq -k^-} \|U^{\mathcal{L}\psi}_{j,k}(t)\|_{L^\infty} \\ &\lesssim \varepsilon_1 \langle t\rangle^{-1+H(q+1,n+1)\delta} 2^{k^-/2} 2^{-N(n+1)k^+ + 2k^+} \min\{1, 2^{2k^-}\langle t\rangle\} \end{aligned} \tag{3.3.13}$$

*and, if $j \geq -k^-$,*

$$\|U^{\mathcal{L}\psi}_{j,k}(t)\|_{L^\infty} \lesssim \varepsilon_1 \langle t\rangle^{-3/2+H(q+1,n+1)\delta} 2^{j/2} 2^{-N(n+1)k^+ + 2k^+}. \tag{3.3.14}$$

*Finally, if $2^{2k^- - 20}\langle t\rangle \geq 1$ and $\mathcal{L} \in \mathcal{V}^q_n$, $n \leq 1$, then*

$$\begin{aligned} &\sum_{2^j \in [2^{-k^-}, 2^{k^- - 20}\langle t\rangle]} \|U^{\mathcal{L}\psi}_{j,k}(t)\|_{L^\infty} \\ &\lesssim \varepsilon_1 \langle t\rangle^{-3/2+H(q+1,n+2)\delta} 2^{-k^-/2} 2^{-N(n+2)k^+ + 5k^+} (\langle t\rangle 2^{2k^-})^{\delta/4}. \end{aligned} \tag{3.3.15}$$

*(ii) In the case $n = 0$ ($\mathcal{L} = Id$) these bounds can be improved slightly. More*

*precisely, assume* $k, J \in \mathbb{Z}$, $t \in [0,T]$, $2^{k^-}\langle t\rangle \geq 2^{20}$, *and* $2^J \in [2^{-k^-}, 2^{-10}\langle t\rangle]$. *Then for any* $H \in \{F, \omega_a, \vartheta_{ab} : a, b \in \{1,2,3\}\}$ *and* $h \in \{h_{\alpha\beta}\}$,

$$\|U^H_{\leq J,k}(t)\|_{L^\infty} + \langle t\rangle^{-\delta}\|U^h_{\leq J,k}(t)\|_{L^\infty} \lesssim \varepsilon_1\langle t\rangle^{-1}2^{k^- - \kappa k^-}2^{-N_0k^+ + 5k^+}. \quad (3.3.16)$$

*Moreover, if* $k, J \in \mathbb{Z}$, $t \in [0,T]$, $2^{2k^-}\langle t\rangle \geq 2^{20}$, *and* $2^J \in [2^{-k^-}, 2^{k^- - 20}\langle t\rangle]$, *then*

$$\|U^\psi_{\leq J,k}(t)\|_{L^\infty} \lesssim \varepsilon_1\langle t\rangle^{-3/2}2^{-k^-/2 + \kappa k^-/20}2^{-N_0k^+ + 5k^+}. \quad (3.3.17)$$

*Proof.* (i) We prove first (3.3.11). We estimate, using just (3.3.4),

$$\|e^{-it\Lambda_{wa}}V^{\mathcal{L}h}_{j,k}(t)\|_{L^\infty} \lesssim 2^{3k/2}\|V^{\mathcal{L}h}_{j,k}(t)\|_{L^2} \lesssim \varepsilon_1 Y(k,t;q,n)2^k2^{-j}(2^{k^-}\langle t\rangle)^{-\gamma}. \quad (3.3.18)$$

This suffices to prove (3.3.11) if $2^k \lesssim \langle t\rangle^{-1}$, by summing over $j \geq -k$. On the other hand, if $2^k \geq 2^{20}\langle t\rangle^{-1}$ then (3.3.18) still suffices to control the sum over $j$ with $2^j \geq 2^{-10}\langle t\rangle$. Finally, if $2^k \geq \langle t\rangle^{-1}$ and $2^j \leq 2^{-10}\langle t\rangle$ then we use (3.2.9) and (3.3.4) to estimate

$$\begin{aligned}\|e^{-it\Lambda_{wa}}V^{\mathcal{L}h}_{j,k}(t)\|_{L^\infty} &\lesssim 2^{3k/2}\langle t\rangle^{-1}2^j\|Q_{j,k}V^{\mathcal{L}h}(t)\|_{L^2}\\ &\lesssim \langle t\rangle^{-1}2^k\varepsilon_1 Y(k,t;q,n)((\langle t\rangle 2^{k^-})^{-\gamma}.\end{aligned}$$

The desired conclusion follows by summation over $j$ with $2^j \in [2^{-k^-}, 2^{-10}\langle t\rangle]$.

For (3.3.12) we use (3.3.3) and (3.2.11). Recalling (3.3.10) we estimate the left-hand side by

$$\begin{aligned}C\sum_{2^j\in[2^{-k^-},2^{-20}\langle t\rangle]} &2^{k^+}\langle t\rangle^{-1}2^{k^-/2}((\langle t\rangle 2^{k^-})^{\delta/8}\|Q_{j,k}V^{\mathcal{L}h}(t)\|_{H^{0,1}_\Omega}\\ &\lesssim \sum_{2^j\in[2^{-k^-},2^{-20}\langle t\rangle]} 2^{k^+}\langle t\rangle^{-1}2^{k^-/2}((\langle t\rangle 2^{k^-})^{\delta/20}\\ &\qquad\times \varepsilon_1\langle t\rangle^{H(q,n+1)\delta}2^{-N(n+1)k^+}2^{k/2}((\langle t\rangle 2^{k^-})^{-\gamma}.\end{aligned}$$

The desired bound (3.3.12) follows.

We prove now (3.3.13). As in (3.3.18) we have

$$\|e^{-it\Lambda_{kg}}V^{\mathcal{L}\psi}_{j,k}(t)\|_{L^\infty} \lesssim 2^{3k/2}\|V^{\mathcal{L}\psi}_{j,k}(t)\|_{L^2} \lesssim \varepsilon_1 Y(k,t;q,n)2^{3k/2}2^{-k^+}2^{-j}. \quad (3.3.19)$$

This suffices to prove the desired bound when $2^{2k} \leq \langle t\rangle^{-1}$. This bound also suffices to control the sum over $j$ with $2^j \geq \langle t\rangle 2^{k^-}2^{-k^+}$ when $2^{2k} \geq \langle t\rangle^{-1}$. On the other hand, if $2^j \leq \langle t\rangle 2^{k^-}2^{-k^+}$ then we use (3.2.13) and (3.3.4) to estimate

$$\begin{aligned}\|e^{-it\Lambda_{kg}}V^{\mathcal{L}\psi}_{j,k}(t)\|_{L^\infty} &\lesssim 2^{3k^+}\langle t\rangle^{-3/2}2^{3j/2}\|Q_{j,k}V^{\mathcal{L}\psi}(t)\|_{L^2}\\ &\lesssim \varepsilon_1 Y(k,t;q,n)2^{2k^+}\langle t\rangle^{-3/2}2^{j/2}.\end{aligned}$$

The desired bound (3.3.13) follows by summation over $j$ with $2^j \leq \langle t\rangle 2^{k^-}2^{-k^+}$.

The bounds (3.3.14) follow from (3.2.13) and (3.3.4). The bounds (3.3.15) follow from (3.2.14) and (3.3.4), once we notice that $\Omega_{ab}Q_{j,k}V^{\mathcal{L}\psi} = Q_{j,k}V^{\Omega_{ab}\mathcal{L}\psi}$, for any rotation vector-field $\Omega_{ab}$.

(ii) To prove (3.3.16) we define $J_0$ by $2^{J_0} = 2^{10}\langle t\rangle^{1/2}2^{-k/2}$ and estimate

$$\|e^{-it\Lambda_{wa}}V^H_{\leq J,k}(t)\|_{L^\infty} \lesssim 2^{2k}\langle t\rangle^{-1}\|\widehat{Q_{\leq Jk}V^H}\|_{L^\infty} \lesssim 2^{2k}\langle t\rangle^{-1}\varepsilon_1 2^{-k^- - \kappa k^-}2^{-N_0k^+},$$

if $J \leq J_0$, using (3.2.12) and (3.3.7). If $J \geq J_0$ then we estimate the remaining contribution by

$$\begin{aligned} C\sum_{j\in[J_0,J]}\|e^{-it\Lambda_{wa}}V^H_{j,k}(t)\|_{L^\infty} &\lesssim \sum_{j\geq J_0} 2^{k^+}\langle t\rangle^{-1}2^{k^-/2}(\langle t\rangle 2^{k^-})^{\delta/20}\|Q_{j,k}V^H(t)\|_{H^{0,1}_\Omega} \\ &\lesssim 2^{-N(2)k^+ + 3k^+}\langle t\rangle^{-3/2+(H(1,2)+1)\delta}2^{k^-/2}, \end{aligned}$$

where we used (3.2.11) and (3.3.8). These two bounds suffice to prove the estimates (3.3.16) for $H$ when $2^k \lesssim \langle t\rangle^{1/(5d)}$; if $2^k \geq \langle t\rangle^{1/(5d)}$ then the desired estimates for $H$ follow by Sobolev embedding from (3.3.3). The bounds for the metric components $h$ follow by a similar argument.

The bounds (3.3.17) follow in a similar way, using just (3.3.3) if $2^k \gtrsim \langle t\rangle^{1/(10d)}$ and (3.3.7) if $2^k \lesssim \langle t\rangle^{-1/2+\kappa/8}$. If $2^k \in [\langle t\rangle^{-1/2+\kappa/8}, \langle t\rangle^{1/(10d)}]$ then we use (3.2.15) and (3.3.7) if $2^J \leq 2^{10}\langle t\rangle^{1/2}$, and (3.2.14) and (3.3.4) to estimate the remaining contribution if $2^J \geq 2^{10}\langle t\rangle^{1/2}$. ☐

*Remark 3.17.* We notice that the last two bounds (3.3.16) and (3.3.17) provide sharp pointwise decay at the rate of $\langle t\rangle^{-1}$ and $\langle t\rangle^{-3/2}$ for some parts of the metric tensor and of the Klein-Gordon field. In all the other pointwise bounds in Lemma 3.16 we allow small $\langle t\rangle^{C\delta}$ losses relative to these sharp decay rates.

We prove now several linear bounds on the profiles $V^{\Omega^a h}$ and $V^{\Omega^a\psi}$. These bounds are slight improvements in certain ranges of the bounds one could derive directly from the bootstrap assumptions. These improvements are important in several nonlinear estimates, and are possible because we use interpolation (Lemma 3.13) to take advantage of the stronger assumptions (3.3.7) we have on the functions $V^h$ and $V^\psi$.

**Lemma 3.18.** *(i) For $a \in [0,3]$ we let $\Omega^a$ denote generic vector-fields of the form $\Omega_{23}^{a_1}\Omega_{31}^{a_2}\Omega_{12}^{a_3}$ with $a_1 + a_2 + a_3 \leq a$. If $t \in [0,T]$ and $2^k \gtrsim \langle t\rangle^{-1}$ then*

$$\begin{aligned} \|P_kV^{\Omega^1 h}(t)\|_{L^2} &\lesssim \varepsilon_1 2^{k/2}2^{-N(1)k^+}\langle t\rangle^{H(0,1)\delta}\cdot\langle t\rangle^{20\delta}2^{-6dk^+}, \\ \|P_kV^{\Omega^2 h}(t)\|_{L^2} &\lesssim \varepsilon_1 2^{k/2}2^{-N(2)k^+}\langle t\rangle^{H(0,2)\delta}\cdot\langle t\rangle^{20\delta}2^{-4dk^+}, \end{aligned} \qquad (3.3.20)$$

*where* $h \in \{h_{\alpha\beta} : \alpha, \beta \in \{0,1,2,3\}\}$ *as before. Moreover,*

$$\begin{aligned} \|P_k V^{\Omega^1 \psi}(t)\|_{L^2} &\lesssim \varepsilon_1 2^{-N(1)k^+} \langle t\rangle^{H(0,1)\delta} \cdot 2^{k^-} \langle t\rangle^{10\delta} 2^{-dk^+}, \\ \|P_k V^{\Omega^2 \psi}(t)\|_{L^2} &\lesssim \varepsilon_1 2^{-N(2)k^+} \langle t\rangle^{H(0,2)\delta} \cdot 2^{k^-/3} \langle t\rangle^{15\delta} 2^{-2dk^+} \end{aligned} \tag{3.3.21}$$

*and, for any* $j \geq -k^-$,

$$\begin{aligned} \|V^{\Omega^1 h}_{j,k}(t)\|_{L^2} &\lesssim \varepsilon_1 2^{k/2} 2^{-N(1)k^+} \langle t\rangle^{H(0,1)\delta} \cdot \langle t\rangle^{50\delta} 2^{dk^+} 2^{-(2/3)(j+k)}, \\ \|V^{\Omega^2 h}_{j,k}(t)\|_{L^2} &\lesssim \varepsilon_1 2^{k/2} 2^{-N(2)k^+} \langle t\rangle^{H(0,2)\delta} \cdot \langle t\rangle^{25\delta} 2^{dk^+} 2^{-(1/3)(j+k)}. \end{aligned} \tag{3.3.22}$$

*(ii) In addition, for any* $J \geq -k^-$,

$$\begin{aligned} \|U^{\Omega^1 h}_{\leq J,k}(t)\|_{L^4} &\lesssim \varepsilon_1 2^{3k^-/4} 2^{-\frac{N(1)+N(2)}{2}k^+} \langle t\rangle^{-1/2+\frac{H(0,1)+H(0,2)}{2}\delta} \cdot \langle t\rangle^{6\delta} 2^{2k^+}, \\ \|U^{\Omega^1 h}_{\leq J,k}(t)\|_{L^6} &\lesssim \varepsilon_1 2^{5k^-/6} 2^{-\frac{2N(1)+N(3)}{3}k^+} \langle t\rangle^{-2/3+\frac{2H(0,1)+H(0,3)}{3}\delta} \cdot \langle t\rangle^{8\delta} 2^{2k^+}, \\ \|U^{\Omega^2 h}_{\leq J,k}(t)\|_{L^3} &\lesssim \varepsilon_1 2^{2k^-/3} 2^{-\frac{N(1)+2N(3)}{3}k^+} \langle t\rangle^{-1/3+\frac{H(0,1)+2H(0,3)}{3}\delta} \cdot \langle t\rangle^{4\delta} 2^{2k^+}. \end{aligned} \tag{3.3.23}$$

*Moreover, if* $2^{2k^-}\langle t\rangle \geq 2^{20}$ *and* $2^J \in [2^{-k^-}, 2^{k^- -20}\langle t\rangle]$ *then*

$$\begin{aligned} \|U^{\Omega^1 \psi}_{\leq J,k}(t)\|_{L^4} &\lesssim \varepsilon_1 2^{-\frac{N(1)+N(2)}{2}k^+} \langle t\rangle^{-3/4+\delta H(0,2)/2} 2^{-k^-/4} \cdot 2^{\kappa k^-/60} 2^{-2k^+}, \\ \|U^{\Omega^1 \psi}_{\leq J,k}(t)\|_{L^6} &\lesssim \varepsilon_1 2^{-\frac{2N(1)+N(3)}{3}k^+} \langle t\rangle^{-1+\delta H(0,3)/3} 2^{-k^-/3} \cdot 2^{\kappa k^-/60} 2^{-2.5k^+}, \\ \|U^{\Omega^2 \psi}_{\leq J,k}(t)\|_{L^3} &\lesssim \varepsilon_1 2^{-\frac{N(1)+2N(3)}{3}k^+} \langle t\rangle^{-1/2+2\delta H(0,3)/3} 2^{-k^-/6} \cdot 2^{\kappa k^-/60} 2^{-1.5k^+}. \end{aligned} \tag{3.3.24}$$

*Proof.* (i) We use the interpolation inequalities in (3.2.66)–(3.2.67). The bounds (3.3.20) (which are relevant only when $2^{dk+} \geq 2^{10}\langle t\rangle^{\delta}$) follow from (3.3.3).

The bounds in the first line of (3.3.21) follow directly from (3.3.5) if $k \leq 0$ (notice that $H(1,1) - H(0,1) = 10$) and from (3.3.3) and (3.2.66) if $k \geq 0$. Similarly, the bounds in the second line of (3.3.21) follow from (3.2.67), (3.3.7) (if $k \leq 0$), and (3.3.3) (if $k \geq 0$).

For the bounds in the first line of (3.3.22) we use (3.2.67) and (3.3.3)–(3.3.4),

$$\begin{aligned} \|V^{\Omega^1 h}_{j,k}(t)\|_{L^2} &\lesssim \|V^h_{j,k}(t)\|^{2/3}_{L^2} \|V^h_{j,k}(t)\|^{1/3}_{H^{0,3}_\Omega} \\ &\lesssim [\varepsilon_1 2^{k/2} \langle t\rangle^{H(1,1)} 2^{-N(1)k^+} 2^{-(j+k)}]^{2/3} [\varepsilon_1 2^{k/2} \langle t\rangle^{H(0,3)} 2^{-N(3)k^+}]^{1/3}, \end{aligned}$$

which gives the desired bounds. The estimates in the second line follow in a similar way.

(ii) To prove (3.3.23) we use (3.2.64)–(3.2.65). Indeed, using also (3.3.11) and (3.3.3),

$$\begin{aligned}\|U^{\Omega^1 h}_{\leq J,k}(t)\|_{L^4} &\lesssim \|U^h_{\leq J,k}(t)\|^{1/2}_{L^\infty}\|U^h_{\leq J,k}(t)\|^{1/2}_{H^{0,2}_\Omega}\\ &\lesssim \varepsilon_1\big(\langle t\rangle^{-1+H(1,1)\delta}2^{k^-}2^{-N(1)k^++2k^+}\big)^{1/2}\big(\langle t\rangle^{H(0,2)\delta}2^{k/2}2^{-N(2)k^+}\big)^{1/2},\end{aligned}$$

which gives the bounds in the first line of (3.3.23) (recall that $H(1,1) = H(0,1)+10$). Similarly,

$$\begin{aligned}\|U^{\Omega^1 h}_{\leq J,k}(t)\|_{L^6} &\lesssim \|U^h_{\leq J,k}(t)\|^{2/3}_{L^\infty}\|U^h_{\leq J,k}(t)\|^{1/3}_{H^{0,3}_\Omega}\\ &\lesssim \varepsilon_1\big(\langle t\rangle^{-1+H(1,1)\delta}2^{k^-}2^{-N(1)k^++2k^+}\big)^{2/3}\big(\langle t\rangle^{H(0,3)\delta}2^{k/2}2^{-N(3)k^+}\big)^{1/3}\end{aligned}$$

and

$$\begin{aligned}\|U^{\Omega^2 h}_{\leq J,k}(t)\|_{L^3} &\lesssim \|U^h_{\leq J,k}(t)\|^{1/3}_{L^\infty}\|U^h_{\leq J,k}(t)\|^{2/3}_{H^{0,3}_\Omega}\\ &\lesssim \varepsilon_1\big(\langle t\rangle^{-1+H(1,1)\delta}2^{k^-}2^{-N(1)k^++2k^+}\big)^{1/3}\big(\langle t\rangle^{H(0,3)\delta}2^{k/2}2^{-N(3)k^+}\big)^{2/3}.\end{aligned}$$

The remaining bounds in (3.3.23) follow. We notice that these bounds can be improved slightly if $2^J \leq \langle t\rangle 2^{-20}$, by using the $L^\infty$ bounds (3.3.12) instead of (3.3.11). This is not useful for us, however, since we will apply these bounds to estimate the contributions of localized profiles corresponding to large $j$.

The bounds (3.3.24) follow in a similar way, using (3.2.64)–(3.2.65), (3.3.17), and (3.3.3):

$$\begin{aligned}&\|U^{\Omega^1\psi}_{\leq J,k}(t)\|_{L^4}\\ &\lesssim \varepsilon_1\big(\langle t\rangle^{-3/2}2^{-k^-/2+\kappa k^-/20}2^{-N(1)k^+-5k^+}\big)^{1/2}\big(\langle t\rangle^{H(0,2)\delta}2^{-N(2)k^+}\big)^{1/2},\\ &\|U^{\Omega^1\psi}_{\leq J,k}(t)\|_{L^6}\\ &\lesssim \varepsilon_1\big(\langle t\rangle^{-3/2}2^{-k^-/2+\kappa k^-/20}2^{-N(1)k^+-5k^+}\big)^{2/3}\big(\langle t\rangle^{H(0,3)\delta}2^{-N(3)k^+}\big)^{1/3},\\ &\|U^{\Omega^2\psi}_{\leq J,k}(t)\|_{L^3}\\ &\lesssim \varepsilon_1\big(\langle t\rangle^{-3/2}2^{-k^-/2+\kappa k^-/20}2^{-N(1)k^+-5k^+}\big)^{1/3}\big(\langle t\rangle^{H(0,3)\delta}2^{-N(3)k^+}\big)^{2/3}.\end{aligned}$$

The bounds in (3.3.24) follow. □

We also record a few additional $L^\infty$ bounds in the Fourier space.

**Lemma 3.19.** *If $k \in \mathbb{Z}$ and $\mathcal{L} \in \mathcal{V}^q_n$, $n \leq 1$, then*

$$\begin{aligned}&\|\widehat{P_k V^{\mathcal{L}h_{\alpha\beta}}}(t)\|_{L^\infty}\\ &\lesssim \varepsilon_1 2^{-k^--\delta k^-/2}\langle t\rangle^{\frac{H(q,n+1)+H(q+1,n+2)}{2}\delta}2^{-N_0k^++(n+3/2)dk^+-3k^+/4}.\end{aligned} \tag{3.3.25}$$

*Moreover, if* $(k,j) \in \mathcal{J}$ *and* $\mathcal{L} \in \mathcal{V}^q_n$, $n \le 1$, *then*

$$\begin{aligned} &2^{3k/2}\big\|\widehat{Q_{j,k}V^{\mathcal{L}}h_{\alpha\beta}}(t)\big\|_{L^\infty} + 2^k\big\|\widehat{Q_{j,k}V^{\mathcal{L}}\psi}(t)\big\|_{L^\infty} \\ &\lesssim \varepsilon_1 2^{-j/2+\delta j/4} Y(k,t;q,n+1)2^{\delta k^+/4}. \end{aligned} \tag{3.3.26}$$

*Proof.* The bounds (3.3.26) follow from (3.2.5), (3.3.4), and (3.3.10). To prove the bounds (3.3.25) we use the estimates

$$\begin{aligned} \|Q_{j,k}V^{\mathcal{L}}h_{\alpha\beta}(t)\|_{H^{0,1}_\Omega} &\lesssim \varepsilon_1 \langle t\rangle^{H(q,n+1)\delta} 2^{k/2} 2^{-N(n+1)k^+}(2^{k^-}\langle t\rangle)^{-\gamma}, \\ 2^{j+k}\|Q_{j,k}V^{\mathcal{L}}h_{\alpha\beta}(t)\|_{H^{0,1}_\Omega} &\lesssim \varepsilon_1 \langle t\rangle^{H(q+1,n+2)\delta} 2^{k/2} 2^{-N(n+2)k^+}(2^{k^-}\langle t\rangle)^{-\gamma}, \end{aligned}$$

which follow from (3.3.3)–(3.3.4), and the bounds (3.2.63). □

# Chapter Four

## The Nonlinearities $\mathcal{N}^h_{\alpha\beta}$ and $\mathcal{N}^\psi$

### 4.1 LOCALIZED BILINEAR ESTIMATES

One of our main goals is to prove good bounds on the various components of the nonlinearities $\mathcal{L}\mathcal{N}^h_{\alpha\beta}$ and $\mathcal{L}\mathcal{N}^\psi$, where $\mathcal{L} \in \mathcal{V}^3_3$. Ideally, we would like to prove that these nonlinearities satisfy bounds of the form

$$\begin{aligned} \sum_{\alpha,\beta\in\{0,1,2,3\}} \|\mathcal{L}\mathcal{N}^h_{\alpha\beta}(t)\| &\lesssim \varepsilon_1 \langle t\rangle^{-1} \sum_{\alpha,\beta\in\{0,1,2,3\}} \|\nabla_{x,t}\mathcal{L}h_{\alpha\beta}(t)\|, \\ \|\mathcal{L}\mathcal{N}^\psi(t)\| &\lesssim \varepsilon_1\langle t\rangle^{-1}\|(\langle\nabla\rangle_x, \partial_t)\mathcal{L}\psi(t)\|, \end{aligned} \tag{4.1.1}$$

in suitable norms. Unfortunately, such optimal bounds do not hold for most of the important components of the nonlinearities. As we will see, we have both derivative loss, due to the quasilinear nature of the system, and loss of decay in time, due to the slower decay of the metric components $h_{\alpha\beta}$. However, we can still prove estimates that are somewhat close to (4.1.1), but with certain losses. To quantify this, we define the acceptable loss function

$$\ell(0,0) := 3, \qquad \ell(0,1) := 13, \qquad \ell(1,1) := 23, \qquad \ell(q,n) := 33 \text{ if } n \geq 2. \tag{4.1.2}$$

Notice that $\ell(q,n) + 7 \leq H(q,n)$ if $n \geq 1$.

#### 4.1.1 Frequency Localized $L^2$ Estimates

In this subsection we prove several bounds on localized bilinear interactions, which are the main building blocks for the estimates on the nonlinearities $\mathcal{N}^h_{\alpha\beta}$ and $\mathcal{N}^\psi$ in the next section. Notice that, as a consequence of the definitions (2.1.49), we have the superlinear inequalities

$$\begin{aligned} H(q_1, n_1+1) + H(q_2,n_2) &\leq H(q_1+q_2, n_1+n_2) + 20, \\ H(q_1, n_1+1) + H(q_2,n_2) &\leq H(q_1+q_2, n_1+n_2) - 40 \qquad \text{if } q_2 \geq 1, \\ H(q_1+1, n_1+1) + H(q_2,n_2) &\leq H(q_1+q_2, n_1+n_2) + 30 \qquad \text{if } q_2 \geq 1, \end{aligned} \tag{4.1.3}$$

which hold when $n_1 \geq \max(1,q_1)$, $n_2 \geq \max(1,q_2)$, $n_1+n_2 \leq 3$.

We start by proving $L^2$ bounds on localized bilinear interactions of the metric components.

**Lemma 4.1.** *Assume that* $\mathcal{L}_1 \in \mathcal{V}^{q_1}_{n_1}$, $\mathcal{L}_2 \in \mathcal{V}^{q_2}_{n_2}$, $n_1+n_2 \leq 3$, $h_1, h_2 \in \{h_{\alpha\beta} :$

*$\alpha,\beta \in \{0,1,2,3\}\}$, and $\iota_1,\iota_2 \in \{+,-\}$. Assume that $m \in \mathcal{M}$ (see (3.2.41)), $I = I_m$ is defined as in (3.2.43), and let*

$$I^{wa,1}_{k,k_1,k_2}(t) := 2^{-k/2}\big\|P_k I[P_{k_1}U^{\mathcal{L}_1 h_1,\iota_1}, P_{k_2}U^{\mathcal{L}_2 h_2,\iota_2}](t)\big\|_{L^2}, \tag{4.1.4}$$

*for any $t \in [0,T]$ and $k,k_1,k_2 \in \mathbb{Z}$. Then*

$$\begin{aligned} I^{wa,1}_{k,k_1,k_2}(t) &\lesssim \varepsilon_1^2 2^{-k/2} 2^{3\min\{k,k_1,k_2\}/2}\big(\langle t\rangle^2 2^{k_1^-+k_2^-}\big)^{-\gamma} 2^{k_1/2+k_2/2} \\ &\quad \times \langle t\rangle^{[H(q_1,n_1)+H(q_2,n_2)]\delta} 2^{-N(n_1)k_1^+-N(n_2)k_2^+}. \end{aligned} \tag{4.1.5}$$

*In addition, assuming $n_1 \le n_2$ (in particular $n_1 \le 1$), we have:*
*(1) if $k = \min\{k,k_1,k_2\}$ and $n_1 = 1$ then*

$$I^{wa,1}_{k,k_1,k_2}(t) \lesssim \varepsilon_1^2 \langle t\rangle^{-1+\delta[H(q_2,n_2)+H(q_1,n_1+1)+1]} 2^{-2k_2^+} 2^{k_2^-/4} 2^{-N(n_2)k^+}; \tag{4.1.6}$$

*(2) if $k = \min\{k,k_1,k_2\}$ and $n_1 = 0$ then*

$$I^{wa,1}_{k,k_1,k_2}(t) \lesssim \varepsilon_1^2 \langle t\rangle^{-1+\delta[H(q_2,n_2)+\ell(q_2,n_2)]} 2^{-2k_2^+} 2^{k_2^-/4} 2^{-N(n_2)k^+}; \tag{4.1.7}$$

*(3) if $k_1 = \min\{k,k_1,k_2\}$ and $n_1 \in \{0,1\}$ then*

$$I^{wa,1}_{k,k_1,k_2}(t) \lesssim \varepsilon_1^2 \langle t\rangle^{-1+\delta[H(q_2,n_2)+H(q_1,n_1+1)+1]} 2^{k_1} 2^{-2k_1^+} 2^{|k|/4} 2^{-N(n_2)k^+}; \tag{4.1.8}$$

*(4) if $k_2 = \min\{k,k_1,k_2\}$ and $n_1 = 1$ then*

$$I^{wa,1}_{k,k_1,k_2}(t) \lesssim \varepsilon_1^2 \langle t\rangle^{-1+\delta[H(q_2,n_2)+H(q_1,n_1+1)+1]} 2^{k_2} 2^{-2k_2^+} 2^{|k|/4} 2^{-N(n_1+1)k^+}; \tag{4.1.9}$$

*(5) if $k_2 = \min\{k,k_1,k_2\}$ and $(n_1,n_2) \in \{(0,0),(0,1)\}$ then*

$$I^{wa,1}_{k,k_1,k_2}(t) \lesssim \varepsilon_1^2 \langle t\rangle^{-1+\delta[H(q_2,n_2)+1]} 2^{k_2} 2^{-2k_2^+} 2^{|k|/4} 2^{-N_0k^+ + k^+}; \tag{4.1.10}$$

*(6) if $k_2 = \min\{k,k_1,k_2\}$ and $(n_1,n_2) \in \{(0,2),(0,3)\}$ then*

$$I^{wa,1}_{k,k_1,k_2}(t) \lesssim \varepsilon_1^2 \langle t\rangle^{-1+\delta[H(q_2,n_2)+\ell(q_2,n_2)]} 2^{k_2} 2^{-2k_2^+} 2^{|k|/4} 2^{-N(1)k^+}. \tag{4.1.11}$$

*Proof.* We remark first that bilinear Wave × Wave interactions appear in the metric nonlinearities $\mathcal{N}^{h_{\alpha\beta}}$, both in semilinear and in quasilinear form. According to the general philosophy described in (4.1.1), we would like to have bounds on the form

$$I^{wa,1}_{k,k_1,k_2}(t) \lesssim \varepsilon_1^2 2^{-|k_1-k_2|} \min(2^k, \langle t\rangle^{-1}) 2^{-N(n_1+n_2)k^+ + k^+}. \tag{4.1.12}$$

The factors $2^{-|k_1-k_2|}$ in the right-hand side are critical, in order to be able to estimate the quasilinear components of the nonlinearities $\mathcal{N}^h_{\alpha\beta}$. We notice that the bounds (4.1.6)–(4.1.11) that we actually prove are variations of the ideal bounds (4.1.12), with small $\langle t\rangle^{C\delta}$ loss of decay and loss of derivative $2^{k^+}$ in some cases. For later use, it is very important to minimize the time decay loss as much as possible.

The estimates follow from a case by case analysis, using Lemmas 3.15 and 3.16, and the bilinear estimates in Lemmas 3.10 and 3.11. We estimate first, using just (3.2.44)

$$I^{wa,1}_{k,k_1,k_2}(t) \lesssim 2^{-k/2}2^{3\min\{k,k_1,k_2\}/2}\|P_{k_1}U^{\mathcal{L}_1h_1,\iota_1}(t)\|_{L^2}\|P_{k_2}U^{\mathcal{L}_2h_2,\iota_2}(t)\|_{L^2}, \tag{4.1.13}$$

which gives (4.1.5) in view of (3.3.3). To prove the rest of the bounds, we consider three cases.

**Step 1.** We prove first (4.1.6) and (4.1.7). Assume that $k=\min\{k,k_1,k_2\}$. We may also assume that $|k_1-k_2|\leq 4$, $2^k\gtrsim\langle t\rangle^{-1}$, and $2^{k_2}\lesssim\langle t\rangle^{1/20}$ (otherwise the bounds follow from (4.1.5)). Let $J_1$ be the largest integer such that $2^{J_1}\leq\langle t\rangle(1+2^{k_1}\langle t\rangle)^{-\delta/20}$ and decompose

$$\begin{aligned} P_{k_1}U^{\mathcal{L}_1h_1,\iota_1}(t) &= U^{\mathcal{L}_1h_1,\iota_1}_{\leq J_1,k_1}(t)+U^{\mathcal{L}_1h_1,\iota_1}_{>J_1,k_1}(t)\\ &= e^{-it\Lambda_{wa,\iota_1}}V^{\mathcal{L}_1h_1,\iota_1}_{\leq J_1,k_1}(t)+e^{-it\Lambda_{wa,\iota_1}}V^{\mathcal{L}_1h_1,\iota_1}_{>J_1,k_1}(t); \end{aligned} \tag{4.1.14}$$

see (3.3.1)–(3.3.2). Using (3.2.54), (3.3.3), and (3.3.4) we estimate

$$\begin{aligned} &2^{-k/2}\big\|P_kI[U^{\mathcal{L}_1h_1,\iota_1}_{\leq J_1,k_1}(t),P_{k_2}U^{\mathcal{L}_2h_2,\iota_2}(t)]\big\|_{L^2}\\ &\quad\lesssim\langle t\rangle^{-1}(1+2^{k_1}\langle t\rangle)^{\delta/10}\big\|Q_{\leq J_1,k_1}V^{\mathcal{L}_1h_1}(t)\big\|_{H^{0,1}_\Omega}\|P_{k_2}U^{\mathcal{L}_2h_2}(t)\|_{L^2}\\ &\quad\lesssim\varepsilon_1^2\langle t\rangle^{-1+\delta[H(q_2,n_2)+H(q_1,n_1+1)+1]}2^{-N(n_2)k_2^+-5k_2^+}2^{k_2^-/2} \end{aligned} \tag{4.1.15}$$

and, using (3.2.44),

$$\begin{aligned} &2^{-k/2}\big\|P_kI[U^{\mathcal{L}_1h_1,\iota_1}_{>J_1,k_1}(t),P_{k_2}U^{\mathcal{L}_2h_2,\iota_2}(t)]\big\|_{L^2}\\ &\quad\lesssim 2^k\big\|U^{\mathcal{L}_1h_1}_{>J_1,k_1}(t)\big\|_{L^2}\|P_{k_2}U^{\mathcal{L}_2h_2}(t)\|_{L^2}\\ &\quad\lesssim\varepsilon_1^2\langle t\rangle^{-1+\delta[H(q_2,n_2)+H(q_1+1,n_1+1)+1]}2^{-N(n_2)k_2^+-5k_2^+}2^{k^-/2}. \end{aligned} \tag{4.1.16}$$

Moreover, if $n_2\leq 2$ then we can use (3.3.4) and (3.3.11) to estimate

$$\begin{aligned} &2^{-k/2}\big\|P_kI[U^{\mathcal{L}_1h_1,\iota_1}_{>J_1,k_1}(t),P_{k_2}U^{\mathcal{L}_2h_2,\iota_2}(t)]\big\|_{L^2}\\ &\quad\lesssim 2^{-k/2}\big\|U^{\mathcal{L}_1h_1}_{>J_1,k_1}(t)\big\|_{L^2}\|P_{k_2}U^{\mathcal{L}_2h_2}(t)\|_{L^\infty}\\ &\quad\lesssim\varepsilon_1^2 2^{-k/2}\langle t\rangle^{-2+\delta'}2^{-N(n_1+1)k_1^+-N(n_2+1)k_2^++5k_2^+}. \end{aligned} \tag{4.1.17}$$

Since $H(1,1)=30$, the bounds (4.1.7) follow from (4.1.15) and (4.1.16) if $n_1=0$ and $n_2\geq 2$. The bounds (4.1.6) follow from (4.1.15) and (4.1.17) if $n_1\geq 1$ (in this case $n_2\leq 2$).

It remains to prove the bounds (4.1.7) when $n_1=0$ and $n_2\leq 1$. The estimates (4.1.17) and (4.1.15) still suffice if $2^{k_2}\leq\langle t\rangle^{-\delta'}$, but are slightly too weak when $2^{k_2} \geq \langle t\rangle^{-\delta'}$. In this case we need a different decomposition: let $J_1'$ be the largest integer such that $2^{J_1'} \leq \langle t\rangle^{1/2}2^{-k_1/2} + 2^{-k}$ and decompose $P_{k_1}U^{h_1,\iota_1}(t) = U^{h_1,\iota_1}_{\leq J_1',k_1}(t) + U^{h_1,\iota_1}_{>J_1',k_1}(t)$ as in (4.1.14). Using (3.2.46), (3.3.3), and (3.3.7) we estimate

$$\begin{aligned}
2^{-k/2}&\big\|P_kI[U^{h_1,\iota_1}_{\leq J_1',k_1}(t), P_{k_2}U^{\mathcal{L}_2h_2,\iota_2}(t)]\big\|_{L^2}\\
&\lesssim \langle t\rangle^{-1}2^{3k_1/2}\big\|\mathcal{F}\{Q_{\leq J_1,k_1}V^{h_1}\}(t)\big\|_{L^\infty}\|P_{k_2}U^{\mathcal{L}_2h_2}(t)\|_{L^2} \qquad (4.1.18)\\
&\lesssim \varepsilon_1^2\langle t\rangle^{-1+\delta[H(q_2,n_2)+1]}2^{-N(n_2)k_2^+-5k_2^+}2^{k_2^-/2}.
\end{aligned}$$

Moreover, using (3.3.4) and (3.3.11) we estimate

$$\begin{aligned}
2^{-k/2}&\big\|P_kI[U^{h_1,\iota_1}_{>J_1',k_1}(t), P_{k_2}U^{\mathcal{L}_2h_2,\iota_2}(t)]\big\|_{L^2}\\
&\lesssim 2^{-k/2}\big\|U^{h_1}_{>J_1',k_1}(t)\big\|_{L^2}\|P_{k_2}U^{\mathcal{L}_2h_2}(t)\|_{L^\infty}\\
&\lesssim \varepsilon_1^2 2^{-k/2}\langle t\rangle^{-1+\delta'}2^{-N(n_2+1)k_2^++5k_2^+}2^{k_2^-/2}\min\{2^{-J_1'}2^{-N(1)k_1^+}, 2^{-N(0)k_1^+}\}. \qquad (4.1.19)
\end{aligned}$$

By analyzing the cases $2^k \geq \langle t\rangle^{\delta'}$, $2^k \in [\langle t\rangle^{-1/2}, \langle t\rangle^{\delta'}]$, and $2^k \leq \langle t\rangle^{-1/2}$ it is easy to see that the right-hand side of (4.1.19) is suitably bounded, as claimed in the right-hand side of (4.1.7). The desired conclusion follows using also (4.1.18).

**Step 2.** We prove now (4.1.8). Assume that $k_1 = \min\{k, k_1, k_2\}$. We may also assume $|k - k_2| \leq 4$ and $2^{k_1} \gtrsim \langle t\rangle^{-1}$ (otherwise the bounds follow from (4.1.5)). We estimate first

$$\begin{aligned}
2^{-k/2}&\big\|P_kI[P_{k_1}U^{\mathcal{L}_1h_1,\iota_1}, P_{k_2}U^{\mathcal{L}_2h_2,\iota_2}](t)\big\|_{L^2}\\
&\lesssim 2^{-k_2/2}\big\|P_{k_1}U^{\mathcal{L}_1h_1}(t)\big\|_{L^\infty}\|P_{k_2}U^{\mathcal{L}_2h_2}(t)\|_{L^2} \qquad (4.1.20)\\
&\lesssim \varepsilon_1^2\langle t\rangle^{-1+[H(q_2,n_2)+H(q_1+1,n_1+1)+1]\delta}2^{-N(n_2)k_2^+}2^{k_1}2^{-5k_1^+},
\end{aligned}$$

using (3.3.3) and (3.3.11). This suffices to prove (4.1.8) if $2^{k_2} \lesssim \langle t\rangle^{-\delta'}$ or $2^{k_2} \gtrsim \langle t\rangle^{\delta'}$. Also, the desired bounds follow directly from (4.1.5) if $2^{k_1} \lesssim \langle t\rangle^{-1+10\delta}$.

It remains to consider the case $\langle t\rangle \gg 1$, $2^{k_2} \in [\langle t\rangle^{-\delta'}, \langle t\rangle^{\delta'}]$, $2^{k_1} \geq \langle t\rangle^{-1+10\delta}$. Let $J_1$ be the largest integer such that $2^{J_1} \leq \langle t\rangle 2^{-30}$ and decompose $P_{k_1}U^{\mathcal{L}_1h_1,\iota_1}$ as in (4.1.14). Let $J_2$ be the largest integer such that $2^{J_2} \leq \langle t\rangle^{1/2}$ and decompose $P_{k_2}U^{\mathcal{L}_2h_2,\iota_2}$ in a similar way. Then

$$\begin{aligned}
2^{-k/2}&\big\|P_kI[U^{\mathcal{L}_1h_1,\iota_1}_{\leq J_1,k_1}(t), P_{k_2}U^{\mathcal{L}_2h_2,\iota_2}(t)]\big\|_{L^2}\\
&\lesssim 2^{-k_2/2}\big\|U^{\mathcal{L}_1h_1}_{\leq J_1,k_1}(t)\big\|_{L^\infty}\|P_{k_2}U^{\mathcal{L}_2h_2}(t)\|_{L^2} \qquad (4.1.21)\\
&\lesssim \varepsilon_1^2\langle t\rangle^{-1+\delta[H(q_2,n_2)+H(q_1,n_1+1)+1]}2^{-N(n_2)k_2^+}2^{k_1}2^{-5k_1^+},
\end{aligned}$$

using (3.3.3) and (3.3.12). Moreover,

$$\begin{aligned} 2^{-k/2}\big\|P_kI[U^{\mathcal{L}_1h_1,\iota_1}_{>J_1,k_1}(t),U^{\mathcal{L}_2h_2,\iota_2}_{>J_2,k_2}(t)]\big\|_{L^2} &\\ \lesssim 2^{-k_2/2}\big\|U^{\mathcal{L}_1h_1}_{>J_1,k_1}(t)\big\|_{L^\infty}\|U^{\mathcal{L}_2h_2}_{>J_2,k_2}(t)\|_{L^2} &\\ \lesssim \varepsilon_1^2\langle t\rangle^{-5/4}2^{-k_2}2^{-N(n_2+1)k_2^+}2^{k_1}2^{-5k_1^+}, & \end{aligned} \tag{4.1.22}$$

using (3.3.4) and (3.3.11). In addition, for any $j_1 > J_1$ we use (3.2.56) and (3.3.3) to estimate

$$\begin{aligned} 2^{-k/2}\big\|P_kI[U^{\mathcal{L}_1h_1,\iota_1}_{j_1,k_1}(t),U^{\mathcal{L}_2h_2,\iota_2}_{\le J_2,k_2}(t)]\big\|_{L^2} &\\ \lesssim 2^{-k_2/2}2^{k_1/2}\langle t\rangle^{-1+3\delta/4}\big\|Q_{j_1,k_1}V^{\mathcal{L}_1h_1}(t)\big\|_{H^{0,1}_\Omega}\|Q_{\le J_2,k_2}V^{\mathcal{L}_2h_2}(t)\|_{L^2} &\\ \lesssim \varepsilon_1^2\langle t\rangle^{-1+3\delta/4+H(q_1,n_1+1)\delta+H(q_2,n_2)\delta}2^{k_1}2^{-5k_1^+}2^{-N(n_2)k_2^+}. & \end{aligned} \tag{4.1.23}$$

Finally, when $2^{j_1}\ge\langle t\rangle^4$ then we just use (3.3.3) and (3.3.4) to estimate

$$\begin{aligned} 2^{-k/2}\big\|P_kI[U^{\mathcal{L}_1h_1,\iota_1}_{j_1,k_1}(t),U^{\mathcal{L}_2h_2,\iota_2}_{\le J_2,k_2}(t)]\big\|_{L^2} &\\ \lesssim 2^{-k_2/2}2^{3k_1/2}\big\|V^{\mathcal{L}_1h_1}_{j_1,k_1}(t)\big\|_{L^2}\|V^{\mathcal{L}_2h_2}_{\le J_2,k_2}(t)\|_{L^2} &\\ \lesssim \varepsilon_1^2 2^{-3j_1/4}2^{k_1}2^{-5k_1^+}2^{-N(n_2)k_2^+}. & \end{aligned} \tag{4.1.24}$$

The desired bounds (4.1.8) follow from (4.1.21)–(4.1.24) if $2^{k_2}\in[\langle t\rangle^{-\delta'},\langle t\rangle^{\delta'}]$.

**Step 3.** Finally, we prove (4.1.9)–(4.1.11). Assume that $k_2=\min\{k,k_1,k_2\}$. We may also assume $|k-k_1|\le 4$, $2^{k_2}\gtrsim\langle t\rangle^{-1}$, and $2^{k_1}\in[\langle t\rangle^{-4/5},\langle t\rangle]$ (otherwise the bounds follow from (4.1.5)). Let $J_1$ be the largest integer with the property that $2^{J_1}\le\langle t\rangle(1+2^{k_1}\langle t\rangle)^{-\delta/20}$ and decompose $P_{k_1}U^{\mathcal{L}_1h_1,\iota_1}(t)$ as in (4.1.14). Using (3.2.54), (3.3.3), and (3.3.4) we estimate

$$\begin{aligned} 2^{-k/2}\big\|P_kI[U^{\mathcal{L}_1h_1,\iota_1}_{\le J_1,k_1}(t),P_{k_2}U^{\mathcal{L}_2h_2,\iota_2}(t)]\big\|_{L^2} &\\ \lesssim 2^{-k_1/2}\langle t\rangle^{-1+\delta/4}2^{k_2/2}\big\|Q_{\le J_1,k_1}V^{\mathcal{L}_1h_1}(t)\big\|_{H^{0,1}_\Omega}\|P_{k_2}U^{\mathcal{L}_2h_2}(t)\|_{L^2} &\\ \lesssim \varepsilon_1^2\langle t\rangle^{-1+\delta[H(q_2,n_2)+H(q_1,n_1+1)+1]}2^{k_2}2^{-5k_2^+}2^{-N(n_1+1)k_1^+} & \end{aligned} \tag{4.1.25}$$

and

$$\begin{aligned} 2^{-k/2}\big\|P_kI[U^{\mathcal{L}_1h_1,\iota_1}_{>J_1,k_1}(t),P_{k_2}U^{\mathcal{L}_2h_2,\iota_2}(t)]\big\|_{L^2} &\\ \lesssim 2^{-k_1/2}2^{3k_2/2}\big\|U^{\mathcal{L}_1h_1}_{>J_1,k_1}(t)\big\|_{L^2}\|P_{k_2}U^{\mathcal{L}_2h_2}(t)\|_{L^2} &\\ \lesssim \varepsilon_1^2\langle t\rangle^{-1+\delta[H(q_2,n_2)+H(q_1+1,n_1+1)+1]}2^{-5k_2^+}2^{2k_2}2^{-k_1}2^{-N(n_1+1)k_1^+}. & \end{aligned} \tag{4.1.26}$$

Since $H(1,1)=30$, these bounds clearly suffice to prove (4.1.11) when $(n_1,n_2)\in\{(0,2),(0,3)\}$.

It remains to consider the cases $(n_1,n_2)\in\{(0,0),(0,1),(1,1),(1,2)\}$. Assume first that $2^k\notin[\langle t\rangle^{-8\delta'},\langle t\rangle^{8\delta'}]$. Then we estimate, using (3.3.3) and (3.3.11),

$$
\begin{aligned}
2^{-k/2}\big\|P_kI[P_{k_1}U^{\mathcal{L}_1h_1,\iota_1}(t),P_{k_2}U^{\mathcal{L}_2h_2,\iota_2}(t)]\big\|_{L^2}
&\lesssim 2^{-k_1/2}\big\|P_{k_1}U^{\mathcal{L}_1h_1}(t)\big\|_{L^2}\|P_{k_2}U^{\mathcal{L}_2h_2}(t)\|_{L^\infty}\\
&\lesssim \varepsilon_1^2\langle t\rangle^{-1+\delta'/2}2^{k_2}2^{-5k_2^+}2^{-N(n_1)k_1^+},
\end{aligned}
\tag{4.1.27}
$$

which suffices to prove (4.1.9)–(4.1.10) when $2^{|k|}\geq\langle t\rangle^{8\delta'}$.

On the other hand, if

$$
(n_1,n_2)\in\{(0,0),(0,1)\}\qquad\text{and}\qquad 2^k\in[\langle t\rangle^{-8\delta'},\langle t\rangle^{8\delta'}],
\tag{4.1.28}
$$

then we would like to use Lemma 3.10. Let $J_1'$ be the largest integer such that $2^{J_1'}\leq\langle t\rangle^{1/2}2^{-k_1/2}$ and decompose $P_{k_1}U^{h_1,\iota_1}(t)$ as in (4.1.14). Using (3.2.46), (3.3.3), and (3.3.7) we estimate

$$
\begin{aligned}
&2^{-k/2}\big\|P_kI[U^{h_1,\iota_1}_{\leq J_1',k_1}(t),P_{k_2}U^{\mathcal{L}_2h_2,\iota_2}(t)]\big\|_{L^2}\\
&\lesssim 2^{-k_1/2+k_2/2}\langle t\rangle^{-1}2^{3k_1/2}\big\|\mathcal{F}\{Q_{\leq J_1',k_1}V^{h_1}\}(t)\big\|_{L^\infty}\|P_{k_2}U^{\mathcal{L}_2h_2}(t)\|_{L^2}\\
&\lesssim \varepsilon_1^2\langle t\rangle^{-1+\delta[H(q_2,n_2)+1]}2^{k_2}2^{-5k_2^+}2^{-\kappa k_1^-}2^{-N_0k_1^++k_1^+}.
\end{aligned}
\tag{4.1.29}
$$

Moreover, using (3.3.4) and (3.3.11) we estimate

$$
\begin{aligned}
2^{-k/2}\big\|P_kI[U^{h_1,\iota_1}_{>J_1',k_1}(t),P_{k_2}U^{\mathcal{L}_2h_2,\iota_2}(t)]\big\|_{L^2}
&\lesssim 2^{-k_1/2}\big\|Q_{>J_1',k_1}V^{h_1}(t)\big\|_{L^2}\|P_{k_2}U^{\mathcal{L}_2h_2}(t)\|_{L^\infty}\\
&\lesssim \varepsilon_1^2\langle t\rangle^{-3/2+\delta'}2^{-k_1/2}2^{k_2}2^{-N(1)k_1^+-5k_2^+}.
\end{aligned}
\tag{4.1.30}
$$

These two bounds clearly suffice to prove (4.1.10) when $2^k\in[\langle t\rangle^{-8\delta'},\langle t\rangle^{8\delta'}]$.

Finally, assume that

$$
(n_1,n_2)\in\{(1,1),(1,2)\}\qquad\text{and}\qquad 2^k\in[\langle t\rangle^{-8\delta'},\langle t\rangle^{8\delta'}].
\tag{4.1.31}
$$

Let $J_1$ be the largest integer such that $2^{J_1}\leq\langle t\rangle(1+2^{k_1}\langle t\rangle)^{-\delta/20}$ as before and notice that (4.1.25) gives suitable bounds for the contributions of $U^{\mathcal{L}_1h_1,\iota_1}_{\leq J_1,k_1}(t)$. Moreover,

$$
\begin{aligned}
2^{-k/2}\big\|P_kI[U^{\mathcal{L}_1h_1,\iota_1}_{>J_1,k_1}(t),P_{k_2}U^{\mathcal{L}_2h_2,\iota_2}(t)]\big\|_{L^2}
&\lesssim 2^{-k_1/2}\big\|Q_{>J_1,k_1}V^{\mathcal{L}_1h_1}(t)\big\|_{L^2}\|P_{k_2}U^{\mathcal{L}_2h_2}(t)\|_{L^\infty}\\
&\lesssim \varepsilon_1^2\langle t\rangle^{-2+\delta'}2^{-5k_2^+}2^{k_2}2^{-k_1}2^{-N(n_1+1)k_1^+},
\end{aligned}
\tag{4.1.32}
$$

using (3.3.4) and (3.3.11). This suffices to complete the proof of (4.1.9). □

We prove now $L^2$ bounds on localized bilinear interactions of the Klein-Gordon field.

**Lemma 4.2.** *Assume that $\mathcal{L}_1 \in \mathcal{V}^{q_1}_{n_1}$, $\mathcal{L}_2 \in \mathcal{V}^{q_2}_{n_2}$, $n_1+n_2 \leq 3$, $n_1 \leq n_2$. Assume also that $m \in \mathcal{M}$ (see (3.2.41)), $I = I_m$ is defined as in (3.2.43), and let*

$$I^{wa,2}_{k,k_1,k_2}(t) := 2^{-k/2}\big\|P_k I[P_{k_1}U^{\mathcal{L}_1\psi,\iota_1}, P_{k_2}U^{\mathcal{L}_2\psi,\iota_2}](t)\big\|_{L^2}, \tag{4.1.33}$$

*for any $t \in [0,T]$, $\iota_1, \iota_2 \in \{+,-\}$, and $k, k_1, k_2 \in \mathbb{Z}$. Then*

$$\begin{aligned} I^{wa,2}_{k,k_1,k_2}(t) &\lesssim \varepsilon_1^2 2^{-k/2} 2^{3\min\{k,k_1,k_2\}/2} 2^{-N(n_1)k_1^+ - N(n_2)k_2^+} \\ &\quad\times \min\big\{\langle t\rangle^{H(q_1,n_1)\delta}, 2^{k_1^-}\langle t\rangle^{H(q_1+1,n_1+1)\delta}\big\} \\ &\quad\times \min\big\{\langle t\rangle^{H(q_2,n_2)\delta}, 2^{k_2^-}\langle t\rangle^{H(q_2+1,n_2+1)\delta}\big\}, \end{aligned} \tag{4.1.34}$$

*where the second factor in the right-hand side is, by definition, $\langle t\rangle^{H(q_2,n_2)\delta}$ if $n_2 = 3$. Moreover,*

$$I^{wa,2}_{k,k_1,k_2}(t) \lesssim \varepsilon_1^2 \langle t\rangle^{-1+\delta[H(q_1+q_2,n_1+n_2)+\ell(q_1+q_2,n_1+n_2)]} 2^{-N(n_1+n_2)k^+} 2^{-k^+/4}. \tag{4.1.35}$$

*Proof.* As in the previous lemma, we remark that bilinear KG × KG interactions appear in the metric nonlinearities $\mathcal{N}^h_{\alpha\beta}$, in semilinear form. According to the general philosophy described in (4.1.1), ideally we would like to have bounds on the form

$$I^{wa,2}_{k,k_1,k_2}(t) \lesssim \varepsilon_1^2 \min(2^k, \langle t\rangle^{-1}) 2^{-N(n_1+n_2)k^+ - k^+/2}. \tag{4.1.36}$$

We notice that the estimates (4.1.34)–(4.1.35) we actually prove are variations of these optimal bounds, with small $\langle t\rangle^{C\delta}$ loss of decay in some cases.

The estimates (4.1.34) follow using just $L^2$ bounds; see (3.2.44), (3.3.3), and (3.3.5). To prove (4.1.35), we will have to consider several cases. We will sometimes use the general bounds

$$\begin{aligned} I^{wa,2}_{k,k_1,k_2}(t) &\lesssim 2^{-k/2}\|P_{k_1}U^{\mathcal{L}_1\psi}(t)\|_{L^\infty}\|P_{k_2}U^{\mathcal{L}_2\psi}(t)\|_{L^2} \\ &\lesssim \varepsilon_1^2 \langle t\rangle^{-1+\delta(H(q_1+1,n_1+1)+H(q_2,n_2))} \\ &\quad\times 2^{-k/2} 2^{k_1^-/2} 2^{-N(n_2)k_2^+} 2^{-N(n_1+1)k_1^+ + 2k_1^+}, \end{aligned} \tag{4.1.37}$$

which follow from (3.3.3) and (3.3.13). Similarly, if $n_2 \leq 2$,

$$\begin{aligned} I^{wa,2}_{k,k_1,k_2}(t) &\lesssim 2^{-k/2}\|P_{k_1}U^{\mathcal{L}_1\psi}(t)\|_{L^2}\|P_{k_2}U^{\mathcal{L}_2\psi}(t)\|_{L^\infty} \\ &\lesssim \varepsilon_1^2 \langle t\rangle^{-1+\delta(H(q_1,n_1)+H(q_2+1,n_2+1))} \\ &\quad\times 2^{-k/2} 2^{k_2^-/2} 2^{-N(n_1)k_1^+} 2^{-N(n_2+1)k_2^+ + 2k_2^+}. \end{aligned} \tag{4.1.38}$$

We can prove one more general bound of this type when $n_2 \leq 2$ by decomposing the profiles. Indeed, let

$$P_{k_1} U^{\mathcal{L}_1\psi,\iota_1} = \sum_{j_1 \geq \max(-k_1,0)} U^{\mathcal{L}_1\psi,\iota_1}_{j_1,k_1}, \qquad P_{k_2} U^{\mathcal{L}_2\psi,\iota_2} = \sum_{j_2 \geq \max(-k_2,0)} U^{\mathcal{L}_2\psi,\iota_2}_{j_2,k_2}, \tag{4.1.39}$$

as in (3.3.1)–(3.3.2). In view of (3.3.4) and (3.3.14), we have

$$\begin{aligned} \|U^{\mathcal{L}_l\psi,\iota_l}_{j_l,k_l}(t)\|_{L^2} &\lesssim 2^{-j_l}\varepsilon_1 Y(k_l,t;q_l,n_l), \\ \|U^{\mathcal{L}_l\psi,\iota_l}_{j_l,k_l}(t)\|_{L^\infty} &\lesssim 2^{3k_l^+}\langle t\rangle^{-3/2}2^{j_l/2}\varepsilon_1 Y(k_l,t;q_l,n_l), \end{aligned} \tag{4.1.40}$$

for $l \in \{1,2\}$, where $Y$ is defined as in (3.3.6). We use the $L^2 \times L^\infty$ estimate for each interaction (as in (4.1.37)–(4.1.38)), and place the factor with the larger $j$ in $L^2$ (in order to gain $2^{-\max(j_1,j_2)}$) and the factor with the smaller $j$ in $L^\infty$. After summation over $j_1, j_2$, it follows that

$$\begin{aligned} I^{wa,2}_{k,k_1,k_2}(t) &\lesssim \varepsilon_1^2 Y(k_1,t;q_1,n_1)Y(k_2,t;q_2,n_2) \\ &\quad \times 2^{3(k_1^+ + k_2^+)}2^{-k/2}\langle t\rangle^{-3/2}2^{\min(k_1^-,k_2^-)/2}. \end{aligned} \tag{4.1.41}$$

**Step 1.** Assume first that $n_1 = 1$, thus $(n_1,n_2) \in \{(1,1),(1,2)\}$. The desired bounds (4.1.35) follow from (4.1.34) if $2^{k_1^- + k_2^- + k^-} \lesssim \langle t\rangle^{-1-\delta'}$. They also follow if $2^{\max\{k_1,k_2\}} \gtrsim \langle t\rangle^{\delta'}$, using (4.1.37)–(4.1.38) if $k \geq 0$ and (4.1.41) if $k \leq 0$. If $2^{\max\{k_1,k_2\}} \leq \langle t\rangle^{\delta'}$ then the bounds (4.1.35) follow from (4.1.41) if $2^k \gtrsim \langle t\rangle^{-1+10\delta'}$. After these reductions, it remains to prove (4.1.35) when $|t| \gg 1$ and

$$2^{k_1}, 2^{k_2} \in [\langle t\rangle^{-8\delta'}, \langle t\rangle^{\delta'}], \qquad 2^k \leq \langle t\rangle^{-1+10\delta'}. \tag{4.1.42}$$

This is a High × High → Low interaction and loss of derivatives is not an issue, so we only need to justify the $\langle t\rangle^{-1+\delta[H(q_1+q_2,n_1+n_2)+33]}$ time decay. The bounds still follow from (4.1.34) if $2^k \lesssim \langle t\rangle^{-1+30\delta}$; see (2.1.53). On the other hand, if $2^k \geq \langle t\rangle^{-1+30\delta}$ then we let $J_1$ be the largest integer such that $2^{J_1} \leq \langle t\rangle^{3/4}$. Using (3.2.60), (3.3.3)–(3.3.5), and (3.3.13) we estimate

$$\begin{aligned} &2^{-k/2}\big\|P_k I[U^{\mathcal{L}_1\psi,\iota_1}_{\leq J_1,k_1}(t), P_{k_2}U^{\mathcal{L}_2\psi,\iota_2}(t)]\big\|_{L^2} \\ &\quad\lesssim \langle t\rangle^{-1+\delta}2^{5k_1^+}2^{-k_1^-}\|Q_{\leq J_1,k_1}V^{\mathcal{L}_1\psi}(t)\|_{H^{0,1}_\Omega}\|P_{k_2}U^{\mathcal{L}_2\psi}(t)\|_{L^2} \\ &\quad\lesssim \varepsilon_1^2\langle t\rangle^{-1+\delta[H(q_1+1,n_1+1)+H(q_2,n_2)+1]} \end{aligned} \tag{4.1.43}$$

and

$$\begin{aligned} &2^{-k/2}\big\|P_k I[U^{\mathcal{L}_1\psi,\iota_1}_{>J_1,k_1}(t), P_{k_2}U^{\mathcal{L}_2\psi,\iota_2}(t)]\big\|_{L^2} \\ &\quad\lesssim 2^{-k/2}\big\|Q_{>J_1,k_1}V^{\mathcal{L}_1\psi}(t)\big\|_{L^2}\|P_{k_2}U^{\mathcal{L}_2\psi}(t)\|_{L^\infty} \\ &\quad\lesssim \varepsilon_1^2 2^{-k/2}\langle t\rangle^{-7/4+\delta'}. \end{aligned} \tag{4.1.44}$$

The desired bounds (4.1.35) follow if $q_2 \geq 1$, using the last inequality in (4.1.3).

We can also repeat these estimates with the roles of the functions $P_{k_1}U^{\mathcal{L}_1\psi,\iota_1}$ and $P_{k_2}U^{\mathcal{L}_2\psi,\iota_2}$ reversed. Thus $I^{wa,2}_{k,k_1,k_2}(t) \lesssim \varepsilon_1^2\langle t\rangle^{-1+\delta[H(q_2+1,n_2+1)+H(q_1,n_1)+1]}$, and the desired conclusion follows if $q_1 \geq 1$.

Finally, $2^{-k_1^-}\|P_{k_1}U^{\mathcal{L}_1\psi}\|_{L^2} \lesssim \varepsilon_1 Y(k_1,t;0,0) \lesssim \varepsilon_1\langle t\rangle^{H(1,1)}2^{-N(1)k_1^+}$ if $q_1 = q_2 = 0$, as a consequence of (3.3.5) and the assumption $n_1 = 1$. Then we estimate, as in (4.1.43),

$$\begin{aligned} &2^{-k/2}\big\|P_kI[P_{k_1}U^{\mathcal{L}_1\psi,\iota_1}(t), U^{\mathcal{L}_2\psi,\iota_2}_{\leq J_2,k_2}(t)]\big\|_{L^2} \\ &\quad\lesssim \langle t\rangle^{-1+\delta}2^{5k_2^+}2^{-k_2^-}\|P_{k_1}U^{\mathcal{L}_1\psi}(t)\|_{L^2}\|Q_{\leq J_2,k_2}V^{\mathcal{L}_2\psi}(t)\|_{H^{0,1}_\Omega} \\ &\quad\lesssim \varepsilon_1^2\langle t\rangle^{-1+\delta[H(1,1)+H(0,n_2+1)+1]}, \end{aligned}$$

where $J_2$ is the largest integer such that $2^{J_2} \leq \langle t\rangle^{3/4}$. Since $H(1,1) = 30$, this is consistent with the desired estimates (4.1.35). The contribution of $U^{\mathcal{L}_2\psi,\iota_2}_{>J_2,k_2}(t)$ can be bounded as in (4.1.44), using an $L^\infty \times L^2$ estimate and (3.3.4). This completes the proof of (4.1.35) when $n_1 = 1$.

**Step 2.** Assume that $n_1 = 0$. It follows from (3.3.7) that $\|P_{k_1}U^{\psi,\iota_1}(t)\|_{L^2} \lesssim \varepsilon_1 2^{k_1^-+\kappa k_1^-}2^{-N_0k_1^++2k_1^+}$. Therefore

$$I^{wa,2}_{k,k_1,k_2}(t) \lesssim \varepsilon_1^2 2^{-k^+/2}2^{\min\{k^-,k_1^-,k_2^-\}}2^{k_1^-+\kappa k_1^-}2^{-N_0k_1^++4k_1^+}2^{-N(n_2)k_2^+}\langle t\rangle^{H(q_2,n_2)\delta}, \tag{4.1.45}$$

as a consequence of (3.3.3) and (3.2.44). The desired bounds (4.1.35) follow if $2^{k_1^-} \lesssim \langle t\rangle^{-1/2+\kappa/8}$ or if ($2^{k_1^-} \geq \langle t\rangle^{-1/2+\kappa/8}2^{20}$ and $2^{k^-+k_1^-} \lesssim \langle t\rangle^{-1}$).

On the other hand, if $\langle t\rangle \geq 2^{1/\delta}$,

$$2^{k_1^-} \geq \langle t\rangle^{-1/2+\kappa/8} \qquad \text{and} \qquad 2^{k^-+k_1^-} \geq \langle t\rangle^{-1} \tag{4.1.46}$$

then we let $J_1$ be the largest integer such that $2^{J_1} \leq 2^{k_1^- -20}\langle t\rangle$ and decompose $P_{k_1}U^{\psi,\iota_1}(t) = e^{-it\Lambda_{kg,\iota_1}}V^{\psi,\iota_1}_{\leq J_1,k_1}(t) + e^{-it\Lambda_{kg,\iota_1}}V^{\psi,\iota_1}_{>J_1,k_1}(t)$ as in (4.1.14). Using (3.3.17), (3.3.3), and (3.3.4) we estimate

$$\begin{aligned} &2^{-k/2}\big\|P_kI[U^{\psi,\iota_1}_{\leq J_1,k_1}(t), P_{k_2}U^{\mathcal{L}_2\psi,\iota_2}(t)]\big\|_{L^2} \\ &\quad\lesssim 2^{-k/2}\|U^{\psi}_{\leq J_1,k_1}(t)\|_{L^\infty}\|P_{k_2}U^{\mathcal{L}_2\psi}(t)\|_{L^2} \\ &\quad\lesssim \varepsilon_1^2\langle t\rangle^{-3/2+\delta[H(q_2,n_2)+1]}2^{-k/2}2^{-k_1^-/2}2^{-N_0k_1^++5k_1^+}2^{-N(n_2)k_2^+} \end{aligned} \tag{4.1.47}$$

and, with $\underline{k} = \min\{k, k_1, k_2\}$,

$$\begin{aligned} &2^{-k/2}\big\|P_kI[U^{\psi,\iota_1}_{>J_1,k_1}(t), P_{k_2}U^{\mathcal{L}_2\psi,\iota_2}(t)]\big\|_{L^2} \\ &\lesssim 2^{-k/2}2^{3\underline{k}/2}\big\|U^{\psi}_{>J_1,k_1}(t)\big\|_{L^2}\|P_{k_2}U^{\mathcal{L}_2\psi}(t)\|_{L^2} \\ &\lesssim \varepsilon_1^2\langle t\rangle^{-1+\delta[H(q_2,n_2)+H(1,1)+1]}2^{-k_1^--k/2}2^{3\underline{k}/2}2^{-N(1)k_1^+}2^{-N(n_2)k_2^+}. \end{aligned} \tag{4.1.48}$$

**Case 2.1.** Assume first that $k = \min(k, k_1, k_2)$ and $2^k \geq \langle t\rangle^{-1}$. Then

$|k_1-k_2|\leq 4$ and the bounds (4.1.47)–(4.1.48) show that

$$\begin{aligned}I^{wa,2}_{k,k_1,k_2}(t)&\lesssim \varepsilon_1^2\langle t\rangle^{-1+\delta[H(q_2,n_2)+1]}2^{-N(n_2)k_2^+-5k_2^+}\\&\quad\times\{2^{-k/2}2^{-k_1^-/2}\langle t\rangle^{-1/2}+2^{k-k_1}\langle t\rangle^{H(1,1)\delta}\}.\end{aligned}\tag{4.1.49}$$

Since $H(1,1)=30$ and $2^{-k/2}2^{-k_1^-/2}\langle t\rangle^{-1/2}\lesssim 1$ (see (4.1.46)), this suffices when $n_2\geq 2$ or when $2^{k-k_1}2^{-4k_2^+}\leq\langle t\rangle^{-\delta'}$. In the remaining case ($n_2\in\{0,1\}$ and $2^{k-k_1}2^{-4k_2^+}\in[\langle t\rangle^{-\delta'},1]$) we need to improve the bounds (4.1.48). Using (3.3.4) and (3.3.13) we estimate

$$\begin{aligned}2^{-k/2}&\big\|P_kI[U^{\psi,\iota_1}_{>J_1,k_1}(t),P_{k_2}U^{\mathcal{L}_2\psi,\iota_2}(t)]\big\|_{L^2}\\&\lesssim 2^{-k/2}\big\|U^{\psi}_{>J_1,k_1}(t)\big\|_{L^2}\|P_{k_2}U^{\mathcal{L}_2\psi}(t)\|_{L^\infty}\\&\lesssim\varepsilon_1^2\langle t\rangle^{-2+\delta'}2^{-k_1^-/2-k/2}.\end{aligned}$$

Since $2^{4k^+}\leq\langle t\rangle^{\delta'}$ we can use also (4.1.47) to complete the proof of (4.1.35).

**Case 2.2.** Assume now that $k_1=\min(k,k_1,k_2)$. Then $|k-k_2|\leq 4$ and the bounds (4.1.47)–(4.1.48) show that

$$\begin{aligned}I^{wa,2}_{k,k_1,k_2}(t)&\lesssim \varepsilon_1^2\langle t\rangle^{-1+\delta[H(q_2,n_2)+1]}2^{-N(n_2)k_2^+-5k_1^+}\\&\quad\times\{2^{-k/2}2^{-k_1^-/2}\langle t\rangle^{-1/2}+2^{(k_1^--k)/2}\langle t\rangle^{H(1,1)\delta}\}.\end{aligned}$$

In view of (4.1.46), this suffices when $n_2\geq 2$ or when $2^{(k_1^--k^-)/2}2^{-k^+/4}\leq\langle t\rangle^{-\delta'}$. In the remaining case ($n_2\in\{0,1\}$ and $2^{(k_1^--k^-)/2}2^{-k^+/4}\in[\langle t\rangle^{-\delta'},1]$) we can use (3.3.4) and (3.3.13) to improve the bounds (4.1.48),

$$\begin{aligned}2^{-k/2}&\big\|P_kI[U^{\psi,\iota_1}_{>J_1,k_1}(t),P_{k_2}U^{\mathcal{L}_2\psi,\iota_2}(t)]\big\|_{L^2}\\&\lesssim 2^{-k/2}\big\|U^{\psi}_{>J_1,k_1}(t)\big\|_{L^2}\|P_{k_2}U^{\mathcal{L}_2\psi}(t)\|_{L^\infty}\lesssim\varepsilon_1^2\langle t\rangle^{-2+\delta'}2^{-k_1^-}.\end{aligned}\tag{4.1.50}$$

Since $2^{k^+}\lesssim\langle t\rangle^{4\delta'}$ and $2^{-k_1^-}\lesssim\langle t\rangle^{1/2}$, the desired bounds (4.1.35) follow.

**Case 2.3.** Finally, assume that $k_2=\min(k,k_1,k_2)$. In proving (4.1.35) we may also assume that $n_2\geq 1$, since the case $n_2=0$ follows from the analysis in Case 2.2 by reversing the roles of the functions $P_{k_1}U^{\psi,\iota_1}$ and $P_{k_2}U^{\psi,\iota_2}$. The bounds (4.1.47)–(4.1.48) show that

$$\begin{aligned}I^{wa,2}_{k,k_1,k_2}(t)&\lesssim \varepsilon_1^2\langle t\rangle^{-1+\delta[H(q_2,n_2)+1]}2^{-4k_2^+-N(1)k_1^+}\\&\quad\{2^{-k/2}2^{-k_1^-/2}\langle t\rangle^{-1/2}2^{-4k_1^+}+2^{(k_2^--k)/2}\langle t\rangle^{H(1,1)\delta}\}.\end{aligned}$$

In view of (4.1.46), this suffices if $n_2\geq 2$ or if ($n_2=1$ and $2^{k_2^--k^-}2^{-k^+/2}\leq\langle t\rangle^{-\delta'}$). In the remaining case ($n_2=1$ and $2^{k_2^--k^-}2^{-k^+/2}\in[\langle t\rangle^{-\delta'},1]$) we use

(3.3.4) and (3.3.13) to prove bounds identical to (4.1.50),

$$2^{-k/2}\big\|P_kI[U^{\psi,\iota_1}_{>J_1,k_1}(t),P_{k_2}U^{\mathcal{L}_2\psi,\iota_2}(t)]\big\|_{L^2}\lesssim\varepsilon_1^2\langle t\rangle^{-2+\delta'}2^{-k_1^-}.$$

The desired bounds (4.1.35) follow in this last case. This completes the proof of the lemma. □

Finally we prove $L^2$ bounds on localized bilinear interactions of the Klein-Gordon field and the metric components.

**Lemma 4.3.** *Assume that $\mathcal{L}_1\in\mathcal{V}^{q_1}_{n_1}$, $\mathcal{L}_2\in\mathcal{V}^{q_2}_{n_2}$, $n_1+n_2\leq 3$. Assume also that $m\in\mathcal{M}$ (see* (3.2.41)*), $I=I_m$ is defined as in* (3.2.43)*, and let*

$$I^{kg}_{k,k_1,k_2}(t):=2^{k_2^+-k_1}\big\|P_kI[P_{k_1}U^{\mathcal{L}_1h,\iota_1},P_{k_2}U^{\mathcal{L}_2\psi,\iota_2}](t)\big\|_{L^2},\tag{4.1.51}$$

*for any $t\in[0,T]$, $\iota_1,\iota_2\in\{+,-\}$, $h\in\{h_{\alpha\beta}\}$, and $k,k_1,k_2\in\mathbb{Z}$. Then*

$$\begin{aligned}I^{kg}_{k,k_1,k_2}(t)&\lesssim\varepsilon_1^2 2^{-k_1/2}2^{3\min\{k,k_1,k_2\}/2}2^{-N(n_1)k_1^+-N(n_2)k_2^++k_2^+}\\&\times\langle t\rangle^{H(q_1,n_1)\delta}(\langle t\rangle 2^{k_1^-})^{-\gamma}\min\big\{\langle t\rangle^{H(q_2,n_2)\delta},2^{k_2^-}\langle t\rangle^{H(q_2+1,n_2+1)\delta}\big\},\end{aligned}\tag{4.1.52}$$

*where the second factor in the right-hand side is, by definition, $\langle t\rangle^{H(q_2,n_2)\delta}$ when $n_2=3$. In addition, we have:*

*(1) if $n_1=0$ and $n_2\geq 0$ then*

$$I^{kg}_{k,k_1,k_2}(t)\lesssim\varepsilon_1^2\langle t\rangle^{-1+\delta[H(q_2,n_2)+33]}2^{-N(n_2)k^++5k^+/4}2^{-2\min\{k_1^+,k_2^+\}};\tag{4.1.53}$$

*(2) if $n_1=n_2=0$ then*

$$I^{kg}_{k,k_1,k_2}(t)\lesssim\begin{cases}\varepsilon_1^2\langle t\rangle^{-1+4\delta}2^{-2k_2^+}2^{-N(0)k_1^+-k_1^+/4} & \text{if } k_1\geq k_2;\\ \varepsilon_1^2\langle t\rangle^{-1+4\delta}2^{-2k_1^+}2^{-N_0k_2^++6k_2^+} & \text{if } k_1\leq k_2;\end{cases}\tag{4.1.54}$$

*(3) if $n_1\geq 1$ and $n_2=0$ then*

$$I^{kg}_{k,k_1,k_2}(t)\lesssim\varepsilon_1^2\langle t\rangle^{-1+\delta(H(q_1,n_1)+\ell(q_1,n_1))}2^{-N(n_1)k^+-k^+/4}2^{-2\min\{k_1^+,k_2^+\}};\tag{4.1.55}$$

*(4) if $n_1\geq 1$ and $n_2\geq 1$ then*

$$I^{kg}_{k,k_1,k_2}(t)\lesssim\varepsilon_1^2\langle t\rangle^{-1+\delta(H(q_1+q_2,n_1+n_2)+33)}2^{-N(n_1+n_2)k^+-k^+/4}2^{-2\min\{k_1^+,k_2^+\}}.\tag{4.1.56}$$

*Proof.* We notice that bilinear Wave × KG interactions appear in the nonlinearities $\mathcal{N}^\psi$, in quasilinear form. As before, ideally we would like to have bounds on the form

$$I^{kg}_{k,k_1,k_2}(t)\lesssim\varepsilon_1^2\min(2^k,\langle t\rangle^{-1})2^{-N(n_1+n_2)k^++k^+},\tag{4.1.57}$$

but we are only able to prove variations of these bounds, with small losses.

The estimates (4.1.52) follow using the $L^2$ bounds (3.2.44), (3.3.3), (3.3.5).

**Step 1.** We observe that we have the following general bounds, which follow from (3.3.3), (3.3.11), and (3.3.13): if $n_1 \leq 2$ then

$$\begin{aligned} I^{kg}_{k,k_1,k_2}(t) &\lesssim 2^{k_2^+-k_1} \|P_{k_1} U^{\mathcal{L}_1 h}(t)\|_{L^\infty} \|P_{k_2} U^{\mathcal{L}_2 \psi}(t)\|_{L^2} \\ &\lesssim \varepsilon_1^2 \langle t\rangle^{-1+\delta(H(q_1+1,n_1+1)+H(q_2,n_2))} 2^{-N(n_2)k_2^+ + k_2^+} 2^{-N(n_1+1)k_1^+ + k_1^+}. \end{aligned} \tag{4.1.58}$$

Similarly, if $n_2 \leq 2$ and $2^{k_1} \gtrsim \langle t\rangle^{-1}$ then

$$\begin{aligned} I^{kg}_{k,k_1,k_2}(t) &\lesssim 2^{k_2^+-k_1} \|P_{k_1} U^{\mathcal{L}_1 h}(t)\|_{L^2} \|P_{k_2} U^{\mathcal{L}_2 \psi}(t)\|_{L^\infty} \\ &\lesssim \varepsilon_1^2 \langle t\rangle^{-1+\delta(H(q_1,n_1)+H(q_2+1,n_2+1))} 2^{-k_1/2} 2^{k_2^-/2} 2^{-N(n_1)k_1^+} 2^{-N(n_2+1)k_2^+ + 3k_2^+}. \end{aligned} \tag{4.1.59}$$

Since $H(1,1) = 30$, the bounds (4.1.53) follow from (4.1.58) if $n_2 \geq 1$. They also follow from (4.1.58) if $n_2 = 0$ and $k_1 \leq k_2$ and from the stronger bounds (4.1.54) if $n_2 = 0$ and $k_1 \geq k_2$.

**Step 2.** Assume now that $n_2 = 0$ and we prove the bounds (4.1.54)–(4.1.55). We have

$$\begin{aligned} I^{kg}_{k,k_1,k_2}(t) &\lesssim 2^{k_2^+-k_1} 2^{3\min\{k_1,k_2\}/2} \|P_{k_1} U^{\mathcal{L}_1 h}(t)\|_{L^2} \|P_{k_2} U^{\psi}(t)\|_{L^2} \\ &\lesssim \varepsilon_1^2 \langle t\rangle^{\delta H(q_1,n_1)} 2^{\min\{k_1^-,k_2^-\}} 2^{k_2^- + \kappa k_2^-} (\langle t\rangle 2^{k_1^-})^{-\gamma} 2^{-N(n_1)k_1^+ - k_1^+/2} 2^{-N_0 k_2^+ + 5k_2^+}, \end{aligned} \tag{4.1.60}$$

using $L^2$ estimates and the inequality $2^{3\min\{k_1,k_2\}/2} \lesssim 2^{k_1^-/2} 2^{\min\{k_1^-,k_2^-\}} 2^{3k_2^+/2}$. This suffices to prove (4.1.54)–(4.1.55) if $2^{\min\{k_1^-,k_2^-\}} 2^{k_2^-} \lesssim \langle t\rangle^{-1}$.

Assume now that $\langle t\rangle^{-1} \leq 2^{\min\{k_1^-,k_2^-\}} 2^{k_2^- - 40}$. Let $J_2$ denote the largest integer such that $2^{J_2} \leq 2^{k_2^- - 20}\langle t\rangle$. Using (3.3.17) and (3.3.3) we estimate

$$\begin{aligned} &2^{k_2^+-k_1} \big\|P_k I[P_{k_1} U^{\mathcal{L}_1 h,\iota_1}, U^{\psi,\iota_2}_{\leq J_2,k_2}](t)\big\|_{L^2} \\ &\lesssim 2^{k_2^+-k_1} \|P_{k_1} U^{\mathcal{L}_1 h}(t)\|_{L^2} \|U^{\psi}_{\leq J_2,k_2}(t)\|_{L^\infty} \\ &\lesssim \varepsilon_1^2 \langle t\rangle^{-3/2+\delta H(q_1,n_1)} 2^{-k_1^-/2} 2^{-k_2^-/2} 2^{-N(n_1)k_1^+ - k_1^+/2} 2^{-N_0 k_2^+ + 6k_2^+} \\ &\lesssim \varepsilon_1^2 \langle t\rangle^{-1+\delta H(q_1,n_1)} 2^{-N(n_1)k_1^+ - k_1^+/2} 2^{-N_0 k_2^+ + 6k_2^+}. \end{aligned} \tag{4.1.61}$$

Using also (3.3.4) and $L^2$ estimates, we also have, with $\underline{k} = \min\{k, k_1, k_2\}$,

$$\begin{aligned} 2^{k_2^+-k_1} &\big\|P_k I[P_{k_1} U^{\mathcal{L}_1 h,\iota_1}, U^{\psi,\iota_2}_{>J_2,k_2}](t)\big\|_{L^2} \\ &\lesssim 2^{k_2^+-k_1} 2^{3\underline{k}/2} \|P_{k_1} U^{\mathcal{L}_1 h}(t)\|_{L^2} \|U^{\psi}_{>J_2,k_2}(t)\|_{L^2} \\ &\lesssim \varepsilon_1^2 \langle t\rangle^{-1+\delta(H(q_1,n_1)+30)} 2^{-N(n_1)k_1^+ - k_1^+/2} 2^{-N(1)k_2^+ + 4k_2^+}. \end{aligned} \tag{4.1.62}$$

The last two bounds suffice to prove (4.1.55) when $n_1 \geq 2$.

On the other hand, if $n_1 \leq 1$ then we can also estimate, using (3.3.11),

$$\begin{aligned}2^{k_2^+-k_1}&\|P_k I[P_{k_1}U^{\mathcal{L}_1 h,\iota_1}, U^{\psi,\iota_2}_{>J_2,k_2}](t)\|_{L^2}\\&\lesssim 2^{k_2^+-k_1}\|P_{k_1}U^{\mathcal{L}_1 h}(t)\|_{L^\infty}\|U^{\psi}_{>J_2,k_2}(t)\|_{L^2}\\&\lesssim \varepsilon_1^2\langle t\rangle^{-2+\delta'}2^{-k_2^-}2^{-N(n_1+1)k_1^++2k_1^+}2^{-N(1)k_2^++2k_2^+}.\end{aligned}\tag{4.1.63}$$

The bounds (4.1.54) and (4.1.55) with $n_1 = 1$ follow from (4.1.61) and (4.1.63) if $2^{k_1^+} + 2^{k_2^+} \lesssim \langle t\rangle^{4\delta'}$. On the other hand, if $k_1 \geq k_2$ and $2^{k_1^+} \geq \langle t\rangle^{4\delta'}$ then the bounds (4.1.54) and (4.1.55) with $n_1 = 1$ follow from (4.1.61) and (4.1.62). Finally, if $k_1 \leq k_2$ and $2^{k_2^+} \geq \langle t\rangle^{4\delta'}$ then the bounds (4.1.54) and (4.1.55) with $n_1 = 1$ follow from (4.1.58).

This completes the proof of the bounds (4.1.54) and (4.1.55) in all cases.

**Step 3.** We prove now the bounds (4.1.56) when $k_2 \leq k_1$. The bounds (4.1.59) give the desired conclusion if $2^{k_1} \gtrsim \langle t\rangle^{\delta'}$. Also, the bounds (4.1.52) give the desired conclusion if $2^{k_1} \lesssim \langle t\rangle^{-1/2-\delta'}$.

In the remaining case

$$k_2 \leq k_1, \qquad 2^{k_1} \in [\langle t\rangle^{-1/2-\delta'}, \langle t\rangle^{\delta'}],\tag{4.1.64}$$

we decompose $P_{k_2}U^{\mathcal{L}_2\psi,\iota_2} = \sum_{j_2\geq\max(-k_2,0)} U^{\mathcal{L}_2\psi,\iota_2}_{j_2,k_2}$ as in (4.1.39). Then we estimate

$$\begin{aligned}2^{k_2^+-k_1}&\|P_k I[P_{k_1}U^{\mathcal{L}_1 h,\iota_1}(t), U^{\mathcal{L}_2\psi,\iota_2}_{j_2,k_2}(t)]\|_{L^2}\\&\lesssim 2^{k_2^+-k_1}\|P_{k_1}U^{\mathcal{L}_1 h}(t)\|_{L^2}\|U^{\mathcal{L}_2\psi}_{j_2,k_2}(t)\|_{L^\infty} \lesssim \varepsilon_1^2\langle t\rangle^{-3/2+2\delta'}2^{j_2/2}2^{-k_1/2},\end{aligned}$$

using (3.3.3) and (3.3.14). This suffices to estimate the contribution of the localized profiles for which $2^{j_2} \lesssim \langle t\rangle^{1/3}$. On the other hand, using (3.3.12) and (3.3.4) we also estimate

$$\begin{aligned}2^{k_2^+-k_1}&\|P_k I[P_{k_1}U^{\mathcal{L}_1 h,\iota_1}(t), U^{\mathcal{L}_2\psi,\iota_2}_{j_2,k_2}(t)]\|_{L^2}\\&\lesssim 2^{k_2^+-k_1}\|P_{k_1}U^{\mathcal{L}_1 h}(t)\|_{L^\infty}\|U^{\mathcal{L}_2\psi}_{j_2,k_2}(t)\|_{L^2} \lesssim \varepsilon_1^2\langle t\rangle^{-1+2\delta'}2^{-j_2},\end{aligned}$$

which suffices to estimate the contribution of the localized profiles for which $2^{j_2} \geq \langle t\rangle^{1/3}$.

**Step 4.** Finally, we prove the bounds (4.1.56) when $k_1 \leq k_2$. The bounds (4.1.58) give the desired conclusion if $2^{k_2} \gtrsim \langle t\rangle^{\delta'}$. Also, the bounds (4.1.52) and (2.1.53) give the desired conclusion if $2^{k_1} \lesssim \langle t\rangle^{-1+40\delta}$.

If $q_2 \geq 1$, the bounds (4.1.56) follow from (4.1.58) and the last inequality in (4.1.3). If $q_2 = 0$ then we let $J_1$ be the largest integer such that $2^{J_1} \leq \langle t\rangle 2^{-30}$ and decompose $P_{k_1}U^{\mathcal{L}_1 h,\iota_1}(t)$ as in (4.1.14). We also decompose $P_{k_2}U^{\mathcal{L}_2 h,\iota_2}(t) = U^{\mathcal{L}_2 h,\iota_2}_{\leq J_2,k_2}(t) + U^{\mathcal{L}_2 h,\iota_2}_{>J_2,k_2}(t)$, where $J_2$ is the largest integer such that $2^{J_2} \leq \langle t\rangle^{4\delta'}$.

Then

$$
\begin{aligned}
&2^{k_2^+-k_1}\big\|P_kI[U^{\mathcal{L}_1h,\iota_1}_{\leq J_1,k_1}(t),P_{k_2}U^{\mathcal{L}_2\psi,\iota_2}(t)]\big\|_{L^2}\\
&\qquad\lesssim 2^{k_2^+-k_1}\big\|U^{\mathcal{L}_1h}_{\leq J_1,k_1}(t)\big\|_{L^\infty}\|P_{k_2}U^{\mathcal{L}_2\psi}(t)\|_{L^2}\\
&\qquad\lesssim \varepsilon_1^2\langle t\rangle^{-1+\delta[H(0,n_2)+H(q_1,n_1+1)+1]}2^{-N(n_2)k_2^++k_2^+}2^{-5k_1^+},
\end{aligned}
$$

using (3.3.3) and (3.3.12). This suffices due to the first inequality in (4.1.3). Also,

$$
\begin{aligned}
2^{k_2^+-k_1}\big\|P_kI[U^{\mathcal{L}_1h,\iota_1}_{>J_1,k_1}(t),U^{\mathcal{L}_2\psi,\iota_2}_{>J_2,k_2}(t)]\big\|_{L^2}&\lesssim 2^{k_2^+-k_1}\big\|U^{\mathcal{L}_1h}_{>J_1,k_1}(t)\big\|_{L^\infty}\|U^{\mathcal{L}_2\psi}_{>J_2,k_2}(t)\|_{L^2}\\
&\lesssim \varepsilon_1^2\langle t\rangle^{-1+\delta[H(1,n_2+1)+H(q_1+1,n_1+1)+1]}2^{-J_2}2^{-N(n_2+1)k_2^++k_2^+}2^{-5k_1^+},
\end{aligned}
$$

using (3.3.4) and (3.3.11). For (4.1.56) it remains to prove that

$$
\begin{aligned}
&2^{k_2^+-k_1}\big\|P_kI[U^{\mathcal{L}_1h,\iota_1}_{>J_1,k_1}(t),U^{\mathcal{L}_2\psi,\iota_2}_{\leq J_2,k_2}(t)]\big\|_{L^2}\\
&\qquad\lesssim \varepsilon_1^2\langle t\rangle^{-1+\delta(H(q_1,n_1+n_2)+33)}2^{-N(n_1+n_2)k^+-k^+/4}2^{-2k_1^+},
\end{aligned}
\tag{4.1.65}
$$

provided that $|t|\gg 1$ and

$$
q_2=0,\qquad \langle t\rangle^{-1+40\delta}\leq 2^{k_1}\leq 2^{k_2}\leq\langle t\rangle^{\delta'},\qquad 2^{J_2}\leq\langle t\rangle^{4\delta'}.\tag{4.1.66}
$$

Let $X:=2^{k_2^+-k_1}\big\|P_kI[U^{\mathcal{L}_1h,\iota_1}_{>J_1,k_1}(t),U^{\mathcal{L}_2\psi,\iota_2}_{\leq J_2,k_2}(t)]\big\|_{L^2}$ denote the expression in the left-hand side of (4.1.65).

**Case 4.1.** Assume first that $q_1=0$. We use the bounds in Lemma 3.18. If $n_1=1$ and $n_2=1$ then we use the $L^4$ bounds in (3.3.23) and (3.3.24) to estimate

$$
\begin{aligned}
X&\lesssim 2^{k_2^+-k_1}\|U^{\mathcal{L}_1h}_{>J_1,k_1}(t)\|_{L^4}\|U^{\mathcal{L}_2\psi}_{\leq J_2,k_2}(t)\|_{L^4}\\
&\lesssim \varepsilon_1^2\langle t\rangle^{-5/4+110\delta}2^{-k_1^-/4}2^{-k_2^-/4}2^{-N(2)k_2^+-2k_2^+}2^{-4k_1^+}.
\end{aligned}
$$

Alternatively, we could use the $L^2$ bounds in (3.3.21)–(3.3.22) to estimate

$$
\begin{aligned}
X&\lesssim 2^{k_2^+-k_1}2^{3k_1/2}\|U^{\mathcal{L}_1h}_{>J_1,k_1}(t)\|_{L^2}\|U^{\mathcal{L}_2\psi}_{\leq J_2,k_2}(t)\|_{L^2}\\
&\lesssim \varepsilon_1^2\langle t\rangle^{-2/3+110\delta}2^{k_1^-/3}2^{k_2^-}2^{-N(1)k_2^+}2^{-4k_1^+}.
\end{aligned}
$$

We use the first estimates if $\langle t\rangle 2^{k_1^-}2^{k_2^-}\geq 1$ and we use the second estimates if $\langle t\rangle 2^{k_1^-}2^{k_2^-}\leq 1$. The desired bounds (4.1.65) follow if $n_1=n_2=1$.

Similarly, if $n_1=1$ and $n_2=2$ then we use the $L^6$ bounds in (3.3.23) and the $L^3$ bounds in (3.3.24) to estimate

$$X \lesssim 2^{k_2^+-k_1}\|U^{\mathcal{L}_1 h}_{>J_1,k_1}(t)\|_{L^6}\|U^{\mathcal{L}_2\psi}_{\le J_2,k_2}(t)\|_{L^3}$$
$$\lesssim \varepsilon_1^2\langle t\rangle^{-7/6+170\delta}2^{-k_1^-/6}2^{-k_2^-/6}2^{-N(3)k_2^+-2k_2^+}2^{-4k_1^+}.$$

Alternatively, using the $L^2$ bounds in (3.3.21)–(3.3.22) we estimate

$$X \lesssim 2^{k_2^+-k_1}2^{3k_1/2}\|U^{\mathcal{L}_1 h}_{>J_1,k_1}(t)\|_{L^2}\|U^{\mathcal{L}_2\psi}_{\le J_2,k_2}(t)\|_{L^2}$$
$$\lesssim \varepsilon_1^2\langle t\rangle^{-2/3+170\delta}2^{k_1^-/3}2^{k_2^-/3}2^{-N(2)k_2^+}2^{-4k_1^+}.$$

As before, the desired bounds (4.1.65) follow if $n_1 = 1$ and $n_2 = 2$ from these two estimates.

Finally, if $n_1 = 2$ and $n_2 = 1$ then we use the $L^3$ bounds in (3.3.23) and the $L^6$ bounds in (3.3.24) to estimate

$$X \lesssim 2^{k_2^+-k_1}\|U^{\mathcal{L}_1 h}_{>J_1,k_1}(t)\|_{L^3}\|U^{\mathcal{L}_2\psi}_{\le J_2,k_2}(t)\|_{L^6}$$
$$\lesssim \varepsilon_1^2\langle t\rangle^{-4/3+160\delta}2^{-k_1^-/3}2^{-k_2^-/3}2^{-N(3)k_2^+-2k_2^+}2^{-4k_1^+}.$$

Using the $L^2$ bounds in (3.3.21)–(3.3.22) we also estimate

$$X \lesssim 2^{k_2^+-k_1}2^{3k_1/2}\|U^{\mathcal{L}_1 h}_{>J_1,k_1}(t)\|_{L^2}\|U^{\mathcal{L}_2\psi}_{\le J_2,k_2}(t)\|_{L^2}$$
$$\lesssim \varepsilon_1^2\langle t\rangle^{-1/3+160\delta}2^{2k_1^-/3}2^{k_2^-}2^{-N(2)k_2^+}2^{-4k_1^+}.$$

The desired bounds (4.1.65) follow if $n_1 = 2$ and $n_2 = 1$ from these estimates.

**Case 4.2.** Assume now that $q_1 \ge 1$ and $q_2 = 0$. Recall that

$$\begin{aligned}&\|U^{\mathcal{L}_1 h}_{>J_1,k_1}(t)\|_{L^2} \lesssim \varepsilon_1 2^{k_1/2}\min\{\langle t\rangle^{H(q_1,n_1)\delta}, \langle t\rangle^{-1}2^{-k_1}\langle t\rangle^{H(q_1+1,n_1+1)\delta}\}2^{-5k_1^+},\\ &\|U^{\mathcal{L}_2\psi}_{\le J_2,k_2}(t)\|_{L^2} \lesssim \varepsilon_1 2^{k_2^-}\langle t\rangle^{H(1,n_2)\delta}2^{-N(n_2)k_2^+},\end{aligned} \tag{4.1.67}$$

as a consequence of (3.3.3)–(3.3.5). Using just $L^2$ estimates we have

$$X \lesssim 2^{k_2^++k_1/2}\|U^{\mathcal{L}_1 h}_{>J_1,k_1}(t)\|_{L^2}\|U^{\mathcal{L}_2\psi}_{\le J_2,k_2}(t)\|_{L^2}$$
$$\lesssim \varepsilon_1^2 2^{k_1^-+k_2^-}\langle t\rangle^{\delta(H(q_1,n_1)+H(1,n_2))}2^{-N(n_2)k_2^++2k_2^+}2^{-4k_1^+}.$$

Since $H(1,n_2) + H(q_1,n_1) \le H(q_1,n_1+n_2) - 150$ (due to the last inequality in (4.1.3) and the assumption $q_1 \ge 1$), this suffices to prove (4.1.65) when $2^{k_1^-+k_2^-} \lesssim \langle t\rangle^{-1+180\delta}$.

Finally, assume that $2^{k_1^-+k_2^-} \ge \langle t\rangle^{-1+180\delta}$. If $n_2 = 1$ then we estimate, using

(4.1.67), Sobolev embedding, and the $L^6$ bounds in (3.3.24),

$$\begin{aligned}X &\lesssim 2^{k_2^+-k_1}\|U^{\mathcal{L}_1 h}_{>J_1,k_1}(t)\|_{L^3}\|U^{\mathcal{L}_2\psi}_{\leq J_2,k_2}(t)\|_{L^6}\\ &\lesssim \varepsilon_1^2\min\{\langle t\rangle^{H(q_1,n_1)\delta},\langle t\rangle^{-1}2^{-k_1}\langle t\rangle^{H(q_1+1,n_1+1)\delta}\}\\ &\quad\times 2^{-5k_1^+}\cdot\langle t\rangle^{-1+50\delta}2^{-k_2^-/3}2^{-N(2)k_2^+-k_2^+}\\ &\lesssim \varepsilon_1^2 2^{-k_2^-/3}\langle t\rangle^{-1/3}2^{-k_1^-/3}\langle t\rangle^{\frac{H(q_1+1,n_1+1)+2H(q_1,n_1)}{3}\delta}2^{-4k_1^+}\langle t\rangle^{-1+50\delta}2^{-N(2)k_2^+-k_2^+}.\end{aligned}$$

Since $H(q_1+1,n_1+1)+2H(q_1,n_1)\leq 3H(q_1,n_1+1)$ (see (2.1.49)), this suffices to prove (4.1.65) if $n_2=1$.

On the other hand, if $n_2=2$ (so necessarily $(q_1,n_1)=(1,1)$) we estimate

$$\begin{aligned}X &\lesssim 2^{k_2^+-k_1}\|U^{\mathcal{L}_1 h}_{>J_1,k_1}(t)\|_{L^6}\|U^{\mathcal{L}_2\psi}_{\leq J_2,k_2}(t)\|_{L^3}\\ &\lesssim \varepsilon_1^2 2^{k_1/2}\min\{\langle t\rangle^{H(q_1,n_1)\delta},\langle t\rangle^{-1}2^{-k_1}\langle t\rangle^{H(q_1+1,n_1+1)\delta}\}\\ &\quad\times 2^{-5k_1^+}\cdot\langle t\rangle^{-1/2+100\delta}2^{-k_2^-/6}2^{-N(3)k_2^+-k_2^+}\\ &\lesssim \varepsilon_1^2 2^{-k_2^-/6}\langle t\rangle^{-2/3}2^{-k_1^-/6}\langle t\rangle^{\frac{2H(q_1+1,n_1+1)+H(q_1,n_1)}{3}\delta}\\ &\quad\times 2^{-4k_1^+}\langle t\rangle^{-1/2+100\delta}2^{-N(3)k_2^+-k_2^+},\end{aligned}$$

using (4.1.67) and the $L^3$ bounds in (3.3.24). Since $[2H(2,2)+H(1,1)]/3\leq H(1,3)-100$ (see (2.1.49)), this suffices to prove (4.1.65) if $n_2=2$. This completes the proof of the lemma. □

### 4.1.2 The Classes of Functions $\mathcal{G}_a$

In most cases, the cubic and higher order nonlinearities can be treated perturbatively, and do not play a significant role in the analysis. To justify this, we need good bounds on the quadratic metric components $g^{\alpha\beta}_{\geq 2}$.

The metric components $g^{\alpha\beta}_{\geq 2}$ satisfy the identities (2.1.8). Therefore they can be represented as infinite sums of monomials of degree $\geq 2$ in the functions $h_{\mu\nu}$. More generally, for integers $a\geq 1$ we define the sets

$$\mathcal{G}_a:=\{G_a=h_1\cdot\ldots\cdot h_a: h_1,\ldots,h_a\in\{h_{\mu\nu}\}\}. \tag{4.1.68}$$

By convention, set $\mathcal{G}_0:=\{1\}$. In this subsection we prove the following bounds:

**Lemma 4.4.** *Assume $\mathcal{L}\in\mathcal{V}^q_n$, $t\in[0,T]$, $l\in\{1,2,3\}$, and $G_a\in\mathcal{G}_a$, $a\geq 2$. Then there is a constant $C_0\geq 1$ such that*

$$\begin{aligned}\|P_k\mathcal{L}G_a(t)\|_{L^2} &\leq (C_0\varepsilon_1)^a\langle t\rangle^{-1+\delta'}2^{-N(n)k^+}2^{-k/2}\min\{1,2^{k^-}\langle t\rangle\}^{1-\delta'},\\ \|P_k\mathcal{L}G_a(t)\|_{L^\infty} &\leq (C_0\varepsilon_1)^a\langle t\rangle^{-2+\delta'}2^{-N(n+1)k^++3k^+}\min\{1,2^{k^-}\langle t\rangle\}^{2-\delta'},\\ \|P_k(x_l\mathcal{L}G_a)(t)\|_{L^2} &\leq (C_0\varepsilon_1)^a(\langle t\rangle+2^{-k^-})^{\delta'}2^{-N(n+1)k^+}2^{-k^-/2},\end{aligned} \tag{4.1.69}$$

*where* $\delta' = 2000\delta$ *(see* (2.1.42)*), and the inequality in the first line holds for all pairs* $(q,n)$ *with* $n \le 3$*, while the inequalities in the last two lines hold for pairs* $(q,n)$ *with* $n \le 2$*. Thus*

$$\begin{aligned}\|P_k\mathcal{L}g^{\alpha\beta}_{\ge 2}(t)\|_{L^2} &\lesssim \varepsilon_1^2\langle t\rangle^{-1+\delta'}2^{-N(n)k^+}2^{-k/2}\min\{1,2^{k^-}\langle t\rangle\}^{1-\delta'},\\ \|P_k\mathcal{L}g^{\alpha\beta}_{\ge 2}(t)\|_{L^\infty} &\lesssim \varepsilon_1^2\langle t\rangle^{-2+\delta'}2^{-N(n+1)k^+ +3k^+}\min\{1,2^{k^-}\langle t\rangle\}^{2-\delta'}, \quad (4.1.70)\\ \|P_k(x_l\mathcal{L}g^{\alpha\beta}_{\ge 2})(t)\|_{L^2} &\lesssim \varepsilon_1^2(\langle t\rangle + 2^{-k^-})^{\delta'}2^{-N(n+1)k^+}2^{-k^-/2}.\end{aligned}$$

*Remark 4.5.* We notice that we prove more than just frequency-localized $L^2$ bounds on the functions $\mathcal{L}g^{\alpha\beta}_{\ge 2}$. In particular, we prove weighted $L^2$ bounds that are important in our bootstrap scheme; see the key estimates (6.1.5) and (6.1.8) in Proposition 6.2.

*Proof.* The bounds (3.3.3), (3.3.11), and (2.1.47) show that

$$\begin{aligned}\|P_k\mathcal{L}h_{\alpha\beta}(t)\|_{L^2} &\lesssim \varepsilon_1\langle t\rangle^{H(q,n)\delta}2^{-N(n)k^+}2^{-k/2}(\langle t\rangle 2^{k^-})^{-\gamma},\\ \|P_k\mathcal{L}h_{\alpha\beta}(t)\|_{L^\infty} &\lesssim \varepsilon_1\langle t\rangle^{-1+H(q+1,n+1)\delta}2^{-N(n+1)k^+ +3k^+}\min\{1,2^{k^-}\langle t\rangle\}^{1-\delta},\\ \|P_k(x_l\mathcal{L}h_{\alpha\beta})(t)\|_{L^2} &\lesssim \varepsilon_1\langle t\rangle^{H(q+1,n+1)\delta}\\ &\quad\times 2^{-N(n+1)k^+}2^{-k^-/2}(2^{-k^-}+\langle t\rangle)(\langle t\rangle 2^{k^-})^{-\gamma},\end{aligned} \quad (4.1.71)$$

for any $\mathcal{L} \in \mathcal{V}^q_n$ and $k \in \mathbb{Z}$, where the first inequality holds for all pairs $(q,n) \le (3,3)$, while the last two inequalities hold for pairs $(q,n) \le (2,2)$. Indeed, for the last bound we estimate first

$$\begin{aligned}\|P_k(x_l\mathcal{L}h_{\alpha\beta})(t)\|_{L^2} \lesssim\, & 2^{-2k^-}\|P_{[k-4,k+4]}U^{\mathcal{L}h_{\alpha\beta}}(t)\|_{L^2}\\ &+ 2^{-k^-}\|\varphi_k(\xi)\partial_{\xi_l}\widehat{U^{\mathcal{L}h_{\alpha\beta}}}(\xi,t)\|_{L^2_\xi}.\end{aligned}$$

Then we recall that $\widehat{U^{\mathcal{L}h_{\alpha\beta}}}(\xi,t) = e^{-it|\xi|}\widehat{V^{\mathcal{L}h_{\alpha\beta}}}(\xi,t)$. Therefore, the right-hand side of the inequality above is bounded by

$$C2^{-k^-}(2^{-k^-}+\langle t\rangle)\|P_{[k-4,k+4]}U^{\mathcal{L}h_{\alpha\beta}}(t)\|_{L^2} + C2^{-k^-}\|\varphi_k(\xi)\partial_{\xi_l}\widehat{V^{\mathcal{L}h_{\alpha\beta}}}(\xi,t)\|_{L^2_\xi}.$$

The estimate in the last line of (4.1.71) follows from (3.3.3) and (3.3.9).

**Step 1.** We consider first the case $a = 2$. Assume $\mathcal{L}_1 \in \mathcal{V}^{q_1}_{n_1}$, $\mathcal{L}_2 \in \mathcal{V}^{q_2}_{n_2}$, $(q_1,n_1)+(q_2,n_2) \le (q,n)$. Assume also $h_1,h_2 \in \{h_{\mu\nu}\}$. If $2^{k^-} \le \langle t\rangle^{-1}$ then we bound, using (4.1.71) and (3.2.44),

$$
\begin{aligned}
&\|P_k(\mathcal{L}_1 h_1 \cdot \mathcal{L}_2 h_2)(t)\|_{L^2} \\
&\quad \lesssim \sum_{(k_1,k_2)\in\mathcal{X}_k} 2^{3\min(k,k_1,k_2)/2} \|P_{k_1}\mathcal{L}_1 h_1(t)\|_{L^2} \|P_{k_2}\mathcal{L}_2 h_2(t)\|_{L^2} \\
&\quad \lesssim \varepsilon_1^2 \langle t\rangle^{\delta'/2} 2^{k^-/2-\delta k^-},
\end{aligned} \tag{4.1.72}
$$

where $\mathcal{X}_k$ are as in (2.1.21). Moreover, if $n_1, n_2 \leq 2$ and $2^{k^-} \geq \langle t\rangle^{-1}$ then

$$
\begin{aligned}
\|P_k(\mathcal{L}_1 h_1 \cdot \mathcal{L}_2 h_2)(t)\|_{L^\infty} &\lesssim \sum_{(k_1,k_2)\in\mathcal{X}_k} \|P_{k_1}\mathcal{L}_1 h_1(t)\|_{L^\infty} \|P_{k_2}\mathcal{L}_2 h_2(t)\|_{L^\infty} \\
&\lesssim \varepsilon_1^2 \langle t\rangle^{-2+\delta'} 2^{-N(n+1)k^+ + 3k^+}.
\end{aligned} \tag{4.1.73}
$$

To estimate $\|P_k(x_l\mathcal{L}_1 h_1 \cdot \mathcal{L}_2 h_2)(t)\|_{L^2}$ we combine the factor $x_l$ with the higher frequency term and estimate it in $L^2$, and estimate the lower frequency term in $L^\infty$. Using (4.1.71) as before, it follows that, for $l \in \{1,2,3\}$,

$$
\begin{aligned}
&\|P_k(x_l\mathcal{L}_1 h_1 \cdot \mathcal{L}_2 h_2)(t)\|_{L^2} \\
&\quad \lesssim \varepsilon_1^2 \langle t\rangle^{-1+\delta'} 2^{-N(n+1)k^+} 2^{-k^-/2} \min(1, 2^{k^-}\langle t\rangle)^{1-\delta'} \cdot (2^{-k^-} + \langle t\rangle).
\end{aligned} \tag{4.1.74}
$$

The $L^\infty$ bounds in (4.1.69) follow from (4.1.73) if $2^{k^-}\langle t\rangle \gtrsim 1$ and from (4.1.72) if $2^{k^-}\langle t\rangle \lesssim 1$. The weighted $L^2$ bounds in the last line of (4.1.69) follow from (4.1.74). The $L^2$ bounds in the first line of (4.1.69) follow from (4.1.72) if $2^{k^-}\langle t\rangle \lesssim 1$. It remains to prove that

$$
\|P_k(\mathcal{L}_1 h_1 \cdot \mathcal{L}_2 h_2)(t)\|_{L^2} \lesssim \varepsilon_1^2 \langle t\rangle^{-1+\delta'} 2^{-k/2} 2^{-N(n)k^+}, \tag{4.1.75}
$$

for any $k \in \mathbb{Z}$ with $2^k \geq \langle t\rangle^{-1}$, and any $\mathcal{L}_1 \in \mathcal{V}_{n_1}^{q_1}$, $\mathcal{L}_2 \in \mathcal{V}_{n_2}^{q_2}$ with $(q_1, n_1) + (q_2, n_2) \leq (q, n)$.

To prove (4.1.75), we assume first that $n_1, n_2 \leq 2$ and estimate

$$
\|P_k(\mathcal{L}_1 h_1 \cdot \mathcal{L}_2 h_2)(t)\|_{L^2} \leq S_1 + S_2 + S_3, \tag{4.1.76}
$$

where, using (3.3.11) and the inequality in the first line of (4.1.71),

$$
\begin{aligned}
S_1 := &\sum_{k_1 \leq k-6,\, |k_2-k|\leq 4} \|P_{k_1}\mathcal{L}_1 h_1(t)\|_{L^\infty} \|P_{k_2}\mathcal{L}_2 h_2(t)\|_{L^2} \\
&\lesssim \varepsilon_1^2 \langle t\rangle^{-1+\delta'/2} 2^{-k/2} 2^{-N(n_2)k^+},
\end{aligned}
$$

$$\begin{aligned}S_2 &:= \sum_{k_2\le k-6,\,|k_1-k|\le 4}\|P_{k_1}\mathcal{L}_1h_1(t)\|_{L^2}\|P_{k_2}\mathcal{L}_2h_2(t)\|_{L^\infty}\\&\lesssim \varepsilon_1^2\langle t\rangle^{-1+\delta'/2}2^{-k/2}2^{-N(n_1)k^+}\end{aligned}$$

and

$$\begin{aligned}S_3 &:= \sum_{k_1,k_2\ge k-6,\,|k_1-k_2|\le 10}\|P_{k_1}\mathcal{L}_1h_1(t)\|_{L^\infty}\|P_{k_2}\mathcal{L}_2h_2(t)\|_{L^2}\\&\lesssim \varepsilon_1^2\langle t\rangle^{-1+\delta'/2}2^{-k/2}2^{-N(n_2)k^+}.\end{aligned}$$

These bounds clearly suffice to prove (4.1.75).

On the other hand if $\max(n_1,n_2)=3$, then we may assume that $n_2=3$ and $n_1=0$. The bounds on $S_1$ and $S_3$ above still hold, but the bounds on $S_2$ fail, because we do not have suitable $L^\infty$ bounds on $P_{k_2}\mathcal{L}_2h_2(t)$. However, we can still use $L^2$ bounds as in (4.1.72) to control the contribution of small frequencies $k_2$, i.e., $2^{k_2}\lesssim\langle t\rangle^{-1}$. For (4.1.75) it remains to show that

$$\|P_k(P_{k_1}h_1\cdot P_{k_2}\mathcal{L}_2h_2)(t)\|_{L^2}\lesssim \varepsilon_1^2\langle t\rangle^{-1+2\delta'/3}2^{-k/2}2^{-N(3)k^+-k^+}, \tag{4.1.77}$$

provided that $|k_1-k|\le 4$, $\langle t\rangle^{-1}\le 2^{k_2}$, $k_2\le k-6$, and $2^k\le\langle t\rangle^{1/10}$.

To prove (4.1.77) we would like to use Lemma 3.11 (i) (the simple idea of directly estimating $P_{k_1}h_1$ in $L^\infty$ and $P_{k_2}\mathcal{L}_2h_2$ in $L^2$ does not work when $k_2$ is small, due to the factor $2^{-k_2/2}$ coming from (4.1.71)). For this we write first

$$h_1(t)=-|\nabla|^{-1}\Im(U^{h_1}(t))=-|\nabla|^{-1}\Im(e^{-it\Lambda_{wa}}V^{h_1}(t));$$

see (2.1.34)). Then we decompose $P_{k_1}V^{h_1,\iota_1}=V^{h_1,\iota_1}_{\le J_1,k_1}+V^{h_1,\iota_1}_{>J_1,k_1}$, where $J_1$ is the largest integer such that $2^{J_1}\le\langle t\rangle(1+2^{k_1}\langle t\rangle)^{-\delta}$ and $\iota_1\in\{+,-\}$. With $I$ defined as in (3.2.41)–(3.2.43), we estimate

$$\begin{aligned}&\|P_kI[e^{-it\Lambda_{wa,\iota_1}}V^{h_1,\iota_1}_{\le J_1,k_1}(t),P_{k_2}\mathcal{L}_2h_2)(t)]\|_{L^2}\\&\quad\lesssim 2^{k_2/2}\langle t\rangle^{-1}(2^{k_1}\langle t\rangle)^{\delta}\|P_{k_2}\mathcal{L}_2h_2(t)\|_{L^2}\|P_{k_1}V^{h_1}\|_{H^{0,1}_\Omega}\\&\quad\lesssim \varepsilon_1^2\langle t\rangle^{-1+\delta'/2}2^{k_1/2}2^{-N(1)k_1^++2k_1^+},\end{aligned}$$

for $\iota_1\in\{+,-\}$, using (3.2.54), (4.1.71), and (3.3.3). We also estimate

$$\begin{aligned}&\|P_kI[e^{-it\Lambda_{wa,\iota_1}}V^{h_1,\iota_1}_{>J_1,k_1}(t),P_{k_2}\mathcal{L}_2h_2)(t)]\|_{L^2}\\&\quad\lesssim 2^{3k_2/2}\|P_{k_2}\mathcal{L}_2h_2(t)\|_{L^2}\|Q_{>J_1,k_1}V^{h_1}(t)\|_{L^2}\\&\quad\lesssim \varepsilon_1^2\langle t\rangle^{-1+\delta'/2}2^{k_1/2}2^{-N(1)k_1^++2k_1^+},\end{aligned}$$

using (4.1.71) and (3.3.4). The estimates (4.1.77) follow from these two bounds, which completes the proof of (4.1.75).

**Step 2.** We prove now the bounds (4.1.69) for $a\ge 3$, by induction over $a$.

The last two bounds in (4.1.69) follow as in **Step 1**, since the bounds satisfied by $(C\varepsilon_1)^{-a}G_a$ in (4.1.69) are stronger than the bounds satisfied by $\varepsilon_1^{-1}h_{\alpha\beta}$ in (4.1.71). As before, it remains to prove that if $h \in \{h_{\alpha\beta}\}$ and $G \in \mathcal{G}_a$ then

$$\|P_k(\mathcal{L}_1 h \cdot \mathcal{L}_2 G)(t)\|_{L^2} \lesssim (C_0\varepsilon_1)^a \varepsilon_1 \langle t\rangle^{-1+\delta'/2} 2^{-k/2} 2^{-N(n)k^+}, \tag{4.1.78}$$

for any $k \in \mathbb{Z}$ with $2^k \geq \langle t\rangle^{-1}$, and any $\mathcal{L}_1 \in \mathcal{V}^{q_1}_{n_1}$, $\mathcal{L}_2 \in \mathcal{V}^{q_2}_{n_2}$ with $(q_1, n_1) + (q_2, n_2) \leq (q, n)$.

The bounds (4.1.78) are similar to the bounds (4.1.75) if $\max\{n_1, n_2\} \leq 2$, using $L^\infty \times L^2$ estimates with the lower frequency measured in $L^\infty$. On the other hand, if $(n_1, n_2) = (0, 3)$, then we have to prove the analogue of the bounds (4.1.77), which is

$$\|P_k(P_{k_1}h \cdot P_{k_2}\mathcal{L}_2 G)(t)\|_{L^2} \lesssim \varepsilon_1 (C_0\varepsilon_1)^a \langle t\rangle^{-1} 2^{-k/2} 2^{-N(3)k^+ - k^+},$$

provided that $|k_1 - k| \leq 4$, $\langle t\rangle^{-1} \leq 2^{k_2}$, $k_2 \leq k - 6$, and $2^k \leq \langle t\rangle^{1/10}$. This is easier now, since we can just use the $L^2$ estimates $\|P_{k_2}\mathcal{L}_2 G(t)\|_{L^2} \lesssim (C_0\varepsilon_1)^a 2^{-k_2/2}\langle t\rangle^{-1+\delta'}$ from the induction hypothesis, and combine them with $L^\infty$ estimates on $P_{k_1}h$. The loss of the factor $2^{-k_2/2}$ is mitigated in this case by the gain of time decay. The proof in the case $(n_1, n_2) = (3, 0)$ is similar, which completes the proof of the lemma. □

In some estimates in sections 4.2 and 4.3 we need slightly different bounds:

**Lemma 4.6.** *Assume $\mathcal{L} \in \mathcal{V}^q_n$, $t \in [0, T]$, $k \in \mathbb{Z}$, and $G_{\geq 1} = \sum_{d\geq 1} a_d g_d$ for some functions $g_d \in \mathcal{G}_d$ (see (4.1.68)) and some coefficients $a_d \in \mathbb{R}$ with $|a_d| \leq C^d$. Then*

$$\begin{aligned}
\|P_k|\nabla|^{-1}\mathcal{L}(G_{\geq 1}\partial_\rho h)(t)\|_{L^2} &\lesssim \varepsilon_1^2 \langle t\rangle^{-1+\delta'} 2^{-N(n)k^+} 2^{-k/2} \min\{1, 2^{k^-}\langle t\rangle\}^{1-\delta'},\\
\|P_k|\nabla|^{-1}\mathcal{L}(G_{\geq 1}\partial_\rho h)(t)\|_{L^\infty} &\lesssim \varepsilon_1^2 \langle t\rangle^{-2+\delta'} 2^{-N(n+1)k^+ + 3k^+} \min\{1, 2^{k^-}\langle t\rangle\}^{2-\delta'},
\end{aligned} \tag{4.1.79}$$

*where $h \in \{h_{\alpha\beta}\}$, $\rho \in \{0, 1, 2, 3\}$, and the bounds in the second line hold only if $n \leq 2$.*

*Proof.* Notice that the bounds (4.1.79) are slightly stronger than the $L^2$ and the $L^\infty$ bounds in (4.1.69), but we do not prove weighted $L^2$ bounds. Low $\times$ High $\to$ High interactions can still be estimated in the same way, but some care is needed to estimate High $\times$ High $\to$ Low interactions, due to the factor $|\nabla|^{-1}$. More precisely, we prove that for any $k \in \mathbb{Z}$, and $\mathcal{L}_1 \in \mathcal{V}^{q_1}_{n_1}$, $\mathcal{L}_2 \in \mathcal{V}^{q_2}_{n_2}$ with

$(q_1, n_1) + (q_2, n_2) \leq (q, n)$, we have

$$\begin{aligned}
\sum_{k_1,k_2\geq k} &\|P_k(P_{k_1}\mathcal{L}_1 G_{\geq 1} \cdot P_{k_2}\partial_\rho \mathcal{L}_2 h)(t)\|_{L^2} \\
&\lesssim \varepsilon_1^2 \langle t\rangle^{-1+\delta'} 2^{-N(n)k^+} 2^{k/2} \min\{1, 2^{k^-}\langle t\rangle\}^{1-\delta'}, \\
\sum_{k_1,k_2\geq k} &\|P_k(P_{k_1}\mathcal{L}_1 G_{\geq 1} \cdot P_{k_2}\partial_\rho \mathcal{L}_2 h)(t)\|_{L^\infty} \\
&\lesssim \varepsilon_1^2 \langle t\rangle^{-2+\delta'} 2^{-N(n+1)k^+ + 3k^+} 2^k \min\{1, 2^{k^-}\langle t\rangle\}^{2-\delta'}.
\end{aligned} \tag{4.1.80}$$

These bounds follow easily when $2^k\langle t\rangle^{1-2\delta} \lesssim 1$ using just $L^2$ estimates as in (4.1.72). They also follow easily when $k \geq 0$, since there is no potential derivative loss in this case. Finally, if $2^k \in [2^{10}\langle t\rangle^{2\delta-1}, 1]$ then the contribution of the components $a_d G_d$, $d \geq 2$, in $G_{\geq 1}$ can be bounded in the same way, using (4.1.69). After these reductions, for (4.1.80) it remains to show that

$$\begin{aligned}
\sum_{k_1,k_2\geq k} 2^{-k_1}\|P_k I[P_{k_1}U^{\mathcal{L}_1 h_1,\iota_1}, P_{k_2}U^{\mathcal{L}_2 h_2,\iota_2}](t)\|_{L^2} &\lesssim \varepsilon_1^2 \langle t\rangle^{-1+\delta'} 2^{k/2}, \\
\sum_{k_1,k_2\geq k} 2^{-k_1}\|P_k I[P_{k_1}U^{\mathcal{L}_1 h_1,\iota_1}, P_{k_2}U^{\mathcal{L}_2 h_2,\iota_2}](t)\|_{L^\infty} &\lesssim \varepsilon_1^2 \langle t\rangle^{-2+\delta'} 2^{k},
\end{aligned} \tag{4.1.81}$$

provided that $h_1, h_2 \in \{h_{\alpha\beta}\}$, $\iota_1, \iota_2 \in \{+,-\}$, $I$ is as in (3.2.43), and $2^k \in [2^{10}\langle t\rangle^{2\delta-1}, 1]$.

To prove (4.1.81), by symmetry, we may assume $n_1 = \min(n_1, n_2) \leq 1$ and decompose $P_{k_1}U^{\mathcal{L}_1 h_1,\iota_1}(t) = U^{\mathcal{L}_1 h_1,\iota_1}_{\leq J_1,k_1}(t) + U^{\mathcal{L}_1 h_1,\iota_1}_{>J_1,k_1}(t)$ as in (4.1.14), where $J_1$ is the largest integer such that $2^{J_1} \leq \langle t\rangle(1 + 2^{k_1}\langle t\rangle)^{-\delta}$. The $L^2$ bounds in (4.1.81) follow using (3.2.54) to estimate the contribution of $U^{\mathcal{L}_1 h_1,\iota_1}_{\leq J_1,k_1}(t)$, and using (3.2.44) and (3.3.4) to bound the contribution of $U^{\mathcal{L}_1 h_1,\iota_1}_{>J_1,k_1}(t)$. For the $L^\infty$ bounds, it suffices to show that

$$\|P_k I[P_{k_1}U^{\mathcal{L}_1 h_1,\iota_1}, P_{k_2}U^{\mathcal{L}_2 h_2,\iota_2}](t)\|_{L^\infty} \lesssim \varepsilon_1^2 2^{k_1} 2^k \langle t\rangle^{-2+3\delta'/4} 2^{-2k_1^+} 2^{-2k_2^+}, \tag{4.1.82}$$

provided that $|k_1 - k_2| \leq 4$, $2^k \in [2^{10}\langle t\rangle^{-1}, 1]$, and $2^{k_1}, 2^{k_2} \in [2^k, \langle t\rangle]$.

We decompose $P_{k_2}U^{\mathcal{L}_2 h_2,\iota_2} = U^{\mathcal{L}_2 h_2,\iota_2}_{\leq J_2,k_2} + U^{\mathcal{L}_2 h_2,\iota_2}_{>J_2,k_2}$, with $J_1 = J_2$, and estimate

$$\begin{aligned}
\big\|P_k I[U^{\mathcal{L}_1 h_1,\iota_1}_{>J_1,k_1}, P_{k_2}U^{\mathcal{L}_2 h_2,\iota_2}](t)\big\|_{L^\infty} + \big\|P_k I[U^{\mathcal{L}_1 h_1,\iota_1}_{\leq J_1,k_1}, U^{\mathcal{L}_2 h_2,\iota_2}_{>J_2,k_2}](t)\big\|_{L^\infty}& \\
\lesssim \varepsilon_1^2 2^{3k/2} 2^{k_1/2} \langle t\rangle^{-2+\delta'/2} 2^{-4k_1^+} 2^{-4k_2^+}&,
\end{aligned}$$

using (3.3.4) and (3.3.11). Finally, using (3.2.16) with $l = k$ we estimate

$$\begin{aligned}
&\|P_k I[U^{\mathcal{L}_1 h_1,\iota_1}_{\leq J_1,k_1}, U^{\mathcal{L}_2 h_2,\iota_2}_{\leq J_2,k_2}](t)\|_{L^\infty} \\
&\lesssim \sum_{|n_1-n_2|\leq 4} \|\mathcal{C}_{n_1,l} U^{\mathcal{L}_1 h_1,\iota_1}_{\leq J_1,k_1}\|_{L^\infty} \|\mathcal{C}_{n_2,l} U^{\mathcal{L}_2 h_2,\iota_2}_{\leq J_2,k_2}(t)\|_{L^\infty} \\
&\lesssim \varepsilon_1^2 \langle t\rangle^{-2+\delta'/2} 2^k 2^{k_1/2} 2^{k_2/2} 2^{-4k_1^+} 2^{-4k_2^+},
\end{aligned}$$

using also (3.3.3) in the last estimate. This completes the proof of (4.1.82). □

## 4.2 BOUNDS ON THE NONLINEARITIES $\mathcal{N}^H_{\alpha\beta}$ AND $\mathcal{N}^\psi$

Recall the decompositions (see (2.1.9)–(2.1.17))

$$\begin{aligned}
\mathcal{N}^h_{\alpha\beta} &= \mathcal{KG}^2_{\alpha\beta} + \mathcal{Q}^2_{\alpha\beta} + \mathcal{S}^2_{\alpha\beta} + \mathcal{N}^{h,\geq 3}_{\alpha\beta}, \\
\mathcal{N}^\psi &= \mathcal{N}^{\psi,2} + \mathcal{N}^{\psi,\geq 3}.
\end{aligned} \tag{4.2.1}$$

In this subsection we prove several frequency-localized bounds on the nonlinear terms $\mathcal{L}\mathcal{KG}^2_{\alpha\beta}$, $\mathcal{L}\mathcal{Q}^2_{\alpha\beta}$, $\mathcal{L}\mathcal{S}^2_{\alpha\beta}$, $\mathcal{L}\mathcal{N}^{h,\geq 3}_{\alpha\beta}$, $\mathcal{L}\mathcal{N}^{\psi,2}$, and $\mathcal{L}\mathcal{N}^{\psi,\geq 3}$. One should think of these as rather general bounds; we could improve some of them slightly, in terms of both differentiability and time decay, but we do not pursue all the possible improvements at this stage.

For $n \geq 0$ we define $\widetilde{\ell}(n)$ (a slightly worse loss function) and $\widetilde{N}(n)$ by

$$\widetilde{\ell}(n) := 35 \text{ if } n \geq 1, \quad \widetilde{\ell}(0) := 25, \qquad \widetilde{N}(n) := N(n) \text{ if } n \geq 1, \quad \widetilde{N}(0) := N_0. \tag{4.2.2}$$

The following proposition is our main result in this section.

**Proposition 4.7.** *For any $k \in \mathbb{Z}$, $t \in [0,T]$, $\mathcal{L} \in \mathcal{V}^q_n$, $q \leq n \leq 3$, and $\alpha,\beta \in \{0,1,2,3\}$,*

$$\|P_k(\mathcal{L}\mathcal{N}^h_{\alpha\beta})(t)\|_{L^2} \lesssim \varepsilon_1^2 2^{k/2} \langle t\rangle^{H(q,n)\delta} 2^{-\widetilde{N}(n)k^+ + 7k^+} \min(2^k \langle t\rangle^{3\delta/2}, \langle t\rangle^{-1+\widetilde{\ell}(n)\delta}) \tag{4.2.3}$$

*and*

$$\|P_k(\mathcal{L}\mathcal{N}^\psi)(t)\|_{L^2} \lesssim \varepsilon_1^2 \langle t\rangle^{H(q,n)\delta} 2^{-\widetilde{N}(n)k^+ + 7k^+} \min(2^k \langle t\rangle^{3\delta/2}, \langle t\rangle^{-1+\widetilde{\ell}(n)\delta}). \tag{4.2.4}$$

*Moreover if $n \leq 2$ and $l \in \{1,2,3\}$ then*

$$\begin{aligned}
\|P_k(\mathcal{L}\mathcal{N}^h_{\alpha\beta})(t)\|_{L^\infty} &\lesssim \varepsilon_1^2 2^k \langle t\rangle^{-1+4\delta'} 2^{-N(n+1)k^+ + 5k^+} \min(2^k, \langle t\rangle^{-1}), \\
\|P_k(x_l \mathcal{L}\mathcal{N}^h_{\alpha\beta})(t)\|_{L^2} &\lesssim \varepsilon_1^2 2^{k/2} \langle t\rangle^{H(q,n)\delta + \widetilde{\ell}(n)\delta} 2^{-N(n+1)k^+ - 2k^+}
\end{aligned} \tag{4.2.5}$$

*and*

$$\begin{aligned}&\|P_k(\mathcal{L}\mathcal{N}^\psi)(t)\|_{L^\infty} \lesssim \varepsilon_1^2 2^{k/2}\langle t\rangle^{-1+4\delta'}2^{-N(n+1)k^+ +5k^+}\min(2^k,\langle t\rangle^{-1}),\\&\|P_k(x_l\mathcal{L}\mathcal{N}^\psi)(t)\|_{L^2} \lesssim \varepsilon_1^2\langle t\rangle^{H(q,n)\delta+\widetilde{\ell}(n)\delta}2^{-N(n+1)k^+ -2k^+}.\end{aligned} \tag{4.2.6}$$

The proposition follows from Lemmas 4.8–4.12 below. We notice again that we prove weighted $L^2$ bounds on the nonlinearities $\mathcal{L}\mathcal{N}^h_{\alpha\beta}$ and $\mathcal{L}\mathcal{N}^\psi$, which will play a critical role in our bootstrap scheme in Proposition 6.2.

### 4.2.1 The Quadratic Nonlinearities

We consider first the nonlinearities $\mathcal{KG}^2_{\alpha\beta}$.

**Lemma 4.8.** *Assume that $t\in[0,T]$, $\mathcal{L}\in\mathcal{V}^q_n$, $q\le n\le 3$, and $k\in\mathbb{Z}$. Then*

$$\begin{aligned}\|P_k\{\mathcal{L}\mathcal{KG}^2_{\alpha\beta}\}(t)\|_{L^2} \lesssim&\varepsilon_1^2 2^{k/2}2^{-N(n)k^+}\langle t\rangle^{H(q,n)\delta}\\&\times\min(2^k\langle t\rangle^{3\delta/2},\langle t\rangle^{-1+\ell(q,n)\delta+\delta/2}).\end{aligned} \tag{4.2.7}$$

*Moreover, if $l\in\{1,2,3\}$ and $n\le 2$ then we also have the bounds*

$$\begin{aligned}&\|P_k\{\mathcal{L}\mathcal{KG}^2_{\alpha\beta}\}(t)\|_{L^\infty} \lesssim \varepsilon_1^2 2^k\langle t\rangle^{-1+2\delta'}2^{-N(n+1)k^+ +3k^+}\min\{2^k,\langle t\rangle^{-1}\},\\&\|P_k\{x_l\mathcal{L}\mathcal{KG}^2_{\alpha\beta}\}(t)\|_{L^2} \lesssim \varepsilon_1^2 2^{k/2}2^{-N(n+1)k^+ -5k^+}\langle t\rangle^{\delta[H(q,n)+\widetilde{\ell}(n)]}.\end{aligned} \tag{4.2.8}$$

*Proof.* **Step 1.** Recall the operators $I$ defined in (3.2.43) and the loss function $\ell$ defined in (4.1.2). In view of (2.1.36), for (4.2.7) it suffices to prove that

$$\begin{aligned}\sum_{(k_1,k_2)\in\mathcal{X}_k}&\|P_kI[P_{k_1}U^{\mathcal{L}_1\psi,\iota_1},P_{k_2}U^{\mathcal{L}_2\psi,\iota_2}](t)\|_{L^2}\\&\lesssim \varepsilon_1^2 2^{k/2}2^{-N(n)k^+}\langle t\rangle^{H(q,n)\delta}\min(2^k\langle t\rangle^{3\delta/2},\langle t\rangle^{-1+\ell(q,n)\delta+\delta/2})\end{aligned} \tag{4.2.9}$$

for any $\iota_1,\iota_2\in\{+,-\}$, $\mathcal{L}_1\in\mathcal{V}^{q_1}_{n_1}$, $\mathcal{L}_2\in\mathcal{V}^{q_2}_{n_2}$, $(q_1,n_1)+(q_2,n_2)\le(q,n)$. This follows easily from the bounds (4.1.34)–(4.1.35).

**Step 2.** To prove the $L^\infty$ bounds in (4.2.8) it suffices to show that

$$\begin{aligned}\sum_{(k_1,k_2)\in\mathcal{X}_k}&\|P_kI[P_{k_1}U^{\mathcal{L}_1\psi,\iota_1},P_{k_2}U^{\mathcal{L}_2\psi,\iota_2}](t)\|_{L^\infty}\\&\lesssim \varepsilon_1^2 2^k\langle t\rangle^{-1+2\delta'}2^{-N(n+1)k^+ +3k^+}\min(2^k,\langle t\rangle^{-1}).\end{aligned} \tag{4.2.10}$$

This follows from (4.2.9) if $2^k\le\langle t\rangle^{-1}$, using just the Cauchy-Schwarz inequality. On the other hand, if $2^k\ge\langle t\rangle^{-1}$ then the contribution of the pairs $(k_1,k_2)$ for which $\min\{k_1,k_2\}\le k+10$ can be bounded easily, using just (3.3.13). These bounds also suffice when $\min\{k_1,k_2\}\ge k+10$, but only if $k\ge 0$. Moreover, using just (3.3.5) we have

$$\|P_kI[P_{k_1}U^{\mathcal{L}_1\psi,\iota_1},P_{k_2}U^{\mathcal{L}_2\psi,\iota_2}](t)\|_{L^\infty}\lesssim 2^{3k}\|P_{k_1}U^{\mathcal{L}_1\psi}(t)\|_{L^2}\|P_{k_2}U^{\mathcal{L}_2\psi}(t)\|_{L^2}$$
$$\lesssim \varepsilon_1^2 2^{3k}2^{k_1}2^{k_2}\langle t\rangle^{\delta'}2^{-4k_1^+}2^{-4k_2^+},$$

and this suffices to bound the contribution of the pairs $(k_1,k_2)$ that satisfy the inequality $2^{2k}2^{k_1^-}2^{k_2^-}\lesssim\langle t\rangle^{-2+\delta'}$.

In the remaining case

$$2^k\in[\langle t\rangle^{-1},1],\qquad \min\{k_1,k_2\}\geq k+10,\qquad 2^{-k}\leq 2^{k_1^-}\langle t\rangle^{1-\delta'/2},\tag{4.2.11}$$

we decompose $P_{k_1}U^{\mathcal{L}_1\psi,\iota_1}=U^{\mathcal{L}_1\psi,\iota_1}_{\leq J_1,k_1}+U^{\mathcal{L}_1\psi,\iota_1}_{>J_1,k_1}$ and $P_{k_2}U^{\mathcal{L}_2\psi,\iota_2}=U^{\mathcal{L}_2\psi,\iota_2}_{\leq J_2,k_2}+U^{\mathcal{L}_2\psi,\iota_2}_{>J_2,k_2}$, where $2^{J_1}=2^{J_2}\approx 2^{k_1^-}\langle t\rangle^{1-\delta'/2}$. Then we estimate

$$\|P_kI[U^{\mathcal{L}_1\psi,\iota_1}_{>J_1,k_1},P_{k_2}U^{\mathcal{L}_2\psi,\iota_2}](t)\|_{L^\infty}\lesssim 2^{3k/2}\|U^{\mathcal{L}_1\psi,\iota_1}_{>J_1,k_1}(t)\|_{L^2}\|P_{k_2}U^{\mathcal{L}_2\psi,\iota_2}(t)\|_{L^\infty}$$
$$\lesssim \varepsilon_1^2 2^{3k/2}2^{-J_1}\langle t\rangle^{-1+\delta'}2^{k_2^-/2}2^{-4k_1^+}2^{-4k_2^+}$$
$$\lesssim \varepsilon_1^2 2^{3k/2}\langle t\rangle^{-2+3\delta'/2}2^{-k_2^-/2}2^{-4k_1^+}2^{-4k_2^+},$$

using (3.3.4) and (3.3.13). Similarly,

$$\|P_kI[U^{\mathcal{L}_1\psi,\iota_1}_{\leq J_1,k_1},U^{\mathcal{L}_2\psi,\iota_2}_{>J_2,k_2}](t)\|_{L^\infty}\lesssim \varepsilon_1^2 2^{3k/2}\langle t\rangle^{-2+3\delta'/2}2^{-k_2^-/2}2^{-4k_1^+}2^{-4k_2^+}.$$

Finally, using (3.2.18) with $l=k$ and recalling the definition (3.2.1), we estimate

$$\big\|P_kI[U^{\mathcal{L}_1\psi,\iota_1}_{\leq J_1,k_1},U^{\mathcal{L}_2\psi,\iota_2}_{\leq J_2,k_2}](t)\big\|_{L^\infty}\lesssim\sum_{|n_1-n_2|\leq 4}\big\|\mathcal{C}_{n_1,l}U^{\mathcal{L}_1\psi,\iota_1}_{\leq J_1,k_1}\big\|_{L^\infty}\big\|\mathcal{C}_{n_2,l}U^{\mathcal{L}_2\psi,\iota_2}_{\leq J_2,k_2}(t)\big\|_{L^\infty}$$
$$\lesssim \varepsilon_1^2\langle t\rangle^{-2+\delta'}2^k2^{-4k_1^+}2^{-4k_2^+},$$

where we also used (3.3.5) in the estimate in the second line. Therefore

$$\|P_kI[P_{k_1}U^{\mathcal{L}_1\psi,\iota_1},P_{k_2}U^{\mathcal{L}_2\psi,\iota_2}](t)\|_{L^\infty}\lesssim \varepsilon_1^2 2^k\langle t\rangle^{-2+3\delta'/2}2^{-4k_1^+}2^{-4k_2^+},\tag{4.2.12}$$

which suffices to bound the contribution of the remaining pairs in (4.2.11).

**Step 3.** To prove the weighted $L^2$ bounds in (4.2.8) it suffices to show that

$$\sum_{(k_1,k_2)\in\mathcal{X}_k,\,k_1\geq k_2}\|P_k\{x_lI[P_{k_1}U^{\mathcal{L}_1\psi,\iota_1},P_{k_2}U^{\mathcal{L}_2\psi,\iota_2}]\}(t)\|_{L^2}$$
$$\lesssim \varepsilon_1^2 2^{k/2}2^{-N(n+1)k^+-5k^+}\langle t\rangle^{\delta[H(q,n)+\widetilde{\ell}(n)]}.\tag{4.2.13}$$

Notice that we may assume that $k_1\geq k_2$ in (4.2.13), due to symmetry. We write $U^{\mathcal{L}_j\psi,\iota_j}(t)=e^{-it\Lambda_{kg,\iota_j}}V^{\mathcal{L}_j\psi,\iota_j}(t)$, $j\in\{1,2\}$. When proving (4.2.13), the $\partial_{\xi_l}$ derivative can hit the multiplier $m(\xi-\eta,\eta)$, or the phase $e^{-it\Lambda_{kg,\iota_1}(\xi-\eta)}$, or the profile $P_{k_1}\widehat{V^{\mathcal{L}_1\psi,\iota_1}}(\xi-\eta)$. In the first two cases, the derivative effectively corresponds to multiplying by factors $\lesssim 2^{-k_1^-}$ or $\lesssim \langle t\rangle$, and the desired bounds are again consequences of Lemma 4.2.

Let $\widehat{U^{\mathcal{L}_j\psi,\iota_j}_{*l,k_j}}(\xi,t) := e^{-it\Lambda_{kg,\iota_j}(\xi)}\partial_{\xi_l}\{\varphi_{k_j}\cdot\widehat{V^{\mathcal{L}_j\psi,\iota_j}}\}(\xi,t)$. In view of (3.3.9) and (3.3.5) we have

$$\|U^{\mathcal{L}_j\psi,\iota_j}_{*l,k_j}(t)\|_{L^2}\lesssim \varepsilon_1\langle t\rangle^{H(q_j+1,n_j+1)\delta}2^{-N(n_j+1)k_j^+}. \tag{4.2.14}$$

Using also the $L^\infty$ bounds (3.3.13) and the $L^2$ bounds (3.3.3) we estimate

$$\begin{aligned}&\|P_k\{I[U^{\mathcal{L}_1\psi,\iota_1}_{*l,k_1},P_{k_2}U^{\mathcal{L}_2\psi,\iota_2}]\}(t)\|_{L^2}\\&\qquad\lesssim \varepsilon_1^2\langle t\rangle^{\delta'}\min(\langle t\rangle^{-1},2^{k_2^-},2^{k^-})2^{-8k_2^+}2^{-N(n_1+1)k_1^+}.\end{aligned}\tag{4.2.15}$$

This suffices to prove (4.2.13), except if there is derivative loss, which happens when

$$n_1=n\quad\text{ and }\quad 2^k\geq\langle t\rangle^{1/100}2^{10}. \tag{4.2.16}$$

In this case the estimates (4.2.15) still suffice to bound the contribution of the pairs $(k_1,k_2)$ as in (4.2.13), unless $k_2\in[-10k,k-10]$. In this last case we make the change of variables $\eta\to\xi-\eta$ in order to move the $\partial_{\xi_l}$ derivative on the low frequency factor, and estimate

$$\|P_k\{I[P_{k_1}U^{\mathcal{L}_1\psi,\iota_1},U^{\mathcal{L}_2\psi,\iota_2}_{*l,k_2}]\}(t)\|_{L^2}\lesssim\varepsilon_1^2\langle t\rangle^{\delta'}2^{-N(n)k_1^+}2^{3k_2^-/2},$$

for any $(k_1,k_2)\in\mathcal{X}_k$ with $k_2\leq k_1$. We remark that the loss of the factor of $\langle t\rangle^{\delta'}$ is mitigated by the gain of derivative and the assumption $2^k\gtrsim\langle t\rangle^{1/100}$. This suffices to bound the remaining contributions as claimed in (4.2.13). □

We consider now the quadratic nonlinearities $\mathcal{Q}^2_{\alpha\beta}$ and $\mathcal{S}^2_{\alpha\beta}$.

**Lemma 4.9.** *Assume that $t\in[0,T]$, $\mathcal{L}\in\mathcal{V}^q_n$, $q\leq n\leq 3$, and $k\in\mathbb{Z}$. Then*

$$\begin{aligned}&\|P_k\{\mathcal{L}\mathcal{Q}^2_{\alpha\beta}\}(t)\|_{L^2}+\|P_k\{\mathcal{L}\mathcal{S}^2_{\alpha\beta}\}(t)\|_{L^2}\\&\qquad\lesssim\varepsilon_1^2 2^{k/2}2^{-\widetilde{N}(n)k^++3k^+}\langle t\rangle^{H(q,n)\delta}\min(2^k\langle t\rangle^{3\delta/2},\langle t\rangle^{-1+\widetilde{\ell}(n)\delta}).\end{aligned}\tag{4.2.17}$$

*Moreover, if $n\leq 2$ and $l\in\{1,2,3\}$ then we also have the bounds*

$$\begin{aligned}&\|P_k\{\mathcal{L}\mathcal{Q}^2_{\alpha\beta}\}(t)\|_{L^\infty}+\|P_k\{\mathcal{L}\mathcal{S}^2_{\alpha\beta}\}(t)\|_{L^\infty}\\&\qquad\lesssim\varepsilon_1^2 2^k\langle t\rangle^{-1+2\delta'}2^{-N(n+1)k^++3k^+}\min\{2^k,\langle t\rangle^{-1}\},\\&\|P_k\{x_l\mathcal{L}\mathcal{Q}^2_{\alpha\beta}\}(t)\|_{L^2}+\|P_k\{x_l\mathcal{L}\mathcal{S}^2_{\alpha\beta}\}(t)\|_{L^2}\\&\qquad\lesssim\varepsilon_1^2 2^{k/2}2^{-N(n+1)k^+-5k^+}\langle t\rangle^{\delta[H(q,n)+\widetilde{\ell}(n)]}.\end{aligned}\tag{4.2.18}$$

*Proof.* We notice that $\mathcal{S}^2_{\alpha\beta}$ is a sum of quadratic expressions of the form $\partial_\mu h_1\cdot$

$\partial_\nu h_2$, where $h_1, h_2 \in \{h_{\alpha\beta}\}$, and $\mathcal{Q}^2_{\alpha\beta}$ is a sum of quadratic expressions of the form $h_1 \cdot \partial_\mu\partial_\nu h_2$, $(\mu,\nu) \neq (0,0)$. The differential operator $\mathcal{L}$ can split between these two factors. Notice that commutation with one vector-field $\partial_\mu$ generates similar terms of the form $\partial_\nu\mathcal{L}'$ (see (2.1.25)), while commutation with two vector-fields $\partial_\mu\partial_\nu$ (which appear in $\mathcal{Q}^2_{\alpha\beta}$) leads to terms of the form $\partial_\alpha\partial_\beta\mathcal{L}'$. If $\alpha = \beta = 0$ then we have to further replace $\partial_t^2\mathcal{L}'h$ with $\Delta\mathcal{L}'h + \mathcal{L}'\mathcal{N}^h$ (as in (2.1.9)), in order to have access to elliptic estimates.

**Step 1.** In view of these considerations, for (4.2.17) it suffices to prove that

$$\begin{aligned}\sum_{(k_1,k_2)\in\mathcal{X}_k} & 2^{|k_2-k_1|}\|P_kI[P_{k_1}U^{\mathcal{L}_1h_1,\iota_1}, P_{k_2}U^{\mathcal{L}_2h_2,\iota_2}](t)\|_{L^2} \\ & \lesssim \varepsilon_1^2 2^{k/2}2^{-\widetilde{N}(n)k^+ + 3k^+}\langle t\rangle^{H(q,n)\delta}\min(2^k\langle t\rangle^{3\delta/2}, \langle t\rangle^{-1+\widetilde{\ell}(n)\delta})\end{aligned} \tag{4.2.19}$$

and

$$\begin{aligned}\sum_{(k_1,k_2)\in\mathcal{X}_k} & 2^{-k_1}\|P_kI[P_{k_1}U^{\mathcal{L}_1h_1,\iota_1}, P_{k_2}\mathcal{L}'\mathcal{N}^h](t)\|_{L^2} \\ & \lesssim \varepsilon_1^2 2^{k/2}2^{-\widetilde{N}(n)k^+ + 3k^+}\langle t\rangle^{H(q,n)\delta}\min(2^k\langle t\rangle^{3\delta/2}, \langle t\rangle^{-1+\widetilde{\ell}(n)\delta}),\end{aligned} \tag{4.2.20}$$

for any $\iota_1, \iota_2 \in \{+,-\}$, $h_1, h_2 \in \{h_{\alpha\beta}\}$, $\mathcal{N}^h \in \{\mathcal{N}^h_{\alpha\beta}\}$, $\mathcal{L}_1 \in \mathcal{V}^{q_1}_{n_1}$, $\mathcal{L}_2 \in \mathcal{V}^{q_2}_{n_2}$, $\mathcal{L}' \in \mathcal{V}^{q'}_{n'}$, $(q_1,n_1) + (q_2,n_2) \leq (q,n)$, $(q_1,n_1) + (q',n') \leq (q,n-1)$.

Without loss of generality, in proving (4.2.19) we may assume that $n_1 \leq n_2$. We use Lemma 4.1. If $2^k\langle t\rangle^{3\delta/2} \lesssim \langle t\rangle^{-1+\widetilde{\ell}(n)\delta}$ then the bounds (4.2.19) follow from (4.1.5) and (2.1.53). On the other hand, if $2^k\langle t\rangle^{3\delta/2} \geq \langle t\rangle^{-1+\widetilde{\ell}(n)\delta}$ then the contribution of the pairs $(k_1,k_2) \in \mathcal{X}_k$ with $\min(k_1,k_2) \geq k$ (thus $|k_1 - k_2| \leq 10$) is bounded as claimed due to (4.1.6), (4.1.7), and the first inequality in (4.1.3). The contribution of the pairs $(k_1,k_2) \in \mathcal{X}_k$ with $k_1 = \min(k,k_1,k_2)$ (thus $|k_2 - k| \leq 4$) is bounded as claimed due to (4.1.8) (this case gives the worst contribution to the growth function $\langle t\rangle^{\delta[H(q,n)+\widetilde{\ell}(n)]}$, when $n_1 = 0$). Finally, the contribution of the pairs $(k_1,k_2) \in \mathcal{X}_k$ with $k_2 = \min(k,k_1,k_2)$ is bounded as claimed due to (4.1.9) if $n_1 = 1$ and (4.1.10)–(4.1.11) if $n_1 = 0$.

To prove (4.2.20) we use only $L^2$ bounds. We may assume, by induction over $n$ (in the case $n = 0$ the left-hand side of (4.2.20) is trivial), that the bounds (4.2.3) hold for $P_{k_2}\mathcal{N}^{\mathcal{L}'h'}(t)$ and estimate the left-hand side of (4.2.20), using also (3.3.3), by

$$\begin{aligned}\sum_{(k_1,k_2)\in\mathcal{X}_k} & 2^{-k_1}2^{3\min\{k,k_1,k_2\}/2}\|P_{k_1}U^{\mathcal{L}_1h_1,\iota_1}(t)\|_{L^2}\|P_{k_2}\mathcal{L}'\mathcal{N}^h(t)\|_{L^2} \\ & \lesssim \varepsilon_1^2 2^{3k/2}2^{-\widetilde{N}(n)k^+}\langle t\rangle^{-1+\delta[H(q_1,n_1)+H(q',n')+\widetilde{\ell}(n')+2]}.\end{aligned} \tag{4.2.21}$$

This suffices to prove (4.2.20), using also (2.1.53).

**Step 2.** Similarly, to prove the $L^\infty$ bounds in (4.2.18) it suffices to show that

$$\sum_{(k_1,k_2)\in\mathcal{X}_k} 2^{|k_2-k_1|}\|P_kI[P_{k_1}U^{\mathcal{L}_1h_1,\iota_1},P_{k_2}U^{\mathcal{L}_2h_2,\iota_2}](t)\|_{L^\infty} \lesssim \varepsilon_1^2 2^k\langle t\rangle^{-1+2\delta'}2^{-N(n+1)k^++3k^+}\min\{2^k,\langle t\rangle^{-1}\} \tag{4.2.22}$$

and

$$\sum_{(k_1,k_2)\in\mathcal{X}_k} 2^{-k_1}\|P_kI[P_{k_1}U^{\mathcal{L}_1h_1,\iota_1},P_{k_2}\mathcal{L}'\mathcal{N}^h](t)\|_{L^\infty} \lesssim \varepsilon_1^2 2^k\langle t\rangle^{-1+2\delta'}2^{-N(n+1)k^++3k^+}\min\{2^k,\langle t\rangle^{-1}\}. \tag{4.2.23}$$

These bounds follow easily from (4.2.19)–(4.2.20) if $2^k\lesssim\langle t\rangle^{-1+\delta'}$. Also, the bounds (4.2.23) follow from (3.3.11) and using Proposition 4.7 inductively.

We prove now the bounds (4.2.22). The contribution of the pairs $(k_1,k_2)$ with $\min\{k_1,k_2\}\leq k+10$ or $\max\{k_1,k_2\}\geq\langle t\rangle$ can be bounded easily using (3.3.11). This also suffices to bound all the contributions if $k\geq 0$. In the remaining case, the desired bounds follow from (4.1.82).

**Step 3.** Similarly, to prove the weighted $L^2$ bounds in (4.2.18) it suffices to show that

$$\sum_{(k_1,k_2)\in\mathcal{X}_k,\,k_2\geq k_1} 2^{k_2-k_1}\|P_k\{x_lI[P_{k_1}U^{\mathcal{L}_1h_1,\iota_1},P_{k_2}U^{\mathcal{L}_2h_2,\iota_2}]\}(t)\|_{L^2} \lesssim \varepsilon_1^2 2^{k/2}2^{-N(n+1)k^+-5k^+}\langle t\rangle^{\delta[H(q,n)+\widetilde{\ell}(n)]} \tag{4.2.24}$$

and

$$\sum_{(k_1,k_2)\in\mathcal{X}_k} 2^{-k_1}\|P_k\{x_lI[P_{k_1}U^{\mathcal{L}_1h_1,\iota_1},P_{k_2}\mathcal{L}'\mathcal{N}^h]\}(t)\|_{L^2} \lesssim \varepsilon_1^2 2^{k/2}2^{-N(n+1)k^+-5k^+}\langle t\rangle^{\delta[H(q,n)+\widetilde{\ell}(n)]}. \tag{4.2.25}$$

Notice that in (4.2.24) it is more convenient to assume that $k_1\leq k_2$ instead of $n_1\leq n_2$.

To prove the bounds (4.2.24) we write $U^{\mathcal{L}_jh_j,\iota_j}(t)=e^{-it\Lambda_{wa,\iota_j}}V^{\mathcal{L}_jh_j,\iota_j}(t)$, $j\in\{1,2\}$ and let $\widehat{U^{\mathcal{L}_jh_j,\iota_j}_{*l,k_j}}(\xi,t):=e^{-it\Lambda_{wa,\iota_j}(\xi)}\partial_{\xi_l}\{\varphi_{k_j}\cdot\widehat{V^{\mathcal{L}_jh_j,\iota_j}}\}(\xi,t)$, as in Lemma 4.8. In view of (3.3.9) and (3.3.3) it follows that, for $j\in\{1,2\}$,

$$\|U^{\mathcal{L}_jh_j,\iota_j}_{*l,k_j}(t)\|_{L^2}\lesssim\varepsilon_1\langle t\rangle^{H(q_j+1,n_j+1)\delta}2^{-k_j/2}2^{-N(n_j+1)k_j^+}(\langle t\rangle 2^{k_j^-})^{-\gamma}. \tag{4.2.26}$$

To prove (4.2.24) we make the change of variables $\eta\to\xi-\eta$ and notice that the $\partial_{\xi_l}$ derivative can hit the multiplier $m(\eta,\xi-\eta)$, or the phase $e^{-it\Lambda_{wa,\iota_2}(\xi-\eta)}$, or the higher frequency profile $P_{k_2}\widehat{V^{\mathcal{L}_2h_2,\iota_2}}(\xi-\eta)$. In the first two cases, the derivative effectively corresponds to multiplying by factors $\lesssim\langle t\rangle+2^{-k_2^-}$, and the desired bounds are then consequences of (4.2.19). In the last case we estimate

$$\begin{aligned}2^{k_2-k_1}&\|P_k\{I[P_{k_1}U^{\mathcal{L}_1h_1,\iota_1},U^{\mathcal{L}_2h_2,\iota_2}_{*l,k_2}]\}(t)\|_{L^2}\\&\lesssim \varepsilon_1^2\langle t\rangle^{\delta'}\min(\langle t\rangle^{-1},2^{k_1^-},2^{k^-})^{9/10}2^{-8k_1^+}2^{-N(n_2+1)k_2^++k_2^+},\end{aligned}\tag{4.2.27}$$

using the $L^2$ bounds (4.2.26) and the bounds (3.3.11) and (3.3.3). As in Lemma 4.8, this suffices except if there is derivative loss, which can happen only when

$$n_2=n\qquad\text{and}\qquad 2^k\geq\langle t\rangle^{1/100}2^{10}.\tag{4.2.28}$$

Assuming that (4.2.28) holds, the sum over $k_1\leq -10k$ or $k_1\geq k-10$ can still be bounded as before, using (4.2.27). To bound the sum over $k_1\in[-10k,k-10]$ we return to the original formula for $I$ (without making the change of variables $\eta\to\xi-\eta$) and notice that the $\partial_{\xi_l}$ derivative now hits the low frequency factor. Then we estimate, using just (4.2.26) and (3.3.3),

$$2^{k_2-k_1}\|P_k\{I[U^{\mathcal{L}_1h_1,\iota_1}_{*l,k_1},P_{k_2}U^{\mathcal{L}_2h_2,\iota_2}]\}(t)\|_{L^2}\lesssim\varepsilon_1^2\langle t\rangle^{\delta'}2^{-N(n)k_2^++2k_2^+}2^{-\delta k_1^-}.$$

As before, the loss of the factor of $\langle t\rangle^{\delta'}$ is compensated by the gain of derivative and the assumption $2^k\gtrsim\langle t\rangle^{1/100}$. This suffices to complete the proof of (4.2.24).

The bounds (4.2.25) are easier: as in (4.2.21) we estimate the expression in the left-hand side by

$$\begin{aligned}\sum_{(k_1,k_2)\in\mathcal{X}_k}&2^{-k_1}2^{3\min\{k,k_1,k_2\}/2}\|P_{k_1}U^{\mathcal{L}_1h_1,\iota_1}(t)\|_{L^2}\\&\times\left[\|P_{k_2}\{x_l\mathcal{L}'\mathcal{N}^h\}(t)\|_{L^2}+2^{-k_2^-}\|P'_{k_2}\mathcal{L}'\mathcal{N}^h(t)\|_{L^2}\right]\\&\lesssim\varepsilon_1^2 2^k2^{-N(n+1)k^+-8k^+}\langle t\rangle^{\delta[H(q_1,n_1)+H(q',n')+\widetilde{\ell}(n')+2]},\end{aligned}$$

where $P'_{k_2}=P_{[k-2,k+2]}$, using (3.3.3) and Proposition 4.7 inductively. This clearly suffices. □

Finally, we consider the quadratic nonlinearities $\mathcal{N}^{\psi,2}$.

**Lemma 4.10.** *Assume that* $t\in[0,T]$, $\mathcal{L}\in\mathcal{V}^q_n$, $n\leq 3$, *and* $k\in\mathbb{Z}$. *Then*

$$\|P_k\{\mathcal{L}\mathcal{N}^{\psi,2}\}(t)\|_{L^2}\lesssim\varepsilon_1^2 2^{-\widetilde{N}(n)k^++7k^+}\langle t\rangle^{H(q,n)\delta}\min(2^k\langle t\rangle^{3\delta/2},\langle t\rangle^{-1+\widetilde{\ell}(n)\delta}).\tag{4.2.29}$$

*Moreover, if* $l\in\{1,2,3\}$ *and* $n\leq 2$ *then we also have the bounds*

$$\begin{aligned}\|P_k\{\mathcal{L}\mathcal{N}^{\psi,2}\}(t)\|_{L^\infty}&\lesssim\varepsilon_1^2 2^{k/2}\langle t\rangle^{-1+2\delta'}2^{-N(n+1)k^++3k^+}\min\{2^k,\langle t\rangle^{-1}\},\\\|P_k\{x_l\mathcal{L}\mathcal{N}^{\psi,2}\}(t)\|_{L^2}&\lesssim\varepsilon_1^2 2^{-N(n+1)k^+-3k^+}\langle t\rangle^{\delta[H(q,n)+\widetilde{\ell}(n)]}.\end{aligned}\tag{4.2.30}$$

*Proof.* We examine the definition (2.1.17) and notice that $\mathcal{N}^{\psi,2}$ is a sum of terms of the form $h\cdot\partial_\mu\partial_\nu\psi$, $(\mu,\nu)\neq(0,0)$, or $h\cdot\psi$, where $h\in\{h_{\mu\nu}\}$. As in Lemma 4.9, we distribute the vector-field $\mathcal{L}$, commute with the derivatives $\partial_\mu\partial_\nu$, and further replace $\partial_t^2\mathcal{L}'\psi$ with $(\Delta-1)\mathcal{L}'\psi+\mathcal{L}'\mathcal{N}^\psi$.

**Step 1.** For (4.2.29) it suffices to prove that

$$\sum_{(k_1,k_2)\in\mathcal{X}_k} 2^{k_2^+-k_1}\|P_kI[P_{k_1}U^{\mathcal{L}_1h,\iota_1},P_{k_2}U^{\mathcal{L}_2\psi,\iota_2}](t)\|_{L^2} \lesssim \varepsilon_1^2 2^{-\widetilde{N}(n)k^++7k^+}\langle t\rangle^{H(q,n)\delta}\min(2^k\langle t\rangle^{3\delta/2},\langle t\rangle^{-1+\widetilde{\ell}(n)\delta}) \tag{4.2.31}$$

and

$$\sum_{(k_1,k_2)\in\mathcal{X}_k} 2^{-k_1}\|P_kI[P_{k_1}U^{\mathcal{L}_1h,\iota_1},P_{k_2}\mathcal{L}'\mathcal{N}^\psi](t)\|_{L^2} \lesssim \varepsilon_1^2 2^{-\widetilde{N}(n)k^++7k^+}\langle t\rangle^{H(q,n)\delta}\min(2^k\langle t\rangle^{3\delta/2},\langle t\rangle^{-1+\widetilde{\ell}(n)\delta}), \tag{4.2.32}$$

for any $\iota_1,\iota_2\in\{+,-\}$, $h\in\{h_{\alpha\beta}\}$, $\mathcal{L}_1\in\mathcal{V}^{q_1}_{n_1}$, $\mathcal{L}_2\in\mathcal{V}^{q_2}_{n_2}$, $\mathcal{L}'\in\mathcal{V}^{q'}_{n'}$, $(q_1,n_1)+(q_2,n_2)\leq(q,n)$, $(q_1,n_1)+(q',n')\leq(q,n-1)$.

To prove (4.2.31) we use Lemma 4.3. The bounds follow from (4.1.52) if $2^k\langle t\rangle^{3\delta/2}\lesssim\langle t\rangle^{-1+\widetilde{\ell}(n)\delta}$. If $2^k\langle t\rangle^{3\delta/2}\geq\langle t\rangle^{-1+\widetilde{\ell}(n)\delta}$ then (4.2.31) follows from (4.1.54) when $n_1=n_2=0$, or from (4.1.53) if $n_1=0$ and $n_2\geq1$, or from (4.1.55) if $n_1\geq1$ and $n_2=0$, or from (4.1.56) if $n_1\geq1$ and $n_2\geq1$.

The bounds (4.2.32) follow easily, using just $L^2$ estimates as in (4.2.21).

**Step 2.** To prove the $L^\infty$ bounds in (4.2.30) it suffices to show that

$$\sum_{(k_1,k_2)\in\mathcal{X}_k} 2^{k_2^+-k_1}\|P_kI[P_{k_1}U^{\mathcal{L}_1h,\iota_1},P_{k_2}U^{\mathcal{L}_2\psi,\iota_2}](t)\|_{L^\infty} \lesssim \varepsilon_1^2 2^{k/2}\langle t\rangle^{-1+2\delta'}2^{-N(n+1)k^++3k^+}\min\{2^k,\langle t\rangle^{-1}\} \tag{4.2.33}$$

and

$$\sum_{(k_1,k_2)\in\mathcal{X}_k} 2^{-k_1}\|P_kI[P_{k_1}U^{\mathcal{L}_1h,\iota_1},P_{k_2}\mathcal{L}'\mathcal{N}^\psi](t)\|_{L^\infty} \lesssim \varepsilon_1^2 2^{k/2}\langle t\rangle^{-1+2\delta'}2^{-N(n+1)k^++3k^+}\min\{2^k,\langle t\rangle^{-1}\}. \tag{4.2.34}$$

These bounds follow easily from (4.2.31)–(4.2.32) if $2^k\lesssim\langle t\rangle^{-1+\delta'}$, using the Cauchy-Schwarz inequality. Also, the bounds (4.2.34) follow from (3.3.11) and using Proposition 4.7 inductively.

We prove now the bounds (4.2.33). The contribution of the pairs $(k_1,k_2)$ with $\min\{k_1,k_2\}\leq k+10$ or $\max\{k_1,k_2\}\geq\langle t\rangle$ can be bounded easily, using just (3.3.11) and (3.3.13). This also suffices to bound all the contributions if

$k \geq 0$. To summarize, it remains to show that

$$2^{-k_1}\|P_kI[P_{k_1}U^{\mathcal{L}_1h_1,\iota_1},P_{k_2}U^{\mathcal{L}_2\psi,\iota_2}](t)\|_{L^\infty} \lesssim \varepsilon_1^2 2^{k/2}\langle t\rangle^{-2+3\delta'/2}2^{-2k_1^+}2^{-2k_2^+}, \tag{4.2.35}$$

provided that $2^k \in [\langle t\rangle^{-1+\delta'},1]$, $|k_1-k_2|\leq 4$, and $k_1,k_2\in[k+10,\langle t\rangle]$.

This is similar to the bounds (4.2.12). First we estimate the left-hand side of (4.2.35) by $2^{3k/2}\varepsilon_1^2\langle t\rangle^{-1+\delta'/2}2^{k_2}2^{-4k_1^+}2^{-4k_2^+}$, using (3.3.5) and (3.3.11). This suffices if $2^k2^{k_2^-}\lesssim\langle t\rangle^{-1+\delta'}$. On the other hand, if $2^k2^{k_2^-}\in[\langle t\rangle^{-1+\delta'},1]$ then we decompose $P_{k_1}U^{\mathcal{L}_1h_1,\iota_1}=U^{\mathcal{L}_1h_1,\iota_1}_{\leq J_1,k_1}+U^{\mathcal{L}_1h_1,\iota_1}_{>J_1,k_1}$ and $P_{k_2}U^{\mathcal{L}_2\psi,\iota_2}=U^{\mathcal{L}_2\psi,\iota_2}_{\leq J_2,k_2}+U^{\mathcal{L}_2\psi,\iota_2}_{>J_2,k_2}$, where $2^{J_1}\approx\langle t\rangle^{1-\delta'/2}$ and $2^{J_2}\approx 2^{k_2^-}\langle t\rangle^{1-\delta'/2}$. Then we estimate

$$\begin{aligned}2^{-k_1}\big\|P_kI[U^{\mathcal{L}_1h_1,\iota_1}_{>J_1,k_1},P_{k_2}U^{\mathcal{L}_2\psi,\iota_2}](t)\big\|_{L^\infty}+2^{-k_1}\big\|P_kI[U^{\mathcal{L}_1h_1,\iota_1}_{\leq J_1,k_1},U^{\mathcal{L}_2\psi,\iota_2}_{>J_2,k_2}](t)\big\|_{L^\infty}\\ \lesssim \varepsilon_1^2 2^{3k/2}2^{-k_2}\langle t\rangle^{-2+3\delta'/2}2^{-4k_1^+}2^{-4k_2^+},\end{aligned}$$

using (3.3.4), (3.3.11), and (3.3.13). Morever, using (3.2.16) and (3.2.18) with $l=k$ we estimate

$$\begin{aligned}&2^{-k_1}\big\|P_kI[U^{\mathcal{L}_1h_1,\iota_1}_{\leq J_1,k_1},U^{\mathcal{L}_2\psi,\iota_2}_{\leq J_2,k_2}](t)\big\|_{L^\infty}\\ &\quad\lesssim 2^{-k_1}\sum_{|n_1-n_2|\leq 4}\big\|C_{n_1,l}U^{\mathcal{L}_1h_1,\iota_1}_{\leq J_1,k_1}\big\|_{L^\infty}\big\|C_{n_2,l}U^{\mathcal{L}_2\psi,\iota_2}_{\leq J_2,k_2}(t)\big\|_{L^\infty}\\ &\quad\lesssim \varepsilon_1^2\langle t\rangle^{-2+\delta'}2^k2^{-k_1/2}2^{-4k_1^+}2^{-4k_2^+},\end{aligned}$$

using also (3.3.3) and (3.3.5) in the estimate in the second line. This completes the proof of (4.2.35).

**Step 3.** As before, to prove the weighted $L^2$ bounds in (4.2.30) it suffices to show that

$$\begin{aligned}\sum_{(k_1,k_2)\in\mathcal{X}_k}2^{k_2^+-k_1}\|P_k\{x_lI[P_{k_1}U^{\mathcal{L}_1h,\iota_1},P_{k_2}U^{\mathcal{L}_2\psi,\iota_2}]\}(t)\|_{L^2}\\ \lesssim\varepsilon_1^2 2^{-N(n+1)k^+-3k^+}\langle t\rangle^{\delta[H(q,n)+\widetilde{\ell}(n)]}\end{aligned}\tag{4.2.36}$$

and

$$\begin{aligned}\sum_{(k_1,k_2)\in\mathcal{X}_k}2^{-k_1}\|P_k\{x_lI[P_{k_1}U^{\mathcal{L}_1h,\iota_1},P_{k_2}\mathcal{L}'\mathcal{N}^\psi]\}(t)\|_{L^2}\\ \lesssim\varepsilon_1^2 2^{-N(n+1)k^+-3k^+}\langle t\rangle^{\delta[H(q,n)+\widetilde{\ell}(n)]}.\end{aligned}\tag{4.2.37}$$

The bounds (4.2.37) are easy. As in the proof of (4.2.25) we estimate the left-hand side by

$$\sum_{(k_1,k_2)\in\mathcal{X}_k} 2^{-k_1}2^{3\min\{k,k_1,k_2\}/2}\|P_{k_1}U^{\mathcal{L}_1h_1,\iota_1}(t)\|_{L^2}$$
$$\times\left[\|P_{k_2}\{x_l\mathcal{L}'\mathcal{N}^\psi\}(t)\|_{L^2}+2^{-k_2^-}\|P'_{k_2}\mathcal{L}'\mathcal{N}^\psi(t)\|_{L^2}\right]$$
$$\lesssim \varepsilon_1^2 2^{k/2}2^{-N(n+1)k^+-8k^+}\langle t\rangle^{\delta[H(q_1,n_1)+H(q',n')+\widetilde{\ell}(n')+2]},$$

using (3.3.3) and Proposition 4.7 inductively. This clearly suffices.

To prove (4.2.36) we write $U^{\mathcal{L}_2\psi,\iota_2}=e^{-it\Lambda_{kg,\iota_2}}V^{\mathcal{L}_2\psi,\iota_2}$ and examine the formula (3.2.43). We make the change of variables $\eta\to\xi-\eta$ and notice that the $\partial_{\xi_l}$ derivative can hit the multiplier $m(\eta,\xi-\eta)$, or the phase $e^{-it\Lambda_{kg,\iota_2}(\xi-\eta)}$, or the profile $P_{k_2}\widehat{V^{\mathcal{L}_2\psi,\iota_2}}(\xi-\eta)$. In the first two cases, the derivative effectively corresponds to multiplying by factors $\lesssim\langle t\rangle$ or $\lesssim 2^{-k_2^-}$, and changing the multiplier $m_1$, in a way that still satisfies (3.2.41). The desired bounds are then consequences of the bounds (4.2.31) (in the case $2^{k_2^-}\lesssim\langle t\rangle^{-1}$ we need to apply (4.1.52) again to control the corresponding contributions).

It remains to consider the case when the $\partial_{\xi_l}$ derivative hits the profile $P_{k_2}\widehat{V^{\mathcal{L}_2\psi,\iota_2}}(\xi-\eta)$. It suffices to prove that

$$\begin{aligned}\sum_{(k_1,k_2)\in\mathcal{X}_k} 2^{k_2^+-k_1}\|P_kI[P_{k_1}U^{\mathcal{L}_1h,\iota_1},U^{\mathcal{L}_2\psi,\iota_2}_{*l,k_2}](t)\|_{L^2}\\ \lesssim \varepsilon_1^2\langle t\rangle^{\delta[H(q,n)+\widetilde{\ell}(n)]}2^{-N(n+1)k^+-4k^+},\end{aligned}\tag{4.2.38}$$

where, as in Lemma 4.8, $\widehat{U^{\mathcal{L}_2\psi,\iota_2}_{*l,k_2}}(\xi,t)=e^{-it\Lambda_{kg,\iota_2}(\xi)}\partial_{\xi_l}\{\varphi_{k_2}\cdot\widehat{V^{\mathcal{L}_2\psi,\iota_2}}\}(\xi,t)$. In view of (4.2.14) we have

$$\|U^{\mathcal{L}_2\psi,\iota_2}_{*l,k_2}(t)\|_{L^2}\lesssim\varepsilon_1\langle t\rangle^{H(q_2+1,n_2+1)\delta}2^{-N(n_2+1)k_2^+}.\tag{4.2.39}$$

Assume first that $n_1\geq 1$. The contribution of the pairs $(k_1,k_2)$ in (4.2.38) for which $k_1\leq k_2+10$ is bounded as claimed, using the $L^\infty$ estimates (3.3.11) and the $L^2$ bounds (4.2.39). Similarly, the contribution of the pairs $(k_1,k_2)$ for which $k_2\leq k_1-10$ (thus $|k-k_1|\leq 4$) is bounded as claimed if $2^k\lesssim\langle t\rangle^{1/100}$. Finally, if $2^k\geq\langle t\rangle^{1/100}$ then the contribution of the pairs $(k_1,k_2)$ for which $k_2\leq k_1-10$ is bounded as claimed using (3.3.3) and (4.2.39).

Assume now that

$$n_1=0\qquad\text{and}\qquad n_2=n.\tag{4.2.40}$$

We need to be slightly more careful than before. If $2^k\lesssim\langle t\rangle^{1/100}$ then we can just use the $L^\infty$ bounds in (3.3.11) and the $L^2$ bounds (4.2.39) to prove (4.2.38). On the other hand, if $2^k\geq\langle t\rangle^{1/100}$ then the contribution of the pairs $(k_1,k_2)$ with $k_1\geq k-10$ or $k_1\leq -10k$ can be estimated as before, using just (4.2.39) and the $L^2$ bounds in (3.3.3).

To estimate the remaining contributions we need to avoid the derivative loss. Going back to (4.2.36), it remains to prove that if $2^k \geq \langle t\rangle^{1/100}$ then

$$\sum_{(k_1,k_2)\in\mathcal{X}_k,\, k_1\in[-10k,k-10]} 2^{k_2^+-k_1}\|P_k\{\partial_{\xi_l}\mathcal{F}\{I[P_{k_1}U^{\mathcal{L}_1h,\iota_1},P_{k_2}U^{\mathcal{L}_2\psi,\iota_2}]\}(t)\|_{L^2_\xi} \lesssim \varepsilon_1^2 2^{-N(n+1)k^+-3k^+}. \tag{4.2.41}$$

We do not make the change of variables $\eta\to\xi-\eta$ now, so the $\partial_{\xi_l}$ derivative hits the low frequency factor or the multiplier in the definition of the operator $I$. The contribution when the derivative $\partial_{\xi_l}$ hits the multiplier can be bounded easily using (3.3.11). Using the $L^2$ bounds (3.3.3) on $\psi$, and $L^2\times L^\infty$ estimates as before, for (4.2.41) it suffices to prove that

$$2^{-k_1}\big\|\mathcal{F}^{-1}\{\partial_{\xi_l}[\widehat{P_{k_1}U^{\mathcal{L}_1h}}(\xi,t)]\}\big\|_{L^\infty} \lesssim \varepsilon_1\langle t\rangle^{\delta'}2^{-2k_1^+}2^{-2\delta k_1^-}. \tag{4.2.42}$$

To prove (4.2.42) we replace $U^{\mathcal{L}_1h}(t)$ by $e^{-it|\nabla|}V^{\mathcal{L}_1h}(t)$ and use either (3.3.11) when the $\partial_{\xi_l}$ derivative hits the factor $\varphi_{k_1}(\xi)e^{-it|\xi|}$ or (3.3.9) when the derivative hits the profile. The desired bounds (4.2.42) follow. This completes the proof of the lemma. □

### 4.2.2 The Cubic and Higher Order Nonlinearities

We now prove bounds on the nonlinearities $\mathcal{LN}^{h,\geq 3}$ and $\mathcal{LN}^{\psi,\geq 3}$.

**Lemma 4.11.** *Assume that $t\in[0,T]$, $\mathcal{L}\in\mathcal{V}_n^q$, $n\leq 2$, $l\in\{1,2,3\}$, and $k\in\mathbb{Z}$. Then*

$$\|P_k(\mathcal{LN}^{h,\geq 3}_{\alpha\beta})(t)\|_{L^2} \lesssim \varepsilon_1^2\langle t\rangle^{-1/2+4\delta'}2^{-\widetilde{N}(n)k^++6k^+}\min(2^k,\langle t\rangle^{-1})^{3/2}, \tag{4.2.43}$$

$$\|P_k(x_l\mathcal{LN}^{h,\geq 3}_{\alpha\beta})(t)\|_{L^2} \lesssim \varepsilon_1^2 2^{k/2}\langle t\rangle^{-1/2+4\delta'}2^{-N(n+1)k^+-3k^+}. \tag{4.2.44}$$

*Moreover, if $n=3$ and $\mathcal{L}\in\mathcal{V}_n^q$ then*

$$\|P_k(\mathcal{LN}^{h,\geq 3}_{\alpha\beta})(t)\|_{L^2} \lesssim \varepsilon_1^2 2^{k/2}\langle t\rangle^{-1/2+4\delta'}2^{-\widetilde{N}(n)k^++6k^+}\min(2^k,\langle t\rangle^{-1}). \tag{4.2.45}$$

*Proof.* We notice that $\mathcal{N}^{h,\geq 3}_{\alpha\beta}$ is a sum of terms of the form $G_{\geq 1}\cdot\mathcal{N}_1$, where $G_{\geq 1}=\sum_{d\geq 1}a_dg_d$ for some functions $g_d\in\mathcal{G}_d$ (see (4.1.68)) and some coefficients $a_d\in\mathbb{R}$ with $|a_d|\leq C^d$, and $\mathcal{N}_1$ is a quadratic term similar to those in $\mathcal{KG}^2_{\alpha\beta}$, $\mathcal{Q}^2_{\alpha\beta}$, $\mathcal{S}^2_{\alpha\beta}$. It suffices to prove that if $\mathcal{L}\in\mathcal{V}_n^q$ then

$$\|P_k(\mathcal{L}(G_{\geq 1}\cdot\mathcal{N}_1))(t)\|_{L^2} \lesssim \varepsilon_1^2\langle t\rangle^{-1/2+4\delta'}2^{-\widetilde{N}(n)k^++6k^+}\min(2^k,\langle t\rangle^{-1})^{3/2}, \tag{4.2.46}$$

$$\|P_k(x_l\mathcal{L}(G_{\geq 1}\cdot\mathcal{N}_1))(t)\|_{L^2} \lesssim \varepsilon_1^2 2^{k/2}\langle t\rangle^{-1/2+4\delta'}2^{-N(n+1)k^+-3k^+}, \qquad (4.2.47)$$

provided that $n \leq 2$. Moreover, if $n = 3$ then

$$\|P_k(\mathcal{L}(G_{\geq 1}\cdot\mathcal{N}_1))(t)\|_{L^2} \lesssim \varepsilon_1^2 2^{k/2}\langle t\rangle^{-1/2+4\delta'}2^{-\widetilde{N}(n)k^+ +6k^+}\min(2^k,\langle t\rangle^{-1}). \qquad (4.2.48)$$

Concerning the functions $G_{\geq 1}$ and $\mathcal{N}_1$, we may assume that (see (4.1.71) and (4.1.70))

$$\begin{aligned}
\|P_k\mathcal{D}G_{\geq 1}(t)\|_{L^2} &\lesssim \varepsilon_1\langle t\rangle^{\delta'}2^{-N(m)k^+}2^{-k/2}2^{-\gamma k^-},\\
\|P_k\mathcal{D}G_{\geq 1}(t)\|_{L^\infty} &\lesssim \varepsilon_1\langle t\rangle^{-1+\delta'}2^{-N(m+1)k^+ +3k^+}\min\{1,2^{k^-}\langle t\rangle\}^{1-\delta}, \qquad (4.2.49)\\
\|P_k(x_l\mathcal{D}G_{\geq 1})(t)\|_{L^2} &\lesssim \varepsilon_1\langle t\rangle^{\delta'}2^{-N(m+1)k^+}2^{-k^-/2}(2^{-k^-}+\langle t\rangle)2^{-\gamma k^-},
\end{aligned}$$

where $l \in \{1,2,3\}$ and $\mathcal{D} \in \mathcal{V}^q_m$, where the inequalities in the first line hold for all pairs $(q,m) \leq (3,3)$, while the inequalities in the last two lines hold only for pairs $(q,n) \leq (2,2)$. We may also assume that $\mathcal{N}_1$ satisfies the bounds (see Lemmas 4.8 and 4.9)

$$\begin{aligned}
\|P_k\mathcal{D}\mathcal{N}_1(t)\|_{L^2} &\lesssim \varepsilon_1^2 2^{k/2}2^{-\widetilde{N}(m)k^+ +3k^+}\langle t\rangle^{-1+\delta'}\min\{1,2^k\langle t\rangle\},\\
\|P_k\mathcal{D}\mathcal{N}_1(t)\|_{L^\infty} &\lesssim \varepsilon_1^2 2^{k}\langle t\rangle^{-2+2\delta'}2^{-N(m+1)k^+ +3k^+}\min\{1,2^k\langle t\rangle\}, \qquad (4.2.50)\\
\|P_k(x_l\mathcal{D}\mathcal{N}_1)(t)\|_{L^2} &\lesssim \varepsilon_1^2 2^{k/2}2^{-N(m+1)k^+ -5k^+}\langle t\rangle^{\delta'}.
\end{aligned}$$

The bounds (4.2.46) follow easily from (4.2.49)–(4.2.50), using $L^2\times L^\infty$ estimates similar to those in Lemma 4.4, with the higher frequency factor placed in $L^2$ and the lower frequency factor in $L^\infty$. The proof of (4.2.47) also follows from (4.2.49)–(4.2.50). The case $2^k \leq \langle t\rangle^{-1}$ follows by $L^2$ estimates as before. If $2^k \geq \langle t\rangle^{-1}$ then we first combine the weight $x_l$ with the higher frequency and place it in $L^2$. Using (4.2.49)–(4.2.50), we have

$$\begin{aligned}
\sum_{(k_1,k_2)\in\mathcal{X}_k,\, k_1\leq k_2+10} &\|P_{k_1}\mathcal{L}_1G_{\geq 1}(t)\|_{L^\infty}\cdot\|P_{k_2}(x_l\mathcal{L}_2\mathcal{N}_1)(t)\|_{L^2}\\
&\lesssim \varepsilon_1^3\langle t\rangle^{-1+3\delta'}2^{-N(n+1)k^+ -4k^+}
\end{aligned}$$

and

$$\begin{aligned}
\sum_{(k_1,k_2)\in\mathcal{X}_k,\, k_2\leq k_1-10} &\|P_{k_1}(x_l\mathcal{L}_1G_{\geq 1})(t)\|_{L^2}\cdot\|P_{k_2}(\mathcal{L}_2\mathcal{N}_1)(t)\|_{L^\infty} \qquad (4.2.51)\\
&\lesssim \varepsilon_1^3\langle t\rangle^{-1+3\delta'}2^{-N(n_1+1)k^+}2^{k^-/2}.
\end{aligned}$$

These bounds suffice in most cases, except when ($n_1 = n$ and $2^k \geq \langle t\rangle^{1/8}$), because of the loss of derivative in (4.2.51). In this case, however, we combine the weight $x_l$ with the lower frequency and use the estimate

$$\sum_{(k_1,k_2)\in\mathcal{X}_k,\, k_2\le k_1-10} \|P_{k_1}(\mathcal{L}_1 G_{\ge 1})(t)\|_{L^2}\cdot 2^{3k_2/2}\|P_{k_2}(x_l\mathcal{L}_2\mathcal{N}_1)(t)\|_{L^2} \lesssim \varepsilon_1^3\langle t\rangle^{3\delta'}2^{-N(n)k^+},$$

which suffices.

The bounds (4.2.48) also follow from (4.2.49)–(4.2.50). We use first $L^2\times L^\infty$ estimates with the lower frequency placed in $L^\infty$. This suffices in most cases, except when all the vector-fields apply to the low frequency factor (so we do not have $L^\infty$ control of this factor). In this case, however, we can reverse the two norms, and still prove (4.2.48) in a similar way. □

**Lemma 4.12.** *Assume that $t\in[0,T]$, $\mathcal{L}\in\mathcal{V}_n^q$, $n\le 3$, and $k\in\mathbb{Z}$. Then*

$$\|P_k(\mathcal{L}\mathcal{N}^{\psi,\ge 3})(t)\|_{L^2}\lesssim \varepsilon_1^2\langle t\rangle^{-0.6}2^{-\widetilde{N}(n)k^+ +6k^+}\min\{2^k,\langle t\rangle^{-1}\}. \tag{4.2.52}$$

*Moreover, if $l\in\{1,2,3\}$ and $n\le 2$ then we also have the bounds*

$$\begin{aligned}\|P_k(\mathcal{L}\mathcal{N}^{\psi,\ge 3})(t)\|_{L^\infty}&\lesssim \varepsilon_1^2 2^{k/2}\langle t\rangle^{-1.6}2^{-N(n+1)k^+ +5k^+}\min\{2^k,\langle t\rangle^{-1}\},\\ \|P_k(x_l\mathcal{L}\mathcal{N}^{\psi,\ge 3})(t)\|_{L^2}&\lesssim \varepsilon_1^2\langle t\rangle^{-0.6}2^{-N(n+1)k^+ -2k^+}.\end{aligned} \tag{4.2.53}$$

*Proof.* We examine the identity (2.1.16). It suffices to prove that if $\Psi'\in\{\psi,\partial_\mu\psi,\partial_\mu\partial_\nu\psi:\ \mu,\nu\in\{0,1,2,3\},\ (\mu,\nu)\neq(0,0)\}$ and $G_{\ge 2}$ is a sum of the form $\sum_{d\ge 2}a_d g_d$ for some functions $g_d\in\mathcal{G}_d$ (see (4.1.68)) and some coefficients $a_d\in\mathbb{R}$ with $|a_d|\le C^d$, then

$$\|P_k\{\mathcal{L}(G_{\ge 2}\cdot\Psi')\}(t)\|_{L^2}\lesssim \varepsilon_1^2\langle t\rangle^{-0.6}2^{-\widetilde{N}(n)k^+ +6k^+}\min\{2^k,\langle t\rangle^{-1}\} \tag{4.2.54}$$

and, assuming that $n\le 2$,

$$\begin{aligned}\|P_k\{\mathcal{L}(G_{\ge 2}\cdot\Psi')\}(t)\|_{L^\infty}&\lesssim \varepsilon_1^2 2^{k/2}\langle t\rangle^{-1.6}2^{-N(n+1)k^+ +5k^+}\min\{2^k,\langle t\rangle^{-1}\},\\ \|P_k\{x_l\mathcal{L}(G_{\ge 2}\cdot\Psi')\}(t)\|_{L^2}&\lesssim \varepsilon_1^2\langle t\rangle^{-0.6}2^{-N(n+1)k^+ -2k^+}.\end{aligned} \tag{4.2.55}$$

Concerning the functions $G_{\ge 2}$ and $\Psi'$, we may assume that (see (4.1.69))

$$\begin{aligned}\|P_k\mathcal{D}G_{\ge 2}(t)\|_{L^2}&\lesssim \varepsilon_1^2\langle t\rangle^{-1+\delta'}2^{-N(m)k^+}2^{-k/2}\min\{1,2^{k^-}\langle t\rangle\}^{1-\delta'},\\ \|P_k\mathcal{D}G_{\ge 2}(t)\|_{L^\infty}&\lesssim \varepsilon_1^2\langle t\rangle^{-2+\delta'}2^{-N(m+1)k^+ +3k^+}\min\{1,2^{k^-}\langle t\rangle\}^{2-\delta'},\\ \|P_k(x_l\mathcal{D}G_{\ge 2})(t)\|_{L^2}&\lesssim \varepsilon_1^2(2^{-k^-}+\langle t\rangle)^{\delta'}2^{-N(m+1)k^+}2^{-k^-/2},\end{aligned} \tag{4.2.56}$$

where $l\in\{1,2,3\}$ and $\mathcal{D}\in\mathcal{V}_m^q$, where the inequalities in the first line hold for all pairs $(q,m)\le(3,3)$, while the inequalities in the last two lines hold only for pairs $(q,m) \leq (2,2)$. Also, as a consequence of (3.3.3), (3.3.13), and (3.3.9),

$$
\begin{aligned}
\|P_k\mathcal{D}\Psi'(t)\|_{L^2} &\lesssim \varepsilon_1\langle t\rangle^{\delta'}2^{-N(m)k^+ + k^+},\\
\|P_k\mathcal{D}\Psi'(t)\|_{L^\infty} &\lesssim \varepsilon_1\langle t\rangle^{-1+\delta'}2^{-N(m+1)k^+ + 3k^+}2^{k^-/2}\min\{1, 2^{2k^-}\langle t\rangle\},\\
\|P_k(x_l\mathcal{D}\Psi')(t)\|_{L^2} &\lesssim \varepsilon_1\langle t\rangle^{\delta'}(1+2^{k^-}\langle t\rangle)2^{-N(m+1)k^+ + k^+},
\end{aligned}
\tag{4.2.57}
$$

As before, we start with $L^2$ estimates,

$$
\begin{aligned}
&\|P_k(\mathcal{L}_1G_{\geq 2}\cdot\mathcal{L}_2\Psi')(t)\|_{L^2}\\
&\lesssim \sum_{(k_1,k_2)\in\mathcal{X}_k} 2^{3\min(k,k_1,k_2)/2}\|P_{k_1}\mathcal{L}_1G_{\geq 2}(t)\|_{L^2}\|P_{k_2}\mathcal{L}_2\Psi'(t)\|_{L^2}\\
&\lesssim \varepsilon_1^3 2^{k^-}\langle t\rangle^{-1+3\delta'}2^{-N(n)k^+ + 2k^+},
\end{aligned}
\tag{4.2.58}
$$

provided that $\mathcal{L}_1 \in \mathcal{V}^{q_1}_{n_1}$, $\mathcal{L}_2 \in \mathcal{V}^{q_2}_{n_2}$, $n_1 + n_2 = n$. This suffices to prove (4.2.54) if $2^k \lesssim \langle t\rangle^{-2/3}$ or if $2^k \gtrsim \langle t\rangle^{1/5}$. Moreover, if $2^k \in [\langle t\rangle^{-2/3}, \langle t\rangle^{1/5}]$ then we use $L^2 \times L^\infty$ estimates and (4.2.56)–(4.2.57), with the lower frequency factor estimated in $L^\infty$ and the higher frequency factor estimated in $L^2$. This suffices in most cases, except when $n = 3$ and all the vector-fields apply to the low frequency factor. In this case, however, we can reverse the two norms, and still prove the desired conclusion (4.2.54).

The $L^\infty$ bounds in (4.2.55) follow from the $L^2$ bounds (4.2.58) if $2^k \lesssim \langle t\rangle^{-0.9}$. On the other hand, if $2^k \geq \langle t\rangle^{-0.9}$ then we just use the $L^2$ and the $L^\infty$ bounds in (4.2.56)–(4.2.57).

The proof of the weighted $L^2$ bounds in (4.2.55) is similar. The case $2^k \leq \langle t\rangle^{-1}$ follows by $L^2$ estimates as before. As in the proof of Lemma 4.11, if $2^k \geq \langle t\rangle^{-1}$ then we first combine the weight $x_l$ with the higher frequency and place it in $L^2$. This gives the desired bounds (4.2.55) in most cases, except when $2^k \geq \langle t\rangle^{1/10}$ and ($n_1 = n$ or $n_2 = n$), because of the loss of derivative. In these cases, however, we combine the weight $x_l$ with the lower frequency factor and use the $L^2$ estimates in the first and third lines of (4.2.56)–(4.2.57). The desired bounds (4.2.55) follow in these cases as well. □

### 4.2.3 Additional Low Frequency Bounds

We prove now some additional linear bounds on the solutions $P_kU^{\mathcal{L}h}$ when $k$ is very small. These bounds are important in some of the energy estimates in the next sections, when the vector-fields hit very low frequency factors.

**Lemma 4.13.** *Assume that $t \in [0,T]$, $k \leq 0$, and $J + k \geq 0$. If $\mathcal{L} = \Gamma_a\mathcal{L}'$, $a \in \{1,2,3\}$, $\mathcal{L}' \in \mathcal{V}^q_n$, $n \leq 2$, and $h \in \{h_{\alpha\beta}\}$ then*

$$
(2^{k^-}\langle t\rangle)^\gamma 2^{-k/2}\|\varphi_{\leq J}(x)\cdot P_kU^{\mathcal{L}h}(t)\|_{L^2} \lesssim \varepsilon_1 2^k(2^J + \langle t\rangle)\cdot\langle t\rangle^{H(q,n)\delta + 2\delta}. \tag{4.2.59}
$$

*In addition, if* $|\alpha| \leq 3$, $\mathcal{L}' \in \mathcal{V}_n^q$, $n + |\alpha| \leq 3$, *then*

$$(2^{k^-}\langle t\rangle)^\gamma 2^{-k/2}\|\varphi_{\leq J}(x)\cdot P_k U^{\Omega^\alpha \mathcal{L}'h}(t)\|_{L^2} \lesssim \varepsilon_1 2^{|\alpha|(J+k)}\cdot \langle t\rangle^{H(q,n)\delta}. \quad (4.2.60)$$

*Proof.* These bounds, which should be compared with (3.3.3), are used only when $2^k \lesssim \langle t\rangle^{-1+\delta'}$. Recall the formula $U^{\mathcal{L}h}(t) = (\partial_t - i\Lambda_{wa})(\mathcal{L}h)(t)$. To prove (4.2.59) we write

$$U^{\mathcal{L}h}(t) = (\partial_t - i\Lambda_{wa})(\Gamma_a\mathcal{L}'h)(t) = \Gamma_a U^{\mathcal{L}'h}(t) + [\partial_t - i\Lambda_{wa}, \Gamma_a](\mathcal{L}'h)(t). \quad (4.2.61)$$

The commutator can be bounded easily, without needing spatial localization,

$$\big\|P_k[\partial_t - i\Lambda_{wa}, \Gamma_a](\mathcal{L}'h)(t)\big\|_{L^2} \lesssim \big\|P_k U^{\mathcal{L}'h}(t)\big\|_{L^2} \lesssim \varepsilon_1(2^{k^-}\langle t\rangle)^{-\gamma}2^{k/2}\langle t\rangle^{H(q,n)\delta};$$

see (3.3.3). For the main term we write $\Gamma_a U^{\mathcal{L}'h}(t) = t\partial_a U^{\mathcal{L}'h}(t) + x_a\partial_t U^{\mathcal{L}'h}(t)$, and estimate

$$\big\|P_k t\partial_a U^{\mathcal{L}'h}(t)\big\|_{L^2} \lesssim \langle t\rangle 2^k \big\|P_k U^{\mathcal{L}'h}(t)\big\|_{L^2} \lesssim \varepsilon_1(2^{k^-}\langle t\rangle)^{-\gamma}2^{k/2}\langle t\rangle^{H(q,n)\delta}\langle t\rangle 2^k.$$

Using the identity $\partial_t U^{\mathcal{L}'h}(t) = -i\Lambda_{wa}U^{\mathcal{L}'h}(t) + \mathcal{L}'\mathcal{N}^h(t)$ and spatial localization,

$$\begin{aligned}
&\|\varphi_{\leq J}(x)\cdot P_k(x_a\partial_t U^{\mathcal{L}'h}(t))\|_{L^2} \lesssim 2^J\|P'_k(\partial_t U^{\mathcal{L}'h}(t))\|_{L^2}\\
&\quad\lesssim 2^{J+k}\|P'_k U^{\mathcal{L}'h}(t)\|_{L^2} + 2^J\|P'_k\mathcal{L}'\mathcal{N}^h(t)\|_{L^2}\\
&\quad\lesssim 2^{J+k}\varepsilon_1(2^{k^-}\langle t\rangle)^{-\gamma}2^{k/2}\langle t\rangle^{H(q,n)\delta} + 2^J\varepsilon_1^2 2^{3k/2}\langle t\rangle^{H(q,n)\delta+3\delta/2},
\end{aligned} \quad (4.2.62)$$

where $P'_k = P_{[k-2,k+2]}$ and we use (4.2.3) in the last line. The desired conclusion (4.2.59) follows from the last three bounds.

The proof of (4.2.60) is easier, since $P_k U^{\Omega^\alpha\mathcal{L}'h}(t) = \Omega^\alpha P_k U^{\mathcal{L}'h}(t)$, and each $\Omega$ vector-field generates a factor of $2^{J+k}$, as in (4.2.62) above. □

### 4.2.4 Additional Bounds on Some Quadratic Nonlinearities

We will also need some slightly stronger bounds on some of the components of the nonlinearities $\mathcal{L}\mathcal{N}^h_{\alpha\beta}$ when $\mathcal{L} \in \mathcal{V}_1^q$.

**Lemma 4.14.** *For any* $k \in \mathbb{Z}$, $t \in [0,T]$, $\mathcal{L} \in \mathcal{V}_n^q$, $n \leq 1$, *and* $\alpha, \beta \in \{0,1,2,3\}$

$$\|P_k(\mathcal{L}\mathcal{KG}^2_{\alpha\beta})(t)\|_{L^2} + \|P_k(\mathcal{L}\mathcal{S}^2_{\alpha\beta})(t)\|_{L^2} \lesssim \varepsilon_1^2 2^{k/2}\langle t\rangle^{-1+H(q,n)\delta+3\delta/2}2^{-N(n)k^+ + k^+}. \quad (4.2.63)$$

*Moreover, if* $2^k \leq \langle t\rangle^{-1/10}$ *then*

$$\|P_k(\mathcal{L}\mathcal{Q}^2_{\alpha\beta})(t)\|_{L^2} \lesssim \varepsilon_1^2 2^{k/2}\langle t\rangle^{-1+H(q,n)\delta+3\delta/2}. \quad (4.2.64)$$

*Proof.* The quadratic nonlinearities $\mathcal{KG}^2_{\alpha\beta}$, $\mathcal{S}^2_{\alpha\beta}$, and $\mathcal{Q}^2_{\alpha\beta}$ are defined in (2.1.12)–

(2.1.15). The point of the lemma is the slightly better estimates in terms of powers of $\langle t\rangle$, when at most one vector-field acts on nonlinearities. We prove the desired bounds in several steps.

**Step 1.** We consider first bilinear interactions of the metric components. For later use we prove slightly stronger frequency-localized estimates. Assume $k, k_1, k_2 \in \mathbb{Z}$, $t \in [0,T]$, $\mathcal{L}_2 \in \mathcal{V}^{q_2}_{n_2}$, $n_2 \leq 1$, $h_1, h_2 \in \{h_{\alpha\beta}\}$, and $\iota_1, \iota_2 \in \{+,-\}$. Assume that $m$ is a multiplier satisfying $\|\mathcal{F}^{-1}m\|_{L^1(\mathbb{R}^6)} \leq 1$ and define $I_m$ as in (3.2.43). Using just $L^2$ estimates we have

$$\begin{aligned}\big\|P_kI_m[P_{k_1}U^{h_1,\iota_1}(t), P_{k_2}U^{\mathcal{L}_2h_2,\iota_2}(t)]\big\|_{L^2}\\ \lesssim \varepsilon_1^2 2^{3\min\{k,k_1,k_2\}/2}\big(\langle t\rangle^2 2^{k_1^-+k_2^-}\big)^{-\gamma}2^{k_1/2+k_2/2}\\ \times \langle t\rangle^{[H(q_2,n_2)+1]\delta}2^{-N(0)k_1^+-N(n_2)k_2^+}.\end{aligned} \tag{4.2.65}$$

Moreover, if $\langle t\rangle \gg 1$, $2^k \geq \langle t\rangle^{-1}$, and $k = \min\{k,k_1,k_2\}$ then we also have the estimates (see (4.1.18)–(4.1.19))

$$\begin{aligned}\big\|P_kI_m[P_{k_1}U^{h_1,\iota_1}(t), P_{k_2}U^{\mathcal{L}_2h_2,\iota_2}(t)]\big\|_{L^2}\\ \lesssim \varepsilon_1^2 2^{k/2}\langle t\rangle^{-1+\delta[H(q_2,n_2)+1]}2^{-N(n_2)k_2^+-5k_2^+}2^{k_2^-/4}.\end{aligned} \tag{4.2.66}$$

On the other hand, if $\langle t\rangle \gg 1$, $2^k \geq \langle t\rangle^{-1}$, and $k_1 = \min\{k,k_1,k_2\}$ then $2^k \approx 2^{k_2}$ and

$$\big\|P_kI_m[P_{k_1}U^{h_1,\iota_1}(t), P_{k_2}U^{\mathcal{L}_2h_2,\iota_2}(t)]\big\|_{L^2} \lesssim \varepsilon_1^2 2^{-|k_1|}\langle t\rangle^{-1+\delta'}2^{k/2}2^{-N(n_2)k^+}, \tag{4.2.67}$$

using the bounds (3.3.11) and (3.3.3). If, in addition, $\langle t\rangle^{-8\delta'} \leq 2^{k_1} \leq 2^{k_2} \leq \langle t\rangle^{8\delta'}$ then we let $J_1$ be the largest integer such that $2^{J_1} \leq \langle t\rangle 2^{-30}$, decompose $P_{k_1}U^{h_1,\iota_1}(t) = U^{h_1,\iota_1}_{\leq J_1,k_1}(t) + U^{h_1,\iota_1}_{>J_1,k_1}(t)$ as in (4.1.14), and use (3.3.4) and (3.3.11), together with the stronger bounds (3.3.16) on $\|U^{h_1,\iota_1}_{\leq J_1,k_1}(t)\|_{L^\infty}$, to show that

$$\begin{aligned}\big\|P_kI_m[P_{k_1}U^{h_1,\iota_1}(t), P_{k_2}U^{\mathcal{L}_2h_2,\iota_2}(t)]\big\|_{L^2}\\ \lesssim \varepsilon_1^2 2^{-|k_1|/2}\langle t\rangle^{-1+\delta[H(q_2,n_2)+1]}2^{k/2}2^{-N(n_2)k^+}.\end{aligned} \tag{4.2.68}$$

Finally, if $\langle t\rangle \gg 1$, $2^k \geq \langle t\rangle^{-1}$, and $k_2 = \min\{k,k_1,k_2\}$ then $2^k \approx 2^{k_1}$ and, using (4.1.10),

$$\begin{aligned}\big\|P_kI_m[P_{k_1}U^{h_1,\iota_1}(t), P_{k_2}U^{\mathcal{L}_2h_2,\iota_2}(t)]\big\|_{L^2}\\ \lesssim \varepsilon_1^2\langle t\rangle^{-1+\delta[H(q_2,n_2)+1]}2^{-|k_2|}2^{|k|/4}2^{-N_0k^++2k^+}.\end{aligned} \tag{4.2.69}$$

We examine the identities (2.1.14)–(2.1.15) and estimate

$$\|P_k(\mathcal{L}\mathcal{S}^2_{\alpha\beta})(t)\|_{L^2} \lesssim \sum_{k_1,k_2,\iota_1,\iota_2,h_1,h_2,\mathcal{L}_2\in\mathcal{V}^q_n} \big\|P_kI_m[P_{k_1}U^{h_1,\iota_1}(t),P_{k_2}U^{\mathcal{L}_2h_2,\iota_2}(t)]\big\|_{L^2}. \tag{4.2.70}$$

The desired bounds in (4.2.63) follow from (4.2.65) if $\langle t\rangle \lesssim 1$ or if $2^k \lesssim \langle t\rangle^{-1}$. On the other hand, if $\langle t\rangle \gg 1$ and $2^k \geq \langle t\rangle^{-1}$ then the High $\times$ High $\to$ Low interactions in (4.2.70) can be estimated as claimed using (4.2.66). The Low $\times$ High $\to$ High interactions in (4.2.70) can be estimated using (4.2.67)–(4.2.68) if $k_1 \leq k_2$ or if $n = 0$ (thus $\mathcal{L}_2 = Id$); they can also be estimated using (4.2.69) if $k_2 \leq k_1$ and $n = 1$. This completes the proof of the bounds on $\|P_k(\mathcal{L}\mathcal{S}^2_{\alpha\beta})(t)\|_{L^2}$ in (4.2.63).

Similarly, using the identities (2.1.13) we estimate

$$\begin{aligned}\|P_k(\mathcal{L}\mathcal{Q}^2_{\alpha\beta})(t)\|_{L^2} &\lesssim \sum_{k_1,k_2,\iota_1,\iota_2,h_1,h_2,\mathcal{L}_2\in\mathcal{V}^q_n} 2^{|k_1-k_2|}\big\|P_kI_m[P_{k_1}U^{h_1,\iota_1}(t),P_{k_2}U^{\mathcal{L}_2h_2,\iota_2}(t)]\big\|_{L^2}\\ &+ \sum_{k_1,k_2,\iota_1,h_1,\mathcal{N}^h_2} 2^{-k_1}\big\|P_kI_m[P_{k_1}U^{h_1,\iota_1}(t),P_{k_2}\mathcal{N}^h_2(t)]\big\|_{L^2},\end{aligned} \tag{4.2.71}$$

where $\mathcal{N}^h_2 \in \{\mathcal{N}^h_{\alpha\beta},\,\alpha,\beta\in\{0,1,2,3\}\}$. The terms in the last line of (4.2.71) are generated by commuting $\mathcal{L}$ and derivatives, and then replacing $\partial_0^2 h$ with $\Delta h + \mathcal{N}^h$, as in the proof of Lemma 4.9. The desired bounds (4.2.64) follow as before, using (4.2.65), (4.2.66), (4.2.67), and (4.2.69) to control the terms in the second line of (4.2.71) (recall that $2^k \leq \langle t\rangle^{-1/10}$), and the $L^2$ estimates (4.2.3) and (3.3.3) to control the terms in the last line.

**Step 2.** We consider now bilinear interactions of the Klein-Gordon field, and prove again slightly stronger frequency-localized estimates. Assume $k,k_1,k_2 \in \mathbb{Z}$, $t\in[0,T]$, $\mathcal{L}_2 \in \mathcal{V}^{q_2}_{n_2}$, $n_2 \leq 1$, $\iota_1,\iota_2 \in \{+,-\}$, and $m$ satisfies $\|\mathcal{F}^{-1}m\|_{L^1(\mathbb{R}^6)} \leq 1$. Using $L^2\times L^\infty$ estimates (with the higher frequency in $L^2$), and the bounds (3.3.13), (3.3.3), and (3.3.5), we have

$$\begin{aligned}&\big\|P_kI_m[P_{k_1}U^{\psi,\iota_1}(t),P_{k_2}U^{\mathcal{L}_2\psi,\iota_2}(t)]\big\|_{L^2}\\ &\qquad\lesssim \varepsilon_1^2\langle t\rangle^{-1+\delta'}2^{\min\{k_1^-,k_2^-\}/2}2^{\max\{k_1^-,k_2^-\}}2^{-N(n_2)\max\{k_1^+,k_2^+\}}.\end{aligned} \tag{4.2.72}$$

Assume that $2^{k_1^-} \geq \langle t\rangle^{-1/2}2^{20}$. Then we write $P_{k_1}U^{\psi,\iota_1}(t) = U^{\psi,\iota_1}_{\leq J_1,k_1}(t) + U^{\psi,\iota_1}_{>J_1,k_1}(t)$ as in (3.3.1)–(3.3.2), where $J_1$ is the largest integer satisfying $2^{J_1} \leq$

$2^{k_1^- - 20}\langle t\rangle$. Using (3.3.17), (3.3.13), and (3.3.3)–(3.3.4) we estimate

$$\begin{aligned}&\big\|P_k I_m[P_{k_1}U^{\psi,\iota_1}(t), P_{k_2}U^{\mathcal{L}_2\psi,\iota_2}(t)]\big\|_{L^2}\\&\quad\lesssim \|U^{\psi,\iota_1}_{\leq J_1,k_1}(t)\|_{L^\infty}\|P_{k_2}U^{\mathcal{L}_2\psi,\iota_2}(t)\|_{L^2} + \|U^{\psi,\iota_1}_{>J_1,k_1}(t)\|_{L^2}\|P_{k_2}U^{\mathcal{L}_2\psi,\iota_2}(t)\|_{L^\infty}\\&\quad\lesssim \varepsilon_1^2\langle t\rangle^{-3/2+H(q_2,n_2)\delta}2^{-k_1^-/2+\kappa k_1^-/20}2^{-N(1)k_1^+}2^{-5k_2^+}.\end{aligned} \tag{4.2.73}$$

Finally, since $\|P_{k_1}U^{\psi,\iota_1}(t)\|_{L^2} \lesssim 2^{k_1^- + \kappa k_1^-}2^{-N_0 k_1^+ + 2k_1^+}$ (see (3.3.7)) we can use just $L^2$ bounds to estimate

$$\begin{aligned}&\big\|P_k I_m[P_{k_1}U^{\psi,\iota_1}(t), P_{k_2}U^{\mathcal{L}_2\psi,\iota_2}(t)]\big\|_{L^2}\\&\quad\lesssim \varepsilon_1^2 2^{3\min\{k^-,k_1^-,k_2^-\}/2}2^{k_1^- + \kappa k_1^-}2^{-N(1)k_1^+}2^{-N(n_2)k_2^+}\langle t\rangle^{H(q_2,n_2)\delta}.\end{aligned} \tag{4.2.74}$$

We can now complete the proof of (4.2.63). We estimate first

$$\|P_k(\mathcal{L}\mathcal{K}\mathcal{G}^2_{\alpha\beta})(t)\|_{L^2} \lesssim \sum_{k_1,k_2,\iota_1,\iota_2,\mathcal{L}_2\in\mathcal{V}^q_n}\big\|P_k I_m[P_{k_1}U^{\psi,\iota_1}(t), P_{k_2}U^{\mathcal{L}_2\psi,\iota_2}(t)]\big\|_{L^2}. \tag{4.2.75}$$

The bounds claimed in (4.2.63) follow using only (4.2.72) if $2^k \gtrsim \langle t\rangle^{\delta'}$. If $2^k \in [1, \langle t\rangle^{\delta'}]$ then the desired bounds follow using (4.2.73) for the contribution of the pairs $(k_1,k_2) \in \mathcal{X}_k$ with $2^{k_1^-} \geq \langle t\rangle^{-1/2}2^{20}$, and (4.2.74) for the other pairs.

Assume now that $2^k \leq 1$. First we use (4.2.74) to bound the contribution of the pairs $(k_1,k_2)$ for which $2^{k_1^- + k^-} \lesssim \langle t\rangle^{-1}$. On the other hand if $2^{k_1^- + k^-} \geq \langle t\rangle^{-1}2^{100}$ then we use (4.2.73) if $2^{k_1^-} \geq \langle t\rangle^{-1/2}2^{20}$ and (4.2.72) for the remaining pairs with $2^{k_1^-} \leq \langle t\rangle^{-1/2}2^{20} \leq 2^{k^-}$. This completes the proof of (4.2.63). □

## 4.3 DECOMPOSITIONS OF THE MAIN NONLINEARITIES

### 4.3.1 The Variables $F^{\mathcal{L}}, \underline{F}^{\mathcal{L}}, \rho^{\mathcal{L}}, \omega_j^{\mathcal{L}}, \Omega_j^{\mathcal{L}}, \vartheta_{jk}^{\mathcal{L}}$

Recall that these variables were defined in (2.1.26) and (2.1.30). The harmonic gauge condition (1.2.7) gives

$$m^{\alpha\beta}\partial_\alpha h_{\beta\mu} - \frac{1}{2}m^{\alpha\beta}\partial_\mu h_{\alpha\beta} = E^{\geq 2}_\mu := -g^{\alpha\beta}_{\geq 1}\partial_\alpha h_{\beta\mu} + \frac{1}{2}g^{\alpha\beta}_{\geq 1}\partial_\mu h_{\alpha\beta}. \tag{4.3.1}$$

These identities and the identities (2.1.2) can be used to derive elliptic equations for the variables $F, \underline{F}, \rho, \omega_j, \Omega_j, \vartheta_{jk}$. More precisely, let $R_0 := |\nabla|^{-1}\partial_t$ and

$$\tau = \tau[H] := (1/2)[\delta_{jk}H_{jk} + R_jR_kH_{jk}] = -(1/2)\delta_{jk}\vartheta[H]_{jk}. \tag{4.3.2}$$

We show below that the variables $\rho^{\mathcal{L}}$ and $\Omega_j^{\mathcal{L}}$ can be expressed in terms of the other variables, up to quadratic errors, while the variables $\tau^{\mathcal{L}} := \tau[\mathcal{L}h]$ are in fact quadratic. More precisely:

**Lemma 4.15.** *Assume that $\mathcal{L} \in \mathcal{V}_3^3$ and define*

$$E^{com}_{\mathcal{L},\mu} := m^{\alpha\beta}[\partial_\alpha, \mathcal{L}]h_{\beta\mu} - \frac{1}{2}m^{\alpha\beta}[\partial_\mu, \mathcal{L}]h_{\alpha\beta}. \tag{4.3.3}$$

*Then the variables $\rho^{\mathcal{L}}, \Omega_j^{\mathcal{L}}$ (defined in (2.1.30)) satisfy the elliptic-type identities*

$$\begin{aligned} \rho^{\mathcal{L}} &= R_0\underline{F}^{\mathcal{L}} + R_0\tau^{\mathcal{L}} + |\nabla|^{-1}\mathcal{L}E_{\overline{0}}^{\geq 2} + |\nabla|^{-1}E^{com}_{\mathcal{L},0}, \\ \Omega_j^{\mathcal{L}} &= R_0\omega_j^{\mathcal{L}} + |\nabla|^{-1} \in_{jlk} R_l\mathcal{L}E_{\overline{k}}^{\geq 2} + |\nabla|^{-1} \in_{jlk} R_l E^{com}_{\mathcal{L},k}. \end{aligned} \tag{4.3.4}$$

*The variables $\tau^{\mathcal{L}}$ satisfy the identities*

$$\begin{aligned} 2|\nabla|^2\tau^{\mathcal{L}} &= \partial_\alpha\mathcal{L}E_{\overline{\alpha}}^{\geq 2} + \partial_\alpha E^{com}_{\mathcal{L},\alpha} + \underline{F}[\mathcal{L}\mathcal{N}^h] + \tau[\mathcal{L}\mathcal{N}^h], \\ 2|\nabla|\partial_0\tau^{\mathcal{L}} &= -|\nabla|\mathcal{L}E_{\overline{0}}^{\geq 2} + R_k\partial_0\mathcal{L}E_{\overline{k}}^{\geq 2} - |\nabla|E^{com}_{\mathcal{L},0} + \partial_0 R_k E^{com}_{\mathcal{L},k} + \rho[\mathcal{L}\mathcal{N}^h]. \end{aligned} \tag{4.3.5}$$

*Proof.* As a consequence of (4.3.1) we have

$$\begin{aligned} m^{\alpha\beta}\partial_\alpha\mathcal{L}h_{\beta\mu} - \frac{1}{2}m^{\alpha\beta}\partial_\mu\mathcal{L}h_{\alpha\beta} &= \mathcal{L}E_{\overline{\mu}}^{\geq 2} + m^{\alpha\beta}[\partial_\alpha, \mathcal{L}]h_{\beta\mu} - \frac{1}{2}m^{\alpha\beta}[\partial_\mu, \mathcal{L}]h_{\alpha\beta} \\ &= \mathcal{L}E_{\overline{\mu}}^{\geq 2} + E^{com}_{\mathcal{L},\mu}. \end{aligned} \tag{4.3.6}$$

Notice that $m^{\alpha\beta}\mathcal{L}h_{\alpha\beta} = 2\tau^{\mathcal{L}} - 2F^{\mathcal{L}}$. Therefore (4.3.6) with $\mu = 0$ gives

$$-\partial_0(F^{\mathcal{L}} + \underline{F}^{\mathcal{L}}) + |\nabla|\rho^{\mathcal{L}} - \partial_0(\tau^{\mathcal{L}} - F^{\mathcal{L}}) = \mathcal{L}E_{\overline{0}}^{\geq 2} + E^{com}_{\mathcal{L},0}.$$

This simplifies to

$$-\partial_0\underline{F}^{\mathcal{L}} + |\nabla|\rho^{\mathcal{L}} - \partial_0\tau^{\mathcal{L}} = \mathcal{L}E_{\overline{0}}^{\geq 2} + E^{com}_{\mathcal{L},0}. \tag{4.3.7}$$

Similarly, using (4.3.6) with $\mu = k \in \{1, 2, 3\}$ gives

$$-\partial_0\mathcal{L}h_{0k} + \partial_j\mathcal{L}h_{jk} - \frac{1}{2}\partial_k(m^{\alpha\beta}\mathcal{L}h_{\alpha\beta}) = \mathcal{L}E_{\overline{k}}^{\geq 2} + E^{com}_{\mathcal{L},k}.$$

Taking the divergence and the curl, and using (2.1.26), this gives

$$-\partial_0\rho^{\mathcal{L}} - |\nabla|\underline{F}^{\mathcal{L}} + |\nabla|\tau^{\mathcal{L}} = R_k\mathcal{L}E_{\overline{k}}^{\geq 2} + R_k E^{com}_{\mathcal{L},k} \tag{4.3.8}$$

and

$$-\partial_0\omega_j^{\mathcal{L}} + |\nabla|\Omega_j^{\mathcal{L}} = \in_{jlk} R_l\mathcal{L}E_{\overline{k}}^{\geq 2} + \in_{jlk} R_l E^{com}_{\mathcal{L},k}. \tag{4.3.9}$$

The identities (4.3.4) follow from (4.3.7) and (4.3.9).

We can now use (4.3.7) and (4.3.8) to derive the identities (4.3.5) for $\tau^{\mathcal{L}}$. Indeed, applying $\partial_0$ to the first equation and $|\nabla|$ to the second equation and adding up we have

$$-(\partial_0^2+|\nabla|^2)\underline{F}^{\mathcal{L}}-(\partial_0^2+|\nabla|^2)\tau^{\mathcal{L}}+2|\nabla|^2\tau^{\mathcal{L}}$$
$$=\partial_0\mathcal{L}E_0^{\geq 2}+|\nabla|R_kE_k^{\geq 2}+\partial_0E^{com}_{\mathcal{L},0}+|\nabla|R_kE^{com}_{\mathcal{L},k}.$$

Similarly, applying $-|\nabla|$ to the first equation and $\partial_0$ to the second equation and adding up,

$$-(\partial_0^2+|\nabla|^2)\rho^{\mathcal{L}}+2|\nabla|\partial_0\tau^{\mathcal{L}}=-|\nabla|\mathcal{L}E_0^{\geq 2}+R_k\partial_0\mathcal{L}E_k^{\geq 2}-|\nabla|E^{com}_{\mathcal{L},0}+\partial_0R_kE^{com}_{\mathcal{L},k}.$$

The desired identities (4.3.5) follow since $(\partial_0^2+|\nabla|^2)G^{\mathcal{L}}=G[\mathcal{L}\mathcal{N}^h]$, $G\in\{\underline{F},\tau,\rho\}$. □

### 4.3.2 Energy Disposable Nonlinearities

To prove energy estimate in the next section we need to bound the contribution of spacetime integrals. Many resulting terms can be estimated easily, without normal form analysis, using just decay properties. In this subsection we identify these terms. We start with a definition.

**Definition 4.16.** *Assume that $(q,n)\leq(3,3)$. A function $L:\mathbb{R}^3\times[0,T]\to\mathbb{C}$ will be called "wave-disposable of order $(q,n)$" if, for any $t\in[0,T]$,*

$$\big\||\nabla|^{-1/2}L(t)\big\|_{H^{N(n)}}\lesssim\varepsilon_1^2\langle t\rangle^{-1+H(q,n)\delta-\delta/2}. \tag{4.3.10}$$

*Similarly, a function $L:\mathbb{R}^3\times[0,T]\to\mathbb{C}$ will be called "KG-disposable of order $(q,n)$" if*

$$\big\|L(t)\big\|_{H^{N(n)}}\lesssim\varepsilon_1^2\langle t\rangle^{-1+H(q,n)\delta-\delta/2}. \tag{4.3.11}$$

We identify now suitable classes of cubic terms that are energy disposable. More precisely:

**Definition 4.17.** *We define two sets of quadratic and higher order expressions $\mathcal{QU}_0\subseteq\mathcal{QU}$*

$$\begin{aligned}\mathcal{QU}&:=\big\{\partial_\alpha h_1\partial_\beta h_2,\,G_{\geq1}\cdot\partial_\alpha h_1\partial_\beta h_2,\,\mathcal{KG}^2_{\alpha\beta},\,G_{\geq1}\cdot\mathcal{KG}^2_{\alpha\beta},\,G_{\geq1}\cdot\partial_\alpha\partial_\beta h_2\big\},\\ \mathcal{QU}_0&:=\big\{\partial_\alpha h_1\partial_\beta h_2,\,G_{\geq1}\cdot\partial_\alpha h_1\partial_\beta h_2,\,\mathcal{KG}^2_{\alpha\beta},\,G_{\geq1}\cdot\mathcal{KG}^2_{\alpha\beta}\big\},\end{aligned} \tag{4.3.12}$$

*where $\alpha,\beta\in\{0,1,2,3\}$, $h_1,h_2\in\{h_{\mu\nu}\}$, $\mathcal{KG}^2_{\alpha\beta}$ are defined in (2.1.12), and $G_{\geq1}=\sum_{d\geq1}a_dg_d$ for some functions $g_d\in\mathcal{G}_d$ (see (4.1.68)) and some coefficients $a_d\in\mathbb{R}$ with $|a_d|\leq C^d$.*

*A "semilinear cubic remainder of order $(q,n)$" is a finite sum of expressions of the form*

$$I[|\nabla|^{-1}\mathcal{L}_1\mathcal{N}, \partial_\alpha\mathcal{L}_2 h_{\mu\nu}] \quad or \quad I[|\nabla|^{-1}\mathcal{L}_1\mathcal{N}, |\nabla|^{-1}\mathcal{L}_2\mathcal{N}'] \quad or \quad I[\mathcal{L}_1\mathcal{N}_0, \mathcal{L}_2 h_{\mu\nu}], \tag{4.3.13}$$

*where $I = I_m$, $m \in \mathcal{M}$, is as in (3.2.43), $\mathcal{N}, \mathcal{N}' \in \mathcal{QU}$, $\mathcal{N}_0 \in \mathcal{QU}_0$, $\mathcal{L}_i \in \mathcal{V}^{q_i}_{n_i}$, and $(q_1, n_1) + (q_2, n_2) \le (q, n)$.*

For example, the functions $\mathcal{N}^h_{\mu\nu}$, $\partial_\mu E^{\ge 2}_{\nu}$, $|\nabla|\partial_\mu\tau$ can be written as sums of terms of the form $R^a\mathcal{N}$ for $\mathcal{N} \in \mathcal{QU}$, where $R^a = R_1^{a_1}R_2^{a_2}R_3^{a_3}$ (see (4.3.1) and (4.3.5)).

We show that functions in $\mathcal{QU}$ satisfy quadratic-type bounds similar to $\mathcal{N}^h_{\alpha\beta}$:

**Lemma 4.18.** *If $\mathcal{N} \in \mathcal{QU}$, $\mathcal{L} \in \mathcal{V}^q_n$, $n \le 3$, $t \in [0,T]$, and $k \in \mathbb{Z}$ then*

$$\|P_k\{\mathcal{L}\mathcal{N}\}(t)\|_{L^2} \lesssim \varepsilon_1^2 2^{k/2}\langle t\rangle^{H(q,n)\delta}2^{-\widetilde{N}(n)k^+ + 7k^+}\min(2^k\langle t\rangle^{3\delta/2}, \langle t\rangle^{-1+\widetilde{\ell}(n)\delta}). \tag{4.3.14}$$

*Moreover, if $n \le 2$ and $l \in \{1,2,3\}$ then we also have the bounds*

$$\begin{aligned}
&\|P_k\{\mathcal{L}\mathcal{N}\}(t)\|_{L^\infty} \lesssim \varepsilon_1^2 2^k\langle t\rangle^{-1+4\delta'}2^{-N(n+1)k^+ + 5k^+}\min\{2^k, \langle t\rangle^{-1}\},\\
&\|P_k\{x_l\mathcal{L}\mathcal{N}\}(t)\|_{L^2} \lesssim \varepsilon_1^2 2^{k/2}\langle t\rangle^{\delta'}2^{-N(n+1)k^+ - 2k^+}.
\end{aligned} \tag{4.3.15}$$

*Finally, if $\mathcal{L} \in \mathcal{V}^q_n$, $n \le 3$, $t \in [0,T]$, $k \in \mathbb{Z}$, and $2^k \lesssim \langle t\rangle^{-\delta'}$ then*

$$\|P_k\{\mathcal{L}\mathcal{N}\}(t)\|_{L^2} \lesssim \varepsilon_1^2 2^{k/2}\langle t\rangle^{H(q,n)\delta}\langle t\rangle^{-1+\ell(q,n)\delta+\delta/2}. \tag{4.3.16}$$

*Proof.* The bounds (4.3.14)–(4.3.15) follow from the proofs of Lemmas 4.8, 4.9, 4.11. The bounds (4.3.16) follow from (4.2.7) if $\mathcal{N} = \mathcal{K}\mathcal{G}^2_{\alpha\beta}$ and from (4.2.46) if $\mathcal{N}$ is a cubic term. It remains to prove that

$$\|P_k\{\mathcal{L}\mathcal{Q}^2_{\alpha\beta}\}(t)\|_{L^2} + \|P_k\{\mathcal{L}\mathcal{S}^2_{\alpha\beta}\}(t)\|_{L^2} \lesssim \varepsilon_1^2 2^{k/2}\langle t\rangle^{H(q,n)\delta}\langle t\rangle^{-1+\ell(q,n)\delta+\delta/2}$$

if $2^k \lesssim \langle t\rangle^{-\delta'}$, where $\mathcal{Q}^2_{\alpha\beta}$ and $\mathcal{S}^2_{\alpha\beta}$ are as in (2.1.13)–(2.1.15). This can be proved as in Lemma 4.9, using Lemma 4.1. Most of the terms gain a factor of $2^{k/2}$, which is more than enough to give the additional time decay. The only exception are the High × High → Low interactions that are controlled using (4.1.5)–(4.1.7); however, in these interactions we already have the time decay factor $\langle t\rangle^{H(q,n)\delta}\langle t\rangle^{-1+\ell(q,n)\delta+\delta/2}$ as claimed. □

We show first that most cubic and higher order terms in $\mathcal{L}\mathcal{N}^h_{\alpha\beta}$ and $\mathcal{L}\mathcal{N}^\psi$ are disposable.

**Lemma 4.19.** *If $(q,n) \le (3,3)$ then any semilinear cubic remainder of order $(q,n)$ (see Definition 4.17) is wave-disposable. In addition, if $I = I_m$, $m \in \mathcal{M}$,* *is as in* (3.2.43) *and* $h_1 \in \{h_{\mu\nu}\}$ *then terms of the form*

$$I[|\nabla|^{-1}\mathcal{L}_1(G_{\geq 1}\partial_\rho h_1), \mathcal{L}_2\psi] \tag{4.3.17}$$

*are KG-disposable of order* $(q,n)$ *for all* $\mathcal{L}_1 \in \mathcal{V}^{q_1}_{n_1}$, $\mathcal{L}_2 \in \mathcal{V}^{q_2}_{n_2}$ *with* $(q_1,n_1) + (q_2,n_2) \leq (q,n)$.

*Moreover, if* $n_2 < n$, *then terms of the form*

$$I[|\nabla|^{-1}\mathcal{L}_1(G_{\geq 1}\partial_\rho h_1), \partial_\alpha\partial_\beta\mathcal{L}_2 h_2] \tag{4.3.18}$$

*are wave-disposable of order* $(q,n)$, *while terms of the form*

$$I[|\nabla|^{-1}\mathcal{L}_1(G_{\geq 1}\partial_\rho h_1), \partial_\alpha\partial_\beta\mathcal{L}_2\psi] \tag{4.3.19}$$

*are KG-disposable of order* $(q,n)$.

*Proof.* Time decay is not an issue in this lemma, since we are considering cubic and higher order terms, but we need to be careful to avoid possible derivative loss. For any $(q,n)$ with $(q,n) \leq (3,3)$ and $t \in [0,T]$ we define the frequency envelopes $\{b_k\}_{k\in\mathbb{Z}} = \{b_k(q,n;t)\}_{k\in\mathbb{Z}}$ by

$$\begin{aligned} b^0_k(q,n;t) &:= \sup_{\mathcal{K}\in\mathcal{V}^q_n,\, \alpha,\beta\in\{0,1,2,3\}} \langle t\rangle^{-H(q,n)\delta} \\ &\quad \times \big\{\big\|P_k\{(\langle t\rangle|\nabla|_{\leq 1})^\gamma|\nabla|^{-1/2}U^{\mathcal{K}h_{\alpha\beta}}\}(t)\big\|_{H^{N(n)}} + \|P_kU^{\mathcal{K}\psi}(t)\|_{H^{N(n)}}\big\}, \\ b_k(q,n;t) &:= \varepsilon_1 2^{-\gamma|k|/4} + \sum_{k'\in\mathbb{Z}} 2^{-\gamma|k-k'|/4} b^0_{k'}(q,n;t). \end{aligned} \tag{4.3.20}$$

In view of the bootstrap assumption (2.1.46) we have

$$\sum_{k\in\mathbb{Z}} (b_k(q,n;t))^2 \lesssim \varepsilon_1^2 \quad \text{and} \quad b_k(q,n;t) \leq b_{k'}(q,n;t)2^{\gamma|k-k'|/4} \text{ for any } k,k' \in \mathbb{Z}. \tag{4.3.21}$$

The main point of the definition is that we have the slightly better $L^2$ bounds

$$\begin{aligned} \big\|P_kU^{\mathcal{K}h_{\alpha\beta}}(t)\big\|_{L^2} &\lesssim b_k(q,n;t)\langle t\rangle^{H(q,n)\delta}2^{k/2}2^{-N(n)k^+}(\langle t\rangle 2^{k^-})^{-\gamma}, \\ \big\|P_kU^{\mathcal{K}\psi}(t)\big\|_{L^2} &\lesssim b_k(q,n;t)\langle t\rangle^{H(q,n)\delta}2^{-N(n)k^+}, \end{aligned} \tag{4.3.22}$$

for any $k \in \mathbb{Z}$, $\mathcal{K} \in \mathcal{V}^q_n$, and $\alpha,\beta \in \{0,1,2,3\}$ (compare with (3.3.3)).

To prove the conclusions we need two more quadratic bounds: if $\mathcal{N}_0 \in \mathcal{QU}_0$ and $\mathcal{N} \in \mathcal{QU}$ (see (4.3.12)), $k \geq 0$, $t \in [0,T]$, and $\mathcal{K} \in \mathcal{V}^{q'}_{n'}$ then

$$\begin{aligned} \|P_k\{\mathcal{K}\mathcal{N}_0\}(t)\|_{L^2} &\lesssim \varepsilon_1 b_k(q',n';t)2^{k/2}2^{-N(n')k}\langle t\rangle^{-1+\delta'/2}, \\ \|P_k\{\mathcal{K}\mathcal{N}\}(t)\|_{L^2} &\lesssim \varepsilon_1 b_k(q',n';t)2^{3k/2}2^{-N(n')k}\langle t\rangle^{-1+\delta'/2}. \end{aligned} \tag{4.3.23}$$

Notice that these bounds are improvements over the general bounds (4.3.14) when $k \geq 0$.

**Step 1.** We assume first that the bounds (4.3.23) hold and show how to prove the conclusions of the lemma. Recall also the bounds (4.1.79). In view of the definitions it suffices to show that

$$\begin{aligned} \|P_kI[|\nabla|^{-1}\mathcal{L}_1\mathcal{N}, U^{\mathcal{L}_2h_2,\iota_2}](t)\|_{L^2} &\lesssim \varepsilon_1 b_k(q,n;t)2^{k/2}2^{-N(n)k^+}\langle t\rangle^{-1},\\ \|P_kI[|\nabla|^{-1}\mathcal{L}_1\mathcal{N}, |\nabla|^{-1}\mathcal{L}_2\mathcal{N}'](t)\|_{L^2} &\lesssim \varepsilon_1 b_k(q,n;t)2^{k/2}2^{-N(n)k^+}\langle t\rangle^{-1},\\ \|P_kI[\mathcal{L}_1\mathcal{N}_0, |\nabla|^{-1}U^{\mathcal{L}_2h_2,\iota_2}](t)\|_{L^2} &\lesssim \varepsilon_1 b_k(q,n;t)2^{k/2}2^{-N(n)k^+}\langle t\rangle^{-1}, \end{aligned} \tag{4.3.24}$$

and

$$\|P_kI[|\nabla|^{-1}\mathcal{L}_1(G_{\geq 1}\partial_\rho h_1), \langle\nabla\rangle^{-1}U^{\mathcal{L}_2\psi,\iota_2}](t)\|_{L^2} \lesssim \varepsilon_1^2 2^{-\delta|k|}2^{-N(n)k^+}\langle t\rangle^{-1}, \tag{4.3.25}$$

for any $k \in \mathbb{Z}$, $t \in [0,T]$, $\mathcal{N},\mathcal{N}' \in \mathcal{QU}$, $\mathcal{N}_0 \in \mathcal{QU}_0$, $\mathcal{L}_1 \in \mathcal{V}^{q_1}_{n_1}$, $\mathcal{L}_2 \in \mathcal{V}^{q_2}_{n_2}$, $(q_1,n_1)+(q_2,n_2) \leq (q,n)$. Moreover, if $n_2 < n$ and $\mathcal{N}^h \in \{\mathcal{N}^h_{\alpha\beta}\}$, then we also have to prove the bounds

$$\begin{aligned} \|P_kI[|\nabla|^{-1}\mathcal{L}_1(G_{\geq 1}\partial_\rho h_1), |\nabla|U^{\mathcal{L}_2h_2,\iota_2}](t)\|_{L^2} &\lesssim \varepsilon_1^2 2^{-\delta|k|}2^{k/2}2^{-N(n)k^+}\langle t\rangle^{-1},\\ \|P_kI[|\nabla|^{-1}\mathcal{L}_1(G_{\geq 1}\partial_\rho h_1), \mathcal{L}_2\mathcal{N}^h](t)\|_{L^2} &\lesssim \varepsilon_1^2 2^{-\delta|k|}2^{k/2}2^{-N(n)k^+}\langle t\rangle^{-1} \end{aligned} \tag{4.3.26}$$

and

$$\begin{aligned} \|P_kI[|\nabla|^{-1}\mathcal{L}_1(G_{\geq 1}\partial_\rho h_1), \langle\nabla\rangle U^{\mathcal{L}_2\psi,\iota_2}](t)\|_{L^2} &\lesssim \varepsilon_1^2 2^{-\delta|k|}2^{-N(n)k^+}\langle t\rangle^{-1},\\ \|P_kI[|\nabla|^{-1}\mathcal{L}_1(G_{\geq 1}\partial_\rho h_1), \mathcal{L}_2\mathcal{N}^\psi](t)\|_{L^2} &\lesssim \varepsilon_1^2 2^{-\delta|k|}2^{-N(n)k^+}\langle t\rangle^{-1}, \end{aligned} \tag{4.3.27}$$

in order to show that the expressions in (4.3.18)–(4.3.19) are also disposable.

The proofs of (4.3.24)–(4.3.27) rely on $L^2 \times L^\infty$ estimates, as in Lemmas 4.4-4.6. In most cases we place the high frequency factor in $L^2$ and the low frequency factor in $L^\infty$, except when the low frequency factor carries all the three vector-fields. We provide all the details only for the proof of the harder estimates in the first line of (4.3.24), which require the frequency envelopes.

The functions $|\nabla|^{-1}\mathcal{L}_1\mathcal{N}$ and $U^{\mathcal{L}_2h_2,\iota_2}$ satisfy the bounds

$$\begin{aligned} \|P_{k_1}(|\nabla|^{-1}\mathcal{L}_1\mathcal{N})(t)\|_{L^2} &\lesssim \varepsilon_1 b_{k_1}(q,n;t)2^{-N(n_1)k_1^+ + k_1^+/2}\min(2^{k_1},\langle t\rangle^{-1})^{1/2-\delta'},\\ \|P_{k_1}(|\nabla|^{-1}\mathcal{L}_1\mathcal{N})(t)\|_{L^\infty} &\lesssim \varepsilon_1^2\langle t\rangle^{-1+4\delta'}2^{-N(n_1+1)k_1^+ + 5k_1^+}\min(2^{k_1},\langle t\rangle^{-1}) \end{aligned} \tag{4.3.28}$$

(see (4.3.23), (4.3.14), and (4.3.15)) and

$$\begin{aligned} \|P_{k_2}(U^{\mathcal{L}_2h_2,\iota_2})(t)\|_{L^2} &\lesssim b_{k_2}(q,n;t)2^{-N(n_2)k_2^+ + k_2^+/2}2^{k_2^-/2-\delta k_2^-}\langle t\rangle^{\delta'},\\ \|P_{k_2}(U^{\mathcal{L}_2h_2,\iota_2})(t)\|_{L^\infty} &\lesssim \varepsilon_1\langle t\rangle^{-1+\delta'}2^{k_2^-}2^{-N(n_2+1)k_2^+ + 2k_2^+} \end{aligned} \tag{4.3.29}$$

(see (4.3.22) and (3.3.11)). As before, the $L^\infty$ bounds in the second lines of (4.3.28) and (4.3.29) hold only if $n_1 \leq 2$ and $n_2 \leq 2$. We estimate first

$$\begin{aligned}
&\|P_k I[|\nabla|^{-1}\mathcal{L}_1\mathcal{N}, U^{\mathcal{L}_2 h_2,\iota_2}](t)\|_{L^2} \\
&\quad\lesssim \sum_{(k_1,k_2)\in\mathcal{X}_k} 2^{3k/2}\|P_{k_1}(|\nabla|^{-1}\mathcal{L}_1\mathcal{N})(t)\|_{L^2}\|P_{k_2}(U^{\mathcal{L}_2 h_2,\iota_2})(t)\|_{L^2} \\
&\quad\lesssim \varepsilon_1^3\langle t\rangle^{-1/2+4\delta'}2^{3k/2},
\end{aligned}$$

which suffices to prove (4.3.24) if $2^k \lesssim \langle t\rangle^{-0.51}$. Moreover, if $2^k \geq \langle t\rangle^{-0.51}$ and $n_1, n_2 \in [0,2]$ then we estimate, using also (4.3.21),

$$\|P_k I[|\nabla|^{-1}\mathcal{L}_1\mathcal{N}, U^{\mathcal{L}_2 h_2,\iota_2}](t)\|_{L^2} \lesssim S_1 + S_2,$$

where

$$\begin{aligned}
S_1 &:= \sum_{(k_1,k_2)\in\mathcal{X}_k,\, k_1\geq k-8} \|P_{k_1}(|\nabla|^{-1}\mathcal{L}_1\mathcal{N})(t)\|_{L^2}\|P_{k_2}(U^{\mathcal{L}_2 h_2,\iota_2})(t)\|_{L^\infty} \\
&\qquad\lesssim \varepsilon_1^2 b_k(q,n;t)2^{-N(n_1)k^+ + k^+/2}\langle t\rangle^{-3/2+4\delta'}, \\
S_2 &:= \sum_{(k_1,k_2)\in\mathcal{X}_k,\, k_1\leq k-8} \|P_{k_1}(|\nabla|^{-1}\mathcal{L}_1\mathcal{N})(t)\|_{L^\infty}\|P_{k_2}(U^{\mathcal{L}_2 h_2,\iota_2})(t)\|_{L^2} \\
&\qquad\lesssim \varepsilon_1^2 b_k(q,n;t)2^{-N(n_2)k^+ + k^+/2}\langle t\rangle^{-5/3}.
\end{aligned}$$

Finally, if $n_1 = 3$ (thus $n_2 = 0$, $n = 3$), then we estimate

$$\begin{aligned}
&\|P_k I[|\nabla|^{-1}\mathcal{L}_1\mathcal{N}, U^{\mathcal{L}_2 h_2,\iota_2}](t)\|_{L^2} \\
&\quad\lesssim \sum_{(k_1,k_2)\in\mathcal{X}_k} \|P_{k_1}(|\nabla|^{-1}\mathcal{L}_1\mathcal{N})(t)\|_{L^2}\|P_{k_2}(U^{\mathcal{L}_2 h_2,\iota_2})(t)\|_{L^\infty} \\
&\quad\lesssim \varepsilon_1^2 b_k(q,n;t)2^{-N(n)k^+ + k^+/2}\langle t\rangle^{-3/2+4\delta'},
\end{aligned}$$

while if $n_2 = 3$ (thus $n_1 = 0$, $n = 3$), then

$$\begin{aligned}
&\|P_k I[|\nabla|^{-1}\mathcal{L}_1\mathcal{N}, U^{\mathcal{L}_2 h_2,\iota_2}](t)\|_{L^2} \\
&\quad\lesssim \sum_{(k_1,k_2)\in\mathcal{X}_k} \|P_{k_1}(|\nabla|^{-1}\mathcal{L}_1\mathcal{N})(t)\|_{L^\infty}\|P_{k_2}(U^{\mathcal{L}_2 h_2,\iota_2})(t)\|_{L^2} \\
&\quad\lesssim \varepsilon_1^2 b_k(q,n;t)2^{-N(n)k^+ + k^+/2}\langle t\rangle^{-5/3}.
\end{aligned}$$

These bounds suffice to prove (4.3.24) when $2^k \geq \langle t\rangle^{-0.51}$.

The estimates (4.3.26)–(4.3.27) are slightly easier, because we do not need to carry the frequency envelopes. The functions $|\nabla|^{-1}\mathcal{L}_1(G_{\geq 1}\partial_\rho h_1)$ satisfy the bounds in Lemma 4.6, while the functions $\mathcal{N}^{\mathcal{L}_2 h_2}$ and $\mathcal{N}^{\mathcal{L}_2\psi}$ satisfy the bounds in Proposition 4.7. Since $n_2 < n$ there is no derivative loss, and the estimates

(4.3.26)–(4.3.27) follow in the same way as (4.3.24).

**Step 2.** We prove now the bounds (4.3.23). We examine the definition (4.3.12). If $\mathcal{N}_0^1 := \partial_\alpha h_1 \cdot \partial_\beta h_2$ and $\mathcal{N}_0^2 := G_{\geq 1}\partial_\alpha h_1 \cdot \partial_\beta h_2$, $\alpha, \beta \in \{0,1,2,3\}$, $h_1, h_2 \in \{h_{\mu\nu}\}$, then

$$\|P_k\{\mathcal{K}\mathcal{N}_0^1\}(t)\|_{L^2} + \|P_k\{\mathcal{K}\mathcal{N}_0^2\}(t)\|_{L^2} \lesssim \varepsilon_1 b_k(q', n'; t)2^{k/2}2^{-N(n')k}\langle t\rangle^{-1+\delta'}, \tag{4.3.30}$$

for any $k \geq 0$, $t \in [0,T]$, and $\mathcal{K} \in \mathcal{V}_{n'}^{q'}$. This is easy to see just using $L^2 \times L^\infty$ estimates and the bounds (4.3.22), (3.3.11), and (4.2.49). Similarly, if $\mathcal{N}_0^3 := \mathcal{K}\mathcal{G}_{\alpha\beta}^2$ (see (2.1.12)) for some $\alpha, \beta \in \{0,1,2,3\}$ and $\mathcal{N}_0^4 := G_{\geq 1} \cdot \mathcal{K}\mathcal{G}_{\alpha\beta}^2$ then

$$\|P_k\{\mathcal{K}\mathcal{N}_0^3\}(t)\|_{L^2} + \|P_k\{\mathcal{K}\mathcal{N}_0^4\}(t)\|_{L^2} \lesssim \varepsilon_1^2 2^{-N(n')k}\langle t\rangle^{-1+\delta'}, \tag{4.3.31}$$

for any $k \geq 0$, $t \in [0,T]$, and $\mathcal{K} \in \mathcal{V}_{n'}^{q'}$. Finally, if $(\alpha, \beta) \neq (0,0)$, then

$$\|P_k\{\mathcal{K}(G_{\geq 1} \cdot \partial_\alpha\partial_\beta h_2)\}(t)\|_{L^2} \lesssim \varepsilon_1 b_k(q', n'; t)2^{-N(n')k+3k/2}\langle t\rangle^{-1+\delta'}, \tag{4.3.32}$$

for any $k \geq 0$, $t \in [0,T]$, and $\mathcal{K} \in \mathcal{V}_{n'}^{q'}$, using again $L^2 \times L^\infty$ estimates as before (and Proposition 4.7 to bound the commutator term $[\mathcal{K}, \partial_\alpha\partial_\beta]h_2$). As a consequence of the last three bounds and the identitites (2.1.9), the metric nonlinearities $\mathcal{N}_{\mu\nu}^h$ satisfy bounds similar to (4.3.32),

$$\|P_k\{\mathcal{K}\mathcal{N}_{\mu\nu}^h\}(t)\|_{L^2} \lesssim \varepsilon_1 b_k(q', n', t)2^{-N(n')k+3k/2}\langle t\rangle^{-1+\delta'}.$$

Therefore, we can use the equation $\partial_0^2 h = \Delta h + \mathcal{N}^h$ to prove the bounds (4.3.32) for $\alpha = \beta = 0$ as well. This completes the proof of (4.3.23). ☐

We show now that all the quadratic terms arising as commutators are also energy disposable.

**Lemma 4.20.** *Assume that* $(q,n) \leq (3,3)$, $\mathcal{L}_1 \in \mathcal{V}_{n_1}^{q_1}$, $\mathcal{L}_2 \in \mathcal{V}_{n_2}^{q_2}$, $(q_1, n_1) + (q_2, n_2) \leq (q, n-1)$. *If* $h_1, h_2 \in \{h_{\mu\nu}\}$, $\rho, \alpha, \beta \in \{0,1,2,3\}$ *then quadratic terms of the form*

$$I[U^{\mathcal{L}_1\psi,\iota_1}, U^{\mathcal{L}_2\psi,\iota_2}] \quad \text{or} \quad I[R_\rho\mathcal{L}_1 h_1, \partial_\alpha\partial_\beta\mathcal{L}_2 h_2] \quad \text{or} \quad I[\partial_\rho\mathcal{L}_1 h_1, \partial_\beta\mathcal{L}_2 h_2] \tag{4.3.33}$$

*are wave-disposable of order* $(q,n)$, *where* $\iota_1, \iota_2 \in \{+,-\}$. *Moreover, terms of the form*

$$I[R_\rho\mathcal{L}_1 h_1, \partial_\alpha\partial_\beta\mathcal{L}_2\psi] \tag{4.3.34}$$

*are KG-disposable of order* $(q,n)$.

Terms such as those in (4.3.33) and (4.3.34) will be called "wave (respectively KG) commutator remainders" of order $(q,n)$.

*Proof.* Derivative loss is not an issue in this lemma, since $n_1 + n_2 \leq n-1$, but we need to be careful with the time decay. We show first for any $k, k_1, k_2 \in \mathbb{Z}$ and $t \in [0, T]$

$$2^{-k/2}\|P_k I[P_{k_1}U^{\mathcal{L}_1\psi,\iota_1}, P_{k_2}U^{\mathcal{L}_2\psi,\iota_2}](t)\|_{L^2} \lesssim \varepsilon_1^2\langle t\rangle^{-1+\delta[H(q,n-1)+\ell(q,n-1)+1]}2^{-N(n-1)k^+}2^{-\gamma(|k|+|k_1|+|k_2|)/4}. \tag{4.3.35}$$

Indeed, these bounds follow from (4.1.34) and (2.1.53) if $2^{\min\{k,k_1,k_2\}} \lesssim \langle t\rangle^{-1}$ (see also the definitions (4.1.2)). On the other hand, if $2^{\min\{k,k_1,k_2\}} \geq \langle t\rangle^{-1}$ then (4.3.35) follows from the bounds (4.1.35). Since

$$H(q, n-1) + \ell(q, n-1) + 12 \leq H(q, n), \tag{4.3.36}$$

the bounds (4.3.35) suffice to show that $I[U^{\mathcal{L}_1\psi,\iota_1}, U^{\mathcal{L}_2\psi,\iota_2}]$ is wave-disposable.

Moreover, we also have

$$2^{-k/2}2^{|k_1-k_2|}\|P_k I[P_{k_1}U^{\mathcal{L}_1h_1,\iota_1}, P_{k_2}U^{\mathcal{L}_2h_2,\iota_2}](t)\|_{L^2} \lesssim \varepsilon_1^2\langle t\rangle^{-1+\delta[H(q,n)-4]}2^{-N(n)k^+-2k^+}2^{-\gamma(|k|+|k_1|+|k_2|)/4}, \tag{4.3.37}$$

for any $k, k_1, k_2 \in \mathbb{Z}$, $t \in [0, T]$, $h_1, h_2 \in \{h_{\alpha\beta}\}$, $\iota_1, \iota_2 \in \{+, -\}$. These bounds follow from (4.1.5) if $2^{\min\{k,k_1,k_2\}} \lesssim \langle t\rangle^{-1}$ and from (4.1.6)–(4.1.11) (see also (4.1.3) and use (4.1.10) instead of (4.1.8) when $n_1 = n_2 = 0$) if $2^{\min\{k,k_1,k_2\}} \geq \langle t\rangle^{-1}$. Thus terms of the form $I[R_\rho\mathcal{L}_1h_1, \partial_\alpha\partial_\beta\mathcal{L}_2h_2]$ and $I[\partial_\rho\mathcal{L}_1h_1, \partial_\mu\mathcal{L}_2h_2]$ are also wave-disposable of order $(q, n)$ (in the case $(\alpha, \beta) = (0, 0)$ we replace first $\partial_0^2\mathcal{L}_2h_2$ with $\Delta\mathcal{L}_2h_2 + \mathcal{L}_2\mathcal{N}^h$ and use (4.2.3) and (4.2.5) to bound the nonlinear contribution).

Finally, we can use Lemma 4.3 in a similar way to show that

$$2^{k_2^+-k_1}\|P_k I[P_{k_1}U^{\mathcal{L}_1h_1,\iota_1}, P_{k_2}U^{\mathcal{L}_2\psi,\iota_2}](t)\|_{L^2} \lesssim \varepsilon_1^2\langle t\rangle^{-1+\delta[H(q,n)-4]}2^{-N(n)k^+-2k^+}2^{-\gamma(|k|+|k_1|+|k_2|)/4}, \tag{4.3.38}$$

for any $k, k_1, k_2 \in \mathbb{Z}$, $t \in [0, T]$, $h_1 \in \{h_{\alpha\beta}\}$, $\iota_1, \iota_2 \in \{+, -\}$. Therefore expressions of the form $I[R_\rho\mathcal{L}_1h_1, \partial_\alpha\partial_\beta\mathcal{L}_2\psi]$ are KG-disposable of order $(q, n)$. This completes the proof. □

### 4.3.3 Null Structures

In the analysis of the wave nonlinearities $\mathcal{N}^h$ it is important to identify null components, for which we prove better estimates. We start with a definition.

**Definition 4.21.** *We define two classes of null multipliers $\mathcal{M}_+^{null}$ and $\mathcal{M}_-^{null}$,*

$$\mathcal{M}_\pm^{null} := \{n : (\mathbb{R}^3 \setminus 0)^2 \to \mathbb{C} : n(x, y) = (x_i/|x| \mp y_i/|y|)m(x, y) \text{ for some } m \in \mathcal{M} \text{ and } i \in \{1, 2, 3\}\}. \tag{4.3.39}$$

*For any $(q,n)$ with $0 \le q \le n \le 3$ we define the set of "semilinear null forms of order $(q,n)$" as the set of finite sums of expressions $I^{null}_{n_{\iota_1\iota_2}}[U^{\mathcal{L}_1 h_1,\iota_1}, U^{\mathcal{L}_2 h_2,\iota_2}]$ defined by*

$$\mathcal{F}\{I^{null}_{n_{\iota_1\iota_2}}[U^{\mathcal{L}_1 h_1,\iota_1}, U^{\mathcal{L}_2 h_2,\iota_2}]\}(\xi) \\ := \frac{1}{8\pi^3}\int_{\mathbb{R}^3} n_{\iota_1\iota_2}(\xi-\eta,\eta)\widehat{U^{\mathcal{L}_1 h_1,\iota_1}}(\xi-\eta)\widehat{U^{\mathcal{L}_2 h_2,\iota_2}}(\eta)\,d\eta, \tag{4.3.40}$$

*where $\iota_1,\iota_2 \in \{+,-\}$, $h_1,h_2 \in \{h_{\alpha\beta}\}$, $\mathcal{L}_1 \in \mathcal{V}^{q_1}_{n_1}$, $\mathcal{L}_2 \in \mathcal{V}^{q_2}_{n_2}$, $(q_1,n_1)+(q_2,n_2) \le (q,n)$, and $n_{\iota_1\iota_2} \in \mathcal{M}^{null}_{\iota_1\iota_2}$. By convention, $++ = -- = +$ and $+- = -+ = -$.*

Our definition of semilinear null forms contains the classical null forms

$$\partial_\alpha h_1 \partial_\beta h_2 - \partial_\beta h_1 \partial_\alpha h_2 \qquad \text{and} \qquad m^{\alpha\beta}\partial_\alpha h_1 \partial_\beta h_2, \tag{4.3.41}$$

for $h_1,h_2 \in \{h_{\mu\nu}\}$, $\alpha,\beta,\mu,\nu \in \{0,1,2,3\}$. Indeed, since

$$\partial_0 h = (1/2)[U^{h,+} + U^{h,-}], \qquad \partial_j h = (i/2)[R_j U^{h,+} - R_j U^{h,-}],$$

for any $h \in \{h_{\alpha\beta}\}$ (see (2.1.34)), we have, for $a,b \in \{1,2,3\}$,

$$\partial_a h_1 \partial_b h_2 - \partial_b h_1 \partial_a h_2 = \sum_{\iota_1,\iota_2\in\{+,-\}} I^{null}_{n^{a,b}_{\iota_1\iota_2}}[U^{h_1,\iota_1}, U^{h_2,\iota_2}], \\ n^{a,b}_\iota(\theta,\eta) := \frac{\iota}{4}\frac{\theta_a\eta_b - \theta_b\eta_a}{|\theta||\eta|}, \tag{4.3.42}$$

$$\partial_0 h_1 \partial_b h_2 - \partial_b h_1 \partial_0 h_2 = \sum_{\iota_1,\iota_2\in\{+,-\}} \iota_1 I^{null}_{n^{0,b}_{\iota_1\iota_2}}[U^{h_1,\iota_1}, U^{h_2,\iota_2}], \\ n^{0,b}_\iota(\theta,\eta) := \frac{1}{4}\Big[\frac{\theta_b}{|\theta|} - \iota\frac{\eta_b}{|\eta|}\Big], \tag{4.3.43}$$

$$m^{\alpha\beta}\partial_\alpha h_1 \partial_\beta h_2 = \sum_{\iota_1,\iota_2\in\{+,-\}} I^{null}_{\widetilde{n}_{\iota_1\iota_2}}[U^{h_1,\iota_1}, U^{h_2,\iota_2}], \\ \widetilde{n}_\iota(\theta,\eta) := \frac{1}{4}\Big[-1 + \iota\frac{\theta\cdot\eta}{|\theta||\eta|}\Big]. \tag{4.3.44}$$

It is easy to verify that the symbols $n^{a,b}_\iota, n^{0,b}_\iota, \widetilde{n}_\iota$ are in $\mathcal{M}^{null}_\iota$, as defined in (4.3.39) (in fact $n^{a,b}_\iota \in \mathcal{M}^{null}_+ \cap \mathcal{M}^{null}_-$), therefore the classical null forms in (4.3.41) are all semilinear null forms of order $(0,0)$. The vector-fields $\mathcal{L}$ can be incorporated as well, without any difficulty.

We remark that our definition (4.21) of semilinear null forms is slightly more general because we would like to allow forms expressed in terms of the variables $F, \underline{F}, \rho, \omega_j, \Omega_j, \vartheta_{jk}$, which involve the Riesz transforms. For example, expressions of the form

$$R_j f R_k g - R_k f R_j g, \qquad fg - R_j R_k f \cdot R_j R_k g \tag{4.3.45}$$

are semilinear null forms of order $(0,0)$ if $f, g \in \{R_1^{a_1} R_2^{a_2} R_3^{a_3} \partial_\alpha h_{\mu\nu}\}$.

### 4.3.4 The Main Decomposition

We are now ready to prove an important proposition concerning the decomposition of the nonlinearities $\mathcal{N}^{\mathcal{L}h}_{\alpha\beta}$ and $\mathcal{N}^{\mathcal{L}\psi}$.

Given three operators $\mathcal{L}_1 = \Gamma^{a'}\Omega^{b'}$, $\mathcal{L}_2 = \Gamma^{a''}\Omega^{b''}$, $\mathcal{L} = \Gamma^a\Omega^b$ (see (2.1.23)), we say that $\mathcal{L}_1 + \mathcal{L}_2 = \mathcal{L}$ if $a' + a'' = a$ and $b' + b'' = b$. Therefore

$$\mathcal{L}(fg) = \sum_{\mathcal{L}_1+\mathcal{L}_2=\mathcal{L}} c_{\mathcal{L}_1,\mathcal{L}_2} \mathcal{L}_1 f \cdot \mathcal{L}_2 g, \tag{4.3.46}$$

for some coefficients $c_{\mathcal{L}_1,\mathcal{L}_2} \in [0,\infty)$.

**Proposition 4.22.** *If $\alpha, \beta \in \{0,1,2,3\}$, $(q,n) \le (3,3)$, and $\mathcal{L} \in \mathcal{V}^q_n$ then*

$$\begin{aligned}\mathcal{L}\mathcal{N}^h_{\alpha\beta} = &\sum_{\mu,\nu\in\{0,1,2,3\}} \widetilde{g}^{\mu\nu}_{\ge 1} \partial_\mu\partial_\nu(\mathcal{L}h_{\alpha\beta}) + \mathcal{Q}^{\mathcal{L}}_{wa}(h_{\alpha\beta}) \\ &+ \mathcal{S}^{\mathcal{L},1}_{\alpha\beta} + \mathcal{S}^{\mathcal{L},2}_{\alpha\beta} + \mathcal{KG}^{\mathcal{L}}_{\alpha\beta} + \mathcal{R}^{\mathcal{L}h}_{\alpha\beta}\end{aligned} \tag{4.3.47}$$

*and*

$$\mathcal{L}\mathcal{N}^\psi = \sum_{\mu,\nu\in\{0,1,2,3\}} \widetilde{g}^{\mu\nu}_{\ge 1} \partial_\mu\partial_\nu(\mathcal{L}\psi) + h_{00}\mathcal{L}\psi + \mathcal{Q}^{\mathcal{L}}_{kg}(\psi) + \mathcal{R}^{\mathcal{L}\psi}. \tag{4.3.48}$$

*The remainders $\mathcal{R}^{\mathcal{L}h}_{\alpha\beta}$ and $\mathcal{R}^{\mathcal{L}\psi}$ are wave-disposable (respectively KG-disposable) of order $(q,n)$, and the reduced metric components $\widetilde{g}^{\mu\nu}_{\ge 1}$ are defined by*

$$\widetilde{g}^{00}_{\ge 1} := 0, \qquad \widetilde{g}^{0j}_{\ge 1} := (1 - g^{00}_{\ge 1})^{-1} g^{0j}_{\ge 1}, \qquad \widetilde{g}^{jk}_{\ge 1} := (1 - g^{00}_{\ge 1})^{-1}\big[g^{jk}_{\ge 1} + g^{00}_{\ge 1}\delta^{jk}\big]. \tag{4.3.49}$$

- *The terms $\mathcal{Q}^{\mathcal{L}}_{wa}(h_{\alpha\beta})$ and $\mathcal{Q}^{\mathcal{L}}_{kg}(\psi)$ are given by*

$$\begin{aligned}\mathcal{Q}^{\mathcal{L}}_{wa}(h_{\alpha\beta}) := &\sum_{G\in\{F,\underline{F},\omega_n,\vartheta_{mn}\}} \sum_{\iota_1,\iota_2\in\{+,-\}} \sum_{\mathcal{L}_1+\mathcal{L}_2=\mathcal{L},\, \mathcal{L}_2\neq\mathcal{L}} \\ &c_{\mathcal{L}_1,\mathcal{L}_2} I_{\mathfrak{q}^{G,wa}_{\iota_1\iota_2}}\big[|\nabla|^{-1}U^{G^{\mathcal{L}_1},\iota_1}, |\nabla|U^{\mathcal{L}_2 h_{\alpha\beta},\iota_2}\big],\end{aligned} \tag{4.3.50}$$

$$\begin{aligned}\mathcal{Q}^{\mathcal{L}}_{kg}(\psi) := &\sum_{G\in\{F,\underline{F},\omega_n,\vartheta_{mn}\}} \sum_{\iota_1,\iota_2\in\{+,-\}} \sum_{\mathcal{L}_1+\mathcal{L}_2=\mathcal{L},\, \mathcal{L}_2\neq\mathcal{L}} \\ &c_{\mathcal{L}_1,\mathcal{L}_2} I_{\mathfrak{q}^{G,kg}_{\iota_1\iota_2}}\big[|\nabla|^{-1}U^{G^{\mathcal{L}_1},\iota_1}, \langle\nabla\rangle U^{\mathcal{L}_2\psi,\iota_2}\big].\end{aligned} \tag{4.3.51}$$

*The multipliers* $\mathfrak{q}^{F,wa}_{\iota_1\iota_2}, \mathfrak{q}^{\underline{F},wa}_{\iota_1\iota_2}, \mathfrak{q}^{\omega_n,wa}_{\iota_1\iota_2}, \mathfrak{q}^{\vartheta_{mn},wa}_{\iota_1\iota_2} \in \mathcal{M}^{null}_{\iota_1\iota_2}$ *(which are in fact double-null as defined in* (5.2.2)*) and* $\mathfrak{q}^{F,kg}_{\iota_1\iota_2}$, $\mathfrak{q}^{\underline{F},kg}_{\iota_1\iota_2}$, $\mathfrak{q}^{\omega_n,kg}_{\iota_1\iota_2}$, $\mathfrak{q}^{\vartheta_{mn},kg}_{\iota_1\iota_2} \in \mathcal{M}$ *are given explicitly in* (4.3.59)-(4.3.60).

- *The semilinear terms* $\mathcal{S}^{\mathcal{L},1}_{\alpha\beta}$, $\mathcal{S}^{\mathcal{L},2}_{\alpha\beta}$, *and* $\mathcal{KG}^{\mathcal{L}}_{\alpha\beta}$ *are given by*

$$\mathcal{S}^{\mathcal{L},1}_{\alpha\beta} := \sum_{h_1,h_2\in\{h_{\mu\nu}\}} \sum_{\iota_1,\iota_2\in\{+,-\}} \sum_{\mathcal{L}_1+\mathcal{L}_2=\mathcal{L}} I^{null}_{\mathfrak{n}_{\iota_1\iota_2}}[U^{\mathcal{L}_1 h_1,\iota_1}, U^{\mathcal{L}_2 h_2,\iota_2}], \tag{4.3.52}$$

$$\mathcal{S}^{\mathcal{L},2}_{\alpha\beta} := \sum_{\mathcal{L}_1+\mathcal{L}_2=\mathcal{L}} (1/2)c_{\mathcal{L}_1,\mathcal{L}_2} R_pR_q\partial_\alpha\vartheta^{\mathcal{L}_1}_{mn}\cdot R_pR_q\partial_\beta\vartheta^{\mathcal{L}_2}_{mn}, \tag{4.3.53}$$

$$\mathcal{KG}^{\mathcal{L}}_{\alpha\beta} := \sum_{\mathcal{L}_1+\mathcal{L}_2=\mathcal{L}} c_{\mathcal{L}_1,\mathcal{L}_2}(2\partial_\alpha\mathcal{L}_1\psi\cdot\partial_\beta\mathcal{L}_2\psi + m_{\alpha\beta}\mathcal{L}_1\psi\mathcal{L}_2\psi), \tag{4.3.54}$$

*where* $\mathfrak{n}_{\iota_1\iota_2} = \mathfrak{n}_{\iota_1\iota_2}(\mathcal{L}_1,\mathcal{L}_2,h_1,h_2)$ *are null multipliers in* $\mathcal{M}^0_{\iota_1\iota_2}$ *(see* (4.3.39)*).*

*Remark 4.23.* (1) The nonlinearities $\mathcal{LN}^h_{\alpha\beta}$ contain five types of components, in addition to wave-disposable remainders:

(i) The top order terms $\widetilde{g}^{\mu\nu}_{\geq 1}\partial_\mu\partial_\nu(\mathcal{L}h_{\alpha\beta})$, which lead to derivative loss in energy estimates and normal form analysis;

(ii) The terms $\mathcal{Q}^{\mathcal{L}}_{wa}(h_{\alpha\beta})$, which are sums of bilinear interactions of the metric components, with null multipliers $\mathfrak{q}^{G,wa}_{\iota_1\iota_2}$. These interactions are all defined by null multipliers, but are not of the same type as the semilinear null forms. The issue is the additional anti-derivative on the first factor $|\nabla|^{-1}U^{G^{\mathcal{L}_1},\pm}$, which leads to significant difficulties at very low frequencies, particularly when these first factors carry all the vector-fields $\mathcal{L}_1 = \mathcal{L}$;

(iii) Generic semilinear null terms $\mathcal{S}^{\mathcal{L},1}_{\alpha\beta}$;

(iv) The special terms $\mathcal{S}^{\mathcal{L},2}_{\alpha\beta}$, which involve non-null bilinear interactions of the "good" metric components $\vartheta$;

(v) The Klein-Gordon nonlinearities $\mathcal{KG}^{\mathcal{L}}_{\alpha\beta}$. These nonlinearities do not have null structure and, in fact, involve the massive field $\psi$ in undifferentiated form.

(2) The Klein-Gordon nonlinearities $\mathcal{LN}^\psi$ contain two types of quasilinear components, which are somewhat similar to the first two types of metric nonlinearities described above, and a KG-disposable remainder.

*Proof.* **Step 1.** We start with the quasilinear components of the metric nonlinearities, which can be written in the form $\sum_{\mu,\nu\in\{0,1,2,3\}}\widetilde{g}^{\mu\nu}_{\geq 1}\partial_\mu\partial_\nu h_{\alpha\beta}$ (see (2.1.9) and (4.3.49)). Thus

$$\mathcal{L}\{\widetilde{g}^{\mu\nu}_{\geq 1}\partial_\mu\partial_\nu h_{\alpha\beta}\} = \widetilde{g}^{\mu\nu}_{\geq 1}\mathcal{L}(\partial_\mu\partial_\nu h_{\alpha\beta}) + \sum_{\mathcal{L}_1+\mathcal{L}_2=\mathcal{L},\, \mathcal{L}_2\neq\mathcal{L}} c_{\mathcal{L}_1,\mathcal{L}_2}\mathcal{L}_1\widetilde{g}^{\mu\nu}_{\geq 1}\mathcal{L}_2(\partial_\mu\partial_\nu h_{\alpha\beta}). \tag{4.3.55}$$

The first term in the right-hand side can be replaced by $\widetilde{g}^{\mu\nu}_{\geq 1}\partial_\mu\partial_\nu(\mathcal{L}h_{\alpha\beta})$, while the second term can be replaced by $\sum_{\mathcal{L}_1+\mathcal{L}_2=\mathcal{L},\, \mathcal{L}_2\neq\mathcal{L}} c_{\mathcal{L}_1,\mathcal{L}_2}\mathcal{L}_1\widetilde{g}^{\mu\nu}_1\partial_\mu\partial_\nu(\mathcal{L}_2 h_{\alpha\beta})$ up to wave-disposable errors (due to Lemmas 4.19 and 4.20), where $\widetilde{g}^{\mu\nu}_1$ is the linear part of $\widetilde{g}^{\mu\nu}_{\geq 1}$. In view of the formulas (2.1.8) and (4.3.49), we have

$$\widetilde{g}^{00}_1 = 0, \qquad \widetilde{g}^{0j}_1 = h_{0j}, \qquad \widetilde{g}^{jk}_1 = -h_{jk} - h_{00}\delta_{jk}. \tag{4.3.56}$$

Therefore

$$\begin{aligned} \mathcal{L}\{\widetilde{g}^{\mu\nu}_{\geq 1}\partial_\mu\partial_\nu h_{\alpha\beta}\} - \widetilde{g}^{\mu\nu}_{\geq 1}\partial_\mu\partial_\nu(\mathcal{L}h_{\alpha\beta}) = \sum_{\mathcal{L}_1+\mathcal{L}_2=\mathcal{L},\, \mathcal{L}_2\neq\mathcal{L}} c_{\mathcal{L}_1,\mathcal{L}_2} \\ \times \big[2\mathcal{L}_1 h_{0j}\cdot\partial_0\partial_j\mathcal{L}_2 h_{\alpha\beta} - \mathcal{L}_1(h_{jk}+h_{00}\delta_{jk})\cdot\partial_j\partial_k\mathcal{L}_2 h_{\alpha\beta}\big] + \mathcal{R}^{\mathcal{L}h,1}_{\alpha\beta}, \end{aligned} \tag{4.3.57}$$

where $\mathcal{R}^{\mathcal{L}h,1}_{\alpha\beta}$ is wave-disposable of order $(q,n)$. Using (2.1.29) with $H = \mathcal{L}_1 h$, we have

$$\begin{aligned} &2\mathcal{L}_1 h_{0j}\cdot\partial_0\partial_j\mathcal{L}_2 h_{\alpha\beta} - \mathcal{L}_1(h_{jk}+h_{00}\delta_{jk})\cdot\partial_j\partial_k\mathcal{L}_2 h_{\alpha\beta} \\ &= [-2R_j\rho^{\mathcal{L}_1} + 2\in_{jmn} R_m\omega_n^{\mathcal{L}_1}]\cdot\partial_j\partial_0\mathcal{L}_2 h_{\alpha\beta} \\ &+ \big[-\delta_{jk}(F^{\mathcal{L}_1} + \underline{F}^{\mathcal{L}_1}) - R_jR_k(F^{\mathcal{L}_1} - \underline{F}^{\mathcal{L}_1}) \\ &+ (\in_{jlm} R_k + \in_{klm} R_j)R_l\Omega_m^{\mathcal{L}_1} - \in_{jpm}\in_{kqn} R_pR_q\vartheta_{mn}^{\mathcal{L}_1}\big]\cdot\partial_j\partial_k\mathcal{L}_2 h_{\alpha\beta}. \end{aligned}$$

We use now the formulas (4.3.4). After reorganizing the terms, the expression above becomes

$$\begin{aligned} &-[\delta_{jk}+R_jR_k]F^{\mathcal{L}_1}\cdot\partial_j\partial_k\mathcal{L}_2 h_{\alpha\beta} \\ &-\big\{2R_jR_0\underline{F}^{\mathcal{L}_1}\cdot\partial_j\partial_0\mathcal{L}_2 h_{\alpha\beta} + (\delta_{jk}-R_jR_k)\underline{F}^{\mathcal{L}_1}\cdot\partial_j\partial_k\mathcal{L}_2 h_{\alpha\beta}\big\} \\ &+\big\{2\in_{jmn} R_m\omega_n^{\mathcal{L}_1}\cdot\partial_j\partial_0\mathcal{L}_2 h_{\alpha\beta} + (\in_{jln} R_k + \in_{kln} R_j)R_lR_0\omega_n^{\mathcal{L}_1}\cdot\partial_j\partial_k\mathcal{L}_2 h_{\alpha\beta}\big\} \\ &-\in_{jpm}\in_{kqn} R_pR_q\vartheta_{mn}^{\mathcal{L}_1}\cdot\partial_j\partial_k\mathcal{L}_2 h_{\alpha\beta} - 2R_jR_0\tau^{\mathcal{L}_1}\cdot\partial_j\partial_0\mathcal{L}_2 h_{\alpha\beta} + \mathcal{R}^{\mathcal{L}h,2}_{\alpha\beta}. \end{aligned} \tag{4.3.58}$$

Here $\mathcal{R}^{\mathcal{L}h,2}_{\alpha\beta}$ corresponds to the contribution of the error terms in (4.3.4), and is wave-disposable of order $(q,n)$, in view of Lemma 4.19 (see (4.3.18)) and Lemma 4.20. The terms $-2R_jR_0\tau^{\mathcal{L}_1}\cdot\partial_j\partial_0\mathcal{L}_2 h_{\alpha\beta}$ are wave-disposable errors (due to Lemma 4.24 below), while the other terms can be rewritten as claimed in (4.3.50), using the identities

$$\begin{aligned} \partial_0\mathcal{L}_2 h_{\alpha\beta} &= (1/2)(U^{\mathcal{L}_2 h_{\alpha\beta},+} + U^{\mathcal{L}_2 h_{\alpha\beta},-}), \\ |\nabla|\mathcal{L}_2 h_{\alpha\beta} &= (i/2)(U^{\mathcal{L}_2 h_{\alpha\beta},+} - U^{\mathcal{L}_2 h_{\alpha\beta},-}) \end{aligned}$$

(see (2.1.36)) and similar identities for $F^{\mathcal{L}_1}, \underline{F}^{\mathcal{L}_1}, \omega_n^{\mathcal{L}_1}, \vartheta_{mn}^{\mathcal{L}_1}$. The symbols $\mathfrak{q}^{G,wa}_{\iota_1\iota_2}$

are given by

$$
\begin{aligned}
\mathfrak{q}^{F,wa}_{\iota_1\iota_2} &:= \frac{-\iota_1\iota_2}{4}\Big[1-\frac{(\theta\cdot\eta)^2}{|\theta|^2|\eta|^2}\Big],\\
\mathfrak{q}^{\underline{F},wa}_{\iota_1\iota_2} &:= \frac{-\iota_1\iota_2}{4}\Big[1-\iota_1\iota_2\frac{\theta\cdot\eta}{|\theta||\eta|}\Big]^2,\\
\mathfrak{q}^{\omega_n,wa}_{\iota_1\iota_2} &:= \frac{-i\iota_1}{2}\frac{\in_{jpn}\theta_p\eta_j}{|\theta||\eta|}\Big[1-\iota_1\iota_2\frac{\theta\cdot\eta}{|\theta||\eta|}\Big],\\
\mathfrak{q}^{\vartheta_{mn},wa}_{\iota_1\iota_2} &:= \frac{\iota_1\iota_2}{4}\frac{\in_{jpm}\theta_p\eta_j}{|\theta||\eta|}\frac{\in_{kqn}\theta_q\eta_k}{|\theta||\eta|}.
\end{aligned}
\tag{4.3.59}
$$

These multipliers are similar to the classical null multipliers in (4.3.42)–(4.3.44), and thus belong to $\mathcal{M}^{null}_{\iota_1\iota_2}$ as claimed.

**Step 2.** We consider now the Klein-Gordon nonlinearity $\mathcal{N}^\psi$ defined in (2.1.16). The analysis is similar to the analysis of the quasilinear wave nonlinearities. One can first place most of the cubic and higher order terms and the commutator terms in the KG-disposable remainder, due to Lemmas 4.19 and 4.20, thus

$$
\mathcal{L}\mathcal{N}^\psi=\widetilde{g}^{\mu\nu}_{\geq 1}\mathcal{L}(\partial_\mu\partial_\nu\psi)+\mathcal{L}(h_{00}\psi)+\sum_{\mathcal{L}_1+\mathcal{L}_2=\mathcal{L},\,\mathcal{L}_2\neq\mathcal{L}}c_{\mathcal{L}_1,\mathcal{L}_2}\mathcal{L}_1\widetilde{g}^{\mu\nu}_1\mathcal{L}_2(\partial_\mu\partial_\nu\psi)+\mathcal{R}^{\mathcal{L}\psi,1},
$$

where $\mathcal{R}^{\mathcal{L}\psi,1}$ is KG-disposable, and the additional term $\mathcal{L}(h_{00}\psi)$ comes from the term $(1-g^{00}_{\geq 1})^{-1}g^{00}_{\geq 1}\psi$ in (2.1.16). The first two terms in the right-hand side are as claimed in (4.3.48), and we can decompose the remaining term as in (4.3.57)–(4.3.58), with $h_{\alpha\beta}$ replaced by $\psi$. The terms $-2R_jR_0\tau^{\mathcal{L}_1}\cdot\partial_j\partial_0\mathcal{L}_2\psi$ are KG-disposable errors (due to Lemma 4.24 below), while all the other terms, including $\mathcal{L}(h_{00}\psi)$, are accounted for in $\mathcal{Q}^{\mathcal{L}}_{kg}(\psi)$. The resulting symbols $\mathfrak{q}^{G,kg}_{\iota_1\iota_2}$ can be calculated as before, and are given explicitly by

$$
\begin{aligned}
\mathfrak{q}^{F,kg}_{\iota_1\iota_2} &:= \frac{-\iota_1\iota_2}{4}\Big[1-\frac{(\theta\cdot\eta)^2}{|\theta|^2\langle\eta\rangle^2}\Big],\\
\mathfrak{q}^{\underline{F},kg}_{\iota_1\iota_2} &:= \frac{-\iota_1\iota_2}{4}\Big[1-\iota_1\iota_2\frac{\theta\cdot\eta}{|\theta|\langle\eta\rangle}\Big]^2,\\
\mathfrak{q}^{\omega_n,kg}_{\iota_1\iota_2} &:= \frac{-i\iota_1}{2}\frac{\in_{jpn}\theta_p\eta_j}{|\theta|\langle\eta\rangle}\Big[1-\iota_1\iota_2\frac{\theta\cdot\eta}{|\theta|\langle\eta\rangle}\Big],\\
\mathfrak{q}^{\vartheta_{mn},kg}_{\iota_1\iota_2} &:= \frac{\iota_1\iota_2}{4}\frac{\in_{jpm}\theta_p\eta_j}{|\theta|\langle\eta\rangle}\frac{\in_{kqn}\theta_q\eta_k}{|\theta|\langle\eta\rangle}.
\end{aligned}
\tag{4.3.60}
$$

**Step 3.** Finally, we consider the semilinear terms coming from the last two terms in (2.1.9). The cubic terms and the commutators can be safely included in the wave-disposable remainders, due to Lemmas 4.19–4.20. The Klein-Gordon contributions coming from $\mathcal{KG}^2_{\alpha\beta}$ are included in the terms $\mathcal{KG}^{\mathcal{L}}_{\alpha\beta}$ in (4.3.54). The null contributions coming from $Q^2_{\alpha\beta}$ are included in the terms

$\mathcal{S}^{\mathcal{L},1}_{\alpha\beta}$ in (4.3.52) (see (4.3.42)–(4.3.44)). The contributions of the terms $P^2_{\alpha\beta}$ in (2.1.15) are recovered in the terms $\mathcal{S}^{\mathcal{L},1}_{\alpha\beta}$ and $\mathcal{S}^{\mathcal{L},2}_{\alpha\beta}$, due to Lemma 4.25 below. □

In the analysis of the quasilinear terms in Lemma 4.22, we used the fact that certain quadratic expressions involving $\tau$ are energy disposable. We prove this below:

**Lemma 4.24.** *Assume that $(q,n) \leq (3,3)$, $\mathcal{L}_1 \in \mathcal{V}^{q_1}_{n_1}$, $\mathcal{L}_2 \in \mathcal{V}^{q_2}_{n_2}$, $(q_1,n_1)+(q_2,n_2) \leq (q,n)$, and $n_2 < n$ (so $n \geq 1$). If $I = I_m$, $m \in \mathcal{M}$, is as in (3.2.43), $h \in \{h_{\mu\nu}\}$, and $\alpha \in \{0,1,2,3\}$ then quadratic terms of the form*

$$I[R_0\tau^{\mathcal{L}_1}, |\nabla|\partial_\alpha\mathcal{L}_2 h] \qquad \text{and} \qquad I[\tau^{\mathcal{L}_1}, |\nabla|\partial_\alpha\mathcal{L}_2 h] \tag{4.3.61}$$

*are wave-disposable of order $(q,n)$. Similarly, quadratic terms of the form*

$$I[R_0\tau^{\mathcal{L}_1}, \langle\nabla\rangle\partial_\alpha\mathcal{L}_2\psi] \qquad \text{and} \qquad I[\tau^{\mathcal{L}_1}, \langle\nabla\rangle\partial_\alpha\mathcal{L}_2\psi] \tag{4.3.62}$$

*are KG-disposable of order $(q,n)$.*

*Proof.* The main point is that the metric components $\tau^{\mathcal{L}_1}$ have quadratic character, up to lower order terms, due to the identities (4.3.5). At low frequencies, however, the resulting quadratic bounds are not effective, and we need to trivialize one vector-field using Lemma 4.13.

More precisely, with $\overline{|k|} := \max\{|k|,|k_1|,|k_2|\}$ it suffices to prove that

$$\begin{aligned} 2^{k_2}2^{-k/2}\|P_kI[P_{k_1}R_\mu\tau^{\mathcal{L}_1}, P_{k_2}U^{\mathcal{L}_2h,\iota_2}](t)\|_{L^2} \\ \lesssim \varepsilon_1^2\langle t\rangle^{-1+H(q,n)\delta-\delta/2}2^{-N(n)k^+}2^{-\gamma\overline{|k|}/4} \end{aligned} \tag{4.3.63}$$

and

$$2^{k_2^+}\|P_kI[P_{k_1}R_\mu\tau^{\mathcal{L}_1}, P_{k_2}U^{\mathcal{L}_2\psi,\iota_2}](t)\|_{L^2} \lesssim \varepsilon_1^2\langle t\rangle^{-1+H(q,n)\delta-\delta/2}2^{-N(n)k^+}2^{-\gamma\overline{|k|}/4}, \tag{4.3.64}$$

for any $\mu \in \{0,1,2,3\}$, $\iota_2 \in \{+,-\}$, $k,k_1,k_2 \in \mathbb{Z}$, and $t \in [0,T]$. These bounds follow from (4.1.5) if $2^{\min\{k,k_1,k_2\}} \lesssim \langle t\rangle^{-1-4\delta}$ or if $2^{\max\{k,k_1,k_2\}} \gtrsim \langle t\rangle^{2.1}$. In the remaining range

$$\langle t\rangle \geq 2^{\delta^{-1}} \qquad \text{and} \qquad 2^k,\, 2^{k_1},\, 2^{k_2} \in [\langle t\rangle^{-1-4\delta}, \langle t\rangle^{2.1}] \tag{4.3.65}$$

we divide the proof into several steps.

**Step 1.** Assume first that (4.3.65) holds and $k_2 = \min\{k,k_1,k_2\}$. Then

$$\begin{aligned} 2^{k_2}2^{-k/2}&\|P_kI[P_{k_1}R_\mu\tau^{\mathcal{L}_1}, P_{k_2}U^{\mathcal{L}_2h,\iota_2}](t)\|_{L^2} \\ &\lesssim 2^{k_2-k/2}\|P_{k_1}R_\mu\tau^{\mathcal{L}_1}(t)\|_{L^2}\|P_{k_2}U^{\mathcal{L}_2h}(t)\|_{L^\infty} \\ &\lesssim \varepsilon_1^2 2^{-k/2}\langle t\rangle^{-1+\delta'}2^{2k_2^-}2^{-8k_2^+}\cdot 2^{-k_1/2}2^{\gamma|k_1|}2^{-N(n_1)k_1^+} \end{aligned} \tag{4.3.66}$$

and

$$\begin{aligned}
&2^{k_2^+}\|P_k I[P_{k_1}R_\mu\tau^{\mathcal{L}_1}, P_{k_2}U^{\mathcal{L}_2\psi,\iota_2}](t)\|_{L^2}\\
&\quad\lesssim 2^{k_2^+}\|P_{k_1}R_\mu\tau^{\mathcal{L}_1}(t)\|_{L^2}\|P_{k_2}U^{\mathcal{L}_2\psi}(t)\|_{L^\infty}\\
&\quad\lesssim \varepsilon_1^2\langle t\rangle^{-1+\delta'}2^{k_2^-/2}\min\{1,\langle t\rangle 2^{2k_2^-}\}2^{-8k_2^+}\cdot 2^{-k_1/2}2^{\gamma|k_1|}2^{-N(n_1)k_1^+},
\end{aligned} \tag{4.3.67}$$

using the $L^\infty$ estimates (3.3.11) and (3.3.13) on the second factor. The desired bounds (4.3.63)–(4.3.64) follow unless $\langle t\rangle^{-1/2-2\delta'}\lesssim 2^{k_2}\le 2^{k_1}\lesssim\langle t\rangle^{2\delta'}$.

In this case, however, we use (4.3.5) and replace $R_\mu\tau^{\mathcal{L}_1}$ with lower order linear terms of the form $R^a|\nabla|^{-2}\partial_\mu\partial_\nu\mathcal{L}_1'h_{\alpha\beta}$ (coming from the commutator terms) and nonlinear terms of the form $R^a|\nabla|^{-2}\mathcal{L}_1''\mathcal{N}$, where $\mathcal{N}\in\mathcal{QU}$ (see (4.3.12)), $\mathcal{L}_1'\in\mathcal{V}_{n_1-1}^{q_1}$, $\mathcal{L}_1''\in\mathcal{V}_{n_1}^{q_1}$, $\alpha,\beta,\mu,\nu\in\{0,1,2,3\}$, and $R^a=R_1^{a_1}R_2^{a_2}R_3^{a_3}$.

The contributions of the linear commutators $R^a|\nabla|^{-2}\partial_\mu\partial_\nu\mathcal{L}_1'h_{\alpha\beta}$ are suitably bounded as claimed, due to the more general estimates (4.3.37) and (4.3.38). The contributions of the nonlinear terms $R^a|\nabla|^{-2}\mathcal{L}_1''\mathcal{N}$ can be bounded using $L^2\times L^\infty$ estimates, as in (4.3.66)–(4.3.67), and recalling the $L^2$ bounds (4.3.14) and the assumption $\langle t\rangle^{-1/2-2\delta'}\lesssim 2^{k_2}\le 2^{k_1}\lesssim\langle t\rangle^{2\delta'}$.

**Step 2.** Assume now that (4.3.65) holds and $k=\min\{k,k_1,k_2\}$. The bounds (4.3.63)–(4.3.64) follow from (4.1.6)–(4.1.7) and (4.3.67) unless $\langle t\rangle^{-1/2-2\delta'}\lesssim 2^{k_2}\approx 2^{k_1}\lesssim\langle t\rangle^{2\delta'}$. In this case, however, we can again use the identities (4.3.5). As before, we estimate the contributions of the linear commutators using (4.3.37) and (4.3.38), and the contributions of the nonlinear terms $R^a|\nabla|^{-2}\mathcal{L}_1''\mathcal{N}$ using the $L^2$ bounds (4.3.14). The desired bounds (4.3.63)–(4.3.64) follow.

**Step 3.** Assume now that $k_1=\min\{k,k_1,k_2\}$ and $n_1<n$. Recall that $n_2<n$, thus $H(q_1,n_1)+H(q_2,n_2)\le H(q,n)-40$ (see (2.1.53)). The desired bounds follow from (4.1.5) if $2^{k_1}\lesssim\langle t\rangle^{-1+39\delta}$ or $2^{k_2}\gtrsim\langle t\rangle$. On the other hand, if $\langle t\rangle^{-1+39\delta}\le 2^{k_1}\le 2^{k_2}\le\langle t\rangle$ then we use the identities (4.3.5) as before, and replace $R_\mu\tau^{\mathcal{L}_1}$ with lower order linear terms of the form $R^a|\nabla|^{-2}\partial_\mu\partial_\nu\mathcal{L}_1'h_{\alpha\beta}$ and nonlinear terms of the form $R^a|\nabla|^{-2}\mathcal{L}_1''\mathcal{N}$. The contributions of the lower order linear terms can be suitably controlled using (4.3.37)–(4.3.38). Moreover, we estimate

$$\begin{aligned}
&2^{k_2}2^{-k/2}\|P_k I[P_{k_1}R^a|\nabla|^{-2}\mathcal{L}_1''\mathcal{N}, P_{k_2}U^{\mathcal{L}_2h,\iota_2}](t)\|_{L^2}\\
&\quad\lesssim 2^{k/2}2^{3k_1/2}\|P_{k_1}|\nabla|^{-2}\mathcal{L}_1''\mathcal{N}(t)\|_{L^2}\|P_{k_2}U^{\mathcal{L}_2h}(t)\|_{L^2}\\
&\quad\lesssim \varepsilon_1^2 2^k\langle t\rangle^{-1+H(q_1,n_1)\delta+36\delta}\langle t\rangle^{H(q_2,n_2)\delta}2^{-N(n_2)k^+}
\end{aligned} \tag{4.3.68}$$

and

$$\begin{aligned}
&2^{k_2^+}\|P_k I[P_{k_1}R^a|\nabla|^{-2}\mathcal{L}_1''\mathcal{N}, P_{k_2}U^{\mathcal{L}_2\psi,\iota_2}](t)\|_{L^2}\\
&\quad\lesssim 2^{k^+}2^{3k_1/2}\|P_{k_1}|\nabla|^{-2}\mathcal{L}_1''\mathcal{N}(t)\|_{L^2}\|P_{k_2}U^{\mathcal{L}_2\psi}(t)\|_{L^2}\\
&\quad\lesssim \varepsilon_1^2 2^{k^+}\langle t\rangle^{-1+H(q_1,n_1)\delta+36\delta}\langle t\rangle^{H(q_2,n_2)\delta}2^{-N(n_2)k^+},
\end{aligned} \tag{4.3.69}$$

using (4.3.14) and (3.3.3). The desired bounds (4.3.63)–(4.3.64) follow.

**Step 4.** Finally, assume that $k_1 = \min\{k, k_1, k_2\}$, $(q_1, n_1) = (q, n)$, and $(q_2, n_2) = (0, 0)$. The proof is harder in this case mainly because the decomposition (4.3.5) is not effective when $k_1$ is very small, say $2^{k_1} \approx \langle t\rangle^{-1}$. We consider two cases:

**Case 4.1.** Let

$$Y(0,1) := 15, \qquad Y(1,1) := 25, \qquad Y(q,n) := 55 \text{ if } n \geq 2. \tag{4.3.70}$$

If $2^{k_1} \in [\langle t\rangle^{-1-4\delta}, \langle t\rangle^{-1+Y(q,n)\delta}]$ then we prove the more general bounds

$$\begin{aligned}2^{k_2-k_1}2^{-k/2}\|P_k I[P_{k_1}U^{\mathcal{L}_1 h_1,\iota_2}, P_{k_2}U^{h_2,\iota_2}](t)\|_{L^2}\\ \lesssim \varepsilon_1^2\langle t\rangle^{-1+H(q,n)\delta-\delta/2}2^{-N(n)k^+}2^{-\gamma\overline{|k|}/4}\end{aligned} \tag{4.3.71}$$

and

$$\begin{aligned}2^{k_2^+-k_1}\|P_k I[P_{k_1}U^{\mathcal{L}_1 h_1,\iota_1}, P_{k_2}U^{\psi,\iota_2}](t)\|_{L^2}\\ \lesssim \varepsilon_1^2\langle t\rangle^{-1+H(q,n)\delta-\delta/2}2^{-N(n)k^+}2^{-\gamma\overline{|k|}/4},\end{aligned} \tag{4.3.72}$$

for any $\mathcal{L}_1 \in \mathcal{V}^q_n$, $n \geq 1$, $\iota_1, \iota_2 \in \{+,-\}$, $h_1, h_2 \in \{h_{\alpha\beta}\}$, $k, k_1, k_2 \in \mathbb{Z}$, and $t \in [0,T]$.

The bounds (4.3.71)–(4.3.72) follow easily using just $L^2$ estimates unless $2^{|k_2|} \lesssim \langle t\rangle^{\delta'}$ and $\langle t\rangle \gg 1$. In this case we decompose $P_{k_2}U^{h_2,\iota_2}(t) = U^{h_2,\iota_2}_{\leq J_2,k_2}(t) + U^{h_2,\iota_2}_{>J_2,k_2}(t)$ and $P_{k_2}U^{\psi,\iota_2}(t) = U^{\psi,\iota_2}_{\leq J_2,k_2}(t) + U^{\psi,\iota_2}_{>J_2,k_2}(t)$ as in (3.3.1)–(3.3.2), where $J_2$ is the largest integer satisfying $2^{J_2} \leq \langle t\rangle^{1/10}$. The contributions of the functions $U^{h_2,\iota_2}_{>J_2,k_2}(t)$ and $U^{\psi,\iota_2}_{>J_2,k_2}(t)$ to (4.3.71) and (4.3.72) respectively can be bounded easily, using again just $L^2$ estimates.

We now consider the main terms. Let $J_1$ denote the largest integer satisfying $2^{J_1} \leq (2^{-k_1} + \langle t\rangle)\langle t\rangle^{\delta/4}$ and decompose

$$U^*_{1,\leq J_1} := P'_{k_1}(\varphi_{\leq J_1}\cdot P_{k_1}U^{\mathcal{L}_1 h_1,\iota_1}), \quad U^*_{1,>J_1} := P'_{k_1}(\varphi_{>J_1}\cdot P_{k_1}U^{\mathcal{L}_1 h_1,\iota_1}). \tag{4.3.73}$$

Notice that in this case we decompose in the physical space the normalized solutions $P_{k_1}U^{\mathcal{L}_1 h_1,\iota_1}$, not the profiles $P_{k_1}V^{\mathcal{L}_1 h_1,\iota_1}$. The point is that the functions $U^*_{1,\leq J_1}$ satisfy the $L^2$ bounds

$$\|U^*_{1,\leq J_1}(t)\|_{L^2} \lesssim \varepsilon_1 2^{k_1/2}\langle t\rangle^{H(q,n)\delta-7\delta/4}. \tag{4.3.74}$$

These bounds are stronger than (3.3.3) (notice the gain of $\langle t\rangle^{-7\delta/4}$) and follow from Lemma 4.13. Indeed, if $q = 0$ then we use (4.2.60) and the assumption $2^{k_1} \in [\langle t\rangle^{-1-4\delta}, \langle t\rangle^{-1+Y(q,n)\delta}]$, so

$$\|U^*_{1,\leq J_1}(t)\|_{L^2} \lesssim (2^{k_1}\langle t\rangle)^{-\gamma}2^{k_1/2}(1 + 2^{k_1}\langle t\rangle)\langle t\rangle^{\delta/4}\langle t\rangle^{H(0,n-1)\delta}.$$

The desired bounds (4.3.74) follow when $q = 0$ since $H(0, n-1) + Y(0, n) \leq$

$H(0,n)-3$; see (4.3.70). The proof is similar in the case $q \geq 1$, using (4.2.59) instead of (4.2.60).

Using (3.2.46), (4.3.74), and (3.3.7), we find that

$$\begin{aligned} &2^{k_2/2-k_1}\|P_k I[U^*_{1,\leq J_1}, U^{h_2,\iota_2}_{\leq J_2,k_2}](t)\|_{L^2} \\ &\qquad \lesssim 2^{-k_1/2}\langle t\rangle^{-1}2^{2k_2}\|U^*_{1,\leq J_1}(t)\|_{L^2}\|\widehat{P_{k_2}V^{h_2}}(t)\|_{L^\infty} \\ &\qquad \lesssim \varepsilon_1^2\langle t\rangle^{-1+H(q,n)\delta-3\delta/4}2^{-N(n)k^+}2^{k^-/2}2^{-3k^+}. \end{aligned} \tag{4.3.75}$$

Similarly, using (3.3.17) and (3.3.7) we have

$$\begin{aligned} &2^{k_2^+-k_1}\|P_k I[U^*_{1,\leq J_1}, U^{\psi,\iota_2}_{\leq J_2,k_2}](t)\|_{L^2} \\ &\qquad \lesssim 2^{k_2^+-k_1}\|U^*_{1,\leq J_1}(t)\|_{L^2}\min\{\|U^{\psi}_{\leq J_2,k_2}(t)\|_{L^\infty}, 2^{3k_1/2}\|U^{\psi}_{\leq J_2,k_2}(t)\|_{L^2}\} \\ &\qquad \lesssim \varepsilon_1^2\langle t\rangle^{-1+H(q,n)\delta-7\delta/4}2^{-N(n)k^+}2^{\kappa k^-/20}2^{-3k^+}. \end{aligned} \tag{4.3.76}$$

Finally, we claim that the remaining contributions are negligible,

$$\begin{aligned} &2^{k_2/2-k_1}\|P_k I[U^*_{1,>J_1}, U^{h_2,\iota_2}_{\leq J_2,k_2}](t)\|_{L^2} \\ &\qquad + 2^{k_2^+-k_1}\|P_k I[U^*_{1,>J_1}, U^{\psi,\iota_2}_{\leq J_2,k_2}](t)\|_{L^2} \lesssim \varepsilon_1^2\langle t\rangle^{-2}2^{-N(n)k^+}. \end{aligned} \tag{4.3.77}$$

To see this we use approximate finite speed of propagation arguments. Indeed, we observe that

$$\begin{aligned} &P_k I[U^*_{1,>J_1}, U^{h_2,\iota_2}_{\leq J_2,k_2}](x,t) \\ &\quad = C\int_{\mathbb{R}^6}\varphi_{>J_1}(z)P_{k_1}U^{\mathcal{L}h_1,\iota_1}(z)\cdot\varphi_{\leq J_2}(y)P_{k_2}V^{h_2,\iota_2}(y)K(x-y,x-z)\,dydz, \end{aligned}$$

where the kernel is given by

$$\begin{aligned} K(y',z') := &\int_{\mathbb{R}^6} e^{-i\iota_2 t|\eta|}e^{iy'\cdot\eta}e^{iz'\cdot\theta}m(\theta,\eta) \\ &\qquad \times \varphi_k(\theta+\eta)\varphi_{[k_1-2,k_1+2]}(\theta)\varphi_{[k_2-2,k_2+2]}(\eta)\,d\theta d\eta. \end{aligned}$$

The point is that $|K(y',z')|$ is small when $|y'-z'| \gtrsim 2^{J_1}$. Indeed, for any $M \geq 1$ we have

$$|K(y',z')| \lesssim_M 2^{3k_1}2^{3k}\left[1+\frac{|y'|+|z'|}{\langle t\rangle + 2^{-k_1}}\right]^{-M},$$

using integration by parts either in $\eta$ or in $\theta$. Since $|y-z| \gtrsim 2^{J_1} \approx \langle t\rangle^{\delta/4}(\langle t\rangle + 2^{-k_1})$ in the support of the integral, it follows that the first expression in (4.3.77) is negligible as claimed. The second expression can be bounded in the same way. In view of (4.3.75)–(4.3.76), this completes the proof of (4.3.71)–(4.3.72).

**Case 4.2.** Finally we prove the bounds (4.3.63)–(4.3.64) when

$$\begin{aligned}(q_1,n_1)&=(q,n), &\quad (q_2,n_2)&=(0,0),\\ k_1&=\min\{k,k_1,k_2\}, &\quad 2^{k_1}&\geq\langle t\rangle^{-1+Y(q,n)\delta}.\end{aligned}\tag{4.3.78}$$

We use the identities (4.3.5). The contributions of the linear commutators $R^a|\nabla|^{-2}\partial_\mu\partial_\nu\mathcal{L}'_1 h_{\alpha\beta}$ are bounded as claimed, due to the estimates (4.3.37)–(4.3.38). It remains to prove that

$$\begin{aligned}&2^{k_2/2}\|P_kI[P_{k_1}|\nabla|^{-2}\mathcal{L}_1\mathcal{N},P_{k_2}U^{h,\iota_2}](t)\|_{L^2}\\&\qquad\lesssim\varepsilon_1^2\langle t\rangle^{-1+H(q,n)\delta-\delta/2}2^{-N(n)k^+}2^{-\gamma|\overline{k}|/4}\end{aligned}\tag{4.3.79}$$

and

$$\begin{aligned}&2^{k_2^+}\|P_kI[P_{k_1}|\nabla|^{-2}\mathcal{L}_1\mathcal{N},P_{k_2}U^{\psi,\iota_2}](t)\|_{L^2}\\&\qquad\lesssim\varepsilon_1^2\langle t\rangle^{-1+H(q,n)\delta-\delta/2}2^{-N(n)k^+}2^{-\gamma|\overline{k}|/4},\end{aligned}\tag{4.3.80}$$

for any $\mathcal{N}\in\mathcal{QU}$. Using (4.3.14), (3.3.3), (3.3.11), and (3.3.13), these bounds follow if $2^{k_1}\gtrsim\langle t\rangle^{-1/2}$ or if $2^{k_2}\notin[\langle t\rangle^{-\delta'},\langle t\rangle^{\delta'}]$.

On the other hand, if $2^{k_1}\leq\langle t\rangle^{-1/2}$ and $2^{k_2}\in[\langle t\rangle^{-\delta'},\langle t\rangle^{\delta'}]$ then we would like to use (4.3.16) and estimate as in (4.3.75)–(4.3.76). For this we replace first $P_{k_2}U^{h,\iota_2}$ with $U^{h,\iota_2}_{\leq J_2,k_2}$ and $P_{k_2}U^{\psi,\iota_2}$ with $U^{\psi,\iota_2}_{\leq J_2,k_2}$ at the expense of acceptable errors, where $J_2$ is the largest integer satisfying $2^{J_2}\leq\langle t\rangle^{1/10}$. Then we estimate, using (3.2.46), (4.3.16), and (3.3.7),

$$\begin{aligned}&2^{k_2/2}\|P_kI[P_{k_1}|\nabla|^{-2}\mathcal{L}_1\mathcal{N},U^{h_2,\iota_2}_{\leq J_2,k_2}](t)\|_{L^2}\\&\quad\lesssim 2^{k_1/2}\langle t\rangle^{-1}2^{2k_2}\|P_{k_1}|\nabla|^{-2}\mathcal{L}_1\mathcal{N}(t)\|_{L^2}\|\widehat{P_{k_2}V^{h_2}}(t)\|_{L^\infty}\\&\quad\lesssim\varepsilon_1^2 2^{-k_1}\langle t\rangle^{-2+H(q,n)\delta+\ell(q,n)\delta+3\delta/2}2^{-N(n)k^+}2^{k^--\kappa k^-}2^{-3k^+}.\end{aligned}$$

Similarly, using (3.3.17) and (3.3.7) we have

$$\begin{aligned}&2^{k_2^+}\|P_kI[P_{k_1}|\nabla|^{-2}\mathcal{L}_1\mathcal{N},U^{\psi,\iota_2}_{\leq J_2,k_2}](t)\|_{L^2}\\&\quad\lesssim 2^{k_2^+-2k_1}\|P_{k_1}\mathcal{L}_1\mathcal{N}(t)\|_{L^2}\min\{\|U^{\psi}_{\leq J_2,k_2}(t)\|_{L^\infty},2^{3k_1/2}\|U^{\psi}_{\leq J_2,k_2}(t)\|_{L^2}\}\\&\quad\lesssim\varepsilon_1^2 2^{-k_1}\langle t\rangle^{-2+H(q,n)\delta+\ell(q,n)\delta+\delta/2}2^{-N(n)k^+}2^{\kappa k^-/20}2^{-3k^+}.\end{aligned}$$

Since $2^{-k_1}\langle t\rangle^{-1}\lesssim\langle t\rangle^{-Y(q,n)\delta}$ (see (4.3.78)) and $Y(q,n)\geq\ell(q,n)+2$, these bounds suffice to prove (4.3.79)–(4.3.80). This completes the proof of the lemma. □

We show now that the nonlinearities $\mathcal{L}P^2_{\alpha\beta}$ can be written as sums of bilinear expressions involving only the good metric components $\vartheta$, null semilinear forms, and acceptable errors.

**Lemma 4.25.** *If $\mathcal{L} \in \mathcal{V}_n^q$ then, with $\mathcal{S}_{\alpha\beta}^{\mathcal{L},2}$ as in (4.3.53),*

$$\mathcal{L}P_{\alpha\beta}^2 = -\mathcal{S}_{\alpha\beta}^{\mathcal{L},2} + \Pi_{\alpha\beta}^{\mathcal{L}} + \mathcal{R}_{\alpha\beta}^{\mathcal{L}}, \tag{4.3.81}$$

*where $\Pi_{\alpha\beta}^{\mathcal{L}}$ are semilinear null forms of order $(q,n)$ and $\mathcal{R}_{\alpha\beta}^{\mathcal{L}}$ are wave-disposable remainders of order $(q,n)$ (sums of semilinear cubic remainders of the first two types as defined in (4.3.13) and wave commutator remainders as defined in (4.3.33)).*

*Proof.* Using (2.1.15) we write

$$\begin{aligned}\mathcal{L}P_{\alpha\beta}^2 = \sum_{\mathcal{L}_1+\mathcal{L}_2=\mathcal{L}} c_{\mathcal{L}_1,\mathcal{L}_2}\big\{ &- (1/2)\partial_\alpha\mathcal{L}_1 h_{00}\partial_\beta\mathcal{L}_2 h_{00} + \partial_\alpha\mathcal{L}_1 h_{0j}\partial_\beta\mathcal{L}_2 h_{0j} \\ &- (1/2)\partial_\alpha\mathcal{L}_1 h_{jk}\partial_\beta\mathcal{L}_2 h_{jk} \\ &+ (1/4)\partial_\alpha\mathcal{L}_1(-h_{00} + \delta_{jk}h_{jk})\partial_\beta\mathcal{L}_2(-h_{00} + \delta_{j'k'}h_{j'k'})\big\} + \mathcal{T}_{\alpha\beta}^{\mathcal{L},1}\end{aligned}$$

for some wave commutator remainders $\mathcal{T}_{\alpha\beta}^{\mathcal{L},1}$ of order $(q,n)$. Using now (2.1.29) and assuming $\mathcal{L}_i \in \mathcal{V}_{n_i}^{q_i}$, $(q_1,n_1)+(q_2,n_2)=(q,n)$, we rewrite the expression between the brackets as

$$\begin{aligned}&- (1/2)\partial_\alpha(F^{\mathcal{L}_1} + \underline{F}^{\mathcal{L}_1})\partial_\beta(F^{\mathcal{L}_2} + \underline{F}^{\mathcal{L}_2}) \\ &+ \partial_\alpha(-R_j\rho^{\mathcal{L}_1} + \in_{jkl} R_k\omega_l^{\mathcal{L}_1})\partial_\beta(-R_j\rho^{\mathcal{L}_2} + \in_{jk'l'} R_{k'}\omega_{l'}^{\mathcal{L}_2}) \\ &- (1/2)\partial_\alpha[R_jR_k(F^{\mathcal{L}_1} - \underline{F}^{\mathcal{L}_1}) - (\in_{klm} R_j + \in_{jlm} R_k)R_l\Omega_m^{\mathcal{L}_1} \\ &\qquad + \in_{jpm}\in_{kqn} R_pR_q\vartheta_{mn}^{\mathcal{L}_1}] \times \partial_\beta[R_jR_k(F^{\mathcal{L}_2} - \underline{F}^{\mathcal{L}_2}) \\ &\qquad - (\in_{kl'm'} R_j + \in_{jl'm'} R_k)R_{l'}\Omega_{m'}^{\mathcal{L}_2} + \in_{jp'm'}\in_{kq'n'} R_{p'}R_{q'}\vartheta_{m'n'}^{\mathcal{L}_2}] \\ &+ \partial_\alpha(\tau^{\mathcal{L}_1} - F^{\mathcal{L}_1})\partial_\beta(\tau^{\mathcal{L}_2} - F^{\mathcal{L}_2}).\end{aligned}$$

We replace also $\rho^{\mathcal{L}_i}$ with $R_0\underline{F}^{\mathcal{L}_i} + R_0\tau^{\mathcal{L}_i} + |\nabla|^{-1}\mathcal{L}_iE_0^{\geq 2} + |\nabla|^{-1}E_{\mathcal{L}_i,0}^{com}$ and $\Omega_j^{\mathcal{L}_i}$ with $R_0\omega_j^{\mathcal{L}_i} + |\nabla|^{-1} \in_{jlk} R_l\mathcal{L}_iE_k^{\geq 2} + |\nabla|^{-1} \in_{jlk} R_lE_{\mathcal{L}_i,k}^{com}$, according to the harmonic gauge identities (4.3.4). The contributions of the error terms lead to either semilinear cubic remainders of order $(q,n)$ (of the first type described in (4.3.13)) or wave commutator remainders of order $(q,n)$ (of the last type in (4.3.33)). Then we regroup and expand the remaining terms, in the form

$$A_{\alpha\beta}^{FF} + A_{\alpha\beta}^{F\underline{F}} + A_{\alpha\beta}^{F\omega} + A_{\alpha\beta}^{F\vartheta} + A_{\alpha\beta}^{\underline{F}\,\underline{F}} + A_{\alpha\beta}^{\underline{F}\omega} + A_{\alpha\beta}^{\underline{F}\vartheta} + A_{\alpha\beta}^{\omega\omega} + A_{\alpha\beta}^{\omega\vartheta} + A_{\alpha\beta}^{\vartheta\vartheta} + \widetilde{A}_{\alpha\beta}^{\tau}, \tag{4.3.82}$$

where

$$A_{\alpha\beta}^{FF} = (1/2)(\partial_\alpha F^{\mathcal{L}_1}\cdot\partial_\beta F^{\mathcal{L}_2} - R_jR_k\partial_\alpha F^{\mathcal{L}_1}\cdot R_jR_k\partial_\beta F^{\mathcal{L}_2}),$$

$$\begin{aligned}A_{\alpha\beta}^{F\underline{F}} = &-(1/2)(\partial_\alpha F^{\mathcal{L}_1}\cdot\partial_\beta\underline{F}^{\mathcal{L}_2} - R_jR_k\partial_\alpha F^{\mathcal{L}_1}\cdot R_jR_k\partial_\beta\underline{F}^{\mathcal{L}_2}) \\ &- (1/2)(\partial_\beta F^{\mathcal{L}_2}\cdot\partial_\alpha\underline{F}^{\mathcal{L}_1} - R_jR_k\partial_\beta F^{\mathcal{L}_2}\cdot R_jR_k\partial_\alpha\underline{F}^{\mathcal{L}_1}),\end{aligned}$$

$$A^{F\omega}_{\alpha\beta} = \in_{klm} R_kR_j\partial_\alpha F^{\mathcal{L}_1}\cdot R_lR_jR_0\partial_\beta\omega^{\mathcal{L}_2}_m + \in_{klm} R_kR_j\partial_\beta F^{\mathcal{L}_2}\cdot R_lR_jR_0\partial_\alpha\omega^{\mathcal{L}_1}_m,$$

$$A^{F\vartheta}_{\alpha\beta} = -(1/2)\in_{jpm} R_jR_k\partial_\alpha F^{\mathcal{L}_1}\cdot R_p\in_{kqn} R_q\partial_\beta\vartheta^{\mathcal{L}_2}_{mn} - (1/2)\in_{jpm} R_jR_k\partial_\beta F^{\mathcal{L}_2}\cdot R_p\in_{kqn} R_q\partial_\alpha\vartheta^{\mathcal{L}_1}_{mn},$$

$$A^{\underline{F}\,\underline{F}}_{\alpha\beta} = -(1/2)(\partial_\alpha\underline{F}^{\mathcal{L}_1}\cdot\partial_\beta\underline{F}^{\mathcal{L}_2} + R_jR_k\partial_\alpha\underline{F}^{\mathcal{L}_1}\cdot R_jR_k\partial_\beta\underline{F}^{\mathcal{L}_2} - 2R_0R_k\partial_\alpha\underline{F}^{\mathcal{L}_1}\cdot R_0R_k\partial_\beta\underline{F}^{\mathcal{L}_2}),$$

$$A^{\underline{F}\,\omega}_{\alpha\beta} = -\in_{jkl} R_jR_0\partial_\alpha\underline{F}^{\mathcal{L}_1}\cdot R_k\partial_\beta\omega^{\mathcal{L}_2}_l - \in_{jkl} R_jR_0\partial_\beta\underline{F}^{\mathcal{L}_2}\cdot R_k\partial_\alpha\omega^{\mathcal{L}_1}_l - \in_{klm} R_kR_j\partial_\alpha\underline{F}^{\mathcal{L}_1}\cdot R_lR_jR_0\partial_\beta\omega^{\mathcal{L}_2}_m - \in_{klm} R_kR_j\partial_\beta\underline{F}^{\mathcal{L}_2}\cdot R_lR_jR_0\partial_\alpha\omega^{\mathcal{L}_1}_m,$$

$$A^{\underline{F}\vartheta}_{\alpha\beta} = (1/2)\in_{jpm} R_jR_k\partial_\alpha\underline{F}^{\mathcal{L}_1}\cdot R_p\in_{kqn} R_q\partial_\beta\vartheta^{\mathcal{L}_2}_{mn} + (1/2)\in_{jpm} R_jR_k\partial_\beta\underline{F}^{\mathcal{L}_2}\cdot R_p\in_{kqn} R_q\partial_\alpha\vartheta^{\mathcal{L}_1}_{mn},$$

$$\begin{aligned}A^{\omega\omega}_{\alpha\beta} = {} & R_l\partial_\alpha\omega^{\mathcal{L}_1}_m\cdot R_l\partial_\beta\omega^{\mathcal{L}_2}_m - R_jR_0R_l\partial_\alpha\omega^{\mathcal{L}_1}_m\cdot R_jR_0R_l\partial_\beta\omega^{\mathcal{L}_2}_m\\ & - (R_l\partial_\alpha\omega^{\mathcal{L}_1}_m\cdot R_m\partial_\beta\omega^{\mathcal{L}_2}_l - R_jR_0R_l\partial_\alpha\omega^{\mathcal{L}_1}_m\cdot R_jR_0R_m\partial_\beta\omega^{\mathcal{L}_2}_l)\\ & - (1/2)\in_{klm}\in_{jl'm'} R_lR_jR_0\partial_\alpha\omega^{\mathcal{L}_1}_m\cdot R_kR_{l'}R_0\partial_\beta\omega^{\mathcal{L}_2}_{m'}\\ & - (1/2)\in_{jlm}\in_{kl'm'} R_lR_kR_0\partial_\alpha\omega^{\mathcal{L}_1}_m\cdot R_jR_{l'}R_0\partial_\beta\omega^{\mathcal{L}_2}_{m'},\end{aligned}$$

$$A^{\omega\vartheta}_{\alpha\beta} = \in_{jkq} R_jR_lR_0\partial_\alpha\omega^{\mathcal{L}_1}_m\cdot R_kR_l\partial_\beta\vartheta^{\mathcal{L}_2}_{qm} - \in_{jkq} R_jR_lR_0\partial_\alpha\omega^{\mathcal{L}_1}_m\cdot R_kR_m\partial_\beta\vartheta^{\mathcal{L}_2}_{ql} + \in_{jkq} R_jR_lR_0\partial_\beta\omega^{\mathcal{L}_2}_m\cdot R_kR_l\partial_\alpha\vartheta^{\mathcal{L}_1}_{qm} - \in_{jkq} R_jR_lR_0\partial_\beta\omega^{\mathcal{L}_2}_m\cdot R_kR_m\partial_\alpha\vartheta^{\mathcal{L}_1}_{ql},$$

$$A^{\vartheta\vartheta}_{\alpha\beta} = R_pR_q\partial_\alpha\vartheta^{\mathcal{L}_1}_{mn}\cdot R_pR_n\partial_\beta\vartheta^{\mathcal{L}_2}_{mq} - (1/2)R_pR_q\partial_\alpha\vartheta^{\mathcal{L}_1}_{mn}\cdot R_pR_q\partial_\beta\vartheta^{\mathcal{L}_2}_{mn} - (1/2)R_pR_q\partial_\alpha\vartheta^{\mathcal{L}_1}_{mn}\cdot R_mR_n\partial_\beta\vartheta^{\mathcal{L}_2}_{pq},$$

$$\begin{aligned}\widetilde{A}^{\tau}_{\alpha\beta} = {} & \partial_\alpha\tau^{\mathcal{L}_1}\cdot\partial_\beta\tau^{\mathcal{L}_2} + R_0R_j\partial_\alpha\tau^{\mathcal{L}_1}\cdot R_0R_j\partial_\beta\tau^{\mathcal{L}_2} - \partial_\alpha\tau^{\mathcal{L}_1}\cdot\partial_\beta F^{\mathcal{L}_2}\\ & - \partial_\alpha F^{\mathcal{L}_1}\cdot\partial_\beta\tau^{\mathcal{L}_2} + R_0R_j\partial_\alpha\tau^{\mathcal{L}_1}\cdot R_0R_j\partial_\beta\underline{F}^{\mathcal{L}_2} + R_0R_j\partial_\alpha\underline{F}^{\mathcal{L}_1}\cdot R_0R_j\partial_\beta\tau^{\mathcal{L}_2}\\ & - R_0R_j\partial_\alpha\tau^{\mathcal{L}_1}\cdot\in_{jpq} R_p\partial_\beta\omega^{\mathcal{L}_2}_q - R_0R_j\partial_\beta\tau^{\mathcal{L}_2}\cdot\in_{jpq} R_p\partial_\alpha\omega^{\mathcal{L}_1}_q.\end{aligned}$$

All the terms in $A^{FF}_{\alpha\beta}$, $A^{F\underline{F}}_{\alpha\beta}$, $A^{F\vartheta}_{\alpha\beta}$, $A^{\underline{F}\vartheta}_{\alpha\beta}$ are clearly semilinear null forms (see also (4.3.45)). Since $\vartheta$ is divergence free (see (2.1.28)), the terms $R_pR_q\partial_\alpha\vartheta^{\mathcal{L}_1}_{mn}\cdot R_pR_n\partial_\beta\vartheta^{\mathcal{L}_2}_{mq}$ and $R_pR_q\partial_\alpha\vartheta^{\mathcal{L}_1}_{mn}\cdot R_mR_n\partial_\beta\vartheta^{\mathcal{L}_2}_{pq}$ in $A^{\vartheta\vartheta}_{\alpha\beta}$ are also semilinear null forms. The remaining term in $A^{\vartheta\vartheta}_{\alpha\beta}$ generates the terms $\mathcal{S}^{\mathcal{L},2}_{\alpha\beta}$ in (4.3.81).

The terms that contain $R_0$ require a little more care. We notice first that if $\alpha\neq 0$ then $R_0\partial_\alpha = R_\alpha\partial_0$; moreover, if $\alpha = 0$ then we use the identities

$$R_0\partial_0 G^{\mathcal{L}_i} = |\nabla|^{-1}\{\Delta G^{\mathcal{L}_i} + G[\mathcal{L}_i\mathcal{N}^h]\} = R_m\partial_m G^{\mathcal{L}_i} + |\nabla|^{-1}G[\mathcal{L}_i\mathcal{N}^h], \quad (4.3.83)$$

for any $G\in\{F,\underline{F},\omega_m,\vartheta_{mn},\tau\}$. Thus we can express all the terms in $A^{F\omega}_{\alpha\beta}$, $A^{\underline{F}\omega}_{\alpha\beta}$, and $A^{\omega\vartheta}_{\alpha\beta}$ as sums of semilinear null forms and semilinear cubic remainders (of the first type in (4.3.13)).

To deal with the remaining terms, we claim that

$$R^a R_0 \partial_\alpha G_1^{\mathcal{L}_1} \cdot R^b R_0 \partial_\beta G_2^{\mathcal{L}_2} \approx R^a R_j \partial_\alpha G_1^{\mathcal{L}_1} \cdot R^b R_j \partial_\beta G_2^{\mathcal{L}_2} \tag{4.3.84}$$

for any $G_1, G_2 \in \{F, \underline{F}, \omega_m, \vartheta_{mn}, \tau\}$, $\alpha, \beta \in \{0,1,2,3\}$, and $R^a = R_1^{a_1} R_2^{a_2} R_3^{a_3}$, $R^b = R_1^{b_1} R_2^{b_2} R_3^{b_3}$, where the identity holds up to sums of semilinear null forms and semilinear cubic remainders (of the first two types in (4.3.13)). Indeed, if $\alpha, \beta \in \{1,2,3\}$ then this follows from (4.3.44) and the identities $R_0 \partial_\rho = R_\rho \partial_0$, $R_j \partial_\rho = R_\rho \partial_j$, $\rho \in \{\alpha, \beta\}$. If $\alpha = 0$ or $\beta = 0$ then (4.3.84) follows using first (4.3.83), to extract the cubic remainders, and combining with (4.3.42)–(4.3.44).

We examine now all the terms in $A_{\alpha\beta}^{\underline{F}\,\underline{F}}$ and $A_{\alpha\beta}^{\omega\omega}$; as a consequence of (4.3.84) and (4.3.45), they can all be written as sums of semilinear null forms and semilinear cubic remainders.

Similarly, all the terms in $\widetilde{A}_{\alpha\beta}^{\tau}$ are sums of semilinear null forms, semilinear cubic remainders, and wave commutator remainders of order $(q,n)$. Indeed, the main point is that $\partial_\mu \tau^{\mathcal{L}_i}$ can be written as a sum of expressions of the form $R^a |\nabla|^{-1} \mathcal{L}' \mathcal{N}$ and $R^a \partial_\rho \mathcal{L}'' h$, $\mathcal{N} \in \mathcal{QU}$, $h \in \{h_{\alpha\beta}\}$, $\mathcal{L}' \in \mathcal{V}_{n_i}^{q_i}$, $\mathcal{L}'' \in \mathcal{V}_{n_i-1}^{q_i}$, $\rho \in \{0,1,2,3\}$ (due to (4.3.5)). We replace also $R_0 X R_0 Y$ with $R_k X R_k Y$, at the expense of acceptable errors, in three of the terms in $\widetilde{A}_{\alpha\beta}^{\tau}$, and use (4.3.83) for the terms in the last line of $\widetilde{A}_{\alpha\beta}^{\tau}$ if $\alpha = 0$ or $\beta = 0$. It follows that all terms in $\widetilde{A}_{\alpha\beta}^{\tau}$ can be written as a sum of acceptable errors, and the lemma follows. □

# Chapter Five

# Improved Energy Estimates

## 5.1 SETUP AND PRELIMINARY REDUCTIONS

In this chapter we prove the main energy estimates (2.1.50). More precisely:

**Proposition 5.1.** *With the hypothesis in Proposition 2.3, for any $t \in [0,T]$ and $\mathcal{L} \in \mathcal{V}_n^q$, $n \le 3$, we have*

$$\|(\langle t\rangle|\nabla|_{\le 1})^{\gamma}|\nabla|^{-1/2}U^{\mathcal{L}h_{\alpha\beta}}(t)\|_{H^{N(n)}} + \|U^{\mathcal{L}\psi}(t)\|_{H^{N(n)}} \lesssim \varepsilon_0\langle t\rangle^{H(q,n)\delta}. \quad (5.1.1)$$

We notice that these bounds hold when $t=0$ due to the initial-data assumptions (1.2.5) (and, in fact, for all $t \in [0,2]$ due to the stronger bounds (7.1.10) proved in Chapter 7). Our main goal in this section is to prove these energy estimates for all $t \in [0,T]$.

### 5.1.1 Energy Increments

To prove Proposition 5.1 we start by defining suitable energy functionals. We consider the modified metric $\widetilde{g}^{\mu\nu} := m^{\mu\nu} + \widetilde{g}^{\mu\nu}_{\ge 1}$, where $\widetilde{g}^{\mu\nu}_{\ge 1}$ are as in (4.3.49), and the modified wave operator $\widetilde{\Box}_{\widetilde{g}} := \widetilde{g}^{\mu\nu}\partial_\mu\partial_\nu$. Suppose that $\lambda \in \{0,1\}$ and $\phi \in C([0,T] : L^2)$ is a real-valued solution of the equation

$$-\widetilde{\Box}_{\widetilde{g}}\phi + \lambda(1-h_{00})\phi = N.$$

We define

$$\begin{aligned}\mathcal{E}_\lambda(\phi) &:= \frac{1}{2}\int_{\mathbb{R}^3}\{-\widetilde{g}^{00}(\partial_t\phi)^2 + \widetilde{g}^{jk}\partial_j\phi\partial_k\phi + \lambda(1-h_{00})(\phi)^2\}\,dx,\\ \mathcal{B}(\phi) &:= \int_{\mathbb{R}^3}\big[\partial_j\widetilde{g}^{0j}(\partial_t\phi)^2 - (1/2)\partial_t\widetilde{g}^{jk}\partial_j\phi\partial_k\phi\\ &\qquad + \partial_j\widetilde{g}^{jk}\partial_t\phi\partial_k\phi + (\lambda/2)\partial_t h_{00}(\phi)^2\big]\,dx.\end{aligned} \quad (5.1.2)$$

Since $\widetilde{g}^{00} = -1$, we have

$$\frac{d}{dt}\mathcal{E}_\lambda(\phi) = -\mathcal{B}(\phi) + \int_{\mathbb{R}^3}\partial_t\phi\cdot N\,dx. \quad (5.1.3)$$

Using (3.3.11), it is easy to see that if $\varepsilon_1$ is small enough then

$$\mathcal{E}_0(\phi) \le \|\nabla_{x,t}\phi\|_{L^2}^2 \le 4\mathcal{E}_0(\phi), \qquad \mathcal{E}_1(\phi) \le \|\nabla_{x,t}\phi\|_{L^2}^2 + \|\phi\|_{L^2}^2 \le 4\mathcal{E}_1(\phi). \tag{5.1.4}$$

This will form the basis of our energy estimates for both $h$ and $\psi$, as we apply these identities to $\mathcal{L}h_{\alpha\beta}$ (with $\lambda = 0$) and $\mathcal{L}\psi$ (with $\lambda = 1$), with suitable multipliers, and use (4.3.47)–(4.3.48).

The bulk terms $\mathcal{B}$ have the same regularity as the energy, and additional "null structure". Indeed, using the formulas (4.3.56), (2.1.29) (with $H = h$), and (4.3.4) (notice that the commutator errors are trivial in this case), we have

$$\begin{aligned}
&\partial_j\widetilde{g}^{0j}(\partial_t\phi)^2 - (1/2)\partial_t\widetilde{g}^{jk}\partial_j\phi\partial_k\phi + \partial_j\widetilde{g}^{jk}\partial_t\phi\partial_k\phi\\
&= (\partial_t\phi)^2\partial_j h_{0j} + (1/2)\partial_j\phi\partial_k\phi\partial_t(h_{jk} + \delta_{jk}h_{00})\\
&- \partial_t\phi\partial_k\phi\partial_j(h_{jk} + \delta_{jk}h_{00}) + E^{\alpha\beta}_{\ge 2}\partial_\alpha\phi\partial_\beta\phi\\
&= (1/2)\left(\partial_j\phi\cdot\partial_j\phi\cdot\partial_t F + \partial_j\phi\cdot\partial_k\phi\cdot R_jR_k\partial_t F\right)\\
&- \in_{klm}\partial_k\phi\cdot[\partial_j\phi\cdot R_jR_l\partial_t\Omega_m + \partial_t\phi\cdot\partial_l\Omega_m]\\
&+ \frac{1}{2}\left(2(\partial_t\phi)^2\cdot\partial_t\underline{F} + \partial_j\phi\cdot\partial_j\phi\cdot\partial_t\underline{F} - 4\partial_t\phi\cdot\partial_k\phi\cdot\partial_k\underline{F} - \partial_j\phi\cdot\partial_k\phi\cdot R_jR_k\partial_t\underline{F}\right)\\
&+ (1/2)\in_{jpm}\in_{kqn}\partial_j\phi\cdot\partial_k\phi\cdot R_pR_q\partial_t\vartheta_{mn} + (\partial_t\phi)^2\cdot\partial_t\tau + E^{\alpha\beta}_{\ge 2}\cdot\partial_\alpha\phi\cdot\partial_\beta\phi.
\end{aligned} \tag{5.1.5}$$

The coefficients of the last two terms in the last line above are of the form $R^a|\nabla|^{-1}\mathcal{N}$, for some $\mathcal{N} \in \mathcal{QU}$ (see (4.3.5) and Definition 4.17).

We claim now that the main terms in (5.1.5) can be expressed in terms of semilinear null forms. Indeed, if $\phi$ is a wave unknown then we let $U^{\phi,\iota} := (\partial_t - \iota i\Lambda_{wa})\phi$ and rewrite

$$\begin{aligned}
&\frac{1}{2}\int_{\mathbb{R}^3}(\partial_j\phi\cdot\partial_j\phi\cdot\partial_t F + \partial_j\phi\cdot\partial_k\phi\cdot R_jR_k\partial_t F)\,dx\\
&= \sum_{\iota_1,\iota_2,\iota\in\{\pm\}}\mathcal{I}_{p^{F,wa}_{\iota_1,\iota_2,\iota}}[U^{F,\iota_1}, U^{\phi,\iota_2}, U^{\phi,\iota}]
\end{aligned} \tag{5.1.6}$$

where

$$\mathcal{I}_p[F,G,H] := \frac{1}{(8\pi^3)^2}\int_{\mathbb{R}^3\times\mathbb{R}^3} p(\xi,\eta)\widehat{F}(\xi-\eta)\widehat{G}(\eta)\overline{\widehat{H}(\xi)}\,d\xi d\eta, \tag{5.1.7}$$

and

$$p^{F,wa}_{\iota_1,\iota_2,\iota}(\xi,\eta) := \frac{1}{16}\Big[\iota\iota_2\frac{\eta}{|\eta|}\cdot\frac{\xi}{|\xi|} - \iota\iota_2\Big(\frac{\xi-\eta}{|\xi-\eta|}\cdot\frac{\eta}{|\eta|}\Big)\Big(\frac{\xi-\eta}{|\xi-\eta|}\cdot\frac{\xi}{|\xi|}\Big)\Big]. \tag{5.1.8}$$

The other terms in the right-hand side of (5.1.5) can be written in a similar way, as sums of integrals of the form $\mathcal{I}_{p^{G,wa}_{\iota_1,\iota_2,\iota}}[U^{G,\iota_1}, U^{\phi,\iota_2}, U^{\phi,\iota}]$, where $G \in \{\underline{F}, \Omega_m, \vartheta_{mn}\}$, with symbols

$$\begin{aligned} p^{\underline{F},wa}_{\iota_1,\iota_2,\iota}(\xi,\eta) := \frac{1}{16}\Big[2 + \iota\iota_2 \frac{\eta}{|\eta|}\cdot\frac{\xi}{|\xi|} - 2\iota_1\iota_2 \frac{\xi-\eta}{|\xi-\eta|}\cdot\frac{\eta}{|\eta|} \\ - 2\iota_1\iota \frac{\xi-\eta}{|\xi-\eta|}\cdot\frac{\xi}{|\xi|} + \iota\iota_2\Big(\frac{\xi-\eta}{|\xi-\eta|}\cdot\frac{\eta}{|\eta|}\Big)\Big(\frac{\xi-\eta}{|\xi-\eta|}\cdot\frac{\xi}{|\xi|}\Big)\Big], \end{aligned} \tag{5.1.9}$$

$$p^{\Omega_m,wa}_{\iota_1,\iota_2,\iota}(\xi,\eta) := \frac{1}{8}\iota_2 \in_{klm} \frac{(\xi-\eta)_l}{|\xi-\eta|}\frac{\eta_k}{|\eta|}\Big(\iota\frac{\xi-\eta}{|\xi-\eta|}\cdot\frac{\xi}{|\xi|} - \iota_1\Big), \tag{5.1.10}$$

$$p^{\vartheta_{mn},wa}_{\iota_1,\iota_2,\iota}(\xi,\eta) := -\frac{1}{16}\iota\iota_2\Big(\in_{jpm}\frac{(\xi-\eta)_p}{|\xi-\eta|}\frac{\eta_j}{|\eta|}\Big)\Big(\in_{kqn}\frac{(\xi-\eta)_q}{|\xi-\eta|}\frac{\xi_k}{|\xi|}\Big). \tag{5.1.11}$$

The calculation is similar if $\phi$ is a Klein-Gordon variable. In this case we define $U^{\phi,\iota} := (\partial_t - \iota i\Lambda_{kg})\phi$. We add up the term $\partial_t h_{00}\phi^2/2$ from (5.1.2), and rewrite the main terms in the right-hand side of (5.1.5) as sums of integrals of the form $\mathcal{I}_{p^{G,kg}_{\iota_1,\iota_2,\iota}}[U^{G,\iota_1}, U^{\phi,\iota_2}, U^{\phi,\iota}]$, $G \in \{F, \underline{F}, \Omega_m, \vartheta_{mn}\}$, $\iota_1, \iota_2, \iota \in \{+,-\}$. The resulting symbols are

$$p^{F,kg}_{\iota_1,\iota_2,\iota}(\xi,\eta) := \frac{1}{16}\Big[\iota\iota_2\frac{\eta\cdot\xi - 1}{\langle\eta\rangle\langle\xi\rangle} - \iota\iota_2\Big(\frac{\xi-\eta}{|\xi-\eta|}\cdot\frac{\eta}{\langle\eta\rangle}\Big)\Big(\frac{\xi-\eta}{|\xi-\eta|}\cdot\frac{\xi}{\langle\xi\rangle}\Big)\Big], \tag{5.1.12}$$

$$\begin{aligned} p^{\underline{F},kg}_{\iota_1,\iota_2,\iota}(\xi,\eta) := \frac{1}{16}\Big[2 + \iota\iota_2\frac{\eta\cdot\xi - 1}{\langle\eta\rangle\langle\xi\rangle} - 2\iota_1\iota_2\frac{\xi-\eta}{|\xi-\eta|}\cdot\frac{\eta}{\langle\eta\rangle} \\ - 2\iota_1\iota\frac{\xi-\eta}{|\xi-\eta|}\cdot\frac{\xi}{\langle\xi\rangle} + \iota\iota_2\Big(\frac{\xi-\eta}{|\xi-\eta|}\cdot\frac{\eta}{\langle\eta\rangle}\Big)\Big(\frac{\xi-\eta}{|\xi-\eta|}\cdot\frac{\xi}{\langle\xi\rangle}\Big)\Big], \end{aligned} \tag{5.1.13}$$

$$p^{\Omega_m,kg}_{\iota_1,\iota_2,\iota}(\xi,\eta) := \frac{1}{8}\iota_2 \in_{klm} \frac{(\xi-\eta)_l}{|\xi-\eta|}\frac{\eta_k}{\langle\eta\rangle}\Big(\iota\frac{\xi-\eta}{|\xi-\eta|}\cdot\frac{\xi}{\langle\xi\rangle} - \iota_1\Big), \tag{5.1.14}$$

$$p^{\vartheta_{mn},kg}_{\iota_1,\iota_2,\iota}(\xi,\eta) := -\frac{1}{16}\iota\iota_2\Big(\in_{jpm}\frac{(\xi-\eta)_p}{|\xi-\eta|}\frac{\eta_j}{\langle\eta\rangle}\Big)\Big(\in_{kqn}\frac{(\xi-\eta)_q}{|\xi-\eta|}\frac{\xi_k}{\langle\xi\rangle}\Big). \tag{5.1.15}$$

### 5.1.2 The Main Spacetime Bounds

We would like now to use the calculations in the previous subsection to start the proof of Proposition 5.1. Assume $\mathcal{L} \in \mathcal{V}^q_n$ and define

$$P^n_{wa} := \langle\nabla\rangle^{N(n)}|\nabla|^{-1/2}|\nabla|^{\gamma}_{\leq 1}, \qquad P^n_{kg} := \langle\nabla\rangle^{N(n)}. \tag{5.1.16}$$

We define also the associated multipliers $P^n_{wa}(\xi) := \langle\xi\rangle^{N(n)}|\xi|^{-1/2}|\xi|^{\gamma}_{\leq 1}$ and $P^n_{kg}(\xi) := \langle\xi\rangle^{N(n)}$, $\xi \in \mathbb{R}^3$. We first let $\phi := P^n_{wa}(\mathcal{L}h)$, $h \in \{h_{\alpha\beta}\}$, and write

$$|\mathcal{E}_0(P^n_{wa}(\mathcal{L}h))(t)| \lesssim |\mathcal{E}_0(P^n_{wa}(\mathcal{L}h))(0)| + \left|\int_0^t \mathcal{B}(P^n_{wa}(\mathcal{L}h))(s)ds\right| \\ + \left|\int_0^t \int_{\mathbb{R}^3} \partial_s(P^n_{wa}(\mathcal{L}h))(s) \cdot \widetilde{\Box}_{\widetilde{g}}(P^n_{wa}(\mathcal{L}h))(s)ds\right|,$$

for any $t \in [0,T]$, as a consequence of (5.1.3). Since $|\mathcal{E}_0(P^n_{wa}(\mathcal{L}h))(s)| \approx \|P^n_{wa}(U^{\mathcal{L}h}(s))\|^2_{L^2}$ for any $s \in [0,t]$ (due to (5.1.4)), to prove the first inequality in (5.1.1) it suffices to show that

$$\left|\int_0^t \mathcal{B}(P^n_{wa}(\mathcal{L}h))(s)\,ds\right| + \left|\int_0^t \int_{\mathbb{R}^3} P^n_{wa}(\partial_s(\mathcal{L}h))(s) \\ \times \widetilde{\Box}_{\widetilde{g}}(P^n_{wa}(\mathcal{L}h))(s)\,ds\right| \lesssim \varepsilon_1^3\langle t\rangle^{2H(q,n)\delta-2\gamma}, \tag{5.1.17}$$

for any $t \in [0,T]$. Similarly, to prove the second inequality in (5.1.1) it suffices to show that

$$\left|\int_0^t \mathcal{B}(P^n_{kg}(\mathcal{L}\psi))(s)\,ds\right| + \left|\int_0^t \int_{\mathbb{R}^3} P^n_{kg}(\partial_s(\mathcal{L}\psi))(s) \\ \times \big\{ -\widetilde{\Box}_{\widetilde{g}}(P^n_{kg}(\mathcal{L}\psi))(s) + (1-h_{00})P^n_{kg}(\mathcal{L}\psi)(s)\big\}ds\right| \lesssim \varepsilon_1^3\langle t\rangle^{2H(q,n)\delta}. \tag{5.1.18}$$

We decompose the time integrals into dyadic pieces. More precisely, given $t \in [0,T]$, we fix a suitable decomposition of the function $\mathbf{1}_{[0,t]}$, i.e., we fix functions $q_0, \ldots, q_{L+1} : \mathbb{R} \to [0,1]$, $|L - \log_2(2+t)| \leq 2$, with the properties

$$\begin{aligned} &\operatorname{supp} q_0 \subseteq [0,2], \qquad \operatorname{supp} q_{L+1} \subseteq [t-2,t], \\ &\operatorname{supp} q_m \subseteq [2^{m-1}, 2^{m+1}] \text{ for } m \in \{1,\ldots,L\}, \\ &\sum_{m=0}^{L+1} q_m(s) = \mathbf{1}_{[0,t]}(s), \\ &q_m \in C^1(\mathbb{R}) \text{ and } \int_0^t |q'_m(s)|\,ds \lesssim 1 \text{ for } m \in \{1,\ldots,L\}. \end{aligned} \tag{5.1.19}$$

We are now ready to state our main estimates on spacetime contributions.

**Proposition 5.2.** *Assume that* $t \in [0,T]$, $\iota, \iota_1, \iota_2 \in \{+,-\}$, $h, h_1, h_2 \in \{h_{\alpha\beta}\}$, $\mathcal{L} \in \mathcal{V}^q_n$, $\mathcal{L}_i \in \mathcal{V}^{q_i}_{n_i}$, $i \in \{1,2\}$, $\mathcal{L}_1 + \mathcal{L}_2 \leq \mathcal{L}$. *For any bounded symbol* $q$ *let* $I_q$ *denote the associated operator as in* (3.2.43). *Let* $J_m$ *denote the supports of the functions* $q_m$ *defined above. Then:*

*(1) if* $\mathcal{L}_2 \neq \mathcal{L}$ *and* $\mathfrak{q}^{*,wa}_{\iota_1\iota_2}$ *is a symbol of the form* $\mathfrak{q}^{G,wa}_{\iota_1\iota_2}(\theta,\eta)m_0(\theta)m_1(\eta)m_2(\theta+$

*$\eta$), where $G \in \{F, \underline{F}, \omega_n, \vartheta_{mn}\}$, $\mathfrak{q}^{G,wa}_{\iota_1\iota_2}$ are as in (4.3.59), and $m_0, m_1, m_2 \in \mathcal{M}_0$, then*

$$\Big|\int_{J_m}\int_{\mathbb{R}^3} q_m(s)[P^n_{wa}(\xi)]^2\mathcal{F}\{I_{\mathfrak{q}^{*,wa}_{\iota_1\iota_2}}[|\nabla|^{-1}U^{\mathcal{L}_1h_1,\iota_1}, |\nabla|U^{\mathcal{L}_2h_2,\iota_2}]\}(\xi,s) \times\overline{\widehat{U^{\mathcal{L}h,\iota}}(\xi,s)}\,d\xi ds\Big| \lesssim \varepsilon_1^3 2^{2H(q,n)\delta m-2\gamma m}; \tag{5.1.20}$$

*(2) if $\mathfrak{n}_{\iota_1\iota_2} \in \mathcal{M}^{null}_{\iota_1\iota_2}$ then*

$$\Big|\int_{J_m}\int_{\mathbb{R}^3} q_m(s)\mathcal{F}\{P^n_{wa}I_{\mathfrak{n}_{\iota_1\iota_2}}[U^{\mathcal{L}_1h_1,\iota_1}, U^{\mathcal{L}_2h_2,\iota_2}]\}(\xi,s) \times \overline{P^n_{wa}\widehat{U^{\mathcal{L}h,\iota}}(\xi,s)}\,d\xi ds\Big| \lesssim \varepsilon_1^3 2^{-\delta m} \tag{5.1.21}$$

*and*

$$\Big|\int_{J_m}\int_{\mathbb{R}^3} q_m(s)\mathcal{F}\{I_{\mathfrak{n}_{\iota_1\iota_2}}[U^{h_1,\iota_1}, P^n_{wa}U^{\mathcal{L}h_2,\iota_2}]\}(\xi,s) \times \overline{P^n_{wa}\widehat{U^{\mathcal{L}h,\iota}}(\xi,s)}\,d\xi ds\Big| \lesssim \varepsilon_1^3 2^{-\delta m}; \tag{5.1.22}$$

*(3) if $\mathfrak{m} \in \mathcal{M}$ and $\vartheta \in \{\vartheta_{mn}\}$, then*

$$\Big|\int_{J_m}\int_{\mathbb{R}^3} q_m(s)[P^n_{wa}(\xi)]^2\mathcal{F}\{I_{\mathfrak{m}}[U^{\vartheta^{\mathcal{L}_1},\iota_1}, U^{\vartheta^{\mathcal{L}_2},\iota_2}]\}(\xi,s) \times \overline{\widehat{U^{\mathcal{L}h,\iota}}(\xi,s)}\,d\xi ds\Big| \lesssim \varepsilon_1^3 2^{2H(q,n)\delta m-2\gamma m}; \tag{5.1.23}$$

*(4) if $\mu,\nu \in \{0,1,2,3\}$ and $\widetilde{g}^{\mu\nu}_{\geq 1}$ are as in (4.3.49), then*

$$\Big|\int_{J_m}\int_{\mathbb{R}^3} q_m(s)\mathcal{F}\{P^n_{wa}[\widetilde{g}^{\mu\nu}_{\geq 1}\partial_\mu\partial_\nu(\mathcal{L}h_2)] - \widetilde{g}^{\mu\nu}_{\geq 1}\partial_\mu\partial_\nu(P^n_{wa}\mathcal{L}h_2)\}(\xi,s) \times \overline{P^n_{wa}\widehat{U^{\mathcal{L}h,\iota}}(\xi,s)}\,d\xi ds\Big| \lesssim \varepsilon_1^3 2^{2H(q,n)\delta m-2\gamma m}; \tag{5.1.24}$$

*(5) if $\mathfrak{m} \in \mathcal{M}$ then*

$$\Big|\int_{J_m}\int_{\mathbb{R}^3} q_m(s)[P^n_{wa}(\xi)]^2\mathcal{F}\{I_{\mathfrak{m}}[U^{\mathcal{L}_1\psi,\iota_1}, U^{\mathcal{L}_2\psi,\iota_2}]\}(\xi,s) \times \overline{\widehat{U^{\mathcal{L}h,\iota}}(\xi,s)}\,d\xi ds\Big| \lesssim \varepsilon_1^3 2^{2H(q,n)\delta m-2\gamma m}; \tag{5.1.25}$$

*(6) if* $\mathfrak{m} \in \mathcal{M}$ *and* $n_2 < n$ *then*

$$\Big| \int_{J_m} \int_{\mathbb{R}^3} q_m(s) [P_{kg}^n(\xi)]^2 \mathcal{F}\{I_{\mathfrak{m}}[|\nabla|^{-1} U^{\mathcal{L}_1 h_{1,\iota_1}}, \langle\nabla\rangle U^{\mathcal{L}_2 \psi,\iota_2}]\}(\xi, s) \times \overline{\widehat{U^{\mathcal{L}\psi,\iota}}(\xi, s)}\, d\xi ds \Big| \lesssim \varepsilon_1^3 2^{2H(q,n)\delta m}; \tag{5.1.26}$$

*(7) if* $\mu, \nu \in \{0, 1, 2, 3\}$ *and* $\widetilde{g}^{\mu\nu}_{\geq 1}$ *are as in* (4.3.49), *then*

$$\Big| \int_{J_m} \int_{\mathbb{R}^3} q_m(s) \mathcal{F}\{P_{kg}^n [\widetilde{g}^{\mu\nu}_{\geq 1} \partial_\mu \partial_\nu (\mathcal{L}\psi) + h_{00} \mathcal{L}\psi] - [\widetilde{g}^{\mu\nu}_{\geq 1} \partial_\mu \partial_\nu (P_{kg}^n \mathcal{L}\psi) + h_{00} P_{kg}^n \mathcal{L}\psi]\}(\xi, s) \overline{P_{kg}^n \widehat{U^{\mathcal{L}\psi,\iota}}(\xi, s)}\, d\xi ds \Big| \lesssim \varepsilon_1^3 2^{2H(q,n)\delta m}; \tag{5.1.27}$$

*(8) if* $\mathfrak{p}^{*,kg}_{\iota_1,\iota_2,\iota}$ *is of the form* $p^{G,kg}_{\iota_1,\iota_2,\iota}(\xi, \eta) m_0(\xi - \eta) m_1(\eta) m_2(\xi)$, *where* $p^{G,kg}_{\iota_1,\iota_2,\iota}$ *are as in* (5.1.12)–(5.1.15), $G \in \{F, \underline{F}, \Omega_n, \vartheta_{mn}\}$, *and* $m_0, m_1, m_2 \in \mathcal{M}_0$, *then*

$$\Big| \int_{J_m} \int_{\mathbb{R}^3 \times \mathbb{R}^3} q_m(s) \mathfrak{p}^{*,kg}_{\iota_1,\iota_2,\iota}(\xi, \eta) \widehat{U^{h_1,\iota_1}}(\xi - \eta, s) P_{kg}^n \widehat{U^{\mathcal{L}\psi,\iota_2}}(\eta, s) \times \overline{P_{kg}^n \widehat{U^{\mathcal{L}\psi,\iota}}(\xi, s)}\, d\xi d\eta ds \Big| \lesssim \varepsilon_1^3 2^{2H(q,n)\delta m}. \tag{5.1.28}$$

*Proof of Proposition 5.1.* It is easy to see that the bounds (5.1.17)–(5.1.18) follow from Proposition 5.2. Indeed, we start from the identity $-\square(\mathcal{L}h_{\alpha\beta}) = \mathcal{L}\mathcal{N}^h_{\alpha\beta}$, and use (4.3.47). Thus

$$\begin{aligned} -\square(P_{wa}^n \mathcal{L}h_{\alpha\beta}) &- \sum_{\mu,\nu\in\{0,1,2,3\}} \widetilde{g}^{\mu\nu}_{\geq 1} \partial_\mu \partial_\nu (P_{wa}^n \mathcal{L}h_{\alpha\beta}) \\ &= \sum_{\mu,\nu\in\{0,1,2,3\}} \{P_{wa}^n [\widetilde{g}^{\mu\nu}_{\geq 1} \partial_\mu \partial_\nu (\mathcal{L}h_{\alpha\beta})] - \widetilde{g}^{\mu\nu}_{\geq 1} \partial_\mu \partial_\nu (P_{wa}^n \mathcal{L}h_{\alpha\beta})\} \\ &+ P_{wa}^n \mathcal{Q}^{\mathcal{L}}_{wa}(h_{\alpha\beta}) + P_{wa}^n \mathcal{S}^{\mathcal{L},1}_{\alpha\beta} + P_{wa}^n \mathcal{S}^{\mathcal{L},2}_{\alpha\beta} + P_{wa}^n \mathcal{K}\mathcal{G}^{\mathcal{L}}_{\alpha\beta} + P_{wa}^n \mathcal{R}^{\mathcal{L}h}_{\alpha\beta}. \end{aligned} \tag{5.1.29}$$

Therefore the contribution of the second term in the left-hand side of (5.1.17) can be bounded as claimed, as a consequence of the estimates (5.1.20)–(5.1.25) and the definitions.

To estimate the contribution of the first term in the left-hand side of (5.1.17) we examine the formulas (5.1.5)–(5.1.11). The contribution of the last two terms in (5.1.5) is bounded as claimed, due to the $L^\infty$ bounds in (4.3.15). Moreover, we claim that all the other terms are null forms that can be estimated using (5.1.22). Indeed, the symbols $p^{\Omega_m,wa}_{\iota_1,\iota_2,\iota}$ and $p^{\vartheta_{mn},wa}_{\iota_1,\iota_2,\iota}$ are clearly null in the variables $\xi - \eta$

and $\eta$, due to (4.3.42). Moreover, by examining (5.1.8)–(5.1.9) we can write

$$16p^{F,wa}_{\iota_1,\iota_2,\iota}(\xi,\eta) = \Big(\iota_1\frac{\xi-\eta}{|\xi-\eta|}-\iota_2\frac{\eta}{|\eta|}\Big)\cdot\Big(\iota_1\frac{\xi-\eta}{|\xi-\eta|}-\iota\frac{\xi}{|\xi|}\Big) \\ -\frac{1}{4}\Big|\iota_1\frac{\xi-\eta}{|\xi-\eta|}-\iota_2\frac{\eta}{|\eta|}\Big|^2\Big|\iota_1\frac{\xi-\eta}{|\xi-\eta|}-\iota\frac{\xi}{|\xi|}\Big|^2$$

and

$$16p^{\underline{F},wa}_{\iota_1,\iota_2,\iota}(\xi,\eta) = \Big(\iota_1\frac{\xi-\eta}{|\xi-\eta|}-\iota_2\frac{\eta}{|\eta|}\Big)\cdot\Big(\iota_1\frac{\xi-\eta}{|\xi-\eta|}-\iota\frac{\xi}{|\xi|}\Big) \\ +\frac{1}{4}\Big|\iota_1\frac{\xi-\eta}{|\xi-\eta|}-\iota_2\frac{\eta}{|\eta|}\Big|^2\Big|\iota_1\frac{\xi-\eta}{|\xi-\eta|}-\iota\frac{\xi}{|\xi|}\Big|^2,$$

thus $p^{F,wa}_{\iota_1,\iota_2,\iota}$ and $p^{\underline{F},wa}_{\iota_1,\iota_2,\iota}$ are sums of acceptable null symbols in the variables $\xi-\eta$ and $\eta$ multiplied by symbols of $\xi$, as desired.

The estimates (5.1.18) follow in a similar way, using (5.1.26)–(5.1.27) to bound the contributions of the nonlinearities, and using (5.1.28) for the spacetime integral of $\mathcal{B}(P^n_{kg}(\mathcal{L}\psi))(s)$. □

### 5.1.3 Poincaré Normal Forms

We examine now the bounds in Proposition 5.2, and notice that all the spacetime integrals in the left-hand sides do not have derivative loss. In most cases, however, the time decay we have is not enough to allow direct estimates. In such a situation we would like to integrate by parts in time (the method of normal forms) to prove the desired spacetime estimates.

More precisely, assume that we are given a trilinear form

$$\mathcal{G}_{\mathfrak{m}}[f,g,h] = \int_{\mathbb{R}^3\times\mathbb{R}^3}\mathfrak{m}(\xi-\eta,\eta)\widehat{f}(\xi-\eta)\widehat{g}(\eta)\overline{\widehat{h}(\xi)}\,d\xi d\eta, \tag{5.1.30}$$

for a suitable multiplier $\mathfrak{m}$, and consider its associated quadratic phase

$$\Phi_{\sigma\mu\nu}(\xi,\eta) = \Lambda_\sigma(\xi)-\Lambda_\mu(\xi-\eta)-\Lambda_\nu(\eta),$$

where $\sigma,\mu,\nu\in\{(wa,+),(wa,-),(kg,+),(kg,-)\}$ as in (2.1.40). Define

$$\mathcal{H}_{\mathfrak{m}}[f,g,h] := \int_{\mathbb{R}^3\times\mathbb{R}^3}\frac{\mathfrak{m}(\xi-\eta,\eta)}{\Phi_{\sigma\mu\nu}(\xi,\eta)}\widehat{f}(\xi-\eta)\widehat{g}(\eta)\overline{\widehat{h}(\xi)}\,d\xi d\eta. \tag{5.1.31}$$

Using integration by parts in time, for $m\in\{1,\ldots,L\}$ we have

$$-i\int_{\mathbb{R}} q_m(s)\mathcal{G}_{\mathfrak{m}}[f(s),g(s),h(s)]ds = \mathcal{H}^1+\mathcal{H}^2+\mathcal{H}^3+\mathcal{H}^4, \tag{5.1.32}$$

where the functions $q_m$ are defined as in (5.1.19), and

$$\begin{aligned}\mathcal{H}^1 &= \int_{\mathbb{R}} q_m'(s)\mathcal{H}_{\mathfrak{m}}[f(s), g(s), h(s)]ds,\\ \mathcal{H}^2 &= \int_{\mathbb{R}} q_m(s)\mathcal{H}_{\mathfrak{m}}[(\partial_s + i\Lambda_\mu) f(s), g(s), h(s)]ds,\\ \mathcal{H}^3 &= \int_{\mathbb{R}} q_m(s)\mathcal{H}_{\mathfrak{m}}[f(s), (\partial_s + i\Lambda_\nu) g(s), h(s)]ds,\\ \mathcal{H}^4 &= \int_{\mathbb{R}} q_m(s)\mathcal{H}_{\mathfrak{m}}[f(s), g(s), (\partial_s + i\Lambda_\sigma) h(s)]ds.\end{aligned} \tag{5.1.33}$$

In other words, we can estimate integrals like those in the left-hand side of (5.1.32) in terms of integrals such as those in (5.1.33). The point is to use the identities (2.1.38) to gain time integrability. The main issue when applying (5.1.32) is the presence of time-resonances, which are frequencies $(\xi, \eta)$ for which $\Phi_{\sigma\mu\nu}(\xi, \eta) = 0$, and which lead to significant difficulties in estimating the terms $\mathcal{H}_{\mathfrak{m}}[f, g, h]$ in (5.1.31). These resonances are stronger in the case of trilinear wave interactions, where parallel frequencies $(\xi, \eta)$ lead to resonances; in the case of mixed interactions involving two Klein-Gordon fields and one wave component, resonances only occur when the frequency of the wave component vanishes (see Lemmas 3.6 and 3.4).

In some cases we can use simple estimates to control trilinear expressions, such as

$$|\mathcal{G}_{\mathfrak{m}}[P_{k_1}F, P_{k_2}G, P_kH]| \lesssim \|\mathcal{F}^{-1}\{\mathfrak{m} \cdot \varphi_{kk_1k_2}\}\|_{L^1} \|P_{k_1}F\|_{L^{p_1}} \|P_{k_2}G\|_{L^{p_2}} \|P_kH\|_{L^p}, \tag{5.1.34}$$

provided that $k, k_1, k_2 \in \mathbb{Z}$, $p, p_1, p_2 \in [1, \infty]$, $1/p_1 + 1/p_2 + 1/p = 1$, where

$$\varphi_{kk_1k_2}(\theta, \eta) := \varphi_{[k-2,k+2]}(\eta + \theta)\varphi_{[k_1-2,k_1+2]}(\theta)\varphi_{[k_2-2,k_2+2]}(\eta). \tag{5.1.35}$$

These bounds follow from Lemma 3.2 (i). We also have pure $L^2$ bounds

$$\begin{aligned}&|\mathcal{G}_{\mathfrak{m}}[P_{k_1}F, P_{k_2}G, P_kH]|\\ &\quad\lesssim 2^{3\min\{k,k_1,k_2\}/2} \|\mathfrak{m} \cdot \varphi_{kk_1k_2}\|_{L^\infty} \|P_{k_1}F\|_{L^2} \|P_{k_2}G\|_{L^2} \|P_kH\|_{L^2}.\end{aligned} \tag{5.1.36}$$

Trilinear expressions like $\mathcal{H}_{\mathfrak{m}}$ are often estimated using Lemmas 3.6 and 3.4.

### 5.1.4 Paralinearization of the Reduced Wave Operator

The use of normal forms as in (5.1.32) to bound spacetime integrals leads to loss of derivative coming from the quasilinear nature of the nonlinearities. To remove this we use paradifferential calculus, as summarized in section (3.1.3). In this subsection we prove a proposition about paralinearization of our wave and Klein-Gordon operators.

Recall the modified metric $\widetilde{g}^{\mu\nu} := m^{\mu\nu} + \widetilde{g}^{\mu\nu}_{\geq 1}$, where $\widetilde{g}^{\mu\nu}_{\geq 1}$ are as in (4.3.49), and the modified wave operator $\widetilde{\Box}_{\widetilde{g}} := \widetilde{g}^{\mu\nu}\partial_\mu\partial_\nu$. Notice that $\widetilde{g}^{00} = -1$. We first define the main symbols

$$\mathcal{D}_{wa} := (\widetilde{g}^{0j}\zeta_j)^2 + \widetilde{g}^{jk}\zeta_j\zeta_k, \quad \Sigma_{wa} := \sqrt{\mathcal{D}_{wa}} - \widetilde{g}^{0j}\zeta_j, \quad \sigma_{wa} := \sqrt{\mathcal{D}_{wa}} + \widetilde{g}^{0j}\zeta_j, \tag{5.1.37}$$

and the main symbols for the Klein-Gordon components,

$$\mathcal{D}_{kg} := (\widetilde{g}^{0j}\zeta_j)^2 + 1 + \widetilde{g}^{jk}\zeta_j\zeta_k, \quad \Sigma_{kg} := \sqrt{\mathcal{D}_{kg}} - \widetilde{g}^{0j}\zeta_j, \quad \sigma_{kg} := \sqrt{\mathcal{D}_{kg}} + \widetilde{g}^{0j}\zeta_j. \tag{5.1.38}$$

These definitions are related to the paradifferential identities in Proposition 5.3.

Using the formulas (4.3.56) we can extract the linear and higher order components of the symbols $\Sigma_{wa}$ and $\Sigma_{kg}$. Indeed, we have

$$\mathcal{D}_{wa} = |\zeta|^2 - h_{00}|\zeta|^2 - h_{jk}\zeta_j\zeta_k + |\zeta|^2\mathcal{D}^{\geq 2}_{wa},$$

where $\mathcal{D}^{\geq 2}_{wa}$ is a quadratic symbol of order 0. Thus, in view of Lemma 4.4,

$$\begin{aligned}
&\Sigma_{wa} = |\zeta|\big(1 + \Sigma^1_{wa} + \Sigma^{\geq 2}_{wa}\big),\\
&\Sigma^1_{wa} := -(1/2)\big[h_{00} + 2h_{0j}\widehat{\zeta_j} + h_{jk}\widehat{\zeta_j}\widehat{\zeta_k}\big],\\
&\|\Sigma^{\geq 2}_{wa}\|_{\mathcal{L}^\infty_0} + \|\partial_t\Sigma^{\geq 2}_{wa}\|_{\mathcal{L}^\infty_0} \lesssim \varepsilon_1^2\langle t\rangle^{2\delta'-2},
\end{aligned} \tag{5.1.39}$$

where $\widehat{\zeta_j} := \zeta_j/|\zeta|$. Similarly, we have the decomposition of the Klein-Gordon symbol $\Sigma_{kg}$,

$$\begin{aligned}
&\Sigma_{kg} = \langle\zeta\rangle\big(1 + \Sigma^1_{kg} + \Sigma^{\geq 2}_{kg}\big),\\
&\Sigma^1_{kg} := -(1/2)\Big[h_{00}\frac{|\zeta|^2}{\langle\zeta\rangle^2} + 2h_{0j}\widehat{\zeta_j} + h_{jk}\frac{\zeta_j\zeta_k}{\langle\zeta\rangle^2}\Big],\\
&\|\Sigma^{\geq 2}_{kg}\|_{\mathcal{L}^\infty_0} + \|\partial_t\Sigma^{\geq 2}_{kg}\|_{\mathcal{L}^\infty_0} \lesssim \varepsilon_1^2\langle t\rangle^{2\delta'-2}.
\end{aligned} \tag{5.1.40}$$

The main result of this subsection is the following:

**Proposition 5.3.** *For $\mathcal{L} \in \mathcal{V}^q_n$, $(q,n) \leq (3,3)$, we define the "quasilinear variables"*

$$\mathcal{U}^{\mathcal{L}h_{\alpha\beta}} := (\partial_t - iT_{\sigma_{wa}})(\mathcal{L}h_{\alpha\beta}), \qquad \mathcal{U}^{\mathcal{L}\psi} := (\partial_t - iT_{\sigma_{kg}})(\mathcal{L}\psi). \tag{5.1.41}$$

*(i) Then, for any $t \in [0,T]$, $k \in \mathbb{Z}$, and $f \in \{\mathcal{L}h_{\alpha\beta}, \mathcal{L}\psi\}$,*

$$\|P_k(\mathcal{U}^f - U^f)(t)\|_{L^2} \lesssim \varepsilon_1 \min\{1, 2^k\langle t\rangle\}^{1-\delta}\langle t\rangle^{-1+31\delta} \cdot \|P_{[k-2,k+2]}U^f(t)\|_{L^2}. \tag{5.1.42}$$

*(ii) Moreover, we have*

$$
\begin{aligned}
(\partial_t + iT_{\Sigma_{wa}})\mathcal{U}^f &= -\widetilde{g}^{\mu\nu}\partial_\mu\partial_\nu f + \Pi_{wa}[U^f], \qquad \text{for } f \in \{\mathcal{L}h_{\alpha\beta}\},\\
(\partial_t + iT_{\Sigma_{kg}})\mathcal{U}^f &= (-\widetilde{g}^{\mu\nu}\partial_\mu\partial_\nu + 1)f + \Pi_{kg}[U^f], \qquad \text{for } f \in \{\mathcal{L}\psi\},
\end{aligned} \tag{5.1.43}
$$

*where the remainder terms* $\Pi_*[U^f]$ *satisfy, for* $* \in \{wa, kg\}$,

$$
\Pi_*[U^f] = \sum_{h\in\{h_{\alpha\beta}\},\,\iota_1,\iota_2\in\{+,-\}} I_{\mathfrak{m}^{h,*}_{\iota_1\iota_2}}[U^{h,\iota_1}, U^{f,\iota_2}] + \mathcal{C}_*[U^f]. \tag{5.1.44}
$$

*Here* $\mathfrak{m}^{h,*}_{\iota_1\iota_2}$ *are multipliers in* $\mathcal{M}$ *satisfying the additional bounds*

$$
\|\mathcal{F}^{-1}\{\mathfrak{m}^{h,*}_{\iota_1\iota_2}(\theta,\eta)\varphi_{k_1}(\theta)\varphi_{k_2}(\eta)\}\|_{L^1} \lesssim \min\{1, 2^{k_2-k_1}\} \tag{5.1.45}
$$

*for any* $k_1, k_2 \in \mathbb{Z}$, *and the cubic remainders satisfy*

$$
\begin{aligned}
&\|P_k\mathcal{C}_{wa}[U^f](t)\|_{L^2} \lesssim \varepsilon_1^2\langle t\rangle^{-5/4}2^{k/2}2^{-N(n)k^+}b_k(q,n;t) \qquad \text{for } f \in \{\mathcal{L}h_{\alpha\beta}\},\\
&\|P_k\mathcal{C}_{kg}[U^f](t)\|_{L^2} \lesssim \varepsilon_1^2\langle t\rangle^{-5/4}2^{-N(n)k^+}b_k(q,n;t) \qquad \text{for } f \in \{\mathcal{L}\psi\},
\end{aligned} \tag{5.1.46}
$$

*where* $b_k(q,n;t)$ *are the frequency envelope coefficients defined in* (4.3.20).

*(iii) As a consequence, if* $Y'(0) = Y'(1) = 2$ *and* $Y'(2) = Y'(3) = 35$, *then*

$$
\big\| (\partial_t + iT_{\Sigma_{wa}})\, P_k\mathcal{U}^{\mathcal{L}h_{\alpha\beta}}(t)\big\|_{L^2} \lesssim \varepsilon_1\langle t\rangle^{-1+\delta'}2^{k/2}2^{-N(n)k^+}b_k(q,n;t), \tag{5.1.47}
$$

$$
\big\| (\partial_t + iT_{\Sigma_{wa}})\, P_k\mathcal{U}^{\mathcal{L}h_{\alpha\beta}}(t)\big\|_{L^2} \lesssim \varepsilon_1^2\langle t\rangle^{-1+H(q,n)\delta+Y'(n)\delta}2^{k/2}2^{-\widetilde{N}(n)k^++7k^+}, \tag{5.1.48}
$$

$$
\big\| (\partial_t + iT_{\Sigma_{kg}})\, P_k\mathcal{U}^{\mathcal{L}\psi}(t)\big\|_{L^2} \lesssim \varepsilon_1\langle t\rangle^{-1+\delta'}2^{-N(n)k^+}b_k(q,n;t). \tag{5.1.49}
$$

*Proof.* The bounds (5.3)–(5.1.49) illustrate the main gains one can expect by using quasilinear profiles instead of linear profiles: there are no derivative losses in (5.1.47) and (5.1.49), and a smaller loss in terms of time decay when $n \leq 1$ in (5.1.48), compared to the bounds on $(\partial_t + i\Lambda_{wa})P_kU^{\mathcal{L}h_{\alpha\beta}} = \mathcal{L}\mathcal{N}^h_{\alpha\beta}$ and $(\partial_t + i\Lambda_{wa})P_kU^{\mathcal{L}\psi} = \mathcal{L}\mathcal{N}^\psi$.

(i) Using (5.1.39) we have

$$
\sigma_{wa} = \Sigma_{wa} + 2\widetilde{g}^{0j}_{\geq 1}\zeta_j = |\zeta| + |\zeta|\big(\Sigma^1_{wa} + \Sigma^{\geq 2}_{wa} + 2\widetilde{g}^{0j}_{\geq 1}\widehat{\zeta}_j\big).
$$

The bounds (5.1.42) follow using (3.1.54) and (3.3.11) when $f \in \{\mathcal{L}h_{\alpha\beta}\}$. The proof is similar when $f \in \{\mathcal{L}\psi\}$ using (5.1.40) instead of (5.1.39).

(ii) We compute, using the definitions,

$$(\partial_t + iT_{\Sigma_{wa}})(\partial_t - iT_{\sigma_{wa}})f = [\partial_t\partial_t + iT_{\Sigma_{wa}-\sigma_{wa}}\partial_t + T_{\Sigma_{wa}\sigma_{wa}}]f \\ - iT_{\partial_t\sigma_{wa}}f + (T_{\Sigma_{wa}}T_{\sigma_{wa}} - T_{\Sigma_{wa}\sigma_{wa}})f, \tag{5.1.50}$$

and, recalling that $\widetilde{g}^{00} = -1$,

$$-\widetilde{g}^{\mu\nu}\partial_\mu\partial_\nu f = \big[\partial_t\partial_t - 2iT_{\widetilde{g}^{0j}\zeta_j}\partial_t + T_{\widetilde{g}^{jk}\zeta_j\zeta_k}\big]f + E_1 + E_2, \tag{5.1.51}$$

where

$$\widehat{E_1}(\xi) := \frac{2i}{8\pi^3}\int_{\mathbb{R}^3}\Big[\frac{\xi_j+\eta_j}{2}\chi_0\Big(\frac{|\xi-\eta|}{|\xi+\eta|}\Big) - \eta_j\Big]\widehat{\widetilde{g}^{0j}_{\geq 1}}(\xi-\eta)\widehat{\partial_t f}(\eta)d\eta,$$
$$\widehat{E_2}(\xi) := \frac{-1}{8\pi^3}\int_{\mathbb{R}^3}\Big[\frac{(\xi_j+\eta_j)(\xi_k+\eta_k)}{4}\chi_0\Big(\frac{|\xi-\eta|}{|\xi+\eta|}\Big) - \eta_j\eta_k\Big]\widehat{\widetilde{g}^{jk}_{\geq 1}}(\xi-\eta)\widehat{f}(\eta)d\eta.$$

In view of the definitions (5.1.37), we have $\Sigma_{wa}-\sigma_{wa} = -2\widetilde{g}^{0j}\zeta_j$ and $\Sigma_{wa}\cdot\sigma_{wa} = \widetilde{g}^{jk}\zeta_j\zeta_k$. Thus the main terms in (5.1.50)–(5.1.51) are the same, and the identities in the first line of (5.1.43) follow easily by extracting the quadratic terms (which do not have derivative loss and can be written as claimed in (5.1.44)). The cubic and higher order remainders can be bounded as claimed in (5.1.46), using the bounds in Lemma 4.4.

The analysis of the Klein-Gordon terms is similar, using the identities $\Sigma_{kg} - \sigma_{kg} = -2\widetilde{g}^{0j}\zeta_j$ and $\Sigma_{kg}\cdot\sigma_{kg} = 1 + \widetilde{g}^{jk}\zeta_j\zeta_k$, see (5.1.38).

We remark that we could obtain more information on the remainder terms, which consist mostly of null quadratic terms and cubic terms, but this is not necessary for our purpose.

(iii) In view of (2.1.9) and the definitions (4.3.49) we have

$$\begin{aligned} -\widetilde{g}^{\mu\nu}\partial_\mu\partial_\nu\mathcal{L}h_{\alpha\beta} &= (\partial_t^2-\Delta)\mathcal{L}h_{\alpha\beta} - \widetilde{g}^{\mu\nu}_{\geq 1}\partial_\mu\partial_\nu\mathcal{L}h_{\alpha\beta} \\ &= \mathcal{L}(\widetilde{g}^{\mu\nu}_{\geq 1}\partial_\mu\partial_\nu h_{\alpha\beta}) - \widetilde{g}^{\mu\nu}_{\geq 1}\partial_\mu\partial_\nu\mathcal{L}h_{\alpha\beta} + \mathcal{L}\{(1-g^{00}_{\geq 1})^{-1}[\mathcal{KG}_{\alpha\beta} - F^{\geq 2}_{\alpha\beta}(g,\partial g)]\}. \end{aligned} \tag{5.1.52}$$

To prove the estimate (5.1.47) we use the identity in the first line of (5.1.43). Notice that the expression $(1-g^{00}_{\geq 1})^{-1}[\mathcal{KG}_{\alpha\beta} - F^{\geq 2}_{\alpha\beta}(g,\partial g)]$ in the right-hand side or (5.1.52) is a sum of terms in $\mathcal{QU}_0$; see (4.3.12). The contribution of these terms is therefore bounded as claimed, due to (4.3.23). The cubic terms $\mathcal{C}[U^{\mathcal{L}h_{\alpha\beta}}]$ also satisfy acceptable estimates, due to (5.1.46). Therefore it remains to prove that

$$\begin{aligned} \|P_kI[U^{h_1,\iota_1},U^{\mathcal{L}h_2,\iota_2}](t)\|_{L^2} &\lesssim \varepsilon_1\langle t\rangle^{-1+\delta'}2^{k/2}2^{-N(n)k^+}b_k(q,n;t), \\ \|P_kI[\mathcal{L}_1\widetilde{g}^{\mu\nu}_{\geq 1},\partial_\alpha\partial_\beta\mathcal{L}_2h_2](t)\|_{L^2} &\lesssim \varepsilon_1^2\langle t\rangle^{-1+\delta'}2^{k/2}2^{-N(n)k^+}2^{-\delta|k|}, \end{aligned} \tag{5.1.53}$$

for any $h_1,h_2\in\{h_{\alpha\beta}\}$, $\iota_1,\iota_2\in\{+,-\}$, $\mu,\nu,\alpha,\beta\in\{0,1,2,3\}$, $\mathcal{L}_1\in\mathcal{V}^{q_1}_{n_1}$, $\mathcal{L}_2\in$

$\mathcal{V}^{q_2}_{n_2}$, $(q_1, n_1) + (q_2, n_2) \leq (q, n)$, and $n_2 < n$.

The proofs of the bounds (5.1.53) are similar to some of the proofs in section 4.3, such as the proofs of (4.3.24) and (4.3.26). Indeed, for the bounds in the first line, we decompose the input functions dyadically in frequency, then use $L^2 \times L^\infty$ estimates for the Low $\times$ High $\to$ High interactions and for the High $\times$ High $\to$ Low interactions when $k \geq 0$; the remaining High $\times$ High $\to$ Low interactions when $k \leq 0$ can be bounded using (4.1.5) and (4.1.7). The proof of the bounds in the second line of (5.1.53) is similar (compare with (4.3.26)).

For $n \geq 2$, the estimates (5.1.48) follow from the formulas (5.1.43), (5.1.52), and the proofs of the $L^2$ estimates in section 4.2 such as (4.2.9), (4.2.19)–(4.2.20), and (4.2.46). For $n \in \{0, 1\}$ the estimates (5.1.48) follow from the improved estimates in Lemma 4.14 and cubic estimates such as (4.2.46). If $n = 1$ the term $\mathcal{L}(\widetilde{g}^{\mu\nu}_1 \partial_\mu \partial_\nu h_{\alpha\beta}) - \widetilde{g}^{\mu\nu}_1 \partial_\mu \partial_\nu \mathcal{L} h_{\alpha\beta}$ in (5.1.52) can be estimated using the bounds (4.2.69) when the undifferentiated factor carries the vector-field and its frequency is small.

The estimates (5.1.49) are similar. We start from the identities

$$
\begin{aligned}
(-\widetilde{g}^{\mu\nu}\partial_\mu\partial_\nu + 1)\mathcal{L}\psi &= (\partial_t^2 - \Delta + 1)\mathcal{L}\psi - \widetilde{g}^{\mu\nu}_{\geq 1}\partial_\mu\partial_\nu\mathcal{L}\psi \\
&= \mathcal{L}(\widetilde{g}^{\mu\nu}_{\geq 1}\partial_\mu\partial_\nu\psi) - \widetilde{g}^{\mu\nu}_{\geq 1}\partial_\mu\partial_\nu\mathcal{L}\psi - \mathcal{L}\{(1 - g^{00}_{\geq 1})^{-1} g^{00}_{\geq 1}\psi\},
\end{aligned}
$$

which follow from (2.1.16). Then we use the identities in the second line of (5.1.43), and notice that all the resulting terms that need to be estimated are quadratic or higher order, and do not lose derivatives. The desired bounds are similar to some of the bounds we proved in section 4.3, such as (4.3.27) and (4.3.25), using also the inequalities (5.1.45) when $\mathcal{L} = Id$ to compensate for the lower regularity of $U^h$ compared to $U^\psi$. □

## 5.2 PURE WAVE INTERACTIONS

In this section we consider Wave×Wave×Wave interactions, and prove the bounds (5.1.20)–(5.1.24) in Proposition 5.2.

### 5.2.1 Null Interactions

We start with a lemma concerning null interactions:

**Lemma 5.4.** *With the assumptions of Proposition 5.2, we have*

$$
\sum_{k,k_1,k_2 \in \mathbb{Z}} 2^{N(n)k^+ - k/2}(2^{N(n)k_1^+ - k_1/2} + 2^{N(n)k_2^+ - k_2/2} + 2^{N(n)k^+ - k/2})
$$
$$
\times \left| \int_{J_m} q_m(s) \mathcal{G}_{\mathfrak{n}_{\iota_1\iota_2}}\left[P_{k_1}U^{\mathcal{L}_1 h_1, \iota_1}(s), P_{k_2}U^{\mathcal{L}_2 h_2, \iota_2}(s), P_k U^{\mathcal{L}h, \iota}(s)\right] ds \right| \lesssim \varepsilon_1^3 2^{-\delta m} \tag{5.2.1}
$$

*for any* $m \in \{0, \ldots, L+1\}$, *where* $\mathfrak{n}_{\iota_1\iota_2} \in \mathcal{M}^{null}_{\iota_1\iota_2}$ *is a null symbol and the operators* $\mathcal{G}_{\mathfrak{n}_{\iota_1\iota_2}}$ *are defined as in* (5.1.30).

*Moreover, if* $\mathfrak{q}_{\iota_1\iota_2}$ *is a double-null symbol of the form*

$$\mathfrak{q}_{\iota_1\iota_2}(\theta,\eta) = (\iota_1\theta_i/|\theta| - \iota_2\eta_i/|\eta|)(\iota_1\theta_j/|\theta| - \iota_2\eta_j/|\eta|)m_1(\theta,\eta), \tag{5.2.2}$$

*where* $i, j \in \{1,2,3\}$, $m_1 \in \mathcal{M}$, *and if* $n_2 < n$ *then we also have*

$$\sum_{k,k_1,k_2\in\mathbb{Z}} 2^{2N(n)k^+ - k + k_2 - k_1} 2^{2\gamma(m+k^-)} \Big| \int_{J_m} q_m(s) \\ \times \mathcal{G}_{\mathfrak{q}_{\iota_1\iota_2}}\big[P_{k_1}U^{\mathcal{L}_1 h_1,\iota_1}(s), P_{k_2}U^{\mathcal{L}_2 h_2,\iota_2}(s), P_k U^{\mathcal{L}h,\iota}(s)\big]\, ds\Big| \lesssim \varepsilon_1^3 2^{2H(q,n)\delta m}. \tag{5.2.3}$$

It is easy to see that this lemma gives four of the bounds in Proposition 5.2:

**Corollary 5.5.** *The estimates* (5.1.20), (5.1.21), (5.1.22), *and* (5.1.24) *hold.*

*Proof of Corollary 5.5.* The bounds (5.1.21) and (5.1.22) follow from (5.2.1). Notice also that all the symbols in (4.3.59), thus all the symbols $\mathfrak{q}^{*;wa}_{\iota_1\iota_2}$ in (1) of Proposition 5.2, are double-null, so the bounds (5.1.20) follow from (5.2.3).

To prove (5.1.24) we calculate, as in (4.3.57)–(4.3.58),

$$\begin{aligned}
\widetilde{g}^{\mu\nu}_{\geq 1}\partial_\mu\partial_\nu H &= -[\delta_{jk} + R_jR_k]F \cdot \partial_j\partial_k H \\
&- \{2R_jR_0\underline{F}\cdot\partial_j\partial_0 H + (\delta_{jk} - R_jR_k)\underline{F}\cdot\partial_j\partial_k H\} \\
&+ \{2 \in_{jmn} R_m\omega_n \cdot \partial_j\partial_0 H + (\in_{jln} R_k + \in_{kln} R_j)R_lR_0\omega_n \cdot \partial_j\partial_k H\} \\
&- \in_{jpm}\in_{kqn} R_pR_q\vartheta_{mn}\cdot\partial_j\partial_k H - 2R_jR_0\tau\cdot\partial_j\partial_0 H + \sum_{(\mu,\nu)\neq(0,0)} \widetilde{G}^{\mu\nu}_{\geq 2}\cdot\partial_\mu\partial_\nu H,
\end{aligned}$$

where $H \in \{\mathcal{L}h_2, P^n_{wa}\mathcal{L}h_2\}$. The quadratic coefficients $\widetilde{G}^{\mu\nu}_{\geq 2}$ are linear combinations of expressions of the form $R^a|\nabla|^{-1}(G_{\geq 1}\partial_\rho h)$, where $R^a = R_1^{a_1}R_2^{a_2}R_3^{a_3}$ and $G_{\geq 1}$ are as in Definition 4.17. Therefore, as in Proposition 4.22, we have

$$\begin{aligned}
P^n_{wa}[\widetilde{g}^{\mu\nu}_{\geq 1}\partial_\mu\partial_\nu\mathcal{L}h_2] - \widetilde{g}^{\mu\nu}_{\geq 1}\partial_\mu\partial_\nu(P^n_{wa}\mathcal{L}h_2) &= \sum_{G\in\{F,\underline{F},\omega_n,\vartheta_{mn}\}}\ \sum_{\iota_1,\iota_2\in\{+,-\}} \\
&\{P^n_{wa} I_{\mathfrak{q}^{G,wa}_{\iota_1\iota_2}}[|\nabla|^{-1}U^{G,\iota_1}, |\nabla|U^{\mathcal{L}h_2,\iota_2}] - I_{\mathfrak{q}^{G,wa}_{\iota_1\iota_2}}[|\nabla|^{-1}U^{G,\iota_1}, P^n_{wa}|\nabla|U^{\mathcal{L}h_2,\iota_2}]\} \\
&- 2\{P^n_{wa}[R_jR_0\tau\cdot\partial_j\partial_0\mathcal{L}h_2] - R_jR_0\tau\cdot\partial_j\partial_0(P^n_{wa}\mathcal{L}h_2)\} \\
&+ \sum_{(\mu,\nu)\neq(0,0)} \{P^n_{wa}[\widetilde{G}^{\mu\nu}_{\geq 2}\cdot\partial_\mu\partial_\nu\mathcal{L}h_2] - \widetilde{G}^{\mu\nu}_{\geq 2}\cdot\partial_\mu\partial_\nu(P^n_{wa}\mathcal{L}h_2)\},
\end{aligned} \tag{5.2.4}$$

where the null multipliers $\mathfrak{q}_{\iota_1\iota_2}^{G,wa}$ are defined in (4.3.59). We notice that

$$\begin{aligned}&P_{wa}^n I_{\mathfrak{q}_{\iota_1\iota_2}^{G,wa}}[|\nabla|^{-1}U^{G,\iota_1}, |\nabla|U^{\mathcal{L}h_2,\iota_2}] - I_{\mathfrak{q}_{\iota_1\iota_2}^{G,wa}}[|\nabla|^{-1}U^{G,\iota_1}, P_{wa}^n|\nabla|U^{\mathcal{L}h_2,\iota_2}] \\ &= P_{wa}^n I_{\mathfrak{n}_{\iota_1\iota_2}^{1,G}}[U^{G,\iota_1}, U^{\mathcal{L}h_2,\iota_2}] + I_{\mathfrak{n}_{\iota_1\iota_2}^{2,G}}[U^{G,\iota_1}, P_{wa}^n U^{\mathcal{L}h_2,\iota_2}],\end{aligned} \tag{5.2.5}$$

where, for $G \in \{F, \underline{F}, \omega_n, \vartheta_{mn}\}$,

$$\begin{aligned}\mathfrak{n}_{\iota_1\iota_2}^{1,G}(\theta,\eta) &:= \mathfrak{q}_{\iota_1\iota_2}^{G,wa}(\theta,\eta)\Big\{\varphi_{\leq -4}(|\theta|/|\eta|)\frac{P_{wa}^n(\eta+\theta) - P_{wa}^n(\eta)}{P_{wa}^n(\eta+\theta)}\frac{|\eta|}{|\theta|} \\ &\qquad + \varphi_{>-4}(|\theta|/|\eta|)\frac{|\eta|}{|\theta|}\Big\}, \\ \mathfrak{n}_{\iota_1\iota_2}^{2,G}(\theta,\eta) &:= -\mathfrak{q}_{\iota_1\iota_2}^{G,wa}(\theta,\eta)\varphi_{>-4}(|\theta|/|\eta|)\frac{|\eta|}{|\theta|}.\end{aligned} \tag{5.2.6}$$

These are null multipliers as in (2) of Proposition 5.2. The resulting integrals are similar to the integrals (5.1.21)–(5.1.22) and can be bounded using (5.2.1).

We claim now that the remaining terms in (5.2.4) are cubic-type acceptable errors satisfying

$$\big\|P_{wa}^n[H\cdot\partial_\mu\partial_\nu\mathcal{L}h_2](s) - H(s)\cdot\partial_\mu\partial_\nu(P_{wa}^n\mathcal{L}h_2)(s)\big\|_{L^2} \lesssim \varepsilon_1^2\langle s\rangle^{-1} \tag{5.2.7}$$

for $(\mu,\nu)\neq(0,0)$ and $s\in[0,T]$, where $H$ is either $R_jR_0\mathcal{T}$ or $\widetilde{G}_{\geq 2}^{\mu\nu}$. This would clearly suffice to complete the proof of (5.1.24). To prove the bounds (5.2.7) we notice that $|\nabla|H$ is of the form $R^a|\nabla|^{-1}\mathcal{N}$, for some $\mathcal{N}\in\mathcal{QU}$ (see (4.3.12) and use (4.3.5)). We decompose as in (5.2.6). For (5.2.7) it suffices to prove that

$$\big\|P_{wa}^n I[|\nabla|^{-1}\mathcal{N}, U^{\mathcal{L}h_2,\iota_2}](s)\big\|_{L^2} + \big\|I[|\nabla|^{-1}\mathcal{N}, P_{wa}^n U^{\mathcal{L}h_2,\iota_2}](s)\big\|_{L^2} \lesssim \varepsilon_1^2\langle s\rangle^{-1}, \tag{5.2.8}$$

for any $\iota_2\in\{+,-\}$ and bilinear operators $I$ as in (3.2.43). The bound on the first term follows from the first inequality in (4.3.24), while the second term can be bounded easily using (4.3.28) and the assumption $\|P_{wa}^n U^{\mathcal{L}h_2}(s)\|_{L^2}\lesssim\varepsilon_1\langle s\rangle^{\delta'}$. This completes the proof of (5.2.7) and (5.1.24). □

*Proof of Lemma 5.4.* We remark first that the two estimates are somewhat similar, except that the bounds (5.2.3) involve stronger null multipliers (double-null), but we have to deal with an additional anti-derivative, which causes significant difficulties at low frequencies.

The contributions of very small frequencies can sometimes be bounded using just $L^2$ norms,

$$\begin{aligned}
&\left|\mathcal{G}_{\mathfrak{m}}\left[P_{k_1}U^{\mathcal{L}_1h_1,\iota_1}(s), P_{k_2}U^{\mathcal{L}_2h_2,\iota_2}(s), P_kU^{\mathcal{L}h,\iota}(s)\right]\right| \lesssim 2^{\underline{k}}2^{\frac{\underline{k}+k_2+k+k_1}{2}}\\
&\times 2^{-\frac{k_1}{2}}\|P_{k_1}U^{\mathcal{L}_1h_1}(s)\|_{L^2}\cdot 2^{-\frac{k_2}{2}}\|P_{k_2}U^{\mathcal{L}_2h_2}(s)\|_{L^2}\cdot 2^{-\frac{k}{2}}\|P_kU^{\mathcal{L}h}(s)\|_{L^2} \qquad (5.2.9)\\
&\lesssim \varepsilon_1^3 2^{2\underline{k}+\overline{k}}2^{-\gamma(k_1^-+k_2^-+k^-)}2^{-N(n_1)k_1^+-N(n_2)k_2^+-N(n)k^+}\\
&\qquad\times \langle s\rangle^{[H(\mathcal{L}_1)+H(\mathcal{L}_2)+H(\mathcal{L})]\delta-3\gamma},
\end{aligned}$$

provided that $\|\mathfrak{m}\|_{L^\infty} \lesssim 1$, where $\overline{k} := \max\{k, k_1, k_2\}$, $\underline{k} := \min\{k, k_1, k_2\}$, and $H(\mathcal{L}) = H(q, n)$, $H(\mathcal{L}_i) = H(q_i, n_i)$, $i \in \{1, 2\}$. These bounds suffice to prove the desired conclusions if $2^m \lesssim 1$, so in the analysis below we may assume that $m \geq \delta^{-2}$.

Step 1. We consider first the contribution of the triplets $(k, k_1, k_2)$ for which $k \leq \underline{k}+8$. In this case $2^{k_2-k_1} \lesssim 1$ and it suffices to focus on the harder estimates (5.2.1). In fact, we will prove the stronger bounds, for any $k, k_1, k_2 \in \mathbb{Z}$ satisfying $k \leq \underline{k}+8$,

$$\begin{aligned}
&2^{N(n)k^+-k/2}(2^{N(n)k_1^+-k_1/2}+2^{N(n)k_2^+-k_2/2}+2^{N(n)k^+-k/2})\\
&\times\left|\int_{J_m} q_m(s)\mathcal{G}_{\mathfrak{n}_{\iota_1\iota_2}}\left[P_{k_1}U^{\mathcal{L}_1h_1,\iota_1}(s), P_{k_2}U^{\mathcal{L}_2h_2,\iota_2}(s), P_kU^{\mathcal{L}h,\iota}(s)\right]ds\right| \qquad (5.2.10)\\
&\lesssim \varepsilon_1^3 2^{-\delta m}2^{-\delta(|k|+|k_1|+|k_2|)}.
\end{aligned}$$

Without loss of generality, in proving these bounds we may assume that $n_1 \leq n_2$.

**Case 1.1.** Assume first that $\iota_1 = -\iota_2$. In this case, we notice that

$$\|\mathcal{F}^{-1}\{\mathfrak{n}_{\iota_1\iota_2}\cdot\varphi_{kk_1k_2}\}\|_{L^1} \lesssim 2^{k-k_1}, \qquad (5.2.11)$$

where $\varphi_{kk_1k_2}$ is as in (5.1.35), since the multiplier $n_{\iota_1\iota_2}$ contains a small factor of the form $\theta/|\theta|+\eta/|\eta|$. If $\underline{k} \leq -\delta' m$ or if $\underline{k} \geq -\delta' m$ and $\overline{k} \geq \delta' m$, then we can just use (5.1.34) with $(p_1, p_2, p) = (\infty, 2, 2)$ (recall that $n_1 \leq n_2$), and the estimates (3.3.3) and (3.3.11). On the other hand, if $k, k_1, k_2 \in [-\delta' m, \delta' m]$ then we may assume $m \leq L$ (the case $m = L+1$ is easier because $|J_m| \lesssim 1$) and decompose the multiplier into resonant and non-resonant contributions. More precisely, let $q_0 = -8\delta' m$ and

$$\mathfrak{n}_{\iota_1\iota_2} = \mathfrak{n}^r_{\iota_1\iota_2} + \mathfrak{n}^{nr}_{\iota_1\iota_2}, \qquad \mathfrak{n}^r_{\iota_1\iota_2}(\theta,\eta) = \varphi_{\leq q_0}(\Xi_{\iota_1\iota_2}(\theta,\eta))\mathfrak{n}_{\iota_1\iota_2}(\theta,\eta), \qquad (5.2.12)$$

where $\Xi_{\iota_1\iota_2}$ is defined as in (3.1.23).

We bound the contributions of the resonant parts $\mathfrak{n}^r_{\iota_1\iota_2}$ using the null structure, while for the non-resonant parts $\mathfrak{n}^{nr}_{\iota_1\iota_2}$ we use normal forms (the identity (5.1.32)). It follows from (3.3.3), (3.3.11), (4.2.3) (or (4.3.23) if $l \geq 0$), and

(4.2.5) (recalling also (2.1.38)) that

$$\begin{aligned}&\|P_l U^{\mathcal{K}h}(t)\|_{L^2} \lesssim \varepsilon_1 \langle t\rangle^{\delta'/2} 2^{l/2} 2^{-\gamma l^-} 2^{-N(n')l^+},\\ &\|P_l U^{\mathcal{K}h}(t)\|_{L^\infty} \lesssim \varepsilon_1 \langle t\rangle^{\delta'/2-1} 2^{l^-} 2^{-N(n'+1)l^+ + 2l^+} \min\{1, 2^{l^-}\langle t\rangle\}^{1-\delta}\end{aligned} \tag{5.2.13}$$

and

$$\begin{aligned}&\|P_l(\partial_t + i\Lambda_{wa})U^{\mathcal{K}h}(t)\|_{L^2} \lesssim \varepsilon_1 \langle t\rangle^{\delta'/2-1} 2^{l/2} 2^{-N(n')l^+ + l^+} \min\{1, 2^{l^-}\langle t\rangle\},\\ &\|P_l(\partial_t + i\Lambda_{wa})U^{\mathcal{K}h}(t)\|_{L^\infty} \lesssim \varepsilon_1 \langle t\rangle^{4\delta'-2} 2^{l^-} 2^{-N(n'+1)l^+ + 6l^+} \min\{1, 2^{l^-}\langle t\rangle\},\end{aligned} \tag{5.2.14}$$

for any $l \in \mathbb{Z}$, $h \in \{h_{\alpha\beta}\}$, $t \in [0, T]$, and $\mathcal{K} \in \mathcal{V}_{n'}^{q'}$, where the inequalities in the first lines hold for all $n' \leq 3$, while the inequalities in the second lines only hold for $n' \leq 2$. Thus, using (3.1.30) and (5.2.13), and recalling the null structure of the symbols $\mathfrak{n}_{\iota_1\iota_2}$, we have

$$\begin{aligned}&\Big|\mathcal{G}_{\mathfrak{n}^r_{\iota_1\iota_2}}[P_{k_1}U^{\mathcal{L}_1h_1,\iota_1}(s), P_{k_2}U^{\mathcal{L}_2h_2,\iota_2}(s), P_kU^{\mathcal{L}h,\iota}(s)]\Big|\\ &\qquad\lesssim 2^{q_0} 2^{\max(k_1,k_2)-k} \|P_{k_1}U^{\mathcal{L}_1h_1}(s)\|_{L^{p_1}} \|P_{k_2}U^{\mathcal{L}_2h_2}(s)\|_{L^{p_2}} \|P_kU^{\mathcal{L}h}(s)\|_{L^2}\\ &\qquad\lesssim \varepsilon_1^3 2^{q_0} 2^{2\delta' m} \langle s\rangle^{-1+2\delta'} 2^{-N(n)k^+ + k/2} 2^{-N(n)k_2^+ + k_2/2},\end{aligned} \tag{5.2.15}$$

assuming that $2^{-\delta' m} \lesssim 2^k \lesssim 2^{k_1} \approx 2^{k_2} \lesssim 2^{\delta' m}$, where $(p_1, p_2) = (\infty, 2)$. This suffices to estimate the resonant contributions as in (5.2.10).

For the non-resonant symbols $\mathfrak{n}^{nr}_{\iota_1\iota_2}$, we can use the normal form formulas (5.1.32)-(5.1.33) and the bounds (3.1.33). Using (5.2.13)–(5.2.14) and estimating as in (5.2.15) we have

$$\begin{aligned}&\Big|\mathcal{H}_{\mathfrak{n}^{nr}_{\iota_1\iota_2}}[P_{k_1}U^{\mathcal{L}_1h_1,\iota_1}(s), P_{k_2}U^{\mathcal{L}_2h_2,\iota_2}(s), P_kU^{\mathcal{L}h,\iota}(s)]\Big|\\ &\qquad\lesssim \varepsilon_1^3 2^{-3q_0} 2^{-0.9m},\\ &\Big|\mathcal{H}_{\mathfrak{n}^{nr}_{\iota_1\iota_2}}[P_{k_1}U^{\mathcal{L}_1h_1,\iota_1}(s), P_{k_2}U^{\mathcal{L}_2h_2,\iota_2}(s), P_k(\partial_s + i\Lambda_{wa,\iota})U^{\mathcal{L}h,\iota}(s)]\Big|\\ &+\Big|\mathcal{H}_{\mathfrak{n}^{nr}_{\iota_1\iota_2}}[P_{k_1}(\partial_s + i\Lambda_{wa,\iota_1})U^{\mathcal{L}_1h_1,\iota_1}(s), P_{k_2}U^{\mathcal{L}_2h_2,\iota_2}(s), P_kU^{\mathcal{L}h,\iota}(s)]\Big|\\ &+\Big|\mathcal{H}_{\mathfrak{n}^{nr}_{\iota_1\iota_2}}[P_{k_1}U^{\mathcal{L}_1h_1,\iota_1}(s), P_{k_2}(\partial_s + i\Lambda_{wa,\iota_2})U^{\mathcal{L}_2h_2,\iota_2}(s), P_kU^{\mathcal{L}h,\iota}(s)]\Big|\\ &\qquad\lesssim \varepsilon_1^3 2^{-3q_0} 2^{-1.9m},\end{aligned} \tag{5.2.16}$$

for any $s \in J_m$ (recall that $2^{|k|} + 2^{|k_1|} + 2^{|k_2|} \lesssim 2^{\delta' m}$). This completes the proof of (5.2.10).

**Case 1.2.** Assume now that $\iota_1 = \iota_2$. If $k \geq \min\{k_1, k_2\} - 10$ then $2^k \approx 2^{k_1} \approx 2^{k_2}$, the bounds (5.2.11) still hold, and the same proof as before gives the desired bounds (5.2.10).

On the other hand, if $k \le \min\{k_1, k_2\} - 10$ then we have no resonant contributions, i.e., $\Xi_{\iota_1\iota_2}(\theta, \eta) \gtrsim 1$ in the support of the integral. So we can integrate by parts directly using (5.1.32). If $2^k \le 2^{-m/4}$ or if $2^{k_1} \gtrsim 2^{\delta' m}$ then we use only the $L^2$ bounds in the first lines of (5.2.13) and (5.2.14), and estimate as in (5.2.9) using (3.1.31). On the other hand, if $-m/4 \le k \le \min\{k_1, k_2\} - 10 \le \delta' m$ then we can estimate the resulting terms in (5.1.33) as in (5.2.16), using $L^2 \times L^2 \times L^\infty$ bounds and (5.2.13)–(5.2.14). The desired bounds (5.2.10) follow.

**Step 2.** We complete now the proof of (5.2.1) by analyzing the contribution of the triplets $(k, k_1, k_2)$ for which $k \ge \underline{k} + 8$. By symmetry, we may assume that $n_1 \le n_2$.

**Case 2.1.** We assume first that $n_1 \ge 1$. In this case $1 \le n_1 \le n_2 \le 2$, $n_2 < n$, and we can still prove the strong bounds (5.2.10). Indeed, if $\underline{k} \le -2\delta' m$ or if $\underline{k} \ge -2\delta' m$ and $\overline{k} \ge 2\delta' m$, then we can use (5.1.34) (with the lowest frequency placed in $L^\infty$) and the estimates (5.2.13). On the other hand, if $k, k_1, k_2 \in [-2\delta' m, 2\delta' m]$ then we still decompose $\mathfrak{n}_{\iota_1\iota_2} = \mathfrak{n}^r_{\iota_1\iota_2} + \mathfrak{n}^{nr}_{\iota_1\iota_2}$ as in (5.2.12), with $q_0 = -8\delta' m$, and apply (3.1.30) and (3.1.33), together with (5.2.13)–(5.2.14). The desired bounds follow easily by estimating the resonant and the non-resonant contributions as in (5.2.15)–(5.2.16).

**Case 2.2.** Assume now that $n_1 = 0$, $n_2 \ge 1$, and consider the contribution of the triplets $(k, k_1, k_2)$ for which $k_1 \ge k_2$. We can still prove the strong bounds (5.2.10) because there is no derivative loss in this case. If $n_2 \le 2$ then we can still estimate as in (5.2.15)–(5.2.16), using $L^2 \times L^2 \times L^\infty$ bounds with the lowest frequency placed in $L^\infty$. On the other hand, if $n_2 = 3$ (thus $n = 3$), then (5.2.10) follows from (5.2.9) if $k_2 \le -3m/4$; if $k_2 \ge -3m/4$ then we decompose the symbol as in (5.2.12) and estimate as in (5.2.15)–(5.2.16), with the terms corresponding to the frequency $\approx 2^{k_1}$ always placed in $L^\infty$.

**Case 2.3.** Finally, assume that $n_1 = 0$ and consider the contribution of the triplets $(k, k_1, k_2)$ for which $k_1 \le k_2$. The main issue here is the loss of derivative in normal form arguments. Let

$$b_{k,m}(q,n) := \Big\{ \int_{J_m} (b_k(q,n;s))^2 \big[|J_m|^{-1} + |q'_m(s)|\big]\, ds \Big\}^{1/2}, \tag{5.2.17}$$

where the frequency envelope coefficients $b_k(q,n;s)$ are defined in (4.3.20) (if $m = 0$ or $m = L+1$ then we do not include the term $|q'_m(s)|$ in the definition (5.2.17)). We will show that

$$\begin{aligned} &2^{N(n)k^+ - k/2}\big(2^{N(n)k_1^+ - k_1/2} + 2^{N(n)k^+ - k/2}\big) \\ &\times \Big| \int_{J_m} q_m(s) \mathcal{G}_{\mathfrak{n}_{\iota_1\iota_2}}\big[P_{k_1} U^{h_1,\iota_1}(s), P_{k_2} U^{\mathcal{L}h_2,\iota_2}(s), P_k U^{\mathcal{L}h,\iota}(s)\big]\, ds \Big| \\ &\lesssim \varepsilon_1 2^{-\delta m} 2^{-\delta |k_1|} (b_{k,m}(q,n))^2, \end{aligned} \tag{5.2.18}$$

provided that $|k - k_2| \le 4$ and $\mathcal{L} \in \mathcal{V}^q_n$, $(q,n) \le (3,3)$. This is slightly weaker than the bounds (5.2.10) (since $b_{k,m}(q,n) \gtrsim \varepsilon_1 2^{-\delta|k|}$ due to (4.3.20)), but still suffices to complete the proof of (5.2.1) due to the square summability of the coefficients $b_{k,m}(q,n)$ (see (4.3.21)).

To prove (5.2.18) we notice that we can still use the normal form argument as in **Case 2.1** if $2^k \lesssim 2^{m/10}$, or to control the contribution of the factor $2^{N(n)k_1^+ - k_1/2}$ in the left-hand side. If $2^k \gtrsim 2^{m/10}$ and $|k_1| \geq \delta' m$ then we can use just $L^\infty \times L^2 \times L^2$ estimates (with the $L^2$ bounds on the two high frequency terms coming from (4.3.22)) to prove (5.2.18).

In the remaining case when $|k_1| \leq \delta' m$ the resonant contribution can be estimated as in (5.2.15). The bound on the non-resonant contribution requires an additional idea (the use of paradifferential calculus) to avoid the derivative loss in the application of the Poincaré normal form; the desired bounds follow from Lemma 5.7 below.

**Step 3.** We turn now to the proof of (5.2.3). In view of (5.2.1) it suffices to prove that

$$2^{-k_1}\Big|\int_{J_m} q_m(s)\mathcal{G}_{\mathfrak{q}_{\iota_1\iota_2}}\big[P_{k_1}U^{\mathcal{L}_1h_1,\iota_1}(s), P_{k_2}U^{\mathcal{L}_2h_2,\iota_2}(s), P_kU^{\mathcal{L}h,\iota}(s)\big]\,ds\Big| \\ \lesssim \varepsilon_1^3 2^{-2N(n)k^+}2^{2H(\mathcal{L})\delta m}2^{-2\gamma m}2^{-\gamma(|k|+|k_1+m|)/4}, \tag{5.2.19}$$

for any triplet $(k,k_1,k_2)$ for which $k_1 = \underline{k} \leq \overline{k} - 10$. Using first (5.2.9), the left-hand side of (5.2.19) is bounded by

$$C\varepsilon_1^3 2^{k_1+k}2^{-\gamma k_1^- - 2\gamma k^-}2^{-8k_1^+}2^{-N(n)k^+ - N(n_2)k^+}2^{m+\delta m(H(\mathcal{L}_1)+H(\mathcal{L}_2)+H(\mathcal{L}))-3\gamma m}. \tag{5.2.20}$$

Since $N(n_2) \geq N(n) + 10$, this implies (5.2.19) unless

$$2^{k_1+m}2^{-2\gamma(k_1+m)^-}2^{-8k^+}2^{(1-\delta)k^-} \gtrsim 2^{\delta m(H(\mathcal{L})-H(\mathcal{L}_1)-H(\mathcal{L}_2))}. \tag{5.2.21}$$

It remains to prove (5.2.19) for triplets $(k,k_1,k_2)$ for which (5.2.21) holds. In particular, we may assume that $m \geq \delta^{-2}$ in the analysis below.

**Case 3.1.** Assume first that $n_1 < n$. Thus $H(\mathcal{L}) - H(\mathcal{L}_1) - H(\mathcal{L}_2) \geq 40$ (due to (2.1.53)) and $k_1 + m \geq 40\delta m$ (due to (5.2.21)). We may also assume that $m \leq L$, and use (5.1.32). We observe that $|\mathfrak{q}_{\iota_1\iota_2}(\xi-\eta,\eta)(\Phi_{\sigma\mu\nu}(\xi,\eta))^{-1}| \lesssim 2^{-k_1}$, where $(\sigma,\mu,\nu) = ((wa,\iota),(wa,\iota_1),(wa,\iota_2))$ in the support of the integral, due to the double-null assumption (5.2.2) and the bounds (3.1.31). Using a normal form and $L^2$ estimates, the left-hand side of (5.2.19) is bounded by

$$\begin{aligned} C2^{3k_1/2}2^{-2k_1}\sup_{s\in J_m}\big\{&(\|P_{k_1}U^{\mathcal{L}_1h_1}(s)\|_{L^2} \\ &+ 2^m\|P_{k_1}\mathcal{L}_1\mathcal{N}_1^h(s)\|_{L^2})\|P_{k_2}U^{\mathcal{L}_2h_2}(s)\|_{L^2}\|P_kU^{\mathcal{L}h}(s)\|_{L^2} \\ &+ 2^m\|P_{k_1}U^{\mathcal{L}_1h_1}(s)\|_{L^2}\|P_{k_2}\mathcal{L}_2\mathcal{N}_2^h(s)\|_{L^2}\|P_kU^{\mathcal{L}h}(s)\|_{L^2} \\ &+ 2^m\|P_{k_1}U^{\mathcal{L}_1h_1}(s)\|_{L^2}\|P_{k_2}U^{\mathcal{L}_2h_2}(s)\|_{L^2}\|P_k\mathcal{L}\mathcal{N}^h(s)\|_{L^2}\big\}, \end{aligned}$$

where $\mathcal{N}_1^h, \mathcal{N}_2^h, \mathcal{N}^h$ are suitable components of the metric nonlinearities $\mathcal{N}_{\alpha\beta}^h$.

In view of (3.3.3) and (4.2.3), and recalling that $\widetilde{\ell}(n) \leq 35$ (see (4.2.2)), all the terms in the expression above are dominated by

$$C\varepsilon_1^3 2^k 2^{\delta m(H(\mathcal{L}_1)+H(\mathcal{L}_2)+H(\mathcal{L}))+36\delta m} 2^{-2N(n)k^+-2k^+},$$

which suffices to prove (5.2.19) in this case (due to (2.1.53)).

The same argument also proves the desired bounds (5.2.19) when $n_2 = 0$, $(q_1, n_1) = (q, n)$, and $|k| \geq \delta' m$.

**Case 3.2.** Assume now that (5.2.21) holds, $n_2 = 0$, $(q_1, n_1) = (q, n)$, $|k| \leq \delta' m$, and

$$k_1 + m \leq Y(q,n)\delta m, \tag{5.2.22}$$

where $Y(q,n)$ is defined as in (4.3.70). In particular, we may assume that $-2\delta m \leq k_1 + m$, due to (5.2.21). To prove (5.2.19) in this case, we trivialize one vector-field using Lemma 4.13. As in (4.3.73) we define

$$U^*_{1,\leq J_1} := P'_{k_1}(\varphi_{\leq J_1} \cdot P_{k_1} U^{\mathcal{L}_1 h_1,\iota_1}), \quad U^*_{1,>J_1} := P'_{k_1}(\varphi_{>J_1} \cdot P_{k_1} U^{\mathcal{L}_1 h_1,\iota_1}),$$

where $J_1$ denotes the largest integer satisfying $J_1 \leq \max\{-k_1, m\} + \delta m/4$. Then

$$\|U^*_{1,\leq J_1}(s)\|_{L^2} \lesssim \varepsilon_1 2^{k_1/2} 2^{H(q,n)\delta m - 7\delta m/4}; \tag{5.2.23}$$

see (4.3.74). We decompose also $P_{k_2} U^{h_2,\iota_2}(s) = U^{h_2,\iota_2}_{\leq J_2,k_2}(s) + U^{h_2,\iota_2}_{>J_2,k_2}(s)$ as in (3.3.1)–(3.3.2), where $J_2$ is the largest integer satisfying $J_2 \leq m/4$.

With $I_{\mathfrak{q}_{\iota_1\iota_2}}$ as in (3.2.43), the left-hand side of (5.2.19) is bounded by

$$C2^{-k_1}2^m \sup_{s\in J_m} \big\|P_k I_{\mathfrak{q}_{\iota_1\iota_2}}[P_{k_1}U^{\mathcal{L}_1 h_1,\iota_1}(s), P_{k_2}U^{h_2,\iota_2}(s)]\big\|_{L^2} \|P_{[k-2,k+2]}U^{\mathcal{L}h,\iota}(s)\|_{L^2}.$$

In view of (3.3.3), for (5.2.19) it suffices to prove that, for any $s \in J_m$,

$$\begin{aligned} &2^{-k_1}\big\|P_k I_{\mathfrak{q}_{\iota_1\iota_2}}[P_{k_1}U^{\mathcal{L}_1 h_1,\iota_1}, P_{k_2}U^{h_2,\iota_2}](s)\big\|_{L^2} \\ &\quad\lesssim \varepsilon_1^2 2^{-N(n)k^+-k^+} 2^{-m+H(q,n)\delta m} 2^{-\gamma m} 2^{-\gamma|k_1+m|/4}. \end{aligned} \tag{5.2.24}$$

In view of (5.2.21), we may assume that $|k| \leq \delta' m$. Using just $L^2$ estimates, we have

$$2^{-k_1}\big\|P_k I_{\mathfrak{q}_{\iota_1\iota_2}}[P_{k_1}U^{\mathcal{L}_1 h_1,\iota_1}, U^{h_2,\iota_2}_{>J_2,k_2}](s)\big\|_{L^2} \lesssim \varepsilon_1^2 2^{-N_0 k^+} 2^{-m/5} 2^{k_1}. \tag{5.2.25}$$

To bound the contribution of the profile $U^{h_2,\iota_2}_{\leq J_2,k_2}(s)$ we use (3.2.46), (3.3.7), and (5.2.23), so

$$\begin{aligned} &2^{-k_1}\big\|P_k I_{\mathfrak{q}_{\iota_1\iota_2}}[U^*_{1,\leq J_1}, U^{h_2,\iota_2}_{\leq J_2,k_2}](s)\big\|_{L^2} \\ &\quad\lesssim 2^{-k_1/2}2^{-m}\|U^*_{1,\leq J_1}(s)\|_{L^2} 2^{3k_2/2}\|\widehat{P_{k_2}U^{h_2}}\|_{L^\infty} \\ &\quad\lesssim \varepsilon_1^2 2^{-m} 2^{H(q,n)\delta m - 7\delta m/4} 2^{\delta m} 2^{-N_0 k^+ + 2k_+}. \end{aligned} \tag{5.2.26}$$

Finally, the bilinear interaction of $U^{*}_{1,>J_1}(s)$ and $U^{h_2,\iota_2}_{\leq J_2,k_2}(s)$ is negligible,

$$2^{-k_1}\big\|P_k I_{\mathfrak{q}_{\iota_1\iota_2}}[U^{*}_{1,>J_1},U^{h_2,\iota_2}_{\leq J_2,k_2}](s)\big\|_{L^2}\lesssim \varepsilon_1^2 2^{-2m}, \tag{5.2.27}$$

using an approximate finite speed of propagation argument as in the proof of (4.3.77). The desired bounds (5.2.24) follow from (5.2.25)–(5.2.27).

**Step 4.** To prove (5.2.19) in the remaining cases we need to use angular localization and integrate by parts in time. More precisely, we decompose

$$\mathfrak{q}_{\iota_1\iota_2}=\sum_{b\leq 4}\mathfrak{q}^b_{\iota_1\iota_2},\qquad \mathfrak{q}^b_{\iota_1\iota_2}(\theta,\eta)=\varphi_b(\Xi_{\iota_1\iota_2}(\theta,\eta))\mathfrak{q}_{\iota_1\iota_2}(\theta,\eta), \tag{5.2.28}$$

and define the associated operators $\mathcal{G}_{\mathfrak{q}^b_{\iota_1\iota_2}}$ as in (5.1.30). For (5.2.19) it suffices to prove that

$$\begin{aligned}2^{-k_1}\Big|\int_{J_m} q_m(s)\mathcal{G}_{\mathfrak{q}^b_{\iota_1\iota_2}}\big[P_{k_1}U^{\mathcal{L}_1h_1,\iota_1}(s),P_{k_2}U^{h_2,\iota_2}(s),P_kU^{\mathcal{L}h,\iota}(s)\big]\,ds\Big|&\\ \lesssim \varepsilon_1^3 2^{-2N(n)k^+}2^{2H(q,n)\delta m}2^{-3\gamma m}2^{\delta b},&\end{aligned} \tag{5.2.29}$$

provided that $\mathcal{L},\mathcal{L}_1\in\mathcal{V}^q_n$, $b\leq 4$, $m\geq\delta^{-2}$, and

$$k_1\geq -m+Y(q,n)\delta m,\qquad |k|\leq\delta' m,\qquad k_1\leq\min\{k,k_2\}-6. \tag{5.2.30}$$

**Case 4.1.** Assume first that $b\leq -3\delta' m$. We decompose $P_{k_2}U^{h_2,\iota_2}(s)=U^{h_2,\iota_2}_{\leq J_2,k_2}(s)+U^{h_2,\iota_2}_{>J_2,k_2}(s)$ as in (3.3.1)–(3.3.2), where $J_2$ is the largest integer satisfying $J_2\leq m-\delta' m$. Using (3.1.36) and the double-null assumption (5.2.2) we have $\|\mathcal{F}^{-1}(q^b_{\iota_1\iota_2}\cdot\varphi_{kk_1k_2})\|_{L^1}\lesssim 2^b$. Using (3.2.54) and (3.3.3) we estimate

$$\begin{aligned}&2^{-k_1}\Big|\int_{J_m} q_m(s)\mathcal{G}_{\mathfrak{q}^b_{\iota_1\iota_2}}\big[P_{k_1}U^{\mathcal{L}_1h_1,\iota_1}(s),U^{h_2,\iota_2}_{\leq J_2,k_2}(s),P_kU^{\mathcal{L}h,\iota}(s)\big]\,ds\Big|\\ &\lesssim 2^{-k_1}2^m\sup_{s\in J_m}\big\|P_kI_{\mathfrak{q}^b_{\iota_1\iota_2}}\big[P_{k_1}U^{\mathcal{L}_1h_1,\iota_1}(s),U^{h_2,\iota_2}_{\leq J_2,k_2}(s)\big]\big\|_{L^2}\big\|P_{[k-2,k+2]}U^{\mathcal{L}h,\iota}(s)\big\|_{L^2}\\ &\lesssim 2^{-k_1}2^m\cdot 2^b2^{k_1/2}2^{-m+\delta m}\\ &\qquad\times\sup_{s\in J_m}\|P_{k_1}U^{\mathcal{L}_1h_1,\iota_1}(s)\|_{L^2}\|U^{h_2,\iota_2}_{\leq J_2,k_2}(s)\|_{H^{0,1}_\Omega}\|P'_kU^{\mathcal{L}h,\iota}(s)\|_{L^2}\\ &\lesssim\varepsilon_1^3 2^b2^{\delta' m}2^{k_2}2^{-2N(n)k^+}.\end{aligned} \tag{5.2.31}$$

Moreover, using $L^2$ estimates, the double-null assumption (5.2.2), and (3.3.3)–

(3.3.4), we have

$$\begin{aligned}&2^{-k_1}\Big|\int_{J_m} q_m(s)\mathcal{G}_{\mathfrak{q}^b_{\iota_1\iota_2}}\big[P_{k_1}U^{\mathcal{L}_1 h_1,\iota_1}(s), U^{h_2,\iota_2}_{>J_2,k_2}(s), P_kU^{\mathcal{L}h,\iota}(s)\big]\,ds\Big|\\&\lesssim 2^{-k_1}2^m2^{3k_1/2}2^{2b}\sup_{s\in J_m}\|P_{k_1}U^{\mathcal{L}_1h_1,\iota_1}(s)\|_{L^2}\|U^{h_2,\iota_2}_{>J_2,k_2}(s)\|_{L^2}\|P'_kU^{\mathcal{L}h,\iota}(s)\|_{L^2}\\&\lesssim \varepsilon_1^3 2^{2b}2^{k_1^-}2^{2\delta' m}2^{-2N(n)k^+}.\end{aligned}\tag{5.2.32}$$

The desired bounds (5.2.29) follow from (5.2.31)–(5.2.32) if $b\leq -3\delta' m$.

**Case 4.2.** Assume now that the inequalities (5.2.30) hold, and, in addition,

$$b\geq -3\delta' m \qquad \text{and} \qquad k_1\geq -0.6m. \tag{5.2.33}$$

If $m=L+1$ then (5.2.29) follows easily using (5.1.34). On the other hand, if $m\leq L$ then we integrate by parts in time and use (5.1.32)–(5.1.33). Using (3.1.33), the left-hand side of (5.2.29) is bounded by

$$\begin{aligned}C2^{-2k_1}2^{-3b}\sup_{s\in J_m}\big\{&(\|P_{k_1}U^{\mathcal{L}_1h_1}(s)\|_{L^2}\|P_{k_2}U^{h_2}(s)\|_{L^\infty}\|P_kU^{\mathcal{L}h}(s)\|_{L^2}\\&+2^m\|P_{k_1}\mathcal{L}_1\mathcal{N}^h_1(s)\|_{L^2}\|P_{k_2}U^{h_2}(s)\|_{L^\infty}\|P_kU^{\mathcal{L}h}(s)\|_{L^2}\\&+2^m\|P_{k_1}U^{\mathcal{L}_1h_1}(s)\|_{L^2}\|P_{k_2}\mathcal{N}^h_2(s)\|_{L^\infty}\|P_kU^{\mathcal{L}h}(s)\|_{L^2}\\&+2^m\|P_{k_1}U^{\mathcal{L}_1h_1}(s)\|_{L^2}\|P_{k_2}U^{h_2}(s)\|_{L^\infty}\|P_k\mathcal{L}\mathcal{N}^h(s)\|_{L^2}\big\},\end{aligned}\tag{5.2.34}$$

where $\mathcal{N}^h_1,\mathcal{N}^h_2,\mathcal{N}^h$ are suitable components of the metric nonlinearities $\mathcal{N}^h_{\alpha\beta}$. In view of (5.2.13)–(5.2.14), the terms in the expression above are dominated by

$$C\varepsilon_1^3 2^{-3k_1/2}2^{-3b}2^{-m+2\delta' m}\lesssim \varepsilon_1^3 2^{-m/20},$$

where we used the assumptions (5.2.33) in the last inequality. This suffices to prove (5.2.29).

**Case 4.3.** Assume now that the inequalities (5.2.30) hold, and, in addition,

$$k_1\leq -0.6m \qquad \text{and} \qquad b\in[-3\delta' m,-2\delta m]. \tag{5.2.35}$$

We can use the condition $k_1\leq -0.6m$ to improve the argument in **Case 4.1.** Indeed, let $J_2':=-k_1$ and decompose $P_{k_2}U^{h_2,\iota_2}(s)=U^{h_2,\iota_2}_{\leq J_2',k_2}(s)+U^{h_2,\iota_2}_{>J_2',k_2}(s)$ as in (3.3.1)–(3.3.2). Using (3.2.46) (instead of (3.2.54)), (3.1.36), the assumption

(5.2.2), (3.3.3), and (3.3.7), we estimate

$$\begin{aligned}
&2^{-k_1}\Big|\int_{J_m} q_m(s)\mathcal{G}_{\mathfrak{q}^b_{\iota_1\iota_2}}\big[P_{k_1}U^{\mathcal{L}_1h_1,\iota_1}(s), U^{h_2,\iota_2}_{\leq J_2',k_2}(s), P_kU^{\mathcal{L}h,\iota}(s)\big]\,ds\Big|\\
&\lesssim 2^{-k_1}2^m \sup_{s\in J_m}\big\|P_kI_{\mathfrak{q}^b_{\iota_1\iota_2}}\big[P_{k_1}U^{\mathcal{L}_1h_1,\iota_1}(s), U^{h_2,\iota_2}_{\leq J_2',k_2}(s)\big]\big\|_{L^2}\big\|P_{[k-2,k+2]}U^{\mathcal{L}h,\iota}(s)\big\|_{L^2}\\
&\lesssim 2^{-k_1}2^m\cdot 2^b2^{k_1/2}2^{-m}2^{3k_2/2}\\
&\qquad\times\sup_{s\in J_m}\|P_{k_1}U^{\mathcal{L}_1h_1,\iota_1}(s)\|_{L^2}\|\widehat{P_{k_2}U^{h_2,\iota_2}}(s)\|_{L^\infty}\|P_k'U^{\mathcal{L}h,\iota}(s)\|_{L^2}\\
&\lesssim \varepsilon_1^3 2^b2^{2H(q,n)\delta m}2^{\delta m}2^{-2N(n)k^+-2k^+};
\end{aligned}\tag{5.2.36}$$

compare with (5.2.31). The contribution of the profile $U^{h_2,\iota_2}_{>J_2',k_2}(s)$ can be estimated as in (5.2.32), and the bounds (5.2.29) follow if $b\leq -2\delta m$.

**Case 4.4.** Finally assume that the inequalities (5.2.30) hold, and, moreover,

$$k_1\leq -0.6m \qquad\text{and}\qquad b\in[-2\delta m, 4]. \tag{5.2.37}$$

If $\iota\neq\iota_2$ then we can still use normal forms and estimate as in (5.2.34), but with a factor of $2^{-k_1}$ replaced by $2^{-k}$, due to the better lower bound $|\Phi(\xi,\eta)|\gtrsim 2^k$. The desired bounds follow. Also in the case $m=L+1$ there is no loss of $2^m$ and the desired bounds follow easily using (5.1.34).

On the other hand, if $\iota=\iota_2$ and $m\leq L$ then we may assume that $\iota=\iota_2=+$, by taking complex conjugates. The proof in this case is more complicated, as it requires switching to the quasilinear variables $\mathcal{U}^{h_2}$ and $\mathcal{U}^{\mathcal{L}h}$, and is provided in Lemma 5.8 below. □

### 5.2.2 Non-null Semilinear Terms

In this subsection we show how to estimate the remaining Wave×Wave×Wave interactions in Proposition 5.2.

**Lemma 5.6.** *With the assumptions of Proposition 5.2 we have*

$$\begin{aligned}
&\sum_{k,k_1,k_2} 2^{2\gamma k^-}2^{-k}2^{2N(n)k^+}\Big|\int_{J_m} q_m(s)\\
&\quad\times\mathcal{G}_{\mathfrak{m}}[P_{k_1}U^{\vartheta^{\mathcal{L}_1},\iota_1}(s), P_{k_2}U^{\vartheta^{\mathcal{L}_2},\iota_2}(s), P_kU^{\mathcal{L}h,\iota}(s)]\,ds\Big|\lesssim\varepsilon_1^3 2^{2H(q,n)\delta m-2\gamma m},
\end{aligned}\tag{5.2.38}$$

*where $\mathfrak{m}\in\mathcal{M}$ and the operators $\mathcal{G}_{\mathfrak{m}}$ are defined as in* (5.1.30). *Thus the bounds* (5.1.23) *hold.*

*Proof.* In proving (5.2.38) it is important to keep in mind that $\vartheta_{ab}$ are among the “good” components of the metric, which satisfy strong bounds like (3.3.7) uniformly in time.

For suitable values $J \geq \max\{-k_a, 0\}$ we sometimes decompose $P_{k_a}U^{\vartheta^{\mathcal{L}_a},\iota_a} = U^{\vartheta^{\mathcal{L}_a},\iota_a}_{\leq J,k_a} + U^{\vartheta^{\mathcal{L}_a},\iota_a}_{>J,k_a}$, $a \in \{1,2\}$, where

$$\begin{aligned} U^{\vartheta^{\mathcal{L}_a},\iota_a}_{\leq J,k_a} &:= e^{-it\Lambda_{wa,\iota_a}} P'_{k_a}(\varphi_{\leq J} \cdot P_{k_a}V^{\vartheta^{\mathcal{L}_a},\iota_a}),\\ U^{\vartheta^{\mathcal{L}_a},\iota_a}_{>J,k_a} &:= e^{-it\Lambda_{wa,\iota_a}} P'_{k_a}(\varphi_{>J} \cdot P_{k_a}V^{\vartheta^{\mathcal{L}_a},\iota_a}), \end{aligned} \tag{5.2.39}$$

and $V^{\vartheta^{\mathcal{L}_a},\iota_a} := e^{it\Lambda_{wa,\iota_a}}U^{\vartheta^{\mathcal{L}_a},\iota_a}$. We consider several cases.

**Step 1.** We assume first that $\min\{n_1, n_2\} = 0$. By symmetry, we may assume that $n_1 = 0$. Let $J_1 := \infty$ if $|k_1| > \delta' m - 10$ and $J_1 := m - \delta' m$ if $|k_1| \leq \delta' m - 10$, and decompose $P_{k_1}U^{\vartheta,\iota_1} = U^{\vartheta,\iota_1}_{\leq J_1,k_1} + U^{\vartheta,\iota_1}_{>J_1,k_1}$ as in (5.2.39) (in particular, $P_{k_1}U^{\vartheta,\iota_1} = U^{\vartheta,\iota_1}_{\leq J_1,k_1}$ if $|k_1| > \delta' m - 10$).

In this case we prove the stronger bounds

$$\begin{aligned} \sum_{k,k_1,k_2} 2^{2\gamma k^-}2^{-k}2^{2N(n)k^+}\Big| \int_{J_m} q_m(s) \\ \times \mathcal{G}_{\mathfrak{m}}[U^{\vartheta,\iota_1}_{>J_1,k_1}, P_{k_2}U^{\vartheta^{\mathcal{L}},\iota_2}(s), P_kU^{\mathcal{L}h,\iota}(s)]\, ds\Big| \lesssim \varepsilon_1^3, \end{aligned} \tag{5.2.40}$$

and, for any $s \in J_m$ and $k \in \mathbb{Z}$,

$$\begin{aligned} 2^{\gamma(m+k^-)} \sum_{k_1,k_2}{}^{*} \|P_kI[U^{\vartheta,\iota_1}_{\leq J_1,k_1}(s), P_{k_2}U^{\vartheta^{\mathcal{L}},\iota_2}(s)]\|_{L^2} \\ \lesssim \varepsilon_1 b_k(q,n;s)2^{k/2}2^{-N(n)k^+}2^{-m+H(q,n)\delta m}, \end{aligned} \tag{5.2.41}$$

where $I = I_{\mathfrak{m}}$ is as in (3.2.43), the coefficients $b_k(q,n;t)$ are defined in (4.3.20), and $\sum_{k_1,k_2}{}^{*}$ denotes the sum over pairs $(k_1,k_2) \in \mathcal{X}_k$ with the additional assumption $k_1 \leq k_2$ if $\mathcal{L} = \mathrm{Id}$ (thus $n_1 = n_2 = n = 0$). It is clear that (5.2.40)–(5.2.41) would suffice to prove (5.2.38).

Using (4.3.22), (3.3.7), and $L^2$ estimates as in (3.2.44), we have

$$\begin{aligned} 2^{\gamma(m+k^-)}\|P_kI[U^{\vartheta,\iota_1}_{*,k_1}(s), P_{k_2}U^{\vartheta^{\mathcal{L}},\iota_2}(s)]\|_{L^2} \lesssim \varepsilon_1 2^{3\underline{k}/2} b_{k_2}(q,n;s) \\ \times 2^{\gamma(k^- - k_2^-)}2^{k_2/2}2^{-N(n)k_2^+}2^{H(q,n)\delta m} \cdot 2^{k_1^-/2-\kappa k_1^-}2^{-N_0k_1^+ + 2k_1^+}, \end{aligned} \tag{5.2.42}$$

for $* \in \{\leq J_1, > J_1\}$ and any $k, k_1, k_2 \in \mathbb{Z}$.

**Substep 1.1.** We prove first the bounds (5.2.41). Recalling (4.3.21), these bounds follow from (5.2.42) when $2^m \lesssim 1$ or when $2^k \lesssim 2^{-m+\gamma m/2}$. Assume that $m \geq \delta^{-2}$ and $2^k \geq 2^{-m+\gamma m/2}$. We examine first the contribution of the pairs $(k_1,k_2)$ with $|k_1| > \delta' m - 10$, thus $U^{\vartheta,\iota_1}_{\leq J_1,k_1} = P_{k_1}U^{\vartheta,\iota_1}$. In view of (4.3.22) and (3.3.11) we have

$$
\begin{aligned}
&2^{\gamma(m+k^-)}\|P_kI[P_{k_1}U^{\vartheta,\iota_1}(s),P_{k_2}U^{\vartheta^{\mathcal{L}},\iota_2}(s)]\|_{L^2}\lesssim\varepsilon_1b_{k_2}(q,n;s)\\
&\quad\times 2^{\gamma(k^--k_2^-)}2^{k_2/2}2^{-N(n)k_2^+}2^{H(q,n)\delta m}\cdot 2^{k_1^-}2^{-m+\delta'm/2}2^{-N(1)k_1^++2k_1^+}.
\end{aligned}
\tag{5.2.43}
$$

This suffices to bound the contribution of the pairs $(k_1,k_2)\in\mathcal{X}_k$ with $2^{|k_1|}\gtrsim 2^{\delta'm}$ and $k_1\le k-4$. It also suffices to bound the contribution of the pairs $(k_1,k_2)$ with $2^{|k_1|}\gtrsim 2^{\delta'm}$ and $|k-k_1|\le 8$ and $n\ge 2$. The contribution of the pairs $(k_1,k_2)$ with $|k-k_1|\le 8$ and $n\le 1$ can be controlled in a similar way, by estimating $P_{k_1}U^{\vartheta,\iota_1}(s)$ in $L^2$ and $P_{k_2}U^{\vartheta^{\mathcal{L}},\iota_2}(s)$ in $L^\infty$.

The contribution of the pairs $(k_1,k_2)$ with $k_1,k_2\ge k+4$ and $k_1\le -3m/4$ or $k_1\ge m/4$ can be estimated using (5.2.42). Finally, to estimate the contribution of the pairs $(k_1,k_2)$ with $k_1,k_2\ge k+4$ and $k_1\in[-3m/4,m/4]$ we decompose $P_{k_1}U^{\vartheta,\iota_1}=U^{\vartheta,\iota_1}_{\le J_1',k_1}+U^{\vartheta,\iota_1}_{>J_1',k_1}$ with $J_1':=m-\delta'm/4$. Then we use the $L^\infty$ super-localized bounds (3.2.16) to estimate

$$
\begin{aligned}
&2^{\gamma(m+k^-)}\|P_kI[U^{\vartheta,\iota_1}_{\le J_1',k_1}P_{k_1}(s),P_{k_2}U^{\vartheta^{\mathcal{L}},\iota_2}(s)]\|_{L^2}\\
&\lesssim\sum_{|n_1-n_2|\le 4}2^{\gamma(m+k^-)}\|\mathcal{C}_{n_1,k}U^{\vartheta,\iota_1}_{\le J_1',k_1}P_{k_1}(s)\|_{L^\infty}\|\mathcal{C}_{n_2,k}P_{k_2}U^{\vartheta^{\mathcal{L}}}(s)\|_{L^2}\\
&\lesssim\varepsilon_1^2 2^{k_2/2}2^{-N(n)k_2^+}2^{H(q,n)\delta m}\cdot 2^{k_1/2}2^{k/2}2^{-m+\delta'm/2}2^{-8k_1^+}.
\end{aligned}
\tag{5.2.44}
$$

The contribution of the $U^{\vartheta,\iota_1}_{>J_1',k_1}$ can be estimated using just $L^2$ bounds, as in (5.2.42). This suffices to bound the contribution of all the pairs $(k_1,k_2)$ with $|k_1|>\delta'm-10$.

We consider now the pairs $(k_1,k_2)\in\mathcal{X}_k$ with $|k_1|\le\delta'm-10$ and use the more precise estimates (3.2.46). Let $J_1''$ denote the largest integer such that

$$
2^{J_1''}\le 2^{-10}[2^{m/2-k_1/2}+2^{-k}]
\tag{5.2.45}
$$

and apply first (3.2.46) (or (3.2.12) if $k_1=\min\{k,k_1,k_2\}$), and then (3.3.7) to estimate

$$
\begin{aligned}
&2^{\gamma(m+k^-)}\|P_kI[U^{\vartheta,\iota_1}_{\le J_1'',k_1}(s),P_{k_2}U^{\vartheta^{\mathcal{L}},\iota_2}(s)]\|_{L^2}\\
&\lesssim 2^{-m}2^{\min\{k,k_1,k_2\}/2}2^{3k_1/2}\|\widehat{P_{k_1}V^{\vartheta}(s)}\|_{L^\infty}\cdot 2^{\gamma(m+k^-)}\|P_{k_2}U^{\vartheta^{\mathcal{L}}}(s)\|_{L^2}\\
&\lesssim\varepsilon_1 2^{-m+H(q,n)\delta m}2^{k_1^-/4}2^{-(N_0-2)k_1^+}\\
&\qquad\times 2^{\min\{k,k_1,k_2\}/2}2^{k_2/2}2^{\gamma(k^--k_2^-)}b_{k_2}(q,n;s)2^{-N(n_2)k_2^+},
\end{aligned}
\tag{5.2.46}
$$

where $U^{\vartheta,\iota_1}_{\le J_1'',k_1}:=e^{-it\Lambda_{wa,\iota_1}}P'_{k_1}(\varphi_{\le J_1''}\cdot P_{k_1}V^{\vartheta,\iota_1})$ as in (5.2.39). Moreover, using

(3.2.11) and (3.3.4) instead, we estimate

$$\sum_{j_1\in[J_1'',J_1]} 2^{\gamma(m+k^-)}\|P_kI[U^{\vartheta,\iota_1}_{j_1,k_1}(s),P_{k_2}U^{\vartheta^{\mathcal{L}},\iota_2}(s)]\|_{L^2} \lesssim \varepsilon_1^2 2^{-m+\delta' m}2^{-J_1''}2^{k_2/2}2^{-8k_1^+}2^{-N(n_2)k_2^+}. \tag{5.2.47}$$

It is easy to see that these two bounds can be summed over $(k_1,k_2)\in\mathcal{X}_k$ with $|k_1|\leq\delta' m-10$ to complete the proof of (5.2.41). In the (harder) case of pairs $k_1,k_2$ with $k_1,k_2\geq k+4$ one can use (4.3.21) to sum (5.2.46), and the definition (5.2.45) to sum (5.2.47).

**Substep 1.2.** We prove now the bounds (5.2.40). We may assume that $|k_1|\leq\delta' m$, due to the definition of $J_1$. We notice that the bounds (5.2.40) follow using just $L^2$ estimates (similar to (5.2.42)) and (3.3.3)–(3.3.4) if $2^m\lesssim 1$ or $k\leq -2\delta' m$.

Assume now that $m\geq\delta^{-2}$ and $k\geq -2\delta' m$. We would like to integrate by parts in time as in (5.1.32). For this we need to decompose into resonant and non-resonant contributions. As in (5.2.12), with $q_0=-8\delta' m$ and $\Xi_{\iota_1\iota_2}$ as in (3.1.23), we decompose

$$\mathfrak{m}=\mathfrak{m}^r+\mathfrak{m}^{nr},\qquad \mathfrak{m}^r(\theta,\eta)=\varphi_{\leq q_0}(\Xi_{\iota_1\iota_2}(\theta,\eta))\mathfrak{m}(\theta,\eta). \tag{5.2.48}$$

To bound the resonant contribution we use the smallness of Fourier support and Schur's lemma. For this we notice that, with $\varphi_{kk_1k_2}$ defined as in (5.1.35),

$$\begin{aligned}\sup_{\eta\in\mathbb{R}^3}\int_{\mathbb{R}^3}&|\widehat{U^{\vartheta,\iota_1}_{>J_1,k_1}}(\xi-\eta,s)||\mathfrak{m}^r(\xi-\eta,\eta)|\varphi_{kk_1k_2}(\xi-\eta,\eta)\,d\xi\\ &\lesssim\big\|\widehat{U^{\vartheta,\iota_1}_{>J_1,k_1}}(s)\big\|_{L^2}\cdot 2^{3k_1/2}2^{q_0},\\ \sup_{\xi\in\mathbb{R}^3}\int_{\mathbb{R}^3}&|\widehat{U^{\vartheta,\iota_1}_{>J_1,k_1}}(\xi-\eta,s)||\mathfrak{m}^r(\xi-\eta,\eta)|\varphi_{kk_1k_2}(\xi-\eta,\eta)\,d\eta\\ &\lesssim\big\|\widehat{U^{\vartheta,\iota_1}_{>J_1,k_1}}(s)\big\|_{L^2}\cdot 2^{3k_1/2}2^{q_0}2^{k_2-k},\end{aligned}$$

where in the second estimate we bound $\widetilde{\Xi}(\xi-\eta,\xi)\lesssim\widetilde{\Xi}(\xi-\eta,\eta)2^{k_2-k}\lesssim 2^{q_0}2^{k_2-k}$ in the support of the integral (see (3.1.28)). Therefore, using to Schur's test,

$$\begin{aligned}&2^{\gamma(m+k^-)}\|P_kI_{\mathfrak{m}^r}[U^{\vartheta,\iota_1}_{>J_1,k_1}(s),P_{k_2}U^{\vartheta^{\mathcal{L}},\iota_2}(s)]\|_{L^2}\\ &\lesssim 2^{\gamma(m+k^-)}\|P_{k_2}U^{\vartheta^{\mathcal{L}}}(s)\|_{L^2}\big\|\widehat{U^{\vartheta,\iota_1}_{>J_1,k_1}}(s)\big\|_{L^2}\cdot 2^{3k_1/2}2^{q_0}(1+2^{k_2-k}),\\ &\lesssim\varepsilon_1 2^{k_1}2^{-N(1)k_1^+}2^{-J_1}2^{q_0}2^{\delta' m}\cdot 2^{k_2/2}2^{-N(n)k_2^+}b_{k_2}(q,n;s)(1+2^{k_2-k})2^{\gamma(k^- -k_2^-)},\end{aligned}$$

where $I_{\mathfrak{m}^r}$ is as in (3.2.43). Since $|k_1|\leq\delta' m$ and $2^{-J_1}\lesssim 2^{-m+\delta' m}$, this suffices to show that

$$2^{\gamma(m+k^-)}\sum_{k_1,k_2}^{*}\|P_kI_{\mathfrak{m}^r}[U^{\vartheta,\iota_1}_{>J_1,k_1}(s),P_{k_2}U^{\vartheta^{\mathcal{L}},\iota_2}(s)]\|_{L^2}$$
$$\lesssim\varepsilon_1b_k(q,n;s)2^{k/2}2^{-N(n)k^+}2^{-m-\delta'm},\tag{5.2.49}$$

if $2^k\gtrsim 2^{-2\delta'm}$. This is similar to (5.2.41) and implies the required bounds on the resonant contributions.

On the other hand, the non-resonant contributions corresponding to the symbol $\mathfrak{m}^{nr}$ can be treated as in the proof of (5.2.1) in Lemma 5.4. We may assume first that $m\le L$, integrate by parts in time, and notice that estimates like (5.2.16) still hold; such estimates suffice to prove (5.2.40) when $k\le m/10$. In the remaining case when $k$ is very large and $|k_1|\le\delta'm$, the desired conclusion follows using paradifferential calculus from Lemma 5.7 below.

**Step 2.** We prove now (5.2.38) when $n_1,n_2\ge 1$. In this case we prove the strong bounds

$$2^{\gamma(m+k^-)}\sum_{k_1,k_2\in\mathcal{X}_k}\|P_kI[P_{k_1}U^{\vartheta^{\mathcal{L}_1},\iota_1}(s),P_{k_2}U^{\vartheta^{\mathcal{L}_2},\iota_2}(s)]\|_{L^2}$$
$$\lesssim\varepsilon_1^2 2^{-\gamma|k|/4}2^{k/2}2^{-N(n)k^+}2^{-m+H(q,n)\delta m},\tag{5.2.50}$$

for any $s\in J_m$ and $k\in\mathbb{Z}$ (compare with (5.2.41) and recall that $b_k(q,n;s)\ge\varepsilon_12^{-\gamma|k|/4}$).

As in (5.2.42), we recall that $\underline{k}=\min\{k,k_1,k_2\}$ and start with $L^2$ estimates,

$$2^{\gamma(m+k^-)}\|P_kI[P_{k_1}U^{\vartheta^{\mathcal{L}_1},\iota_1}(s),P_{k_2}U^{\vartheta^{\mathcal{L}_2},\iota_2}(s)]\|_{L^2}$$
$$\lesssim\varepsilon_1^2 2^{3\underline{k}/2}2^{k_1/2}2^{k_2/2}2^{-\gamma k_1^- -\gamma k_2^-}2^{-N(n_1)k_1^+}2^{-N(n_2)k_2^+}2^{H(q_1,n_1)\delta m+H(q_2,n_2)\delta m},\tag{5.2.51}$$

for any $k,k_1,k_2\in\mathbb{Z}$. This suffices to prove the bounds (5.2.50) when $2^m\lesssim 1$, or when $k\le -m+35\delta m$, or when $k\ge m/9$ (using (2.1.53) and $N(n)+10\le\min\{N(n_1),N(n_2)\}$).

On the other hand, if $m\ge\delta^{-1}$ and $k\in[-m+35\delta m,m/9]$ then the bounds (5.2.51) suffice to control the contribution of the pairs $(k_1,k_2)\in\mathcal{X}_k$ for which either $\min\{k_1,k_2\}\le -3m/4$ or $\max\{k_1,k_2\}\ge m/9$. Using $L^2\times L^\infty$ estimates as in (5.2.43)–(5.2.44) we can also control the contribution of the pairs $(k_1,k_2)$ for which $\max\{|k_1|,|k_2|\}\ge\delta'm$. In the remaining range $k_1,k_2\in[-\delta'm,\delta'm]$ we consider two cases.

**Case 2.1.** Assume first that $q_1=q_2=0$. Since rotations and Riesz transforms essentially commute (up to multipliers that are accounted for in the multiplier $\mathfrak{m}$), we may replace $U^{\vartheta^{\mathcal{L}_a},\iota_a}$ by $\mathcal{L}_aU^{\vartheta,\iota_a}$, $a\in\{1,2\}$, and use interpolation inequalities. Let

$$J^* = \max\{\frac{m - \max\{k_1, k_2\}}{2}, -k\}. \tag{5.2.52}$$

Assuming that $k \geq \max\{k_1, k_2\} - 10$, we use (3.2.64)–(3.2.65) to estimate

$$\begin{aligned}&\|P_k I[U^{\vartheta^{\mathcal{L}_1},\iota_1}_{\leq J^*,k_1}(s), U^{\vartheta^{\mathcal{L}_2},\iota_2}_{\leq J^*,k_2}(s)]\|_{L^2} \lesssim \|\mathcal{L}_1 U^{\vartheta,\iota_1}_{\leq J^*,k_1}(s)\|_{L^a} \|\mathcal{L}_2 U^{\vartheta,\iota_2}_{\leq J^*,k_2}(s)\|_{L^b} \\ &\lesssim \|U^{\vartheta,\iota_1}_{\leq J^*,k_1}(s)\|_{L^\infty}^{\frac{n-n_1}{n}} \|P_{k_1} V^{\vartheta,\iota_1}\|_{H^{0,n}_\Omega}^{\frac{n_1}{n}} \|U^{\vartheta,\iota_2}_{\leq J^*,k_2}(s)\|_{L^\infty}^{\frac{n-n_2}{n}} \|P_{k_2} V^{\vartheta,\iota_2}\|_{H^{0,n}_\Omega}^{\frac{n_2}{n}},\end{aligned}$$

where $U^{\vartheta^{\mathcal{L}_a},\iota_a}_{\leq J^*,k_a}$ are as in (5.2.39), $n = n_1 + n_2$, and

$$\frac{1}{a} = \frac{n_1}{n}\frac{1}{2}, \qquad \frac{1}{b} = \frac{n_2}{n}\frac{1}{2}. \tag{5.2.53}$$

Using now (3.3.16) and (3.3.3) we have

$$\begin{aligned}&2^{\gamma(m+k^-)} \|P_k I[U^{\vartheta^{\mathcal{L}_1},\iota_1}_{\leq J^*,k_1}(s), U^{\vartheta^{\mathcal{L}_2},\iota_2}_{\leq J^*,k_2}(s)]\|_{L^2} \\ &\quad\lesssim \varepsilon_1^2 2^{-m+H(0,n)\delta m} 2^{k_1/2+k_2/2} 2^{-N(n)k_1^+ - N(n)k_2^+} 2^{-|k_1|/8-|k_2|/8}.\end{aligned} \tag{5.2.54}$$

On the other hand, if $k \leq \max\{k_1, k_2\} - 10$ then we need a bilinear estimate to bring in the small factor $2^{k/2}$. We use Lemma 3.14 followed by (3.2.64)–(3.2.65) to estimate

$$\begin{aligned}&\|P_k I[U^{\vartheta^{\mathcal{L}_1},\iota_1}_{\leq J^*,k_1}(s), U^{\vartheta^{\mathcal{L}_2},\iota_2}_{\leq J^*,k_2}(s)]\|_{L^2} \\ &\lesssim 2^{-m} 2^{k/2} 2^{3k_1/2} \|\mathcal{F}\{\mathcal{L}_1 P_{k_1} U^{\vartheta,\iota_1}\}(s)\|_{L^a} \|\mathcal{F}\{\mathcal{L}_2 P_{k_2} U^{\vartheta,\iota_2}\}(s)\|_{L^b} \\ &\lesssim 2^{-m} 2^{k/2} 2^{3k_1/2} \|\mathcal{F}\{P_{k_1} V^\vartheta\}\|_{L^\infty}^{\frac{n-n_1}{n}} \|P_{k_1} V^\vartheta\|_{H^{0,n}_\Omega}^{\frac{n_1}{n}} \|\mathcal{F}\{P_{k_2} V^\vartheta\}\|_{L^\infty}^{\frac{n-n_2}{n}} \|P_{k_2} V^\vartheta\|_{H^{0,n}_\Omega}^{\frac{n_2}{n}},\end{aligned}$$

where $a, b$ are as in (5.2.53). Using (3.3.7) and (3.3.3), it follows that

$$\begin{aligned}&2^{\gamma(m+k^-)} \|P_k I[U^{\vartheta^{\mathcal{L}_1},\iota_1}_{\leq J^*,k_1}(s), U^{\vartheta^{\mathcal{L}_2},\iota_2}_{\leq J^*,k_2}(s)]\|_{L^2} \\ &\quad\lesssim \varepsilon_1^2 2^{-m+H(0,n)\delta m} 2^{\gamma k^-} 2^{k/2} 2^{-N_0 k_1^+} 2^{3k_1^-/4}.\end{aligned} \tag{5.2.55}$$

For the remaining terms we use (3.3.11), (3.3.4), and $L^2 \times L^\infty$ bounds,

$$\begin{aligned}&2^{\gamma(m+k^-)} \{\|P_k I[U^{\vartheta^{\mathcal{L}_1},\iota_1}_{\leq J^*,k_1}(s), U^{\vartheta^{\mathcal{L}_2},\iota_2}_{> J^*,k_2}(s)]\|_{L^2} \\ &\quad + \|P_k I[U^{\vartheta^{\mathcal{L}_1},\iota_1}_{> J^*,k_1}(s), P_{k_2} U^{\vartheta^{\mathcal{L}_2},\iota_2}(s)]\|_{L^2}\} \\ &\quad\lesssim \varepsilon_1^2 2^{-m+\delta' m} 2^{-J^*} 2^{|k_1|/2+|k_2|/2} 2^{-N(n_1+1)k_1^+} 2^{-N(n_2+1)k_2^+} 2^{2(k_1^+ + k_2^+)}.\end{aligned} \tag{5.2.56}$$

We combine (5.2.54) and (5.2.56) to control the contribution of the pairs $(k_1, k_2)$ for which $k \geq \max\{k_1, k_2\} - 10$; we also combine (5.2.55) and (5.2.56) to control the contribution of the pairs $(k_1, k_2)$ for which $k \leq \max\{k_1, k_2\} - 10$. This completes the proof of (5.2.50).

**Case 2.2.** The case $\max\{q_1, q_2\} \geq 1$ is comparatively easier. Without loss of generality we may assume that $q_2 \geq \max\{q_1, 1\}$. We use Lemma 3.11 and (3.3.3) to estimate

$$\begin{aligned}
&2^{\gamma(m+k^-)}\|P_k I[U^{\vartheta^{\mathcal{L}_1},\iota_1}_{\leq m(1-\delta),k_1}(s), P_{k_2}U^{\vartheta^{\mathcal{L}_2},\iota_2}(s)]\|_{L^2} \\
&\lesssim 2^{\gamma m}2^{-m+\delta m}2^{\min\{k,k_2\}/2}\|P_{k_1}U^{\vartheta^{\mathcal{L}_1}}(s)\|_{H^{0,1}_\Omega}\|P_{k_2}U^{\vartheta^{\mathcal{L}_2}}(s)\|_{L^2} \\
&\lesssim 2^{-m+\delta m}2^{\min\{k,k_2\}/2}2^{k_1/2}2^{k_2/2}2^{-N(n_1+1)k_1^+ - N(n_2)k_2^+}2^{H(q_1,n_1+1)\delta m + H(q_2,n_2)\delta m}.
\end{aligned}$$

This gives acceptable contributions using (4.1.3). On the other hand, using just $L^2 \times L^\infty$ estimates, (3.3.4), and (3.3.11), we have

$$2^{\gamma(m+k^-)}\|P_k I[U^{\vartheta^{\mathcal{L}_1},\iota_1}_{>m(1-\delta),k_1}(s), P_{k_2}U^{\vartheta^{\mathcal{L}_2},\iota_2}(s)]\|_{L^2} \lesssim \varepsilon_1^2 2^{-1.9m},$$

for any $k_1, k_2 \in [-\delta' m, \delta' m]$. The bounds (5.2.50) follow in this case as well, which completes the proof of the lemma. □

### 5.2.3 Second Symmetrization and Paradifferential Calculus

In the case of interactions of vastly different frequencies $|\xi| \approx |\eta| \gg |\xi - \eta|$, $\iota = \iota_2$, the application of normal forms as in (5.1.32) leads to a loss of derivatives due to the quasilinear nature of the nonlinearities. To avoid this loss we use paradifferential calculus and perform a second symmetrization.

Our first lemma concerns the contribution of very high frequencies and applies to conclude the analysis in **Case 2.3** in Lemma 5.4 and **Substep 1.2** in Lemma 5.6.

**Lemma 5.7.** *If $\iota, \iota_1, \iota_2 \in \{+,-\}$, $h, h_1, h_2 \in \{h_{\alpha\beta}\}$, $t \in [0,T]$, $m \in [\delta^{-1}, L+1]$, $|k_1| \leq \delta' m$, $k, k_2 \geq m/10$, $J_1 \geq m - 2\delta' m$, $(q,n) \leq (3,3)$, and $\mathcal{L}, \mathcal{L}_2 \in \mathcal{V}^q_n$ then*

$$\begin{aligned}
\Big|\int_{J_m} q_m(s)\mathcal{G}_{\mathfrak{m}^{nr}}\big[U^{h_1,\iota_1}_{\leq J_1,k_1}(s), P_{k_2}U^{\mathcal{L}_2 h_2,\iota_2}(s), P_k U^{\mathcal{L}h,\iota}(s)\big]\,ds\Big| \\
\lesssim \varepsilon_1 2^{-2N(n)k+k_2^{}} 2^{-2\delta m}(b_{k,m}(q,n))^2,
\end{aligned} \tag{5.2.57}$$

*where $\mathfrak{m}^{nr}(\theta,\eta) = \mathfrak{m}(\theta,\eta)\varphi_{>q_0}(\Xi_{\iota_1\iota_2}(\theta,\eta))$ is the non-resonant part of a symbol $\mathfrak{m} \in \mathcal{M}$, $q_0 = -8\delta' m$, and $b_{k,m}(q,n)$ are defined in* (5.2.17).

*Proof.* Notice that, as a consequence of (4.3.22)–(4.3.23),

$$\|P_{k_2}U^{\mathcal{L}_2 h_2,\iota_2}(s)\|_{L^2} + \|P_k U^{\mathcal{L}h,\iota}(s)\|_{L^2} \lesssim b_k(s)2^{\delta' m}2^{k/2-N(n)k} \tag{5.2.58}$$

and

$$\|(\partial_s + i\Lambda_{wa,\iota_2})P_{k_2}U^{\mathcal{L}_2 h_2,\iota_2}(s)\|_{L^2} + \|(\partial_s + i\Lambda_{wa,\iota})P_k U^{\mathcal{L}h,\iota}(s)\|_{L^2} \lesssim \varepsilon_1 b_k(s) 2^{-m+\delta' m} 2^{3k/2 - N(n)k}, \tag{5.2.59}$$

for any $s \in J_m$, where, for simplicity of notation, we let $b_k(s) := b_k(q, n; s)$. Moreover,

$$\|U^{h_1,\iota_1}_{\leq J_1,k_1}(s)\|_{L^\infty} \lesssim \varepsilon_1 2^{-m+\delta' m} 2^{k_1^-} 2^{-8k_1^+} \tag{5.2.60}$$

and

$$\|(\partial_s + i\Lambda_{wa,\iota_1})U^{h_1,\iota_1}_{\leq J_1,k_1}(s)\|_{L^\infty} \lesssim \varepsilon_1^2 2^{-1.9m}. \tag{5.2.61}$$

The bounds (5.2.60) follow from (3.3.11). The bounds (5.2.61) are slightly harder because of the spatial cutoffs. To prove them we write

$$\begin{aligned}(\partial_s + i\Lambda_{wa})U^{h_{\alpha\beta}}_{\leq J_1,k_1}(s) &= (\partial_s + i\Lambda_{wa})[e^{-is\Lambda_{wa}} P'_{k_1}(\varphi_{\leq J_1} \cdot P_{k_1} V^{h_{\alpha\beta}}(s))] \\ &= e^{-is\Lambda_{wa}} P'_{k_1}(\varphi_{\leq J_1} \cdot \partial_s P_{k_1} V^{h_{\alpha\beta}}(s)) \\ &= e^{-is\Lambda_{wa}} P'_{k_1}(\varphi_{\leq J_1} \cdot P_{k_1} e^{is\Lambda_{wa}} \mathcal{N}^h_{\alpha\beta}(s)),\end{aligned}$$

for any $\alpha, \beta \in \{0, 1, 2, 3\}$. Therefore, for any $x \in \mathbb{R}^3$,

$$\begin{aligned}&(\partial_s + i\Lambda_{wa})U^{h_{\alpha\beta}}_{\leq J_1,k_1}(x, s) \\ &= C \int_{\mathbb{R}^6} e^{ix\cdot\xi} e^{-is|\xi|} \varphi'_{k_1}(\xi) \widehat{\varphi_{\leq J_1}}(\xi - \eta) e^{is|\eta|} \varphi_{k_1}(\eta) \widehat{\mathcal{N}^h_{\alpha\beta}}(\eta, s)\, d\xi d\eta \\ &= C \int_{\mathbb{R}^6} e^{ix\cdot\eta} e^{ix\cdot\theta} e^{-is[|\eta+\theta| - |\eta|]} \varphi'_{k_1}(\eta + \theta) \widehat{\varphi_{\leq J_1}}(\theta) \varphi_{k_1}(\eta) \widehat{\mathcal{N}^h_{\alpha\beta}}(\eta, s)\, d\theta d\eta,\end{aligned}$$

where $\varphi'_{k_1} = \varphi_{[k_1-2,k_1+2]}$. Let

$$L_{x,s}(\eta) := \int_{\mathbb{R}^3} e^{ix\cdot\theta} e^{-is[|\eta+\theta| - |\eta|]} \varphi'_{k_1}(\eta + \theta)(2^{3J_1} \widehat{\varphi}(2^{J_1}\theta))\, d\theta, \tag{5.2.62}$$

so

$$(\partial_s + i\Lambda_{wa})U^{h_{\alpha\beta}}_{\leq J_1,k_1}(x, s) = C \int_{\mathbb{R}^6} e^{i(x-y)\cdot\eta} L_{x,s}(\eta) \varphi_{k_1}(\eta) \mathcal{N}^h_{\alpha\beta}(y, s)\, d\eta dy. \tag{5.2.63}$$

Since $s2^{-J_1} + 2^{|k_1|} \lesssim 2^{2\delta' m}$ (the hypothesis of the lemma), it is easy to see that $|D^\alpha_\eta L_{x,s}(\eta)| \lesssim 2^{4\delta' m|\alpha|}$ for any $x \in \mathbb{R}^3$, $|\eta| \approx 2^{k_1}$, and multi-indices $\alpha$ with $|\alpha| \leq 10$. Therefore

$$|(\partial_s + i\Lambda_{wa})U^{h_{\alpha\beta}}_{\leq J_1,k_1}(x, s)| \lesssim \int_{\mathbb{R}^3} 2^{4\delta' m}(1 + |x - y| 2^{-4\delta' m})^{-4} |\mathcal{N}^h_{\alpha\beta}(y, s)|\, dy,$$

using integration by parts in $\eta$ in (5.2.63). The desired conclusion (5.2.61) follows from (4.3.15).

We divide the rest of the proof of the lemma into several steps.

**Step 1.** We start with some preliminary reductions. First, we may assume that $m \in [\delta^{-1}, L]$ since otherwise $|J_m| \lesssim 1$ and the desired bounds follow using (5.2.58), (5.2.60), and (3.1.30). We may also assume that $\iota = \iota_2$; otherwise

$$|\Phi_{\sigma\mu\nu}(\xi,\eta)| = |\Lambda_{wa,\iota}(\xi) - \Lambda_{wa,\iota_2}(\eta) - \Lambda_{wa,\iota_1}(\xi-\eta)| \gtrsim 2^k$$

in the support of the integral, so the normal form argument (5.1.32) still gives the desired conclusion since the loss of derivatives in (5.2.59) is compensated by the large denominator. By taking complex conjugates, we may actually assume that $\iota_2 = \iota = +$.

To continue we switch to the quasilinear variables defined in (5.1.41),

$$\mathcal{U}^{\mathcal{L}_2 h_2} = (\partial_t - iT_{\sigma_{wa}})(\mathcal{L}_2 h_2) \quad \text{and} \quad \mathcal{U}^{\mathcal{L}h} = (\partial_t - iT_{\sigma_{wa}})(\mathcal{L}h).$$

See subsections 3.1.3–5.1.4 for the definition of the paradifferential operators $T_a$ and the symbols $\sigma_{wa}$. In view of (5.1.42) we have, for any $s \in J_m$,

$$\begin{aligned}\|P_{k_2}(\mathcal{U}^{\mathcal{L}_2 h_2} - U^{\mathcal{L}_2 h_2})(s)\|_{L^2} + \|P_k(\mathcal{U}^{\mathcal{L}h} - U^{\mathcal{L}h})(s)\|_{L^2}& \\ \lesssim 2^{-m+\delta' m} b_k(s) 2^{k/2-N(n)k}&.\end{aligned} \tag{5.2.64}$$

Thus we may replace $P_{k_2}U^{\mathcal{L}_2 h_2}$ with $P_{k_2}\mathcal{U}^{\mathcal{L}_2 h_2}$ and $P_k U^{\mathcal{L}h}$ with $P_k\mathcal{U}^{\mathcal{L}h}$ in the integral in (5.2.57), at the expense of acceptable errors. To summarize, it remains to prove that

$$\begin{aligned}\Big|\int_{J_m} q_m(s)\mathcal{G}_{\mathfrak{m}^{nr}}\big[U^{h_1,\iota_1}_{\leq J_1,k_1}(s), P_{k_2}\mathcal{U}^{\mathcal{L}_2 h_2}(s), P_k\mathcal{U}^{\mathcal{L}h}(s)\big]\,ds\Big|& \\ \lesssim \varepsilon_1 2^{-2N(n)k+k} 2^{-2\delta m} b_{k,m}^2&,\end{aligned} \tag{5.2.65}$$

provided that $m \in [\delta^{-1}, L]$ and, for simplicity, $b_{k,m} := b_{k,m}(q,n)$.

**Step 2.** We integrate by parts in $s$ using (5.1.32). The contributions of the first two terms, when the $d/ds$ derivative hits either the function $q_m(s)$ or the first term $U^{h_1,\iota_1}_{\leq J_1,k_1}(s)$, can be bounded easily, using the $L^\infty$ bounds (5.2.60)–(5.2.61) and the $L^2$ bounds (5.2.58) and (5.2.64) (there are no derivative losses in this case). So it remains to prove that

$$\begin{aligned}\Big|\mathcal{H}_{\mathfrak{m}^{nr}}&\big[U^{h_1,\iota_1}_{\leq J_1,k_1}(s), (\partial_s + i\Lambda_{wa})P_{k_2}\mathcal{U}^{\mathcal{L}_2 h_2}(s), P_k\mathcal{U}^{\mathcal{L}h}(s)\big] \\ &+ \mathcal{H}_{\mathfrak{m}^{nr}}\big[U^{h_1,\iota_1}_{\leq J_1,k_1}(s), P_{k_2}\mathcal{U}^{\mathcal{L}_2 h_2}(s), (\partial_s + i\Lambda_{wa})P_k\mathcal{U}^{\mathcal{L}h}(s)\big]\Big| \\ &\lesssim \varepsilon_1 2^{-2N(n)k+k} 2^{-m-2\delta m} b_k(s)^2,\end{aligned} \tag{5.2.66}$$

for any $s \in J_m$, where the operators $\mathcal{H}_{\mathfrak{m}^{nr}}$ are defined in (5.1.31). Notice that

$$(\partial_s + i\Lambda_{wa})P_k\mathcal{U}^{\mathcal{L}h} = (\partial_s + iT_{\Sigma_{wa}})P_k\mathcal{U}^{\mathcal{L}h} - iT_{\Sigma_{wa}-|\zeta|}P_k\mathcal{U}^{\mathcal{L}h},$$

where the symbols $\Sigma_{wa}$ are defined in (5.1.37), and a similar identity holds for $P_{k_2}\mathcal{U}^{\mathcal{L}_2h_2}$. Therefore we use the bounds (5.1.47) to replace $(\partial_s+i\Lambda_{wa})P_{k_2}\mathcal{U}^{\mathcal{L}_2h_2}$ and $(\partial_s+i\Lambda_{wa})P_k\mathcal{U}^{\mathcal{L}h}$ with $-iT_{\Sigma_{wa}-|\zeta|}P_{k_2}\mathcal{U}^{\mathcal{L}_2h_2}$ and $-iT_{\Sigma_{wa}-|\zeta|}P_k\mathcal{U}^{\mathcal{L}h}$ respectively in (5.2.66), at the expense of acceptable errors. For (5.2.66) it remains to prove that

$$\begin{aligned}\Big|&\mathcal{H}_{\mathfrak{m}^{nr}}\big[U^{h_1,\iota_1}_{\leq J_1,k_1}(s), iT_{\Sigma^{\geq 1}_{wa}}P_{k_2}\mathcal{U}^{\mathcal{L}_2h_2}(s), P_k\mathcal{U}^{\mathcal{L}h}(s)\big]\\ &+\mathcal{H}_{\mathfrak{m}^{nr}}\big[U^{h_1,\iota_1}_{\leq J_1,k_1}(s), P_{k_2}\mathcal{U}^{\mathcal{L}_2h_2}(s), iT_{\Sigma^{\geq 1}_{wa}}P_k\mathcal{U}^{\mathcal{L}h}(s)\big]\Big|\\ &\lesssim \varepsilon_1 2^{-2N(n)k+k}2^{-m-2\delta m}b_k(s)^2,\end{aligned}\tag{5.2.67}$$

for any $s\in J_m$, where $\Sigma^{\geq 1}_{wa}(x,\zeta):=\Sigma_{wa}(x,\zeta)-|\zeta|$.

**Step 3.** We now write explicitly the expression in the left-hand side of (5.2.67) and exploit the cancellation between the two terms to avoid derivative loss. Using the definitions we write

$$\begin{aligned}\mathcal{H}_{\mathfrak{m}^{nr}}\big[U^{h_1,\iota_1}_{\leq J_1,k_1}, iT_{\Sigma^{\geq 1}_{wa}}P_{k_2}\mathcal{U}^{\mathcal{L}_2h_2}, P_k\mathcal{U}^{\mathcal{L}h}\big] &= \frac{i}{8\pi^3}\int_{\mathbb{R}^9}\frac{\mathfrak{m}^{nr}(\xi-\eta,\eta)}{|\xi|-|\eta|-\iota_1|\xi-\eta|}\\ \times\widetilde{\Sigma^{\geq 1}_{wa}}\Big(\eta-\rho,\frac{\eta+\rho}{2}\Big)\chi_0\Big(\frac{|\eta-\rho|}{|\eta+\rho|}\Big)&\widehat{U^{h_1,\iota_1}_{\leq J_1,k_1}}(\xi-\eta)\widehat{P_{k_2}\mathcal{U}^{\mathcal{L}_2h_2}}(\rho)\overline{\widehat{P_k\mathcal{U}^{\mathcal{L}h}}(\xi)}\,d\xi d\eta d\rho,\end{aligned}$$

$$\begin{aligned}\mathcal{H}_{\mathfrak{m}^{nr}}\big[U^{h_1,\iota_1}_{\leq J_1,k_1}, P_{k_2}\mathcal{U}^{\mathcal{L}_2h_2}, iT_{\Sigma^{\geq 1}_{wa}}P_k\mathcal{U}^{\mathcal{L}h}\big] &= \frac{-i}{8\pi^3}\int_{\mathbb{R}^9}\frac{\mathfrak{m}^{nr}(\xi-\eta,\eta)}{|\xi|-|\eta|-\iota_1|\xi-\eta|}\\ \times\overline{\widetilde{\Sigma^{\geq 1}_{wa}}}\Big(\xi-\rho,\frac{\xi+\rho}{2}\Big)\chi_0\Big(\frac{|\xi-\rho|}{|\xi+\rho|}\Big)&\widehat{U^{h_1,\iota_1}_{\leq J_1,k_1}}(\xi-\eta)\widehat{P_{k_2}\mathcal{U}^{\mathcal{L}_2h_2}}(\eta)\overline{\widehat{P_k\mathcal{U}^{\mathcal{L}h}}(\rho)}\,d\xi d\eta d\rho.\end{aligned}$$

The key property that allows symmetrization is the reality of the symbol $\Sigma^{\geq 1}_{wa}$, which shows that $\overline{\widetilde{\Sigma^{\geq 1}_{wa}}}\big(\xi-\rho,\frac{\xi+\rho}{2}\big)=\widetilde{\Sigma^{\geq 1}_{wa}}\big(\rho-\xi,\frac{\xi+\rho}{2}\big)$. Therefore, after changes of variables, we have

$$\begin{aligned}&\mathcal{H}_{\mathfrak{m}^{nr}}\big[U^{h_1,\iota_1}_{\leq J_1,k_1}(s), iT_{\Sigma^{\geq 1}_{wa}}P_{k_2}\mathcal{U}^{\mathcal{L}_2h_2}(s), P_k\mathcal{U}^{\mathcal{L}h}(s)\big]\\ &+\mathcal{H}_{\mathfrak{m}^{nr}}\big[U^{h_1,\iota_1}_{\leq J_1,k_1}(s), P_{k_2}\mathcal{U}^{\mathcal{L}_2h_2}(s), iT_{\Sigma^{\geq 1}_{wa}}P_k\mathcal{U}^{\mathcal{L}h}(s)\big]\\ &= C\int_{\mathbb{R}^9}K_{\mathfrak{m}^{nr}}(\xi,\eta,\rho;s)\widehat{U^{h_1,\iota_1}_{\leq J_1,k_1}}(\xi-\eta-\rho,s)\widehat{P_{k_2}\mathcal{U}^{\mathcal{L}_2h_2}}(\eta,s)\overline{\widehat{P_k\mathcal{U}^{\mathcal{L}h}}(\xi,s)}\,d\xi d\eta d\rho,\end{aligned}\tag{5.2.68}$$

where

$$\begin{aligned}&K_{\mathfrak{m}^{nr}}(\xi,\eta,\rho;s)\\ &:=\frac{\mathfrak{m}^{nr}(\xi-\eta-\rho,\eta+\rho)}{|\xi|-|\eta+\rho|-\iota_1|\xi-\eta-\rho|}\widetilde{\Sigma^{\geq 1}_{wa}}\Big(\rho,\frac{2\eta+\rho}{2},s\Big)\chi_0\Big(\frac{|\rho|}{|2\eta+\rho|}\Big)\\ &\quad-\frac{\mathfrak{m}^{nr}(\xi-\eta-\rho,\eta)}{|\xi-\rho|-|\eta|-\iota_1|\xi-\eta-\rho|}\widetilde{\Sigma^{\geq 1}_{wa}}\Big(\rho,\frac{2\xi-\rho}{2},s\Big)\chi_0\Big(\frac{|\rho|}{|2\xi-\rho|}\Big).\end{aligned}\tag{5.2.69}$$

**Step 4.** To prove (5.2.67) it suffices to show that

$$\Big|\int_{\mathbb{R}^9} K_{\mathfrak{m}^{nr}}(\xi,\eta,\rho;s)\widehat{U^{h_1,\iota_1}_{\leq J_1,k_1}}(\xi-\eta-\rho,s)\widehat{P_{k_2}f_2}(\eta)\overline{\widehat{P_kf}(\xi)}\,d\xi d\eta d\rho\Big| \lesssim \varepsilon_1 2^{-3m/2}\|P_kf\|_{L^2}\|P_{k_2}f_2\|_{L^2}, \tag{5.2.70}$$

for any $s\in J_m$ and $f,f_2\in L^2$. The main issue is the possible loss of derivative, so one should think of $|\xi|,|\eta|\in[2^{k-4},2^{k+4}]$ as large and $|\rho|,|\xi-\eta-\rho|\leq 2^{k-20}$ as small. For $a,b\in[0,1]$ let

$$\begin{aligned}S(\xi,\eta,\rho;s;a,b):=&\frac{\mathfrak{m}(\xi-\eta-\rho,\eta+a\rho)\varphi_{>q_0}(\Xi_{\iota_1+}(\xi-\eta-\rho,\eta+a\rho))}{|\xi-\rho+a\rho|-|\eta+a\rho|-\iota_1|\xi-\eta-\rho|}\\ &\times\widetilde{\Sigma^{\geq 1}_{wa}}\Big(\rho,\frac{2\xi-\rho}{2}-b(\xi-\eta-\rho),s\Big)\chi_0\Big(\frac{|\rho|}{|2\xi-\rho-2b(\xi-\eta-\rho)|}\Big),\end{aligned} \tag{5.2.71}$$

so $K_{\mathfrak{m}}(\xi,\eta,\rho;s)=S(\xi,\eta,\rho;s;1,1)-S(\xi,\eta,\rho;s;0,0)$. It suffices to prove that

$$\Big|\int_{\mathbb{R}^9}\nabla_{a,b}S(\xi,\eta,\rho;s;a,b)\widehat{U^{h_1,\iota_1}_{\leq J_1,k_1}}(\xi-\eta-\rho,s)\widehat{P_{k_2}f_2}(\eta)\overline{\widehat{P_kf}(\xi)}\,d\xi d\eta d\rho\Big| \lesssim \varepsilon_1 2^{-3m/2}\|P_kf\|_{L^2}\|P_{k_2}f_2\|_{L^2}, \tag{5.2.72}$$

for any $a,b\in[0,1]$.

We notice that the symbol $\Sigma^{\geq 1}_{wa}(x,\zeta)=\Sigma_{wa}(x,\zeta)-|\zeta|$ can be decomposed as a sum of symbols of the form $A_{d,l}G_{d,l}(x)\mu_{d,l}(\zeta)$, $d\geq 1$, $l\in\{1,\dots,L(d)\}$, where $G_{d,l}\in\mathcal{G}_d$ (see definition (4.1.68)), $\mu_{d,l}$ are smooth homogeneous multipliers of order 1, and the constants $A_{d,l}$ and $L(d)$ are bounded by $C^d$. Using the $L^\infty$ norms in (4.1.69) and (5.2.60), together with the general estimate (3.1.3), for (5.2.72) it suffices to bound

$$\|\mathcal{F}^{-1}M^{l,l_1,l_3}_{a,b}\|_{L^1(\mathbb{R}^9)}\lesssim 2^{m/4}2^{6l_3^+} \tag{5.2.73}$$

for any $a,b\in[0,1]$ and integers $l\geq m/20$, $l_1\in[-2\delta'm,2\delta'm]$, and $l_3\leq l-20$, where

$$\begin{aligned}&M^{l,l_1,l_3}_{a,b}(\eta,\rho,\theta)\\ &:=\nabla_{a,b}\Big\{\frac{\mathfrak{m}(\theta,\eta+a\rho)\varphi_{>q_0}(\Xi_{\iota_1+}(\theta,\eta+a\rho))}{|\theta+\eta+a\rho|-|\eta+a\rho|-\iota_1|\theta|}A(\eta+(1-b)\theta+\rho/2)\\ &\times\chi_0\Big(\frac{|\rho|}{|2\eta+2(1-b)\theta+\rho|}\Big)\Big\}\varphi_l(\eta)\varphi_{l_1}(\theta)\varphi_{l_3}(\rho).\end{aligned} \tag{5.2.74}$$

This multiplier is obtained from the expression in (5.2.71) by making the change of variables $\xi=\theta+\eta+\rho$, and $A:\mathbb{R}^3\to\mathbb{R}$ is a smooth homogeneous function of order 1.

To prove (5.2.73) we use first (3.1.48), thus

$$\frac{\varphi_{>q_0}(\Xi_{\iota_1+}(\theta,\eta+a\rho))}{|\theta+\eta+a\rho|-|\eta+a\rho|-\iota_1|\theta|} = \frac{|\theta+\eta+a\rho|+|\eta+a\rho|+\iota_1|\theta|}{-\iota_1|\theta||\eta+a\rho|}\frac{\varphi_{>q_0}(\Xi_{\iota_1+}(\theta,\eta+a\rho))}{|\Xi_{\iota_1+}(\theta,\eta+a\rho)|^2}.$$

We use now (3.1.36) and recall that $2^{-q_0} \lesssim 2^{8\delta' m}$. The bounds (5.2.73) follow by examining the terms resulting from taking the derivatives in $a$ or $b$, and recalling the algebra property $\|\mathcal{F}^{-1}(m\cdot m')\|_{L^1} \lesssim \|\mathcal{F}^{-1}(m)\|_{L^1}\|\mathcal{F}^{-1}(m')\|_{L^1}$. □

Finally, we complete the analysis in **Case 4.4** in the proof of Lemma 5.4.

**Lemma 5.8.** *If $\iota_1 \in \{+,-\}$, $t\in[0,T]$, $m\in[\delta^{-2},L]$, $\mathcal{L},\mathcal{L}_1 \in \mathcal{V}_n^q$, $n\geq 1$, and*

$$k_1 \in [-m+Y(q,n)\delta m, -0.6m], \qquad |k|,|k_2|\leq \delta' m, \qquad b\in[-2\delta m, 4], \tag{5.2.75}$$

*then, with $h,h_1,h_2 \in \{h_{\alpha\beta}\}$ and $\mathfrak{q}^b_{\iota_1\iota_2}$ defined as in (5.2.28) and (5.2.2),*

$$2^{-k_1}\Big|\int_{J_m} q_m(s)\mathcal{G}_{\mathfrak{q}^b_{\iota_1+}}[P_{k_1}U^{\mathcal{L}_1h_1,\iota_1}(s), P_{k_2}U^{h_2}(s), P_kU^{\mathcal{L}h}(s)]\,ds\Big| \lesssim \varepsilon_1^3 2^{-2N(n)k^+}2^{2H(q,n)\delta m}2^{-\delta m}.$$

*Proof.* As in the proof of Lemma 5.7 we replace first the solutions $U^{h_2}$ and $U^{\mathcal{L}h}$ with the quasilinear variables $\mathcal{U}^{h_2} = (\partial_t - iT_{\sigma_{wa}})h_2$ and $\mathcal{U}^{\mathcal{L}h} = (\partial_t - iT_{\sigma_{wa}})(\mathcal{L}h)$ defined in (5.1.41), at the expense of acceptable errors that can be estimated as in (5.2.64). It remains to prove that

$$2^{-k_1}\Big|\int_{J_m} q_m(s)\mathcal{G}_{\mathfrak{q}^b_{\iota_1+}}[P_{k_1}U^{\mathcal{L}_1h_1,\iota_1}(s), P_{k_2}\mathcal{U}^{h_2}(s), P_k\mathcal{U}^{\mathcal{L}h}(s)]\,ds\Big| \lesssim \varepsilon_1^3 2^{-2N(n)k^+}2^{2H(q,n)\delta m-\delta m}. \tag{5.2.76}$$

Then we apply the integration by parts identity (5.1.32). With $\mathcal{H}_{\mathfrak{q}^b_{\iota_1+}}$ defined as in (5.1.31), for (5.2.76) it suffices to prove that

$$\big|\mathcal{H}_{\mathfrak{q}^b_{\iota_1+}}[P_{k_1}U^{\mathcal{L}_1h_1,\iota_1}(s), P_{k_2}\mathcal{U}^{h_2}(s), P_k\mathcal{U}^{\mathcal{L}h}(s)]\big| \lesssim \varepsilon_1^3 2^{k_1}2^{-2N(n)k^+}2^{2H(q,n)\delta m-\delta m}, \tag{5.2.77}$$

$$2^m\big|\mathcal{H}_{\mathfrak{q}^b_{\iota_1+}}[(\partial_s + i\Lambda_{wa,\iota_1})P_{k_1}U^{\mathcal{L}_1h_1,\iota_1}(s), P_{k_2}\mathcal{U}^{h_2}(s), P_k\mathcal{U}^{\mathcal{L}h}(s)]\big| \lesssim \varepsilon_1^3 2^{k_1}2^{-2N(n)k^+}2^{2H(q,n)\delta m-\delta m}, \tag{5.2.78}$$

and

$$
\begin{aligned}
2^m \big| &\mathcal{H}_{\mathfrak{q}^b_{\iota_1+}}\big[P_{k_1}U^{\mathcal{L}_1 h_1,\iota_1}(s), (\partial_s + i\Lambda_{wa})P_{k_2}\mathcal{U}^{h_2}(s), P_k\mathcal{U}^{\mathcal{L}h}(s)\big] \\
&+ \mathcal{H}_{\mathfrak{q}^b_{\iota_1+}}\big[P_{k_1}U^{\mathcal{L}_1 h_1,\iota_1}(s), P_{k_2}\mathcal{U}^{h_2}(s), (\partial_s + i\Lambda_{wa})P_k\mathcal{U}^{\mathcal{L}h}(s)\big]\big| \\
&\lesssim \varepsilon_1^3 2^{k_1} 2^{-2N(n)k^+} 2^{2H(q,n)\delta m - \delta m},
\end{aligned}
\tag{5.2.79}
$$

for any $s \in J_m$. We prove these estimate in several steps.

**Step 1.** We start with the easier estimates (5.2.77) and (5.2.78). The main point is that

$$
\begin{aligned}
\|P_{k_1}U^{\mathcal{L}_1 h_1,\iota_1}(s)\|_{L^2} + 2^m\|(\partial_s + i\Lambda_{wa,\iota_1})P_{k_1}U^{\mathcal{L}_1 h_1,\iota_1}(s)\|_{L^2} \\
\lesssim \epsilon_1 2^{k_1/2} 2^{H(q,n)\delta m + Y'(n)\delta m},
\end{aligned}
\tag{5.2.80}
$$

where $Y'(1) := 2$ and $Y'(2) = Y'(3) := 35$. Indeed, these bounds follow from (3.3.3) and (4.2.3) when $n \geq 2$. If $n = 1$ they follow from (3.3.3), Lemma 4.14 (recall $k_1 \leq -0.6m$), and (4.2.43).

Using (3.1.36), (3.1.48), and (5.2.2) we have

$$
\Big\|\mathcal{F}^{-1}\Big\{\frac{\mathfrak{q}^b_{\iota_1+}(\xi-\eta,\eta)}{|\xi| - |\eta| - \iota_1|\xi-\eta|}\varphi_{kk_1k_2}(\xi-\eta,\eta)\Big\}\Big\|_{L^1(\mathbb{R}^6)} \lesssim 2^{-k_1}2^{-b}. \tag{5.2.81}
$$

With $J_2 = -k_1$ we decompose, as in (3.3.1)–(3.3.2),

$$
P_{k_2}\mathcal{U}^{h_2} = P_{k_2}U^{h_2} + P_{k_2}(\mathcal{U}^{h_2} - U^{h_2}) = U^{h_2,+}_{\leq J_2,k_2} + U^{h_2,+}_{>J_2,k_2} + P_{k_2}(\mathcal{U}^{h_2} - U^{h_2}). \tag{5.2.82}
$$

For $G(s) \in \{U^{\mathcal{L}_1 h_1,\iota_1}(s), 2^m(\partial_s + i\Lambda_{wa,\iota_1})P_{k_1}U^{\mathcal{L}_1 h_1,\iota_1}(s)\}$, we estimate, using (3.2.46) and (5.2.81),

$$
\begin{aligned}
&\big|\mathcal{H}_{\mathfrak{q}^b_{\iota_1+}}\big[G(s), U^{h_2,+}_{\leq J_2,k_2}(s), P_k\mathcal{U}^{\mathcal{L}h}(s)\big]\big| \\
&\lesssim 2^{k_1/2}2^{-m}2^{3k/2}2^{-k_1}2^{-b}\|G(s)\|_{L^2}\|\widehat{P_{k_2}U^{h_2}}(s)\|_{L^\infty}\|P_k\mathcal{U}^{\mathcal{L}h}(s)\|_{L^2} \\
&\lesssim \varepsilon_1^3 2^{k^-/2}2^{-2N(n)k^+ - 4k^+}2^{2H(q,n)\delta m}2^{-m+Y'(n)\delta m + \delta m}2^{-b},
\end{aligned}
\tag{5.2.83}
$$

using also (3.3.3), (3.3.7), (5.2.80) in the last line. We also estimate, using just $L^2$ bounds,

$$
\begin{aligned}
&\big|\mathcal{H}_{\mathfrak{q}^b_{\iota_1+}}\big[G(s), U^{h_2,+}_{>J_2,k_2}(s) + P_{k_2}(\mathcal{U}^{h_2} - U^{h_2})(s), P_k\mathcal{U}^{\mathcal{L}h}(s)\big]\big| \\
&\lesssim 2^{3k_1/2}2^{-k_1}2^{-b}\|G(s)\|_{L^2}\|P_k\mathcal{U}^{\mathcal{L}h}(s)\|_{L^2} \\
&\qquad \times \big\{\|U^{h_2,+}_{>J_2,k_2}(s)\|_{L^2} + \|P_{k_2}(\mathcal{U}^{h_2} - U^{h_2})(s)\|_{L^2}\big\} \\
&\lesssim \varepsilon_1^3 2^{2k_1}2^{2\delta' m},
\end{aligned}
\tag{5.2.84}
$$

using also (3.3.3)–(3.3.4), (5.1.42), (5.2.80) in the last line. The desired bounds

(5.2.77)–(5.2.78) follow once we notice that $-m+Y'(n)\delta m+\delta m-b\leq k_1-6\delta m$, due to (5.2.75) and (4.3.70).

**Step 2.** We consider now the estimates (5.2.79). We decompose

$$\begin{aligned}(\partial_s+i\Lambda_{wa})P_{k_2}\mathcal{U}^{h_2}&=P_{k_2}(\partial_s+iT_{\Sigma_{wa}})\mathcal{U}^{h_2}-iP_{k_2}T_{\Sigma_{wa}-|\zeta|}\mathcal{U}^{h_2},\\ P_k(\partial_s+i\Lambda_{wa})\mathcal{U}^{\mathcal{L}h}&=P_k(\partial_s+iT_{\Sigma_{wa}})\mathcal{U}^{\mathcal{L}h}-iP_kT_{\Sigma_{wa}-|\zeta|}\mathcal{U}^{\mathcal{L}h}.\end{aligned}$$

For (5.2.79) it suffices to prove that

$$\begin{aligned}2^m\big|\mathcal{H}_{\mathfrak{q}^b_{\iota_1+}}\big[P_{k_1}U^{\mathcal{L}_1h_1,\iota_1}(s),P_{k_2}(\partial_s+iT_{\Sigma_{wa}})\mathcal{U}^{h_2}(s),P_k\mathcal{U}^{\mathcal{L}h}(s)\big]\big|\\ \lesssim\varepsilon_1^32^{k_1}2^{-2N(n)k^+}2^{2H(q,n)\delta m-\delta m},\end{aligned}\tag{5.2.85}$$

$$\begin{aligned}2^m\big|\mathcal{H}_{\mathfrak{q}^b_{\iota_1+}}\big[P_{k_1}U^{\mathcal{L}_1h_1,\iota_1}(s),P_{k_2}\mathcal{U}^{h_2}(s),P_k(\partial_s+iT_{\Sigma_{wa}})\mathcal{U}^{\mathcal{L}h}(s)\big]\big|\\ \lesssim\varepsilon_1^32^{k_1}2^{-2N(n)k^+}2^{2H(q,n)\delta m-\delta m},\end{aligned}\tag{5.2.86}$$

and

$$\begin{aligned}2^m\big|&\mathcal{H}_{\mathfrak{q}^b_{\iota_1+}}\big[P_{k_1}U^{\mathcal{L}_1h_1,\iota_1}(s),P_{k_2}T_{\Sigma_{wa}-|\zeta|}\mathcal{U}^{h_2}(s),P_k\mathcal{U}^{\mathcal{L}h}(s)\big]\\ &-\mathcal{H}_{\mathfrak{q}^b_{\iota_1+}}\big[P_{k_1}U^{\mathcal{L}_1h_1,\iota_1}(s),P_{k_2}\mathcal{U}^{h_2}(s),P_kT_{\Sigma_{wa}-|\zeta|}\mathcal{U}^{\mathcal{L}h}(s)\big]\big|\\ &\lesssim\varepsilon_1^32^{k_1}2^{-2N(n)k^+}2^{2H(q,n)\delta m-\delta m}.\end{aligned}\tag{5.2.87}$$

We notice that the bounds (5.2.86) follow from (5.1.48), using the decomposition (5.2.82) and estimating as in (5.2.83)–(5.2.84).

**Step 3.** We prove now the bounds (5.2.85). We use the formulas (5.1.43) and (5.1.52) with $\mathcal{L}=Id$. The contribution of the cubic and higher order terms can be bounded easily, proceeding as in (5.2.84). To control the main contributions we will prove that

$$\begin{aligned}2^m\big|\mathcal{Q}_{\mathfrak{p}}[P_{k_1}U^{\mathcal{L}_1h_1,\iota_1}(s),P_{k_3}U^{h_3,\iota_3}(s),P_{k_4}U^{h_4,\iota_4}(s),P_k\mathcal{U}^{\mathcal{L}h}(s)]\big|\\ \lesssim\varepsilon_1^42^{2k_1}2^{-2N(n)k^+}2^{2H(q,n)\delta m-10\delta m}2^{k^-/2}2^{k_3^-/4}2^{k_4^-/2-4k_4^+}\end{aligned}\tag{5.2.88}$$

and

$$\begin{aligned}2^m\big|\mathcal{Q}_{\mathfrak{p}}[P_{k_1}U^{\mathcal{L}_1h_1,\iota_1}(s),P_{k_3}U^{\psi,\iota_3}(s),P_{k_4}U^{\psi,\iota_4}(s),P_k\mathcal{U}^{\mathcal{L}h}(s)]\big|\\ \lesssim\varepsilon_1^42^{2k_1}2^{-2N(n)k^+}2^{2H(q,n)\delta m-10\delta m}2^{k^-/2}2^{k_3^-/4}2^{k_4^-/2-4k_4^+},\end{aligned}\tag{5.2.89}$$

for any $s\in J_m$, $\iota_3,\iota_4\in\{+,-\}$, $h_3,h_4\in\{h_{\alpha\beta}\}$, $k_3\leq k_4\in\mathbb{Z}$. Here

$$\mathcal{Q}_{\mathfrak{p}}[f_1,f_3,f_4,f]:=\int_{(\mathbb{R}^3)^3}\mathfrak{p}(\xi-\eta,\eta-\rho,\rho)\widehat{f_1}(\xi-\eta)\widehat{f_3}(\eta-\rho)\widehat{f_4}(\rho)\overline{\widehat{f}(\xi)}\,d\xi d\eta d\rho,\tag{5.2.90}$$

and $\mathfrak{p}$ is a multiplier satisfying $\|\mathcal{F}^{-1}\mathfrak{p}\|_{L^1(\mathbb{R}^9)} \leq 1$. These bounds clearly suffice to prove (5.2.85); they are in fact stronger than needed because we would like to apply them in the proof of the estimates (5.2.87) as well.

**Substep 3.1.** We prove first the bounds (5.2.88). Since $k_3 \leq k_4$ we may assume that $k_4 \geq k-8$. Using (3.1.3) we estimate first the left-hand side of (5.2.88) by

$$C2^m 2^{3k_1/2}\|P_{k_1}U^{\mathcal{L}_1 h_1,\iota_1}(s)\|_{L^2}\|P_{k_3}U^{h_3,\iota_3}(s)\|_{L^\infty}\|P_{k_4}U^{h_4,\iota_4}(s)\|_{L^2}\|P_k\mathcal{U}^{\mathcal{L}h}(s)\|_{L^2}$$
$$\lesssim \varepsilon_1^4 2^{2k_1}2^{\delta' m}2^{k_3^-}2^{-N(0)k_4^+ + 2k_4^+}2^{k^-/2}2^{-N(n)k^+}, \tag{5.2.91}$$

where we used (3.3.3) and (3.3.11) in the second line. This suffices if either $k_3 \leq -8\delta' m$ or $k_4 \geq 8\delta' m$. On the other hand, if $k_3, k_4 \in [-8\delta' m, 8\delta' m]$ then we fix $J_3 = J_4$ the largest integer smaller than $m/4$ and decompose $P_{k_3}U^{h_3,\iota_3} = U^{h_3,\iota_3}_{\leq J_3,k_3} + U^{h_3,\iota_3}_{>J_3,k_3}$ and $P_{k_4}U^{h_4,\iota_4} = U^{h_4,\iota_4}_{\leq J_4,k_4} + U^{h_4,\iota_4}_{>J_4,k_4}$ as in (3.3.1)–(3.3.2). The contributions of the functions $U^{h_3,\iota_3}_{>J_3,k_3}$ and $U^{h_4,\iota_4}_{>J_4,k_4}$ can be estimated easily, using (3.3.4). After these reductions it remains to prove that

$$\begin{aligned}2^m\big|\mathcal{Q}_{\mathfrak{p}}[P_{k_1}U^{\mathcal{L}_1 h_1,\iota_1}(s), U^{h_3,\iota_3}_{\leq J_3,k_3}(s), U^{h_4,\iota_4}_{\leq J_4,k_4}(s), P_k\mathcal{U}^{\mathcal{L}h}(s)]\big| \\ \lesssim \varepsilon_1^4 2^{2k_1}2^{-2N(n)k^+}2^{2H(q,n)\delta m - 10\delta m}2^{k^-/2}2^{k_3^-/4}2^{k_4^-/2 - 4k_4^+},\end{aligned} \tag{5.2.92}$$

for any $s \in J_m$ and $k_3 \leq k_4 \in [-8\delta' m, 8\delta' m]$.

To prove (5.2.92) we examine the formula (5.2.90) and write

$$\begin{aligned}\mathcal{Q}_{\mathfrak{p}}[f_1, f_3, f_4, f] = C\int_{\mathbb{R}^9}\int_{\mathbb{R}^9} K(x,y,z)e^{-ix\cdot\rho}e^{-iy\cdot(\xi-\rho)}e^{-iz\cdot(\eta-\xi)} \\ \times\widehat{f_1}(\rho)\widehat{f_3}(\xi-\rho)\widehat{f_4}(\eta-\xi)\overline{\widehat{f}(\eta)}\,d\xi d\eta d\rho\,dxdydz\end{aligned}$$

after changes of variables, where $K = \mathcal{F}^{-1}(\mathfrak{p})$. Since $\|K\|_{L^1} \lesssim 1$, we have

$$\begin{aligned}&\big|\mathcal{Q}_{\mathfrak{p}}[f_1, f_3, f_4, f]\big| \\ &\lesssim \sup_{x,y,z\in\mathbb{R}^3}\Big|\int_{\mathbb{R}^9} e^{ix\cdot\xi}(\widehat{f_1}(\rho)e^{iy\rho})\widehat{f_3}(\xi-\rho)\widehat{f_4}(\eta-\xi)\overline{(\widehat{f}(\eta)e^{iz\cdot\eta})}\,d\xi d\eta d\rho\Big| \\ &\lesssim \sup_{y,z\in\mathbb{R}^3}\big\|I[f_1(.-y), f_3]\big\|_{L^2}\big\|I[\overline{f_4}, f(.-z)]\big\|_{L^2},\end{aligned} \tag{5.2.93}$$

where $I$ is defined as in (3.2.43) with the multiplier $m$ equal to 1.

In our case, we apply (5.2.93) and (3.2.46) to estimate the left-hand side of

(5.2.92) by

$$
\begin{aligned}
&C2^m(2^{k_1/2}2^{-m}2^{3k_3/2}\|P_{k_1}U^{\mathcal{L}_1h_1}(s)\|_{L^2}\|\widehat{P_{k_3}U^{h_3}}(s)\|_{L^\infty})\\
&\quad\times(2^{k/2}2^{-m}2^{3k_4/2}\|P_k\mathcal{U}^{\mathcal{L}h}(s)\|_{L^2}\|\widehat{P_{k_4}U^{h_4}}(s)\|_{L^\infty})\\
&\quad\lesssim \varepsilon_1^4 2^{-m}2^{2H(q,n)\delta m+2\delta m}2^{k_1}2^{-N(n)k^++2k^+}2^{-N_0k_4^++2k_4^+}2^{k^-/2}2^{k_3^-/4}2^{k_4^-/2},
\end{aligned}
$$

where we used (3.3.3) and (3.3.7) in the last line. This gives the claimed bounds (5.2.92) once we recall that $k_4 \geq k-8$ and $2^{-m+Y(q,n)\delta m} \leq 2^{k_1}$; see (5.2.75) and recall (4.3.70).

**Substep 3.2.** We prove now the bounds (5.2.89). Estimating as in (5.2.91), this is easy using the $L^\infty$ estimates (3.3.13) unless $k_3, k_4 \in [-8\delta' m, 8\delta' m]$. In this case we fix $J_3 = J_4$ the largest integer smaller than $m/4$, as before, and reduce matters to proving that

$$
\begin{aligned}
&2^m\big|\mathcal{Q}_{\mathfrak{p}}[P_{k_1}U^{\mathcal{L}_1h_1,\iota_1}(s), U^{\psi,\iota_3}_{\leq J_3,k_3}(s), U^{\psi,\iota_4}_{\leq J_4,k_4}(s), P_k\mathcal{U}^{\mathcal{L}h}(s)]\big|\\
&\quad\lesssim \varepsilon_1^4 2^{2k_1}2^{-2N(n)k^+}2^{2H(q,n)\delta m-10\delta m}2^{k^-/2}2^{k_3^-/4}2^{k_4^-/2-4k_4^+}
\end{aligned}
$$

for any $s \in J_m$ and $k_3 \leq k_4 \in [-8\delta' m, 8\delta' m]$. This follows easily using (3.3.17).

**Step 4.** Finally we prove the bounds (5.2.87). We write $\Sigma_{wa} - |\zeta| = |\zeta|\Sigma^1_{wa} + |\zeta|\Sigma^{\geq 2}_{wa}$, as in (5.1.39). The contribution of the symbol $\Sigma^{\geq 2}_{wa}$ leads to higher order terms that can be estimated using just $L^2$ bounds. To bound the main term we write, as in (5.2.68)–(5.2.69),

$$
\begin{aligned}
&\mathcal{H}_{\mathfrak{q}^b_{\iota_1+}}[P_{k_1}U^{\mathcal{L}_1h_1,\iota_1}(s), P_{k_2}T_{|\zeta|\Sigma^1_{wa}}\mathcal{U}^{h_2}(s), P_k\mathcal{U}^{\mathcal{L}h}(s)]\\
&-\mathcal{H}_{\mathfrak{q}^b_{\iota_1+}}[P_{k_1}U^{\mathcal{L}_1h_1,\iota_1}(s), P_{k_2}\mathcal{U}^{h_2}(s), P_kT_{|\zeta|\Sigma^1_{wa}}\mathcal{U}^{\mathcal{L}h}(s)]\\
&= C\int_{\mathbb{R}^9} A(\xi,\eta,\rho)P_{k_1}\widehat{U^{\mathcal{L}_1h_1,\iota_1}}(\xi-\eta-\rho,s)\widehat{P'_{k_2}\mathcal{U}^{h_2}}(\eta,s)\overline{\widehat{P'_k\mathcal{U}^{\mathcal{L}h}}(\xi,s)}\,d\xi d\eta d\rho,
\end{aligned}
\tag{5.2.94}
$$

where $P'_k = P_{[k-2,k+2]}$, $P'_{k_2} = P_{[k_2-2,k_2+2]}$, and

$$
\begin{aligned}
A(\xi,\eta,\rho) := &\frac{\mathfrak{q}^b_{\iota_1+}(\xi-\eta-\rho,\eta+\rho)\varphi_{k_2}(\eta+\rho)}{|\xi|-|\eta+\rho|-\iota_1|\xi-\eta-\rho|}\\
&\times(\widetilde{|\zeta|\Sigma^1_{wa}})\Big(\rho,\frac{2\eta+\rho}{2}\Big)\chi_0\Big(\frac{|\rho|}{|2\eta+\rho|}\Big)\varphi_k(\xi)\\
&-\frac{\mathfrak{q}^b_{\iota_1+}(\xi-\eta-\rho,\eta)\varphi_{k_2}(\eta)}{|\xi-\rho|-|\eta|-\iota_1|\xi-\eta-\rho|}\\
&\times(\widetilde{|\zeta|\Sigma^1_{wa}})\Big(\rho,\frac{2\xi-\rho}{2}\Big)\chi_0\Big(\frac{|\rho|}{|2\xi-\rho|}\Big)\varphi_k(\xi-\rho).
\end{aligned}
\tag{5.2.95}
$$

The formula (5.1.39) shows that $(\widetilde{|\zeta|\Sigma^1_{wa}})(\rho,v)$ is a sum of expressions of the form $\widehat{h_3}(\rho)g(v)$, where $h_3 \in \{h_{\alpha\beta}\}$ and $g(v)$ is either $|v|$ or $|v|\widehat{v_j}$, or $|v|\widehat{v_j}\widehat{v_k}$. As in (5.2.71), for $x, y \in [0,1]$ let

$$B(\xi,\eta,\rho;x,y) := \frac{\mathfrak{q}_{\iota_1+}(\xi-\eta-\rho,\eta+x\rho)\varphi_b(\Xi_{\iota_1+}(\xi-\eta-\rho,\eta+x\rho))}{|\xi-\rho+x\rho|-|\eta+x\rho|-\iota_1|\xi-\eta-\rho|} \times \varphi_{k_2}(\eta+x\rho)g\Big(\frac{2\xi-\rho}{2}-y(\xi-\eta-\rho)\Big) \times \chi_0\Big(\frac{|\rho|}{|2\xi-\rho-2y(\xi-\eta-\rho)|}\Big)\varphi_k(\xi-\rho+x\rho) \tag{5.2.96}$$

such that $A(\xi,\eta,\rho)$ is a sum over $h_3, g$ of expressions of the form $[B(\xi,\eta,\rho;1,1)-B(\xi,\eta,\rho;0,0)]\widehat{h_3}(\rho)$. Using these identities, for (5.2.87) it suffices to show that

$$2^m \sum_{k_3\leq k-10}\Big|\int_{\mathbb{R}^9}\nabla_{x,y}B(\xi,\eta,\rho;x,y)|\rho|^{-1}P_{k_1}\widehat{U^{\mathcal{L}_1h_1,\iota_1}}(\xi-\eta-\rho,s)P_{k_3}\widehat{U^{h_3,\iota_3}}(\rho,s) \times \widehat{P'_{k_2}\mathcal{U}^{h_2}}(\eta,s)\overline{\widehat{P'_k\mathcal{U}^{\mathcal{L}h}}(\xi,s)}\,d\xi d\eta d\rho\Big| \lesssim \varepsilon_1^3 2^{k_1}2^{-2N(n)k^+}2^{2H(q,n)\delta m-\delta m}, \tag{5.2.97}$$

for any $x, y \in [0,1]$ and $\iota_3 \in \{+,-\}$.

Let $B'(\xi,\eta,\rho) := \nabla_{x,y}B(\xi,\eta,\rho;x,y)$, for $x, y \in [0,1]$. To prove (5.2.97) we notice that

$$\big\|\mathcal{F}^{-1}\{B'(\xi,\eta,\rho)|\rho|^{-1}\varphi_{l_3}(\rho)\varphi_{l_1}(\xi-\eta-\rho)\varphi_{l_2}(\eta)\varphi_l(\xi)\}\big\|_{L^1(\mathbb{R}^9)} \lesssim 2^{-2b}(2^{-l_3}+2^{-l_1}) \tag{5.2.98}$$

for any $x, y \in [0,1]$, where $l, l_1, l_3, l_2 \in \mathbb{Z}$, $l, l_2 \in [-2\delta' m, 2\delta' m]$, $|l-l_2| \leq 4$, $l_1 \leq -0.6m+4$, $l_3 \leq l-10$. Indeed, one can think of $2^l$ and $2^{l_2}$ as large and comparable, and $2^{l_1}, 2^{l_3}$ as small. Using (3.1.48) we rewrite

$$B(\xi,\eta,\rho;x,y)|\rho|^{-1} = \frac{\mathfrak{q}_{\iota_1+}(\xi-\eta-\rho,\eta+x\rho)\varphi_b(\Xi_{\iota_1+}(\xi-\eta-\rho,\eta+x\rho))}{-\iota_1|\Xi_{\iota_1+}(\xi-\eta-\rho,\eta+x\rho)|^2} \times \frac{[|\xi-\rho+x\rho|+|\eta+x\rho|+\iota_1|\xi-\eta-\rho|]\varphi_{k_2}(\eta+x\rho)\varphi_k(\xi-\rho+x\rho)}{|\rho||\eta+x\rho|\cdot|\xi-\eta-\rho|} \times g\Big(\frac{2\xi-\rho}{2}-y(\xi-\eta-\rho)\Big)\chi_0\Big(\frac{|\rho|}{|2\xi-\rho-2y(\xi-\eta-\rho)|}\Big). \tag{5.2.99}$$

Using (3.1.36) and the double-null assumption (5.2.2), it is easy to see that

$$\big\|\mathcal{F}^{-1}\{B(\xi,\eta,\rho;x,y)|\rho|^{-1}\varphi_{l_3}(\rho)\varphi_{l_1}(\xi-\eta-\rho)\varphi_{l_2}(\eta)\varphi_l(\xi)\}\big\|_{L^1(\mathbb{R}^9)} \lesssim 2^{-b}2^l2^{-l_1-l_3} \tag{5.2.100}$$

for any $x, y \in [0,1]$ and $l, l_1, l_3, l_2$ as above. Taking $x$ derivatives generates factors $\lesssim 2^{-b}2^{l_3-l}$ from the terms in the first two lines of (5.2.99), while taking $y$ derivatives generates factors $\lesssim 2^{l_1-l}$ from the terms in the third line of (5.2.99). Combining these estimates yields (5.2.98).

We can now complete the proof of (5.2.97). In view of (5.2.98) it suffices to show that

$$\begin{aligned}2^m2^{-2b}\sum_{k_3\le k-10}(2^{-k_1}+2^{-k_3})\Big|\int_{\mathbb{R}^9}\mathfrak{p}'(\rho,\xi-\eta-\rho,\eta)P_{k_1}\widehat{U^{\mathcal{L}_1h_1,\iota_1}}(\xi-\eta-\rho,s)\\ \times P_{k_3}\widehat{U^{h_3,\iota_3}}(\rho,s)\widehat{P'_{k_2}\mathcal{U}^{h_2}}(\eta,s)\overline{\widehat{P'_k\mathcal{U}^{\mathcal{L}h}}(\xi,s)}\,d\xi d\eta d\rho\Big|\\ \lesssim \varepsilon_1^3 2^{k_1}2^{-2N(n)k^+}2^{2H(q,n)\delta m-\delta m},\end{aligned} \tag{5.2.101}$$

provided that $\mathfrak{p}'$ is a multiplier satisfying $\|\mathcal{F}^{-1}\mathfrak{p}'\|_{L^1(\mathbb{R}^9)}\le 1$. The sum over $k_3\ge k_1$ is bounded as claimed due to (5.2.88), while the sum over $k_3\le k_1$ can be estimated easily as in (5.2.91). This completes the proof of the lemma. □

## 5.3 MIXED WAVE-KLEIN-GORDON INTERACTIONS

We consider now the interactions of the metric components and the Klein-Gordon field, and prove the bounds (5.1.25)–(5.1.28) in Proposition 5.2.

We start with the semilinear estimates.

**Lemma 5.9.** *With the assumptions of Proposition 5.2, for any $m\in\{0,\dots,L+1\}$ we have*

$$\begin{aligned}\sum_{k,k_1,k_2\in\mathbb{Z}}2^{2N(n)k^+-k}2^{2\gamma(m+k^-)}\Big|\int_{J_m}q_m(s)\\ \times\mathcal{G}_{\mathfrak{m}}[P_{k_1}U^{\mathcal{L}_1\psi,\iota_1}(s),P_{k_2}U^{\mathcal{L}_2\psi,\iota_2}(s),P_kU^{\mathcal{L}h,\iota}(s)]\,ds\Big|\lesssim\varepsilon_1^3 2^{2H(q,n)\delta m},\end{aligned} \tag{5.3.1}$$

*where $\mathfrak{m}\in\mathcal{M}^*$ (see (3.2.42)) and the operators $\mathcal{G}_{\mathfrak{m}}$ are as in (5.1.30). Moreover,*

$$\begin{aligned}\sum_{k,k_1,k_2\in\mathbb{Z}}2^{2N(n)k^+}2^{k_2^+-k_1}\Big|\int_{J_m}q_m(s)\\ \times\mathcal{G}_{\mathfrak{m}}[P_{k_1}U^{\mathcal{L}_1h_1,\iota_1}(s),P_{k_2}U^{\mathcal{L}_2\psi,\iota_2}(s),P_kU^{\mathcal{L}\psi,\iota}(s)]\,ds\Big|\lesssim\varepsilon_1^3 2^{2H(q,n)\delta m}\end{aligned} \tag{5.3.2}$$

*if $n_2<n$. Therefore the bounds (5.1.25)–(5.1.26) hold.*

*Proof.* The proofs are similar to some of the proofs in section 5.2, using mainly $L^2$ or $L^\infty$ estimates on the frequency-localized solutions. In some cases we integrate by parts in time, using (5.1.32)–(5.1.33) and the bounds in Lemma 3.4 on the resulting multipliers.

Recall some of the $L^2$ estimates we proved earlier,

$$\begin{aligned}(2^{l^-}2^m)^\gamma 2^{-l/2}\|P_l U^{\mathcal{K}h}(s)\|_{L^2} &\lesssim \varepsilon_1 2^{H(\mathcal{K})\delta m}2^{-N(n')l^+},\\ \|P_l U^{\mathcal{K}\psi}(s)\|_{L^2} &\lesssim \varepsilon_1 2^{H(\mathcal{K})\delta m}2^{-N(n')l^+}\end{aligned} \tag{5.3.3}$$

and

$$\begin{aligned}2^m 2^{-l/2}\|P_l(\partial_s + i\Lambda_{wa})U^{\mathcal{K}h}(s)\|_{L^2} &\lesssim \varepsilon_1^2 2^{H(\mathcal{K})\delta m}2^{-\widetilde{N}(n')l^+ + 7l^+}\cdot 2^{35\delta m},\\ 2^m\|P_l(\partial_s + i\Lambda_{kg})U^{\mathcal{K}\psi}(s)\|_{L^2} &\lesssim \varepsilon_1^2 2^{H(\mathcal{K})\delta m}2^{-\widetilde{N}(n')l^+ + 7l^+}\cdot 2^{35\delta m},\end{aligned} \tag{5.3.4}$$

for any $\mathcal{K} \in \mathcal{V}_{n'}^{q'}$, $l \in \mathbb{Z}$, and $s \in J_m$. See (3.3.3) and (4.2.3)–(4.2.4).

Letting $\underline{k} = \min\{k_1, k_2, k_3\}$ and $\overline{k} = \max\{k_1, k_2, k_3\}$ as before we find that

$$\begin{aligned}&2^{2N(n)k^+ - k}2^{2\gamma(m+k^-)}\big|\mathcal{G}_{\mathfrak{m}}[P_{k_1}U^{\mathcal{L}_1\psi,\iota_1}(s), P_{k_2}U^{\mathcal{L}_2\psi,\iota_2}(s), P_k U^{\mathcal{L}h,\iota}(s)]\big|\\ &\lesssim \varepsilon_1^3 2^{\underline{k}}2^{\gamma(m+k^-)}2^{(\underline{k}-k)/2}2^{[H(\mathcal{L}_1)+H(\mathcal{L}_2)+H(\mathcal{L})]\delta m}2^{-N(n_1)k_1^+ - N(n_2)k_2^+ + N(n)k^+},\end{aligned} \tag{5.3.5}$$

using just the $L^2$ bounds (5.3.3), and

$$\begin{aligned}&2^{2N(n)k^+ + k_2^+ - k_1}\big|\mathcal{G}_{\mathfrak{m}}[P_{k_1}U^{\mathcal{L}_1 h_1,\iota_1}(s), P_{k_2}U^{\mathcal{L}_2\psi,\iota_2}(s), P_k U^{\mathcal{L}\psi,\iota}(s)]\big|\\ &\quad\lesssim \varepsilon_1^3 2^{\underline{k}}2^{-\gamma(k_1^- + m)}2^{(\underline{k}-k_1)/2}\\ &\quad\times 2^{[H(\mathcal{L}_1)+H(\mathcal{L}_2)+H(\mathcal{L})]\delta m}2^{-N(n_1)k_1^+ - N(n_2)k_2^+ + k_2^+ + N(n)k^+}.\end{aligned} \tag{5.3.6}$$

We prove the main estimates in several steps.

**Step 1.** We consider first the case when $\min\{n_1, n_2, n\} \geq 1$. The bounds (5.3.5)–(5.3.6) already suffice in this case to bound the contributions of the sums over the triples $(k, k_1, k_2)$ with $\min\{k, k_1, k_2\} \leq -m$ or $\max\{k_1, k_2, k_3\} \geq m/4$, due to (2.1.53). They also suffice to bound the entire sums when $|J_m| \lesssim 1$.

For the remaining contributions we integrate by parts using (5.1.32)–(5.1.33). Using also Lemma 3.4 (i) each term in the sum in the left-hand side of (5.3.1) is bounded by

$$\begin{aligned}C2^{2N(n)k^+ - k}2^{2\gamma(m+k^-)}2^{3\underline{k}/2}2^{-k}2^{2\overline{k}^+}\sup_{s\in J_m}\big\{&\big[\|P_{k_1}U^{\mathcal{L}_1\psi}(s)\|_{L^2}\\ &+ 2^m\|P_{k_1}(\partial_s + i\Lambda_{kg})U^{\mathcal{L}_1\psi}(s)\|_{L^2}\big]\|P_{k_2}U^{\mathcal{L}_2\psi}(s)\|_{L^2}\|P_k U^{\mathcal{L}h}(s)\|_{L^2}\\ &+ 2^m\|P_{k_1}U^{\mathcal{L}_1\psi}(s)\|_{L^2}\|P_{k_2}(\partial_s + i\Lambda_{kg})U^{\mathcal{L}_2\psi}(s)\|_{L^2}\|P_k U^{\mathcal{L}h}(s)\|_{L^2}\\ &+ 2^m\|P_{k_1}U^{\mathcal{L}_1\psi}(s)\|_{L^2}\|P_{k_2}U^{\mathcal{L}_2\psi}(s)\|_{L^2}\|P_k(\partial_s + i\Lambda_{wa})U^{\mathcal{L}h}(s)\|_{L^2}\big\}.\end{aligned}$$

In view of (5.3.3)–(5.3.4) this is bounded by

$$C\varepsilon_1^3 2^{3(\underline{k}-k)/2} 2^{[H(\mathcal{L}_1)+H(\mathcal{L}_2)+H(\mathcal{L})]\delta m} \times 2^{-N(n_1)k_1^+ - N(n_2)k_2^+ + N(n)k^+} 2^{36\delta m} 2^{9\max\{k^+,k_1^+,k_2^+\}}.$$

Since $n > \max\{n_1, n_2\}$ and recalling the bounds (2.1.53), this suffices to complete the proof of (5.3.1). Similarly, each term in the sum in the left-hand side of (5.3.2) is bounded by

$$C\varepsilon_1^3 2^{3(\underline{k}-k_1)/2} 2^{[H(\mathcal{L}_1)+H(\mathcal{L}_2)+H(\mathcal{L})]\delta m} 2^{-N(n_1)k_1^+ - N(n_2)k_2^+ + k_2^+ + N(n)k^+} 2^{36\delta m} 2^{9\overline{k}^+}$$

(compare with (5.3.6)) and the desired bounds (5.3.2) follow.

**Step 2.** We consider now the case when $\min\{n_1, n_2, n\} = 0$. By symmetry, the possibilities are ($n_1 = 0$, $n \geq n_2 \geq 0$) in (5.3.1) or ($n_2 = 0$, $n \geq n_1 \geq 1$) in (5.3.2). The two possibilities are similar, by changes of variables. More precisely, assume that $0 \leq q \leq n \leq 3$ and $\mathcal{L}, \mathcal{L}_2 \in \mathcal{V}_n^q$, $t \in [0, T]$, and $m \in \{0, \ldots, L+1\}$. For any $k, k_1, k_2 \in \mathbb{Z}$ and $\iota, \iota_1, \iota_2 \in \{+, -\}$ let

$$\mathcal{I}_{m;k,k_1,k_2} := \int_{J_m} q_m(s) \mathcal{G}_{\mathfrak{m}}[P_{k_1} U^{\psi,\iota_1}(s), P_{k_2} U^{\mathcal{L}_2\psi,\iota_2}(s), P_k U^{\mathcal{L}h,\iota}(s)]\, ds, \quad (5.3.7)$$

where $\|\mathcal{F}^{-1}(\mathfrak{m})\|_{L^1} \leq 1$. We will show that

$$\sum_{k,k_1,k_2 \in \mathbb{Z}} 2^{2\gamma(m+k^-)} 2^{2N(n)k^+} 2^{-k} |\mathcal{I}_{m;k,k_1,k_2}| \lesssim \varepsilon_1^3 2^{2H(q,n)\delta m} \quad (5.3.8)$$

and, if $n \geq 1$,

$$\sum_{k,k_1,k_2 \in \mathbb{Z}} 2^{k_1^+} 2^{2N(n)k_2^+} 2^{-k} |\mathcal{I}_{m;k,k_1,k_2}| \lesssim \varepsilon_1^3 2^{2H(q,n)\delta m}. \quad (5.3.9)$$

These two bounds would clearly suffice to complete the proof of (5.3.1)–(5.3.2).

Using just the $L^2$ bounds (3.3.3) and (3.3.7), we have

$$2^{-k}|\mathcal{I}_{m;k,k_1,k_2}| \lesssim \varepsilon_1^3 |J_m| 2^{2H(q,n)\delta m} 2^{\underline{k}+k_1^-} 2^{(\underline{k}-k)/2} \cdot 2^{-\gamma(m+k^-)} 2^{\kappa k_1^-} \times 2^{-N(n)k^+ - N(n)k_2^+} 2^{-(N_0-2)k_1^+}. \quad (5.3.10)$$

Using the $L^\infty$ bounds (3.3.13) on the $P_{k_1} U^{\psi,\iota_1}$ component we also have

$$2^{-k}|\mathcal{I}_{m;k,k_1,k_2}| \lesssim \varepsilon_1^3 |J_m| 2^{-k/2} 2^{-m+\delta' m/2} 2^{k_1^-/2} 2^{-\gamma(m+k^-)} \times 2^{-N(n)k^+ - N(n)k_2^+} 2^{-(N(1)-2)k_1^+}. \quad (5.3.11)$$

The bounds (5.3.8)–(5.3.9) follow if $|J_m| \lesssim 1$. Indeed, the bounds (5.3.10) suffice to estimate the contribution of the triplets $(k, k_1, k_2)$ with $|k|+|k_1|+|k_2| \geq$

$\delta' m$ (in the case $n = 0$ we use also a similar bound with the roles of $k_1$ and $k_2$ reversed). On the other hand, the contribution of the triplets $(k, k_1, k_2)$ with $|k| + |k_1| + |k_2| \leq \delta' m$ can be bounded using (5.3.11).

On the other hand, if $|J_m| \approx 2^m \gg 1$ (thus $m \in [\delta^{-1}, L]$) the bounds (5.3.10) still suffice to bound the contribution of triplets $(k, k_1, k_2)$ for which either $\underline{k} \leq -m$ or $\max\{k, k_1, k_2\} \geq 4m$. For the remaining contributions, we consider several cases.

**Step 3.** We show first that if $m \in [1/\delta, L]$ then

$$\begin{aligned} \sum_{k,k_1,k_2 \in \mathbb{Z},\, \underline{k} \geq -m,\, k \leq -0.6m} 2^{2\gamma(m+k)} 2^{-k} |\mathcal{I}_{m;k,k_1,k_2}| &\lesssim \varepsilon_1^3 2^{2H(q,n)\delta m}, \\ \sum_{k,k_1,k_2 \in \mathbb{Z},\, \underline{k} \geq -m,\, k \leq -0.6m} 2^{2N(n)k_2^+} 2^{k_1^+} 2^{-k} |\mathcal{I}_{m;k,k_1,k_2}| &\lesssim \varepsilon_1^3 2^{2H(q,n)\delta m}, \qquad n \geq 1. \end{aligned} \tag{5.3.12}$$

This is the case of small frequencies $2^k$. The estimates (5.3.10) cleary suffice to control the contribution of the triplets $(k, k_1, k_2)$ for which $k \leq -0.6m$ and $k_1 \leq -0.4m$. They also suffice to control the contribution of the triplets $(k, k_1, k_2)$ for which $k \leq -0.6m$ and $(1+\gamma)(m+k) + k_1^-(1+\kappa) - 6.5k_1^+ \leq 0$.

It remains to bound the contribution of the triplets $(k, k_1, k_2)$ for which

$$k \in [-m, -0.6m] \qquad \text{and} \qquad (m+k) + k_1^- \geq 6k_1^+. \tag{5.3.13}$$

In particular, $k_1 \geq -m/2 + 100$. Let $J_1 := k_1^- + m - 40$ and decompose $P_{k_1} U^{\psi,\iota_1} = U^{\psi,\iota_1}_{\leq J_1,k_1} + U^{\psi,\iota_1}_{>J_1,k_1}$ as in (3.3.1)–(3.3.2). Let

$$\begin{aligned} \mathcal{I}^1_{m;k,k_1,k_2} &:= \int_{J_m} q_m(s) \mathcal{G}_{\mathfrak{m}}[U^{\psi,\iota_1}_{\leq J_1,k_1}(s), P_{k_2} U^{\mathcal{L}_2\psi,\iota_2}(s), P_k U^{\mathcal{L}h,\iota}(s)]\, ds, \\ \mathcal{I}^2_{m;k,k_1,k_2} &:= \int_{J_m} q_m(s) \mathcal{G}_{\mathfrak{m}}[U^{\psi,\iota_1}_{>J_1,k_1}(s), P_{k_2} U^{\mathcal{L}_2\psi,\iota_2}(s), P_k U^{\mathcal{L}h,\iota}(s)]\, ds. \end{aligned} \tag{5.3.14}$$

Using (3.3.3) and (3.3.17) we estimate

$$\begin{aligned} 2^{-k} |\mathcal{I}^1_{m;k,k_1,k_2}| &\lesssim 2^m 2^{-k} \sup_{s \in J_m} \|P_k U^{\mathcal{L}h,\iota}(s)\|_{L^2} \|U^{\psi,\iota_1}_{\leq J_1,k_1}(s)\|_{L^\infty} \|P_{k_2} U^{\mathcal{L}_2\psi,\iota_2}(s)\|_{L^2} \\ &\lesssim \varepsilon_1^3 2^{2H(q,n)\delta m} (2^m 2^k 2^{k_1^-})^{-1/2} 2^{\kappa k_1^-/20} 2^{-\gamma(m+k)} 2^{-N(n)k_2^+} 2^{-(N_0-5)k_1^+}. \end{aligned}$$

Therefore, for $(k, k_1, k_2)$ as in (5.3.13),

$$\begin{aligned} 2^{-k} 2^{2\gamma(m+k)} |\mathcal{I}^1_{m;k,k_1,k_2}| &\lesssim \varepsilon_1^3 2^{2H(q,n)\delta m} (2^m 2^k 2^{k_1^-})^{-1/2+\gamma} 2^{-\delta' |k_1|}, \\ 2^{-k} 2^{2N(n)k_2^+} 2^{k_1^+} |\mathcal{I}^1_{m;k,k_1,k_2}| &\lesssim \varepsilon_1^3 2^{2H(q,n)\delta m} (2^m 2^k 2^{k_1^-})^{-1/2} \\ &\quad \times 2^{-\gamma(m+k)} 2^{N(n)k_2^+ - (N_0-6)k_1^+}. \end{aligned} \tag{5.3.15}$$

Similarly, using (3.3.3), (3.3.4), and $L^2$ bounds we estimate

$$\begin{aligned}2^{-k}|\mathcal{I}^2_{m;k,k_1,k_2}| &\lesssim 2^m 2^{k/2}\sup_{s\in I_m}\|P_k U^{\mathcal{L}h,\iota}(s)\|_{L^2}\|U^{\psi,\iota_1}_{>J_1,k_1}(s)\|_{L^2}\|P_{k_2}U^{\mathcal{L}_2\psi,\iota_2}(s)\|_{L^2}\\ &\lesssim \varepsilon_1^3 2^{2H(q,n)\delta m}2^{-\gamma(m+k)}2^{30\delta m}2^{k-k_1^-}2^{-N(n)k_2^+}2^{-N(1)k_1^+}.\end{aligned}$$

Therefore, for $(k,k_1,k_2)$ as in (5.3.13),

$$\begin{aligned}2^{-k}2^{2\gamma(m+k)}|\mathcal{I}^2_{m;k,k_1,k_2}| &\lesssim \varepsilon_1^3 2^{2H(q,n)\delta m}2^{\gamma(m+k)}2^{30\delta m}2^{k-k_1^-}2^{-k_1^+},\\ 2^{-k}2^{2N(n)k_2^+}2^{k_1^+}|\mathcal{I}^2_{m;k,k_1,k_2}| &\lesssim \varepsilon_1^3 2^{2H(q,n)\delta m}2^{-\gamma(m+k)}2^{30\delta m}\\ &\quad\times 2^{k-k_1^-}2^{N(n)k_2^+}2^{-N(1)k_1^++k_1^+}.\end{aligned}\tag{5.3.16}$$

It is easy to see that (5.3.15)–(5.3.16) suffice to bound the remaining contribution of the triplets $(k,k_1,k_2)$ as in (5.3.13). The desired estimates (5.3.12) follow.

**Step 4.** We show now that if $m\in[1/\delta,L]$ then

$$\begin{aligned}\sum_{k,k_1,k_2\in\mathbb{Z},\,k\geq-0.6m,\,\overline{k}\leq 8\delta' m} 2^{2\gamma(m+k^-)}2^{2N(n)k^+}2^{-k}|\mathcal{I}_{m;k,k_1,k_2}| &\lesssim \varepsilon_1^3 2^{2H(q,n)\delta m},\\ \sum_{k,k_1,k_2\in\mathbb{Z},\,k\geq-0.6m,\,\overline{k}\leq 8\delta' m} 2^{2N(n)k_2^++k_1^+}2^{-k}|\mathcal{I}_{m;k,k_1,k_2}| &\lesssim \varepsilon_1^3 2^{2H(q,n)\delta m}.\end{aligned}\tag{5.3.17}$$

For this we use normal forms; see (5.1.32)–(5.1.33). Using also Lemma 3.4 (ii) we estimate

$$\begin{aligned}2^{-k}|\mathcal{I}_{m;k,k_1,k_2}| \lesssim 2^{-2k}2^{4\overline{k}^+}\sup_{s\in J_m}&\{\|P_{k_1}U^{\psi}(s)\|_{L^\infty}\|P_{k_2}U^{\mathcal{L}_2\psi}(s)\|_{L^2}\|P_kU^{\mathcal{L}h}(s)\|_{L^2}\\ &+2^m\|P_{k_1}(\partial_s+i\Lambda_{kg})U^{\psi}(s)\|_{L^\infty}\|P_{k_2}U^{\mathcal{L}_2\psi}(s)\|_{L^2}\|P_kU^{\mathcal{L}h}(s)\|_{L^2}\\ &+2^m\|P_{k_1}U^{\psi}(s)\|_{L^\infty}\|P_{k_2}(\partial_s+i\Lambda_{kg})U^{\mathcal{L}_2\psi}(s)\|_{L^2}\|P_kU^{\mathcal{L}h}(s)\|_{L^2}\\ &+2^m\|P_{k_1}U^{\psi}(s)\|_{L^\infty}\|P_{k_2}U^{\mathcal{L}_2\psi}(s)\|_{L^2}\|P_k(\partial_s+i\Lambda_{wa})U^{\mathcal{L}h}(s)\|_{L^2}\}.\end{aligned}$$

Using the $L^2$ bounds (5.3.3)–(5.3.4) and the $L^\infty$ bounds (3.3.13) and (4.2.6) we then estimate

$$2^{-k}|\mathcal{I}_{m;k,k_1,k_2}| \lesssim \varepsilon_1^3 2^{-|k_1|/4}2^{-3k/2}2^{-m+8\delta' m}2^{-4\overline{k}^+},$$

and the bounds (5.3.17) follow.

**Step 5.** We finally bound the contribution of high frequencies: if $m\in$

$[1/\delta, L]$ then

$$\sum_{k,k_1,k_2\in\mathbb{Z},\, k\geq -0.6m,\, \overline{k}\geq 8\delta' m} 2^{2N(n)k^+}2^{2\gamma(m+k^-)}2^{-k}|\mathcal{I}_{m;k,k_1,k_2}| \lesssim \varepsilon_1^3,$$
$$\sum_{k,k_1,k_2\in\mathbb{Z},\, k\geq -0.6m,\, \overline{k}\geq 8\delta' m} 2^{2N(n)k_2^+ + k_1^+}2^{-k}|\mathcal{I}_{m;k,k_1,k_2}| \lesssim \varepsilon_1^3, \qquad n\geq 1. \tag{5.3.18}$$

Assume first that $n \leq 2$. Using (3.3.3), (3.3.11), and (3.3.13), with the lowest frequency placed in $L^\infty$, for triplets $(k, k_1, k_2)$ as in (5.3.18) we have

$$2^{-k}|\mathcal{I}_{m;k,k_1,k_2}| \lesssim \varepsilon_1^3 2^{2\delta' m}2^{-N(n)k_2^+}2^{-N(0)k_1^+} \qquad \text{if } k = \underline{k},$$
$$2^{-k}|\mathcal{I}_{m;k,k_1,k_2}| \lesssim \varepsilon_1^3 2^{2\delta' m}2^{k_1^-/2-4k_1^+}2^{-N(n)k_2^+}2^{-N(n)k^+-k/2} \qquad \text{if } k_1 = \underline{k},$$
$$2^{-k}|\mathcal{I}_{m;k,k_1,k_2}| \lesssim \varepsilon_1^3 2^{2\delta' m}2^{k_2^-/2-4k_2^+}2^{-N(n)k^+-k/2}2^{-N(0)k_1^+} \qquad \text{if } k_2 = \underline{k}.$$

These bounds suffice to prove (5.3.18), due to the gain of high derivative in all cases.

Assume now that $n = 3$. The bounds (5.3.11) suffice to bound the contribution of the triplets $(k, k_1, k_2)$ as in (5.3.18) for which $k \geq 0$. On the other hand, if $k \leq 0$ (thus $k = \underline{k}$, $k_1, k_2 \geq 8\delta' m - 6$), then we let $J_1 = m - 40$ and decompose $P_{k_1}U^{\psi,\iota_1} = U^{\psi,\iota_1}_{\leq J_1,k_1} + U^{\psi,\iota_1}_{>J_1,k_1}$ and $\mathcal{I}_{m;k,k_1,k_2} = \mathcal{I}_{m;k,k_1,k_2} + \mathcal{I}_{m;k,k_1,k_2}$ as in (5.3.14). Then we estimate

$$\begin{aligned}2^{-k}|\mathcal{I}^1_{m;k,k_1,k_2}| &\lesssim 2^m 2^{-k} \sup_{s\in J_m} \|P_k U^{\mathcal{L}h,\iota}(s)\|_{L^2}\|U^{\psi,\iota_1}_{\leq J_1,k_1}(s)\|_{L^\infty}\|P_{k_2}U^{\mathcal{L}_2\psi,\iota_2}(s)\|_{L^2}\\ &\lesssim \varepsilon_1^3 2^{2\delta' m}2^{-k/2}2^{-m/2}2^{-N(n)k_2^+}2^{-(N_0-5)k_1^+},\end{aligned}$$

using (3.3.3) and (3.3.17). Also

$$\begin{aligned}2^{-k}|\mathcal{I}^2_{m;k,k_1,k_2}| &\lesssim 2^m 2^{k/2} \sup_{s\in I_m} \|P_k U^{\mathcal{L}h,\iota}(s)\|_{L^2}\|U^{\psi,\iota_1}_{>J_1,k_1}(s)\|_{L^2}\|P_{k_2}U^{\mathcal{L}_2\psi,\iota_2}(s)\|_{L^2}\\ &\lesssim \varepsilon_1^3 2^{2\delta' m}2^{k}2^{-N(n)k_2^+}2^{-N(1)k_1^+},\end{aligned}$$

using (3.3.3)–(3.3.4). Therefore, $2^{-k}|\mathcal{I}_{m;k,k_1,k_2}| \lesssim \varepsilon_1^3 2^{2\delta' m}2^{-N(n)k_2^+}2^{-N(1)k_1^+}$ if $k \in [-0.6m, 0]$, which suffices to bound the remaining contributions over $k \leq 0$ in (5.3.18). □

We can now finally complete the proof of Proposition 5.2.

**Lemma 5.10.** *With the assumptions of Proposition 5.2, for any $m \in \{0, \dots, L+$*

1\} *we have*

$$\sum_{k,k_1,k_2\in\mathbb{Z}} 2^{N(n)k^+}(2^{N(n)k_2^+}+2^{k_2^+-k_1^+}2^{N(n)k_1^+})(2^{-k^+}+2^{-k_2^+})\Big|\int_{J_m} q_m(s) \\ \times \mathcal{G}_{\mathfrak{m}}[P_{k_1}U^{h_1,\iota_1}(s),P_{k_2}U^{\mathcal{L}_2\psi,\iota_2}(s),P_kU^{\mathcal{L}\psi,\iota}(s)]\,ds\Big| \lesssim \varepsilon_1^3 2^{2H(q,n)\delta m}, \tag{5.3.19}$$

*where* $\mathcal{L},\mathcal{L}_2\in\mathcal{V}_n^q$, $n\leq 3$, *and* $\mathfrak{m}\in\mathcal{M}^*$. *Moreover, if* $\mathfrak{n}_{\iota_1\iota_2}\in\mathcal{M}^0_{\iota_1\iota_2}$ *is a null multiplier then*

$$\sum_{k,k_1,k_2\in\mathbb{Z}} 2^{N(n)k^+}(2^{N(n)k_2^+}+2^{k_2^+-k_1^+}2^{N(n)k_1^+})\Big|\int_{J_m} q_m(s) \\ \times \mathcal{G}_{\mathfrak{n}_{\iota_1\iota_2}}[P_{k_1}U^{h_1,\iota_1}(s),P_{k_2}U^{\mathcal{L}_2\psi,\iota_2}(s),P_kU^{\mathcal{L}\psi,\iota}(s)]\,ds\Big| \lesssim \varepsilon_1^3 2^{2H(q,n)\delta m}. \tag{5.3.20}$$

*As a consequence, the bounds* (5.1.27)–(5.1.28) *hold.*

*Proof.* Using the $L^\infty$ bounds (3.3.11) and the $L^2$ bounds (3.3.3) we have the general estimates

$$\Big|\int_{J_m} q_m(s)\mathcal{G}_{\mathfrak{m}}[P_{k_1}U^{h_1,\iota_1}(s),P_{k_2}U^{\mathcal{L}_2\psi,\iota_2}(s),P_kU^{\mathcal{L}\psi,\iota}(s)]\,ds\Big| \\ \lesssim \varepsilon_1^3|J_m|2^{2H(q,n)\delta m}2^{-m+30\delta m}2^{k_1^-}2^{-N(n)k^+-N(n)k_2^+}2^{-(N(1)-2)k_1^+}. \tag{5.3.21}$$

As before, we divide the proof into several steps.

**Step 1.** We consider the contribution of large frequencies $k_1$, and show that

$$\sum_{k,k_1,k_2\in\mathbb{Z},\,k_1\geq\overline{k}-8} 2^{N(n)k^+}(2^{N(n)k_2^+}+2^{k_2^+-k_1^+}2^{N(n)k_1^+})\Big|\int_{J_m} q_m(s) \\ \times \mathcal{G}_{\mathfrak{m}}[P_{k_1}U^{h_1,\iota_1}(s),P_{k_2}U^{\mathcal{L}_2\psi,\iota_2}(s),P_kU^{\mathcal{L}\psi,\iota}(s)]\,ds\Big| \lesssim \varepsilon_1^3 2^{2H(q,n)\delta m}, \tag{5.3.22}$$

where $\overline{k}=\max\{k,k_1,k_2\}$. Clearly, $2^{N(n)k_2^+}\lesssim 2^{k_2^+-k_1^+}2^{N(n)k_1^+}$ for $(k,k_1,k_2)$ as in (5.3.22). By symmetry we may assume $k_2\leq k$ (the harder case).

If $n\leq 2$ then we can use the $L^\infty$ bounds (3.3.13) and the $L^2$ bounds (3.3.3) to estimate

$$\Big|\int_{J_m} q_m(s)\mathcal{G}_{\mathfrak{m}}[P_{k_1}U^{h_1,\iota_1}(s),P_{k_2}U^{\mathcal{L}_2\psi,\iota_2}(s),P_kU^{\mathcal{L}\psi,\iota}(s)]\,ds\Big| \\ \lesssim \varepsilon_1^3|J_m|2^{2H(q,n)\delta m}2^{k_2^-/2}2^{-4k_2^+}2^{-m+\delta' m/2}2^{k_1^+/2}2^{-N(0)k_1^+-N(n)k^+}. \tag{5.3.23}$$

The bounds (5.3.21) (used for $n = 3$) and (5.3.23) (used for $n \leq 2$) show that the contribution of the triples $(k, k_1, k_2)$ in (5.3.22) for which $\max\{|k|, |k_1|, |k_2|\} \geq 2\delta' m$ is bounded as claimed.

Assume now that $\max\{|k|, |k_1, |k_2|\} \leq 2\delta' m$. If $|J_m| \lesssim 1$ then the bounds (5.3.21) and (5.3.23) still suffice to control the contribution of these triples. On the other hand, if $m \in [\delta^{-1}, L]$ then we use normal forms (see (5.1.32)–(5.1.33)) and Lemma 3.4 (ii) to estimate

$$\begin{aligned} &\Big| \int_{J_m} q_m(s) \mathcal{G}_{\mathfrak{m}}[P_{k_1} U^{h_1,\iota_1}(s), P_{k_2} U^{\mathcal{L}_2\psi,\iota_2}(s), P_k U^{\mathcal{L}\psi,\iota}(s)]\, ds \Big| \\ &\lesssim 2^{-k_1} 2^{4\overline{k}^+} \sup_{s\in J_m} \{ \|P_{k_1} U^{h_1}(s)\|_{L^\infty} \|P_{k_2} U^{\mathcal{L}_2\psi}(s)\|_{L^2} \|P_k U^{\mathcal{L}\psi}(s)\|_{L^2} \\ &+ 2^m \|P_{k_1}(\partial_s + i\Lambda_{wa}) U^{h_1}(s)\|_{L^\infty} \|P_{k_2} U^{\mathcal{L}_2\psi}(s)\|_{L^2} \|P_k U^{\mathcal{L}\psi}(s)\|_{L^2} \\ &+ 2^m \|P_{k_1} U^{h_1}(s)\|_{L^\infty} \|P_{k_2}(\partial_s + i\Lambda_{kg}) U^{\mathcal{L}_2\psi}(s)\|_{L^2} \|P_k U^{\mathcal{L}\psi}(s)\|_{L^2} \\ &+ 2^m \|P_{k_1} U^{h_1}(s)\|_{L^\infty} \|P_{k_2} U^{\mathcal{L}_2\psi}(s)\|_{L^2} \|P_k(\partial_s + i\Lambda_{kg}) U^{\mathcal{L}\psi}(s)\|_{L^2} \}. \end{aligned} \tag{5.3.24}$$

In view of the $L^2$ bounds (3.3.3) and (4.2.4) and the $L^\infty$ bounds (3.3.11) and (4.2.5), all the terms in the right-hand side of (5.3.24) are bounded by $C2^{-m/2}$. This suffices to bound the remaining contributions in (5.3.22).

**Step 2.** We complete now the proof of (5.3.19) by showing that

$$\begin{aligned} &\sum_{k,k_1,k_2\in\mathbb{Z},\, k_1\leq \overline{k}-8} 2^{N(n)k^+} 2^{N(n)k_2^+} (2^{-k^+} + 2^{-k_2^+}) \Big| \int_{J_m} q_m(s) \\ &\times \mathcal{G}_{\mathfrak{m}}[P_{k_1} U^{h_1,\iota_1}(s), P_{k_2} U^{\mathcal{L}_2\psi,\iota_2}(s), P_k U^{\mathcal{L}\psi,\iota}(s)]\, ds \Big| \lesssim \varepsilon_1^3 2^{2H(q,n)\delta m}. \end{aligned} \tag{5.3.25}$$

Indeed, we may assume that $2^k \approx 2^{k_2}$ and use (5.3.21) to bound the contribution of the triples $(k, k_1, k_2)$ in (5.3.25) for which $\max\{|k|, |k_1, |k_2|\} \geq 2\delta' m$. For the remaining triples $(k, k_1, k_2)$ with $\max\{|k|, |k_1|, |k_2|\} \leq 2\delta' m$ we use normal forms and estimate as in (5.3.24).

**Step 3.** Finally, we complete the proof of (5.3.20) by showing that

$$\begin{aligned} &\sum_{k,k_1,k_2\in\mathbb{Z},\, k_1\leq \overline{k}-8} 2^{N(n)k^+} 2^{N(n)k_2^+} \Big| \int_{J_m} q_m(s) \\ &\times \mathcal{G}_{\mathfrak{n}_{\iota_1\iota_2}}[P_{k_1} U^{h_1,\iota_1}(s), P_{k_2} U^{\mathcal{L}_2\psi,\iota_2}(s), P_k U^{\mathcal{L}\psi,\iota}(s)]\, ds \Big| \lesssim \varepsilon_1^3 2^{2H(q,n)\delta m}. \end{aligned} \tag{5.3.26}$$

This is more subtle, essentially due to the possible loss of high derivative, and we need to exploit the null structure of the symbol $\mathfrak{n}_{\iota_1\iota_2}$. We start by estimating some of the easier contributions. Recall the coefficients $b_k(s) = b_k(q, n; s)$ defined in (4.3.20) and the time averages $b_{k,m} = b_{k,m}(q, n)$ defined in

(5.2.17). Using (3.3.11) and (4.3.22) we have

$$\begin{aligned}&\Big|\int_{J_m} q_m(s)\mathcal{G}_{\mathfrak{n}_{\iota_1\iota_2}}[P_{k_1}U^{h_1,\iota_1}(s), P_{k_2}U^{\mathcal{L}_2\psi,\iota_2}(s), P_kU^{\mathcal{L}\psi,\iota}(s)]\,ds\Big|\\&\qquad\lesssim \varepsilon_1|J_m|2^{2H(q,n)\delta m}2^{-m+30\delta m}2^{k_1^-}2^{-4k_1^+}2^{-N(n)k^+-N(n)k_2^+}b_{k,m}^2\end{aligned}\tag{5.3.27}$$

if $|k-k_2|\leq 8$. In view of (4.3.21) this suffices to control the contribution of the triplets $(k,k_1,k_2)$ in (5.3.26) for which $|k_1|\geq \delta' m$.

Assume that $|k_1|\leq \delta' m$. The bounds (5.3.27) still suffice to control the sum if $|J_m|\lesssim 1$. On the other hand, if $m\in[\delta^{-1},L]$ then we can use normal forms as in (5.3.24) to control the contribution of the triplets $(k,k_1,k_2)$ for which $\max\{k,k_2\}\leq 90\delta' m$.

After these reductions, we may assume that

$$m\in[\delta^{-1},L],\qquad |k_1|\leq\delta' m,\qquad k,k_2\geq 80\delta' m.\tag{5.3.28}$$

We can further dispose of the resonant part of the multipliers. As in (5.2.12) we decompose

$$\mathfrak{n}_{\iota_1\iota_2}=\mathfrak{n}^r_{\iota_1\iota_2}+\mathfrak{n}^{nr}_{\iota_1\iota_2},\qquad \mathfrak{n}^r_{\iota_1\iota_2}(\theta,\eta)=\varphi_{\leq q_0}(\Xi_{\iota_1\iota_2}(\theta,\eta))\mathfrak{n}_{\iota_1\iota_2}(\theta,\eta),\tag{5.3.29}$$

where $q_0:=-2\delta' m$ and $\Xi_{\iota_1\iota_2}$ is defined as in (3.1.23). In view of (3.1.30),

$$\begin{aligned}&\Big|\int_{J_m} q_m(s)\mathcal{G}_{\mathfrak{n}^r_{\iota_1\iota_2}}[P_{k_1}U^{h_1,\iota_1}(s), P_{k_2}U^{\mathcal{L}_2\psi,\iota_2}(s), P_kU^{\mathcal{L}\psi,\iota}(s)]\,ds\Big|\\&\qquad\lesssim \varepsilon_1 2^{q_0}2^{2H(q,n)\delta m}2^{-m+30\delta m}2^{k_1^-}2^{-4k_1^+}2^{-N(n)k^+-N(n)k_2^+}b_{k,m}^2,\end{aligned}\tag{5.3.30}$$

for $k,k_1,k_2$ as in (5.3.28). The key factor $2^{q_0}$ in the right-hand side is due to the nullness assumption on the multiplier $\mathfrak{n}_{\iota_1\iota_2}$. The bounds (5.3.30) suffice to control the contributions of the resonant interactions. To summarize, for (5.3.26) it remains to show that

$$\begin{aligned}&\Big|\int_{J_m} q_m(s)\mathcal{G}_{\mathfrak{n}^{nr}_{\iota_1\iota_2}}[P_{k_1}U^{h_1,\iota_1}(s), P_{k_2}U^{\mathcal{L}_2\psi,\iota_2}(s), P_kU^{\mathcal{L}\psi,\iota}(s)]\,ds\Big|\\&\qquad\lesssim \varepsilon_1 2^{-2N(n)k^+}b_{k,m}^2,\end{aligned}\tag{5.3.31}$$

provided that $m,k,k_1,k_2$ satisfy (5.3.28).

**Step 4.** The bounds (5.3.31) are similar to the bounds (5.2.57) in Lemma 5.7. Recall that

$$\begin{aligned}&\|P_{k_2}U^{\mathcal{L}_2\psi,\iota_2}(s)\|_{L^2}+\|P_kU^{\mathcal{L}\psi,\iota}(s)\|_{L^2}\lesssim b_k(s)2^{\delta' m}2^{-N(n)k},\\&2^m\|P_{k_1}U^{h_1,\iota_1}(s)\|_{L^\infty}+2^{2m}\|(\partial_s+i\Lambda_{wa,\iota_1})P_{k_1}U^{h_1,\iota_1}(s)\|_{L^\infty}\lesssim\varepsilon_1 2^{2\delta' m}\end{aligned}\tag{5.3.32}$$

(see (4.3.22)) and

$$\begin{aligned}\|(\partial_s + i\Lambda_{kg,\iota_2})P_{k_2}U^{\mathcal{L}_2\psi,\iota_2}(s)\|_{L^2} + \|(\partial_s + i\Lambda_{kg,\iota})P_kU^{\mathcal{L}\psi,\iota}(s)\|_{L^2} \\ \lesssim \varepsilon_1 b_k(s)2^{-m+\delta' m}2^{-N(n)k+k}.\end{aligned} \tag{5.3.33}$$

These bounds follow easily using $L^2 \times L^\infty$ estimates from the formulas (2.1.38) and (2.1.16).

To prove (5.3.31) it is convenient for us to apply the wave operator $\Lambda_{wa}$ to the Klein-Gordon variables $U^{\mathcal{L}_2\psi}$ and $U^{\mathcal{L}\psi}$, instead of the more natural Klein-Gordon operator $\Lambda_{kg}$. The reason is to be able to reuse some of the estimates in Lemma 5.7, such as (5.2.70), and also use some of the bounds in Lemma 3.6 that do not involve derivative loss, such as (3.1.36). We notice that the two operators are not too different at high frequency,

$$\|(\Lambda_{wa} - \Lambda_{kg})P_l f\|_{L^2} \lesssim 2^{-l}\|P_l f\|_{L^2}, \qquad l \in \{k, k_2\}. \tag{5.3.34}$$

As in **Step 1** in the proof of Lemma 5.7, the estimates (5.3.31) follow easily if $\iota = -\iota_2$, using normal forms and the ellipticity of the phase $\Lambda_{wa,\iota}(\xi) - \Lambda_{wa,\iota_2}(\eta) - \Lambda_{\iota_1}(\xi - \eta)$. On the other hand, if $\iota = \iota_2$ then we may assume that $\iota = \iota = +$ and replace the variables $U^{\mathcal{L}_2\psi,\iota_2}$ and $U^{\mathcal{L}\psi,\iota}$ with the quasilinear variables $\mathcal{U}^{\mathcal{L}_2\psi}$ and $\mathcal{U}^{\mathcal{L}\psi}$, at the expense of acceptable errors (as bounded in (5.1.42)). We then apply normal forms and notice that

$$\Big\|\mathcal{F}^{-1}\Big\{\frac{\mathfrak{n}^{nr}_{\iota_1+}(\xi-\eta,\eta)}{|\xi| - |\eta| - \iota_1|\xi - \eta|}\varphi_{kk_1k_2}(\xi - \eta, \eta)\Big\}\Big\|_{L^1(\mathbb{R}^6)} \lesssim 2^{-k_1}2^{-3q_0} \lesssim 2^{8\delta' m}, \tag{5.3.35}$$

as a consequence of (3.1.36) and (3.1.48) (notice that there is no high derivative loss in these bounds). Therefore we can estimate the first two terms in (5.1.33) using just (5.3.32).

To control the remaining two terms and prove (5.3.31) it remains to show that

$$\begin{aligned}\Big|\mathcal{H}_{\mathfrak{n}^{nr}_{\iota_1+}}\big[P_{k_1}U^{h_1,\iota_1}(s), (\partial_s + i\Lambda_{wa})P_{k_2}\mathcal{U}^{\mathcal{L}_2\psi}(s), P_k\mathcal{U}^{\mathcal{L}\psi}(s)\big] \\ + \mathcal{H}_{\mathfrak{n}^{nr}_{\iota_1+}}\big[P_{k_1}U^{h_1,\iota_1}(s), P_{k_2}\mathcal{U}^{\mathcal{L}_2\psi}(s), (\partial_s + i\Lambda_{wa})P_k\mathcal{U}^{\mathcal{L}\psi}(s)\big]\Big| \\ \lesssim \varepsilon_1 2^{-2N(n)k}2^{-m}b_k(s)^2,\end{aligned} \tag{5.3.36}$$

for any $s \in J_m$, where the operators $\mathcal{H}_{\mathfrak{n}^{nr}_{\iota_1+}}$ are defined in (5.1.31). We decompose

$$(\partial_s + i\Lambda_{wa})P_k\mathcal{U}^{\mathcal{L}\psi} = (\partial_s + iT_{\Sigma_{kg}})P_k\mathcal{U}^{\mathcal{L}\psi} - iT_{\Sigma_{kg}-\Sigma_{wa}}P_k\mathcal{U}^{\mathcal{L}\psi} - iT_{\Sigma^{\geq 1}_{wa}}P_k\mathcal{U}^{\mathcal{L}\psi},$$

where $\Sigma_{wa}, \Sigma_{kg}$ are defined in (5.1.37)-(5.1.38) and $\Sigma^{\geq 1}_{wa}(x,\zeta) = \Sigma_{wa}(x,\zeta) - |\zeta|$. Notice that $\|\Sigma_{kg} - \Sigma_{wa}\|_{\mathcal{L}^\infty_{-1}} \lesssim 1$; see the definition (3.1.52). Therefore, using

(3.1.54) and (5.1.49) we have

$$\|(\partial_s + iT_{\Sigma_{kg}})P_k\mathcal{U}^{\mathcal{L}\psi}(s)\|_{L^2} \lesssim \varepsilon_1 2^{-m+\delta' m}2^{-N(n)k}b_k(s),$$
$$\|T_{\Sigma_{kg}-\Sigma_{wa}}P_k\mathcal{U}^{\mathcal{L}\psi}\|_{L^2} \lesssim 2^{\delta' m}2^{-N(n)k-k}b_k(s),$$

for any $s \in J_m$. Therefore, recalling that $k \geq 80\delta' m$ and the bounds (5.3.32), (5.3.35), these contributions to the second term in the left-hand side of (5.3.36) can be bounded as claimed. Similarly, we can replace $(\partial_s + i\Lambda_{wa})P_{k_2}\mathcal{U}^{\mathcal{L}_2\psi}$ with $-iT_{\Sigma_{wa}^{\geq 1}}P_{k_2}\mathcal{U}^{\mathcal{L}_2\psi}$ in the first term, at the expense of acceptable errors. For (5.3.36) it remains to prove that, for any $s \in J_m$,

$$\begin{aligned}&\Big|\mathcal{H}_{\mathfrak{n}^{nr}_{\iota_1+}}\left[P_{k_1}U^{h_1,\iota_1}(s), iT_{\Sigma_{wa}^{\geq 1}}P_{k_2}\mathcal{U}^{\mathcal{L}_2\psi}(s), P_k\mathcal{U}^{\mathcal{L}\psi}(s)\right]\\ &+\mathcal{H}_{\mathfrak{n}^{nr}_{\iota_1+}}\left[P_{k_1}U^{h_1,\iota_1}(s), P_{k_2}\mathcal{U}^{\mathcal{L}_2\psi}(s), iT_{\Sigma_{wa}^{\geq 1}}P_k\mathcal{U}^{\mathcal{L}\psi}(s)\right]\Big|\\ &\quad\lesssim \varepsilon_1 2^{-2N(n)k}2^{-m}b_k(s)^2.\end{aligned} \tag{5.3.37}$$

The bounds (5.3.37) are similar to the bounds (5.2.67) (with $J_1 = \infty$). They follow by the same argument, first rewriting the expression as in (5.2.68)–(5.2.69), and then using the general bounds (5.2.70). This completes the proof of (5.3.37), and the desired bounds (5.3.20) follow.

**Step 5.** Finally we prove the bounds (5.1.27)–(5.1.28). Notice that

$$\frac{\rho_j}{|\rho|} - \frac{\rho_j}{\langle\rho\rangle} = \frac{\rho_j}{|\rho|}\frac{1}{\langle\rho\rangle(|\rho|+\langle\rho\rangle)}, \tag{5.3.38}$$

for any $\rho \in \mathbb{R}^3$ and $j \in \{1,2,3\}$. We examine now the multipliers $p^{G,kg}_{\iota_1,\iota_2,\iota}$ in the formulas (5.1.12)–(5.1.15), and compare them with the multipliers $p^{G,wa}_{\iota_1,\iota_2,\iota}$ in (5.1.8)–(5.1.11). The multipliers $p^{G,wa}_{\iota_1,\iota_2,\iota}$ are null in the variables $\xi-\eta$ and $\eta$, as shown in the proof of Proposition 5.1 in section 5.1, and the difference between $p^{G,kg}_{\iota_1,\iota_2,\iota} - p^{G,wa}_{\iota_1,\iota_2,\iota}$ gains at least one derivative either in $\xi$ or in $\eta$, due to (5.3.38). The bounds (5.1.28) follow from (5.3.19)–(5.3.20).

To prove (5.1.27) we use identities similar to (5.2.4)–(5.2.6),

$$\begin{aligned}&P^n_{kg}[\widetilde{g}^{\mu\nu}_{\geq 1}\partial_\mu\partial_\nu\mathcal{L}\psi + h_{00}\mathcal{L}\psi] - [\widetilde{g}^{\mu\nu}_{\geq 1}\partial_\mu\partial_\nu(P^n_{kg}\mathcal{L}\psi) + h_{00}P^n_{kg}\mathcal{L}\psi]\\ &\quad= \sum_{G\in\{F,\underline{F},\omega_n,\vartheta_{mn}\}}\ \sum_{\iota_1,\iota_2\in\{+,-\}}\{P^n_{kg}I_{\mathfrak{q}^{G,kg}_{\iota_1\iota_2}}[|\nabla|^{-1}U^{G,\iota_1},\langle\nabla\rangle U^{\mathcal{L}\psi,\iota_2}]\\ &\qquad\qquad - I_{\mathfrak{q}^{G,kg}_{\iota_1\iota_2}}[|\nabla|^{-1}U^{G,\iota_1}, P^n_{kg}\langle\nabla\rangle U^{\mathcal{L}\psi,\iota_2}]\}\\ &\quad - 2\{P^n_{kg}[R_jR_0\tau\cdot\partial_j\partial_0\mathcal{L}\psi] - R_jR_0\tau\cdot\partial_j\partial_0(P^n_{kg}\mathcal{L}\psi)\}\\ &\quad + \sum_{(\mu,\nu)\neq(0,0)}\{P^n_{kg}[\widetilde{G}^{\mu\nu}_{\geq 2}\cdot\partial_\mu\partial_\nu\mathcal{L}\psi] - \widetilde{G}^{\mu\nu}_{\geq 2}\cdot\partial_\mu\partial_\nu(P^n_{kg}\mathcal{L}\psi)\},\end{aligned}$$

where the null multipliers $\mathfrak{q}^{G,kg}_{\iota_1\iota_2}$ are defined in (4.3.60) and, as in (5.2.4), $\widetilde{G}^{\mu\nu}_{\geq 2}$ are linear combinations of expressions of the form $R^a|\nabla|^{-1}(G_{\geq 1}\partial_\rho h)$. The contributions of the cubic and higher order terms in the last two lines can be bounded as in (5.2.7).

To estimate the main terms we decompose, for $G \in \{F, \underline{F}, \omega_n, \vartheta_{mn}\}$,

$$\begin{aligned} &P_{kg}^n I_{\mathfrak{q}_{\iota_1\iota_2}^{G,kg}}[|\nabla|^{-1}U^{G,\iota_1}, \langle\nabla\rangle U^{\mathcal{L}\psi,\iota_2}] - I_{\mathfrak{q}_{\iota_1\iota_2}^{G,kg}}[|\nabla|^{-1}U^{G,\iota_1}, P_{kg}^n\langle\nabla\rangle U^{\mathcal{L}\psi,\iota_2}] \\ &= P_{kg}^n I_{\mathfrak{m}_{\iota_1\iota_2}^{1,G}}[U^{G,\iota_1}, U^{\mathcal{L}\psi,\iota_2}] + P_{kg}^n I_{\mathfrak{m}_{\iota_1\iota_2}^{2,G}}[U^{G,\iota_1}, U^{\mathcal{L}\psi,\iota_2}] \\ &+ I_{\mathfrak{m}_{\iota_1\iota_2}^{3,G}}[U^{G,\iota_1}, P_{kg}^n U^{\mathcal{L}\psi,\iota_2}], \end{aligned} \tag{5.3.39}$$

where

$$\begin{aligned} \mathfrak{m}_{\iota_1\iota_2}^{1,G}(\theta,\eta) &:= \varphi_{\leq 0}(\eta)\mathfrak{q}_{\iota_1\iota_2}^{G,kg}(\theta,\eta)\frac{P_{kg}^n(\eta+\theta) - P_{kg}^n(\eta)}{P_{kg}^n(\eta+\theta)}\frac{\langle\eta\rangle}{|\theta|}, \\ \mathfrak{m}_{\iota_1\iota_2}^{2,G}(\theta,\eta) &:= \varphi_{>1}(\eta)\mathfrak{q}_{\iota_1\iota_2}^{G,kg}(\theta,\eta)\Big\{\varphi_{>-4}(|\theta|/|\eta|)\frac{\langle\eta\rangle}{|\theta|} \\ &\qquad + \varphi_{\leq -4}(|\theta|/|\eta|)\frac{P_{kg}^n(\eta+\theta) - P_{kg}^n(\eta)}{P_{kg}^n(\eta+\theta)}\frac{\langle\eta\rangle}{|\theta|}\Big\}, \\ \mathfrak{m}_{\iota_1\iota_2}^{3,G}(\theta,\eta) &:= -\varphi_{>1}(\eta)\mathfrak{q}_{\iota_1\iota_2}^{G,kg}(\theta,\eta)\varphi_{>-4}(|\theta|/|\eta|)\frac{\langle\eta\rangle}{|\theta|}. \end{aligned} \tag{5.3.40}$$

It is easy to see that, for $a \in \{1,2,3\}$ and $G \in \{F, \underline{F}, \omega_n, \vartheta_{mn}\}$

$$\|\mathcal{F}^{-1}\{\mathfrak{m}_{\iota_1\iota_2}^{a,G}(\theta,\eta)\varphi_{kk_1k_2}(\theta,\eta)\}\|_{L^1(\mathbb{R}^6)} \lesssim \min\{1, 2^{k_2^+ - k_1^+}\} \quad \text{for any } k, k_1, k_2 \in \mathbb{Z}.$$

Therefore the contribution of the multipliers $\mathfrak{m}_{\iota_1\iota_2}^{1,G}$ can be controlled using just (5.3.19). To bound the contributions of the multipliers $\mathfrak{m}_{\iota_1\iota_2}^{2,G}$ and $\mathfrak{m}_{\iota_1\iota_2}^{3,G}$ we replace first the symbols $\mathfrak{q}_{\iota_1\iota_2}^{G,kg}$ with $\mathfrak{q}_{\iota_1\iota_2}^{G,wa}$, and gain a factor of $\langle\eta\rangle^{-1}$ (compare (4.3.60) and (4.3.59), and use (5.3.38)). After this substitution we can thus use (5.3.19) to estimate the contributions of the smoothing components, and (5.3.20) for the contributions of the null components associated to the null symbols $\mathfrak{q}_{\iota_1\iota_2}^{G,wa}$. This completes the proof of the bounds (5.1.27)–(5.1.28). □

# Chapter Six

## Improved Profile Bounds

### 6.1 WEIGHTED BOUNDS

In this section we prove the main bounds (2.1.51). Recall the identities (2.1.38). We also need a key identity that connects the vector-fields $\Gamma_l$ with weighted norms on profiles.

**Lemma 6.1.** *Assume* $\mu \in \{wa, kg\}$ *and*

$$(\partial_t + i\Lambda_\mu)U = \mathcal{N}, \tag{6.1.1}$$

*on* $\mathbb{R}^3 \times [0, T]$. *If* $V(t) = e^{it\Lambda_\mu}U(t)$ *and* $l \in \{1, 2, 3\}$ *then, for any* $t \in [0, T]$,

$$\widehat{\Gamma_l U}(\xi, t) = i(\partial_{\xi_l}\widehat{\mathcal{N}})(\xi, t) + e^{-it\Lambda_\mu(\xi)}\partial_{\xi_l}[\Lambda_\mu(\xi)\widehat{V}(\xi, t)]. \tag{6.1.2}$$

*Proof.* We calculate

$$\begin{aligned}
\widehat{\Gamma_l U}(\xi, t) &= \mathcal{F}\{x_l\partial_t U + t\partial_l U\}(\xi, t) \\
&= i(\partial_{\xi_l}\widehat{\mathcal{N}})(\xi, t) + \partial_{\xi_l}[\Lambda_\mu(\xi)\widehat{U}(\xi, t)] + it\xi_l\widehat{U}(\xi, t) \\
&= i(\partial_{\xi_l}\widehat{\mathcal{N}})(\xi, t) + e^{-it\Lambda_\mu(\xi)}\partial_{\xi_l}[\Lambda_\mu(\xi)\widehat{V}(\xi, t)] \\
&\quad - it(\partial_{\xi_l}\Lambda_\mu)(\xi)e^{-it\Lambda_\mu(\xi)}\Lambda_\mu(\xi)\widehat{V}(\xi, t) + it\xi_l\widehat{U}(\xi, t).
\end{aligned}$$

This gives (6.1.2) since $(\partial_{\xi_l}\Lambda_\mu)(\xi)\Lambda_\mu(\xi) = \xi_l$. □

Our main result in this section is the following proposition:

**Proposition 6.2.** *With the hypothesis in Proposition 2.3, for any* $t \in [0, T]$, $\mathcal{L} \in \mathcal{V}_n^q$, $n \leq 2$, $l \in \{1, 2, 3\}$, *and* $k \in \mathbb{Z}$ *we have*

$$\begin{aligned}
2^{k/2}(2^{k^-}\langle t\rangle)^\gamma \|P_k(x_l V^{\mathcal{L}h_{\alpha\beta}})(t)\|_{L^2} + 2^{k^+}\|P_k(x_l V^{\mathcal{L}\psi})(t)\|_{L^2} \\
\lesssim \varepsilon_0 \langle t\rangle^{H(q+1,n+1)\delta} 2^{-N(n+1)k^+}.
\end{aligned} \tag{6.1.3}$$

*Proof.* The identities (6.1.2) and (2.1.38) for $U^{\mathcal{L}h_{\alpha\beta}}$ give

$$\Gamma_l \widehat{U^{\mathcal{L}h_{\alpha\beta}}}(\xi, t) = i(\partial_{\xi_l}\widehat{\mathcal{L}\mathcal{N}^h_{\alpha\beta}})(\xi, t) + e^{-it\Lambda_{wa}(\xi)}\partial_{\xi_l}[\Lambda_{wa}(\xi)\widehat{V^{\mathcal{L}h_{\alpha\beta}}}(\xi, t)],$$

for $l \in \{1,2,3\}$. Therefore

$$e^{-it\Lambda_{wa}(\xi)}\Lambda_{wa}(\xi)(\partial_{\xi_l}\widehat{V^{\mathcal{L}h_{\alpha\beta}}})(\xi) = \widehat{\Gamma_l U^{\mathcal{L}h_{\alpha\beta}}}(\xi) - i(\partial_{\xi_l}\widehat{\mathcal{L}\mathcal{N}^h_{\alpha\beta}})(\xi) \\ - e^{-it\Lambda_{wa}(\xi)}(\xi_l/|\xi|)\widehat{V^{\mathcal{L}h_{\alpha\beta}}}(\xi).$$

We multiply all the terms by $2^{-k/2}(2^{k^-}\langle t\rangle)^\gamma\varphi_k(\xi)$ and take $L^2$ norms, so

$$\begin{aligned} 2^{k/2}(2^{k^-}\langle t\rangle)^\gamma\|\varphi_k(\xi)(\partial_{\xi_l}\widehat{V^{\mathcal{L}h_{\alpha\beta}}})(\xi)\|_{L^2_\xi} &\lesssim 2^{-k/2}(2^{k^-}\langle t\rangle)^\gamma\|\varphi_k(\xi)\widehat{\Gamma_l U^{\mathcal{L}h_{\alpha\beta}}}(\xi)\|_{L^2_\xi} \\ &+ 2^{-k/2}(2^{k^-}\langle t\rangle)^\gamma\|\varphi_k(\xi)(\partial_{\xi_l}\widehat{\mathcal{N}^{\mathcal{L}h_{\alpha\beta}}})(\xi)\|_{L^2_\xi} \\ &+ 2^{-k/2}(2^{k^-}\langle t\rangle)^\gamma\|\varphi_k(\xi)\widehat{V^{\mathcal{L}h_{\alpha\beta}}}(\xi)\|_{L^2_\xi}. \end{aligned} \tag{6.1.4}$$

To control the first term in the left-hand side of (6.1.3) it suffices to prove that

$$2^{-k/2}(2^{k^-}\langle t\rangle)^\gamma\|\varphi_k(\xi)(\partial_{\xi_l}\widehat{\mathcal{L}\mathcal{N}^h_{\alpha\beta}})(\xi)\|_{L^2_\xi} \lesssim \varepsilon_1^2\langle t\rangle^{H(q+1,n+1)\delta}2^{-N(n+1)k^+} \tag{6.1.5}$$

and

$$\begin{aligned} 2^{-k/2}(2^{k^-}\langle t\rangle)^\gamma\|\varphi_k(\xi)\widehat{\Gamma_l U^{\mathcal{L}h_{\alpha\beta}}}(\xi)\|_{L^2_\xi} &\lesssim \varepsilon_0\langle t\rangle^{H(q+1,n+1)\delta}2^{-N(n+1)k^+}, \\ 2^{-k/2}(2^{k^-}\langle t\rangle)^\gamma\|\varphi_k(\xi)\widehat{V^{\mathcal{L}h_{\alpha\beta}}}(\xi)\|_{L^2_\xi} &\lesssim \varepsilon_0\langle t\rangle^{H(q+1,n+1)\delta}2^{-N(n+1)k^+}. \end{aligned} \tag{6.1.6}$$

The bounds (6.1.6) follow from the main improved energy estimates in Proposition 5.1, and the commutation identities

$$\Gamma_l U^{\mathcal{L}h_{\alpha\beta}} - U^{\Gamma_l\mathcal{L}h_{\alpha\beta}} = -i\partial_l\Lambda_{wa}^{-1}U^{\mathcal{L}h_{\alpha\beta}}. \tag{6.1.7}$$

The nonlinear estimates (6.1.5) follow from (4.2.5) and the observation $H(q,n)+\widetilde{\ell}(n) \leq H(q+1,n+1)-4$ (see (2.1.49)).

The inequality for the Klein-Gordon component in (6.1.3) follows similarly, using the identity (6.1.2) for $\mu = kg$, the improved energy estimates in Proposition 5.1, and the nonlinear bounds

$$\|\varphi_k(\xi)(\partial_{\xi_l}\widehat{\mathcal{L}\mathcal{N}^\psi})(\xi)\|_{L^2_\xi} \lesssim \varepsilon_1^2\langle t\rangle^{H(q+1,n+1)\delta}2^{-N(n+1)k^+}. \tag{6.1.8}$$

These nonlinear bounds follow from (4.2.6). □

## 6.2 $Z$-NORM CONTROL OF THE KLEIN-GORDON FIELD

In this section we prove the bounds (2.1.52) for the Klein-Gordon field. We notice that, unlike the energy norms, the $Z$ norm of the Klein-Gordon profile

$V^\psi$ is not allowed to grow slowly in time. Because of this we need to renormalize the profile $V^\psi$.

### 6.2.1 Renormalization

We start from the equation $\partial_t V^\psi = e^{it\Lambda_{kg}}\mathcal{N}^\psi$ in (2.1.39), where $\mathcal{N}^\psi$ is the nonlinearity defined in (2.1.4). To extract the nonlinear phase correction we examine only the quadratic part of the nonlinearity, which is (see (2.1.17))

$$\mathcal{N}^{\psi,2} := -h_{00}(\Delta\psi - \psi) + 2h_{0j}\partial_0\partial_j\psi - h_{jk}\partial_j\partial_k\psi. \tag{6.2.1}$$

The formulas in the second line of (2.1.36) show that

$$-\widehat{\partial_j\partial_k\psi}(\rho) = \rho_j\rho_k \frac{i\widehat{U^{\psi,+}}(\rho) - i\widehat{U^{\psi,-}}(\rho)}{2\Lambda_{kg}(\rho)}, \qquad \widehat{\partial_0\partial_j\psi}(\rho) = i\rho_j \frac{\widehat{U^{\psi,+}}(\rho) + \widehat{U^{\psi,-}}(\rho)}{2}.$$

Therefore, using also the identitities $\widehat{U^{\psi,\pm}}(\rho,t) = e^{\mp it\Lambda_{kg}(\rho)}\widehat{V^{\psi,\pm}}(\rho,t)$,

$$\begin{aligned} &e^{it\Lambda_{kg}(\xi)}\widehat{\mathcal{N}^{\psi,2}}(\xi,t) \\ &= \frac{1}{(2\pi)^3}\sum_{\pm}\int_{\mathbb{R}^3} ie^{it\Lambda_{kg}(\xi)}e^{\mp it\Lambda_{kg}(\xi-\eta)}\widehat{V^{\psi,\pm}}(\xi-\eta,t)\mathfrak{q}_{kg,\pm}(\xi-\eta,\eta,t)\,d\eta, \end{aligned} \tag{6.2.2}$$

where

$$\mathfrak{q}_{kg,\pm}(\rho,\eta,t) := \pm\widehat{h_{00}}(\eta,t)\frac{\Lambda_{kg}(\rho)}{2} + \widehat{h_{0j}}(\eta,t)\rho_j \pm \widehat{h_{jk}}(\eta,t)\frac{\rho_j\rho_k}{2\Lambda_{kg}(\rho)}. \tag{6.2.3}$$

We would like to eliminate the resonant bilinear interaction between $h_{\alpha\beta}$ and $V^{\psi,+}$ in (6.2.2) corresponding to $|\eta| \ll 1$. For this we define the Klein-Gordon phase correction (justified heuristically by the approximate formulas (2.2.25))

$$\begin{aligned} \Theta_{kg}(\xi,t) := \int_0^t \Big\{ & h_{00}^{low}(s\xi/\Lambda_{kg}(\xi),s)\frac{\Lambda_{kg}(\xi)}{2} \\ &+ h_{0j}^{low}(s\xi/\Lambda_{kg}(\xi),s)\xi_j + h_{jk}^{low}(s\xi/\Lambda_{kg}(\xi),s)\frac{\xi_j\xi_k}{2\Lambda_{kg}(\xi)}\Big\}\,ds, \end{aligned} \tag{6.2.4}$$

where, for any $h \in \{h_{\alpha\beta}\}$ the low frequency component $h^{low}$ is defined by

$$\widehat{h^{low}}(\rho,s) := \varphi_{\le 0}(\langle s\rangle^{p_0}\rho)\widehat{h}(\rho,s), \qquad p_0 := 0.68. \tag{6.2.5}$$

The reason for this choice of $p_0$, slightly bigger than $2/3$, will become clear later, in the proof of Lemma 6.7. Finally, we define the nonlinear (modified)

Klein-Gordon profile $V_*^\psi$ by

$$\widehat{V_*^\psi}(\xi,t) := e^{-i\Theta_{kg}(\xi,t)}\widehat{V^\psi}(\xi,t). \tag{6.2.6}$$

Recall that the functions $h^{low}$ are real-valued, thus $\Theta_{kg}$ is real-valued. Let $h^{high}_{\alpha\beta} := h_{\alpha\beta} - h^{low}_{\alpha\beta}$ and recall the definitions (6.2.3). For $X \in \{low,\, high\}$ let

$$\mathfrak{q}^X_{kg,\pm}(\rho,\eta,t) := \pm\widehat{h^X_{00}}(\eta,t)\frac{\Lambda_{kg}(\rho)}{2} + \widehat{h^X_{0j}}(\eta,t)\rho_j \pm \widehat{h^X_{jk}}(\eta,t)\frac{\rho_j\rho_k}{2\Lambda_{kg}(\rho)}. \tag{6.2.7}$$

The formula (6.2.2) and the equation $\partial_t V^\psi = e^{it\Lambda_{kg}}\mathcal{N}^\psi$ show that

$$\partial_t\widehat{V_*^\psi}(\xi,t) = e^{-i\Theta_{kg}(\xi,t)}\{\partial_t\widehat{V^\psi}(\xi,t) - i\widehat{V^\psi}(\xi,t)\dot{\Theta}_{kg}(\xi,t)\} = \sum_{a=1}^{4}\mathcal{R}^\psi_a(\xi,t), \tag{6.2.8}$$

where $\dot{\Theta}_{kg}(\xi,t) := (\partial_t\Theta_{kg})(\xi,t)$,

$$\begin{aligned}\mathcal{R}^\psi_1(\xi,t) := \frac{e^{-i\Theta_{kg}(\xi,t)}}{(2\pi)^3}\int_{\mathbb{R}^3} & ie^{it\Lambda_{kg}(\xi)}e^{it\Lambda_{kg}(\xi-\eta)}\widehat{V^{\psi,-}}(\xi-\eta,t)\\ & \times \mathfrak{q}^{low}_{kg,-}(\xi-\eta,\eta,t)\,d\eta,\end{aligned} \tag{6.2.9}$$

$$\begin{aligned}\mathcal{R}^\psi_2(\xi,t) := \frac{e^{-i\Theta_{kg}(\xi,t)}}{(2\pi)^3}\int_{\mathbb{R}^3} & i\{e^{it(\Lambda_{kg}(\xi)-\Lambda_{kg}(\xi-\eta))}\widehat{V^\psi}(\xi-\eta,t)\mathfrak{q}^{low}_{kg,+}(\xi-\eta,\eta,t)\\ & - e^{it(\xi\cdot\eta)/\Lambda_{kg}(\xi)}\widehat{V^\psi}(\xi,t)\mathfrak{q}^{low}_{kg,+}(\xi,\eta,t)\}\,d\eta,\end{aligned} \tag{6.2.10}$$

$$\begin{aligned}\mathcal{R}^\psi_3(\xi,t) := \frac{e^{-i\Theta_{kg}(\xi,t)}}{(2\pi)^3}\sum_{\iota\in\{+,-\}}\int_{\mathbb{R}^3} & ie^{it\Lambda_{kg}(\xi)}e^{-it\Lambda_{kg,\iota}(\xi-\eta)}\widehat{V^{\psi,\iota}}(\xi-\eta,t)\\ & \times \mathfrak{q}^{high}_{kg,\iota}(\xi-\eta,\eta,t)\,d\eta,\end{aligned} \tag{6.2.11}$$

and

$$\mathcal{R}^\psi_4(\xi,t) := e^{-i\Theta_{kg}(\xi,t)}e^{it\Lambda_{kg}(\xi)}[\widehat{\mathcal{N}^\psi}(\xi,t) - \widehat{\mathcal{N}^{\psi,2}}(\xi,t)]. \tag{6.2.12}$$

### 6.2.2 Improved Control

In the rest of this section we prove our $Z$-norm estimate for the profile $V^\psi$.

**Proposition 6.3.** *We have, for any $t \in [0,T]$,*

$$\|V^\psi(t)\|_{Z_{kg}} \lesssim \varepsilon_0. \tag{6.2.13}$$

Since $|\widehat{V^\psi}(\xi,t)| = |\widehat{V_*^\psi}(\xi,t)|$, in view of the definition (2.1.45) it suffices to prove that

$$\|\varphi_k(\xi)\{\widehat{V_*^\psi}(\xi,t_2) - \widehat{V_*^\psi}(\xi,t_1)\}\|_{L^\infty_\xi} \lesssim \varepsilon_0 2^{-\delta m/2} 2^{-k^-/2+\kappa k^-} 2^{-N_0 k^+} \tag{6.2.14}$$

for any $k \in \mathbb{Z}$, $m \geq 1$, and $t_1, t_2 \in [2^m - 2, 2^{m+1}] \cap [0,T]$. We prove these bounds in several steps. We start with the contribution of very low and very high frequencies.

**Lemma 6.4.** *The bounds* (6.2.14) *hold if* $k \leq -\kappa m$ *or if* $k \geq \delta' m - 10$.

*Proof.* **Step 1.** We start with the case of large $k \geq \delta' m - 10$. Notice that

$$\|Q_{j,k}V^\psi(t)\|_{L^2} \lesssim \varepsilon_0 \langle t\rangle^{\delta'} 2^{-N(0)k^+}, \qquad 2^{j+k}\|Q_{j,k}V^\psi(t)\|_{H^{0,1}_\Omega} \lesssim \varepsilon_0 \langle t\rangle^{\delta'} 2^{-N(2)k^+},$$

for any $j \geq -k^-$, due to Propositions 5.1 and 6.2, and Lemma 3.3 (i). Using (3.2.62) we have

$$\begin{aligned}\|\widehat{P_k V^\psi}(t)\|_{L^\infty} &\lesssim 2^{-3k/2}\varepsilon_0 \langle t\rangle^{\delta'} \cdot 2^{-N(0)k^+(1-\delta)/4} 2^{-N(2)k^+(3+\delta)/4} \\ &\lesssim \varepsilon_0 \langle t\rangle^{\delta'} 2^{-3k/2} 2^{-N_0 k^+} 2^{-dk^+},\end{aligned}$$

for any $t \in [0,T]$ and $k \in \mathbb{Z}$ (recall (2.1.43)). The bounds (6.2.14) follow if $2^k \gtrsim 2^{\delta' m}$.

**Step 2.** It remains to show that if $k \leq -\kappa m$ and $t \in [2^m - 2, 2^{m+1}] \cap [0,T]$ then

$$\|\varphi_k(\xi)\widehat{V^\psi}(\xi,t)\|_{L^\infty_\xi} \lesssim \varepsilon_0 2^{-\delta m/2} 2^{-k/2+\kappa k}. \tag{6.2.15}$$

It follows from Proposition 6.2 that

$$2^{k^+}\|\varphi_k(\xi)(\partial_{\xi_l}\widehat{V^{\mathcal{L}\psi}})(\xi,t)\|_{L^2_\xi} \lesssim \varepsilon_0 \langle t\rangle^{H(q+1,n+1)\delta} 2^{-N_0 k^+ + (n+1)dk^+},$$

for any $t \in [0,T]$, $k \in \mathbb{Z}$, $l \in \{1,2,3\}$, and $\mathcal{L} \in \mathcal{V}^q_n$, $n \in [0,2]$. Using Lemma 3.3 (i), we have

$$\sup_{j \geq -k^-} 2^j \|Q_{j,k}V^{\mathcal{L}\psi}(t)\|_{L^2} \lesssim \varepsilon_0 \langle t\rangle^{H(q+1,n+1)\delta} 2^{-N_0 k^+ - k^+ + (n+1)dk^+}. \tag{6.2.16}$$

We use now (6.2.16) and (3.2.63) to estimate

$$\|\widehat{P_k V^\psi}(t)\|_{L^\infty} \lesssim 2^{-3k/2}\{\sup_{j\geq -k^-}\|Q_{j,k}V^\psi(t)\|_{H^{0,1}_\Omega}\}^{(1-\delta)/2}\{\varepsilon_0\langle t\rangle^{H(1,2)\delta}2^k\}^{(1+\delta)/2}.$$

Therefore, recalling that $\kappa^2 \geq 4\delta'$, for (6.2.15) it suffices to prove that

$$\|P_k V^\psi(t)\|_{H^{0,1}_\Omega} \lesssim \varepsilon_0 \langle t\rangle^{2\delta'} 2^{k+10\kappa k}$$

if $k \leq -\kappa m$ and $t \in [2^m - 2, 2^{m+1}] \cap [0, T]$. In view of (3.1.7), for this it suffices to prove that

$$\|P_l V^{\Omega\psi}(t)\|_{L^2} + \sum_{a=1}^{3} \|\varphi_l(\xi)(\partial_{\xi_a} \widehat{V^{\Omega\psi}})(\xi, t)\|_{L^2} \lesssim \varepsilon_0 \langle t \rangle^{2\delta'} 2^{10\kappa l}, \tag{6.2.17}$$

for any $l \in \mathbb{Z}$ and $t \in [0, T]$, where $\Omega \in \{Id, \Omega_{23}, \Omega_{31}, \Omega_{12}\}$.

The bound on the first term in the left-hand side of (6.2.17) follows from (6.2.16). To bound the remaining terms we use the identities (6.1.2). For (6.2.17) it suffices to prove that

$$\|P_l \Gamma_a U^{\Omega\psi}(t)\|_{L^2} + \|\varphi_l(\xi)(\partial_{\xi_a} \widehat{\Omega\mathcal{N}^\psi})(\xi, t)\|_{L^2} \lesssim \varepsilon_0 \langle t \rangle^{2\delta'} 2^{10\kappa l}, \tag{6.2.18}$$

for any $l \in \mathbb{Z}$, $t \in [0, T]$, and $a \in \{1, 2, 3\}$. The term $\|P_l \Gamma_a U^{\Omega\psi}(t)\|_{L^2}$ is bounded as claimed due to (6.2.16) and the commutation identities (6.1.7) (with $\Lambda_{wa}$ replaced by $\Lambda_{kg}$ and $h_{\alpha\beta}$ replaced by $\psi$). Therefore, it remains to prove that

$$\begin{aligned} &\|P_k \Omega\mathcal{N}^\psi(\xi, t)\|_{L^2} \lesssim \varepsilon_0 \langle t \rangle^{2\delta'} 2^{10\kappa k} \min\{\langle t \rangle^{-1}, 2^{k^-}\} 2^{-2k^+} \\ &\|\varphi_k(\xi)(\partial_{\xi_a} \widehat{\Omega\mathcal{N}^\psi})(\xi, t)\|_{L^2} \lesssim \varepsilon_0 \langle t \rangle^{2\delta'} 2^{10\kappa k} \end{aligned} \tag{6.2.19}$$

for any $k \in \mathbb{Z}$, $t \in [0, T]$, $\Omega \in \{Id, \Omega_{23}, \Omega_{31}, \Omega_{12}\}$, and $a \in \{1, 2, 3\}$.

**Step 3.** The bounds (6.2.19) are similar to the bounds in Lemmas 4.10 and 4.12. We may assume $k \leq 0$ and the only issue is to gain the factors $2^{10\kappa k}$ and we are allowed to lose small powers $\langle t \rangle^{2\delta'}$. Notice that the cubic components $\Omega\mathcal{N}^{\psi,\geq 3}$ satisfy stronger bounds (this follows from (4.2.52)–(4.2.53) if $2^k \gtrsim \langle t \rangle^{-1}$ and the $L^2$ estimates (4.1.69) and (3.3.5) if $2^k \lesssim \langle t \rangle^{-1}$). After these reductions, with $I$ as in (3.2.41)–(3.2.43), for (6.2.19) it suffices to prove that

$$\begin{aligned} \sum_{(k_1,k_2) \in \mathcal{X}_k} & 2^{k_2^+ - k_1} \|P_k I[P_{k_1} U^{\mathcal{L}_1 h, \iota_1}, P_{k_2} U^{\mathcal{L}_2 \psi, \iota_2}](t)\|_{L^2} \\ & \lesssim \varepsilon_1^2 \langle t \rangle^{2\delta'} 2^{10\kappa k} \min(\langle t \rangle^{-1}, 2^k) \end{aligned} \tag{6.2.20}$$

and

$$\begin{aligned} \sum_{(k_1,k_2) \in \mathcal{X}_k} & 2^{k_2^+ - k_1} \|\varphi_k(\xi)(\partial_{\xi_a} \mathcal{F}\{I[P_{k_1} U^{\mathcal{L}_1 h, \iota_1}, P_{k_2} U^{\mathcal{L}_2 \psi, \iota_2}]\})(\xi, t)\|_{L^2_\xi} \\ & \lesssim \varepsilon_1^2 \langle t \rangle^{2\delta'} 2^{10\kappa k}, \end{aligned} \tag{6.2.21}$$

for any $k \leq 0$, $h \in \{h_{\alpha\beta}\}$, $\iota_1, \iota_2 \in \{+, -\}$, $a \in \{1, 2, 3\}$, $\mathcal{L}_1 \in \mathcal{V}^0_{n_1}$, $\mathcal{L}_2 \in \mathcal{V}^0_{n_2}$, $n_1 + n_2 \leq 1$.

**Substep 3.1.** We prove first (6.2.20). These bounds easily follow from (4.1.52) when $2^k \lesssim \langle t \rangle^{-1}$. On the other hand, if $2^k \geq \langle t \rangle^{-1}$ then we estimate, using (3.3.5) and (3.3.11),

$$\begin{aligned}
&2^{k_2^+-k_1}\|P_kI[P_{k_1}U^{\mathcal{L}_1h,\iota_1},P_{k_2}U^{\mathcal{L}_2\psi,\iota_2}](t)\|_{L^2}\\
&\quad\lesssim 2^{k_2^+-k_1}\|P_{k_1}U^{\mathcal{L}_1h}(t)\|_{L^\infty}\|P_{k_2}U^{\mathcal{L}_2\psi}(t)\|_{L^2}\\
&\quad\lesssim \varepsilon_1^2 2^{k_2^-}\langle t\rangle^{-1+\delta'}\min(1,2^{k_1^-}\langle t\rangle)^{1/2}2^{-4(k_1^++k_2^+)}.
\end{aligned}$$

This suffices to bound the contribution of the pairs $(k_1,k_2)\in\mathcal{X}_k$ in (6.2.20) for which $2^{k_2}\lesssim 2^{k/10}$. For the remaining pairs we have $\min(k_1,k_2)\geq k/10+10$, thus $|k_1-k_2|\leq 4$. Let $J_1$ denote the largest integer such that $2^{J_1}\leq\langle t\rangle(1+2^{k_1}\langle t\rangle)^{-\delta}$ and decompose $P_{k_1}U^{\mathcal{L}_1h,\iota_1}(t)=U^{\mathcal{L}_1h,\iota_1}_{\leq J_1,k_1}(t)+U^{\mathcal{L}_1h,\iota_1}_{>J_1,k_1}(t)$, as in (4.1.14). We use first Lemma 3.11. Therefore

$$\begin{aligned}
&2^{k_2^+-k_1}\big\|P_kI[U^{\mathcal{L}_1h,\iota_1}_{\leq J_1,k_1}(t),P_{k_2}U^{\mathcal{L}_2\psi,\iota_2}(t)]\big\|_{L^2}\\
&\quad\lesssim 2^{k_2^+-k_1}2^{k/2}(1+2^{k_1}\langle t\rangle)^{\delta}\langle t\rangle^{-1}\|Q_{\leq J_1,k_1}V^{\mathcal{L}_1h}(t)\|_{H^{0,1}_\Omega}\|P_{k_2}U^{\mathcal{L}_2\psi}(t)\|_{L^2}\\
&\quad\lesssim \varepsilon_1^2 2^{k/2}\langle t\rangle^{-1+\delta'}2^{-k_1^-/2}2^{-4(k_1^++k_2^+)}.
\end{aligned}$$

Moreover, using (3.3.4) and (3.3.13),

$$\begin{aligned}
&2^{k_2^+-k_1}\big\|P_kI[U^{\mathcal{L}_1h,\iota_1}_{>J_1,k_1}(t),P_{k_2}U^{\mathcal{L}_2\psi,\iota_2}(t)]\big\|_{L^2}\\
&\quad\lesssim 2^{k_2^+-k_1}\|U^{\mathcal{L}_1h}_{>J_1,k_1}(t)\|_{L^2}\|P_{k_2}U^{\mathcal{L}_2\psi}(t)\|_{L^\infty}\\
&\quad\lesssim \varepsilon_1^2\langle t\rangle^{-1+\delta'}2^{-3k_1^-/2}2^{-J_1}2^{-4(k_1^++k_2^+)}.
\end{aligned}$$

Therefore, recalling that $2^k\in[\langle t\rangle^{-1},1]$ and $\min(k_1,k_2)\geq k/10+10$, we have

$$2^{k_2^+-k_1}\big\|P_kI[U^{\mathcal{L}_1h,\iota_1}_{\leq J_1,k_1}(t),P_{k_2}U^{\mathcal{L}_2\psi,\iota_2}(t)]\big\|_{L^2}\lesssim\varepsilon_1^2 2^{k/2}\langle t\rangle^{-1+\delta'}2^{-k_1^-/2}2^{-4(k_1^++k_2^+)}.$$

This suffices to complete the proof of (6.2.20).

**Substep 3.2.** To prove (6.2.21) we write $U^{\mathcal{L}_2\psi,\iota_2}=e^{-it\Lambda_{kg,\iota_2}}V^{\mathcal{L}_2\psi,\iota_2}$ and examine the formula (3.2.43). We make the change of variables $\eta\to\xi-\eta$ and notice that the $\partial_{\xi_a}$ derivative can hit the multiplier $m(\eta,\xi-\eta)$, or the phase $e^{-it\Lambda_{kg,\iota_2}(\xi-\eta)}$, or the profile $P_{k_2}\widehat{V^{\mathcal{L}_2\psi,\iota_2}}(\xi-\eta)$. In the first two cases, the derivative effectively corresponds to multiplying by factors $\lesssim\langle t\rangle$ or $\lesssim 2^{-k_2^-}$. The desired bounds then follow from (6.2.20) (in the case $2^{k_2^-}\lesssim\langle t\rangle^{-1}$ we need to apply (4.1.52) again to control the corresponding contributions).

Finally, assume that the $\partial_{\xi_a}$ derivative hits the profile $P_{k_2}\widehat{V^{\mathcal{L}_2\psi,\iota_2}}(\xi-\eta)$. Letting (as in Lemma 4.10) $\widehat{U^{\mathcal{L}_2\psi,\iota_2}_{*a,k_2}}(\xi,t)=e^{-it\Lambda_{kg,\iota_2}(\xi)}\partial_{\xi_a}\{\varphi_{k_2}\cdot\widehat{V^{\mathcal{L}_2\psi,\iota_2}}\}(\xi,t)$ it suffices to prove that

$$\sum_{(k_1,k_2)\in\mathcal{X}_k}2^{k_2^+-k_1}\|P_kI[P_{k_1}U^{\mathcal{L}_1h,\iota_1},U^{\mathcal{L}_2\psi,\iota_2}_{*a,k_2}](t)\|_{L^2}\lesssim\varepsilon_1^2\langle t\rangle^{2\delta'}2^{10\kappa k}. \tag{6.2.22}$$

This follows easily using the $L^2$ bounds (4.2.39) and (3.3.3). □

We return now to the main estimates (6.2.14). In the remaining range $-\kappa m \leq k \leq \delta' m - 10$, they follow from the identity (6.2.8) and the bounds (which are proved in Lemmas 6.5–6.8 below)

$$\Big\|\varphi_k(\xi)\int_{t_1}^{t_2}\mathcal{R}_a^{\psi}(\xi,s)\,ds\Big\|_{L^\infty_\xi} \lesssim \varepsilon_1^2 2^{-100\delta' m} \tag{6.2.23}$$

for $a\in\{1,2,3,4\}$, $m\geq 100$, and $t_1\leq t_2\in[2^{m-1},2^{m+1}]\cap[0,T]$.

In some estimates we need to use integration by parts in time (normal forms). For $\mu\in\{(kg,+),(kg,-)\}$, $\nu\in\{(wa,+),(wa,-)\}$, and $s\in[0,T]$ we define the operators $T_{\mu\nu}^{kg}$ by

$$T_{\mu\nu}^{kg}[f,g](\xi,s) := \int_{\mathbb{R}^3}\frac{e^{is\Phi_{(kg,+)\mu\nu}(\xi,\eta)}}{\Phi_{(kg,+)\mu\nu}(\xi,\eta)} m(\xi-\eta,\eta)\widehat{f}(\xi-\eta,s)\widehat{g}(\eta,s)\,d\eta, \tag{6.2.24}$$

where $\Phi_{(kg,+)\mu\nu}(\xi,\eta) = \Lambda_{kg}(\xi)-\Lambda_\mu(\xi-\eta)-\Lambda_\nu(\eta)$ (see (2.1.41)) and $m\in\mathcal{M}$ (see (3.2.41)).

**Lemma 6.5.** *The bounds* (6.2.23) *hold for* $m\geq 100$, $k\in[-\kappa m,\delta' m]$, *and* $a=4$.

*Proof.* We use the bounds (3.2.63), combined with either (4.2.4) and (4.2.6) or (4.2.52)–(4.2.53) (in both cases $n=1$). It follows that

$$\begin{aligned}\|\widehat{P_k\mathcal{N}^\psi}(t)\|_{L^\infty} &\lesssim \varepsilon_1^2\langle t\rangle^{-1/2+\delta'}2^{-k^-}2^{-N(2)k^+-2k^+},\\ \|\widehat{P_k\mathcal{N}^{\psi,\geq 3}}(t)\|_{L^\infty} &\lesssim \varepsilon_1^2\langle t\rangle^{-1.1+\delta'}2^{-k^-}2^{-N(2)k^+-2k^+},\end{aligned} \tag{6.2.25}$$

for any $t\in[0,T]$ and $k\in\mathbb{Z}$. The estimates (6.2.23) follow from definitions. □

**Lemma 6.6.** *The bounds* (6.2.23) *hold for* $m\geq 100$, $k\in[-\kappa m,\delta' m]$, *and* $a=1$.

*Proof.* We examine the formula (6.2.9), substitute $h=i\Lambda_{wa}^{-1}(U^{h,+}-U^{h,-})/2$, $h\in\{h_{\alpha\beta}\}$, and decompose the input functions dyadically in frequency. For $h\in\{h_{\alpha\beta}\}$ let

$$\widehat{U_{low}^{h,\iota_2}}(\xi,s) := \varphi_{\leq 0}(\langle s\rangle^{p_0}\xi)\widehat{U^{h,\iota_2}}(\xi,s), \qquad \widehat{V_{low}^{h,\iota_2}}(\xi,s) := \varphi_{\leq 0}(\langle s\rangle^{p_0}\xi)\widehat{V^{h,\iota_2}}(\xi,s). \tag{6.2.26}$$

With $I$ defined as in (3.2.41)–(3.2.43), it suffices to prove that, for $\iota_2 \in \{+,-\}$,

$$\sum_{(k_1,k_2)\in\mathcal{X}_k} 2^{k_1^+-k_2}\Big\|\varphi_k(\xi)\int_{t_1}^{t_2} e^{is\Lambda_{kg}(\xi)-i\Theta_{kg}(\xi,s)} \times \mathcal{F}\{I[P_{k_1}U^{\psi,-}, P_{k_2}U^{h,\iota_2}_{low}]\}(\xi,s)\,ds\Big\|_{L^\infty_\xi} \lesssim \varepsilon_1^2 2^{-\kappa m}. \tag{6.2.27}$$

We estimate first, using just (3.3.7) and (3.3.3),

$$\begin{aligned} 2^{k_1^+-k_2}\|\mathcal{F}\{I[P_{k_1}U^{\psi,-}, P_{k_2}U^{h,\iota_2}_{low}]\}(\xi,s)\|_{L^\infty_\xi} &\lesssim 2^{k_1^+-k_2}\|\widehat{P_{k_1}U^{\psi}}(s)\|_{L^\infty}\|\widehat{P_{k_2}U^{h}_{low}}(s)\|_{L^1} \\ &\lesssim \varepsilon_1^2 2^{-k_1^-/2+\kappa k_1^-}2^{-N_0k_1^++k_1^+}2^{k_2-\delta k_2}2^{\delta' m}. \end{aligned} \tag{6.2.28}$$

This suffices to bound the contribution of the pairs $(k_1,k_2)$ for which $k_2 \leq -1.1m-10$. It remains to prove that

$$\Big|\int_{t_1}^{t_2}\int_{\mathbb{R}^3} e^{is\Lambda_{kg}(\xi)-i\Theta_{kg}(\xi,s)} m(\xi-\eta,\eta)e^{is\Lambda_{kg}(\xi-\eta)}\widehat{P_{k_1}V^{\psi,-}}(\xi-\eta,s) \times e^{-is\Lambda_{wa,\iota_2}(\eta)}\widehat{P_{k_2}V^{h,\iota_2}_{low}}(\eta,s)\,d\eta ds\Big| \lesssim \varepsilon_1^2 2^{-\kappa m-2\delta' m}2^{k_2}, \tag{6.2.29}$$

for any $\xi$ with $|\xi|\in[2^{k_1-4},2^{k_1+4}]$, provided that

$$k_2 \in [-1.1m-10, -p_0m+10], \qquad k_1 \in [-\kappa m-10, \delta' m+10]. \tag{6.2.30}$$

To prove (6.2.29) we integrate by parts in time. Letting $\sigma = (kg,+)$, $\mu = (kg,-)$, $\nu = (wa,\iota_2)$, we notice that $\Phi_{\sigma\mu\nu}(\xi,\eta) \gtrsim 1$ in the support of the integral. Here it is important that $\mu \neq (kg,+)$, so the phase is nonresonant, as the nonlinear correction (6.2.6) was done precisely to weaken the corresponding resonant contribution of the profile $V^{kg,+}$.

The left-hand side of (6.2.29) is dominated by $C(I_{kg}(\xi)+II_{kg}(\xi)+III_{kg}(\xi))$, where, with $T^{kg}_{\mu\nu}$ defined as in (6.2.24),

$$\begin{aligned} I_{kg}(\xi) &:= \Big(1+\int_{t_1}^{t_2}|\dot\Theta_{kg}(\xi,s)|\,ds\Big)\sup_{s\in[t_1,t_2]}|T^{kg}_{\mu\nu}[P_{k_1}V^{\psi,-}, P_{k_2}V^{h,\iota_2}_{low}](\xi,s)|, \\ II_{kg}(\xi) &:= \int_{t_1}^{t_2}|T^{kg}_{\mu\nu}[\partial_s(P_{k_1}V^{\psi,-}), P_{k_2}V^{h,\iota_2}_{low}](\xi,s)|\,ds, \\ III_{kg}(\xi) &:= \int_{t_1}^{t_2}|T^{kg}_{\mu\nu}[P_{k_1}V^{\psi,-}, \partial_s(P_{k_2}V^{h,\iota_2}_{low})](\xi,s)|\,ds. \end{aligned} \tag{6.2.31}$$

As in (6.2.28), assuming $k_1, k_2$ as in (6.2.30), we estimate for $s \in [t_1, t_2]$

$$|T^{kg}_{\mu\nu}[P_{k_1}V^{\psi,-}, P_{k_2}V^{h,\iota_2}_{low}](\xi, s)| \lesssim \varepsilon_1^2 2^{-k_1^-/2} 2^{2k_2} 2^{4\delta m}.$$

The definition (6.2.4) gives $|\dot{\Theta}_{kg}(\xi, s)| \lesssim 2^{k_1^+} \sup_{\alpha,\beta} \|h_{\alpha\beta}\|_{L^\infty} \lesssim 2^{k_1^+} 2^{-m+\delta' m}$. Therefore

$$I_{kg}(\xi) \lesssim \varepsilon_1^2 2^{-k_1^-/2} 2^{2k_2} 2^{2\delta' m}. \tag{6.2.32}$$

Similarly, using (6.2.25) and the bounds $\|\widehat{P_{k_2}V^{h,\iota_2}_{low}}(s)\|_{L^1} \lesssim \varepsilon_1 2^{2k_2} 2^{4\delta m}$, we have

$$II_{kg}(\xi) \lesssim \varepsilon_1^2 2^{m/2+2\delta' m} 2^{-k_1^-} 2^{2k_2}.$$

Finally, using also (4.2.3),

$$\|\mathcal{F}\{\partial_s(P_{k_2}V^{h,\iota_2}_{low})(s)\}\|_{L^1} \lesssim 2^{3k_2/2} \|\partial_s(P_{k_2}V^h_{low})(s)\|_{L^2} \lesssim \varepsilon_1 2^{-m+\delta' m/2} 2^{2k_2}, \tag{6.2.33}$$

and it follows that $III_{kg}(\xi)$ can be bounded as in (6.2.32). Therefore

$$I_{kg}(\xi) + II_{kg}(\xi) + III_{kg}(\xi) \lesssim \varepsilon_1^2 2^{m/2+2\delta' m} 2^{-k_1^-} 2^{2k_2}.$$

The desired bounds (6.2.29) follow, recalling that $p_0 = 0.68$ and the frequency restrictions (6.2.30). □

**Lemma 6.7.** *The bounds* (6.2.23) *hold for* $m \geq 100$, $k \in [-\kappa m, \delta' m]$, *and* $a = 2$.

*Proof.* We decompose $V^\psi = \sum_{(k_1,j_1)\in\mathcal{J}} V^{\psi,+}_{j_1,k_1}$ as in (3.3.1) and examine the definition (6.2.10). For (6.2.23) it suffices to prove that

$$\begin{aligned}\varphi_k(\xi)\Big| \int_{\mathbb{R}^3} \{ & e^{it(\Lambda_{kg}(\xi)-\Lambda_{kg}(\xi-\eta))} \widehat{V^{\psi,+}_{j_1,k_1}}(\xi-\eta, t) \mathfrak{q}^{low}_{kg,+}(\xi-\eta, \eta, t) \\ & - e^{it(\xi\cdot\eta)/\Lambda_{kg}(\xi)} \widehat{V^{\psi,+}_{j_1,k_1}}(\xi, t) \mathfrak{q}^{low}_{kg,+}(\xi, \eta, t)\} \, d\eta \Big| \lesssim \varepsilon_1^2 2^{-m-\kappa m} 2^{-\delta j_1},\end{aligned}$$

provided that $|k_1 - k| \leq 10$ and $t \in [2^{m-1}, 2^{m+1}]$. Using also the definitions (6.2.7) it suffices to prove that for any multiplier $m \in \mathcal{M}_0$ (see (3.2.40)) and $\alpha, \beta \in \{0, 1, 2, 3\}$ we have

$$\begin{aligned}\varphi_k(\xi)\Big| \int_{\mathbb{R}^3} \widehat{h^{low}_{\alpha\beta}}(\eta, t) \{ & e^{it(\Lambda_{kg}(\xi)-\Lambda_{kg}(\xi-\eta))} \widehat{V^{\psi,+}_{j_1,k_1}}(\xi-\eta, t) m(\xi-\eta)\langle\xi-\eta\rangle \\ & - e^{it(\xi\cdot\eta)/\Lambda_{kg}(\xi)} \widehat{V^{\psi,+}_{j_1,k_1}}(\xi, t) m(\xi)\langle\xi\rangle\} \, d\eta \Big| \lesssim \varepsilon_1^2 2^{-m-\kappa m} 2^{-\delta j_1}.\end{aligned} \tag{6.2.34}$$

Recall that $\|\widehat{V^{\psi,+}_{j_1,k_1}}(t)\|_{L^\infty} \lesssim \varepsilon_1 2^{-k_1} 2^{-j_1/2+\delta j_1/2} 2^{\delta' m} 2^{-4k_1^+}$; see (3.3.26). Therefore, without using the cancellation of the two terms in the integral, the left-hand side of (6.2.34) is bounded by

$$C\varepsilon_1 2^{-j_1/2+\delta j_1}2^{\delta' m}2^{-k_1^-}\|\widehat{h^{low}_{\alpha\beta}}(t)\|_{L^1}\lesssim \varepsilon_1^2 2^{-j_1/2+\delta j_1}2^{2\delta' m}2^{-k_1^-}2^{-p_0 m}.$$

This suffices to prove (6.2.34) when $j_1$ is large, i.e., $2^{j_1/2}\gtrsim 2^{(1.01-p_0)m}$.

On the other hand, if $2^m\gg 1$ and $j_1/2\leq (1.01-p_0)m=0.33m$ then we estimate

$$\begin{aligned}&\big|e^{it(\Lambda_{kg}(\xi)-\Lambda_{kg}(\xi-\eta))}-e^{it(\xi\cdot\eta)/\Lambda_{kg}(\xi)}\big|\lesssim 2^{-2p_0 m+m},\\&\big|\widehat{V^{\psi,+}_{j_1,k_1}}(\xi-\eta,t)m(\xi-\eta)\langle\xi-\eta\rangle-\widehat{V^{\psi,+}_{j_1,k_1}}(\xi,t)m(\xi)\langle\xi\rangle\big|\\&\qquad\lesssim \varepsilon_1 2^{j_1/2}2^{2\delta' m}2^{-2k_1^-}2^{-p_0 m},\end{aligned}\tag{6.2.35}$$

provided that $|\xi|\approx 2^k$ and $|\eta|\lesssim 2^{-p_0 m}$. Indeed, the first bound follows from the observation that $\nabla\Lambda_{kg}(\xi)=\xi/\Lambda_{kg}(\xi)$. The second bound follows from (3.3.26) once we notice that taking $\partial_\xi$ derivative of the localized profiles $\widehat{V^{\psi,+}_{j_1,k_1}}$ corresponds essentially to multiplication by a factor of $2^{j_1}$. If $j_1/2\leq 0.33m$ it follows that the left-hand side of (6.2.34) is bounded by

$$C\varepsilon_1 2^{-0.34m}\|\widehat{h^{low}_{\alpha\beta}}(t)\|_{L^1}\lesssim \varepsilon_1^2 2^{-p_0 m-0.34m+\delta' m}.$$

This suffices to prove (6.2.34) when $j_1/2\leq 0.33m$, which completes the proof of the lemma. □

**Lemma 6.8.** *The bounds* (6.2.23) *hold for* $m\geq 100$, $k\in[-\kappa m,\delta' m]$, *and* $a=3$.

*Proof.* We examine the formula (6.2.11). Let $U^{h,\iota_2}_{high}:=U^{h,\iota_2}-U^{h,\iota_2}_{low}$, $V^{h,\iota_2}_{high}:=V^{h,\iota_2}-V^{h,\iota_2}_{low}$; see (6.2.26). As in the proof of Lemma 6.6, after simple reductions it suffices to prove that

$$\begin{aligned}2^{k_1^+-k_2}\Big\|\varphi_k(\xi)\int_{t_1}^{t_2}&e^{is\Lambda_{kg}(\xi)-i\Theta(\xi,s)}\\&\times\mathcal{F}\{I[P_{k_1}U^{\psi,\iota_1},P_{k_2}U^{h,\iota_2}_{high}]\}(\xi,s)\,ds\Big\|_{L^\infty_\xi}\lesssim\varepsilon_1^2 2^{-101\delta' m}\end{aligned}\tag{6.2.36}$$

for $\iota_1,\iota_2\in\{+,-\}$, and $(k_1,k_2)\in\mathcal{X}_k$, $k_1,k_2\in[-p_0 m-10,m/100]$.

As in the proof of Lemma 6.6 we integrate by parts in time to estimate

$$\begin{aligned}\Big|\int_{t_1}^{t_2}e^{is\Lambda_{kg}(\xi)-i\Theta_{kg}(\xi,s)}&\mathcal{F}\{I[P_{k_1}U^{\psi,\iota_1},P_{k_2}U^{h,\iota_2}_{high}]\}(\xi,s)\,ds\Big|\\&\lesssim I'_{kg}(\xi)+II'_{kg}(\xi)+III'_{kg}(\xi),\end{aligned}$$

where, with $\mu = (kg, \iota_1)$ and $\nu = (wa, \iota_2)$ and $T^{kg}_{\mu\nu}$ defined as in (6.2.24),

$$I'_{kg}(\xi) := \Big(1 + \int_{t_1}^{t_2} |\dot{\Theta}_{kg}(\xi, s)|\, ds\Big) \sup_{s\in[t_1,t_2]} |T^{kg}_{\mu\nu}[P_{k_1}V^{\psi,\iota_1}, P_{k_2}V^{h,\iota_2}_{high}](\xi, s)|,$$

$$II'_{kg}(\xi) := \int_{t_1}^{t_2} |T^{kg}_{\mu\nu}[\partial_s(P_{k_1}V^{\psi,\iota_1}), P_{k_2}V^{h,\iota_2}_{high}](\xi, s)|\, ds,$$

$$III'_{kg}(\xi) := \int_{t_1}^{t_2} |T^{kg}_{\mu\nu}[P_{k_1}V^{\psi,\iota_1}, \partial_s(P_{k_2}V^{h,\iota_2}_{high})](\xi, s)|\, ds.$$

Notice that $|\dot{\Theta}_{kg}(\xi, s)| \lesssim 2^{-m+4\delta' m}$, as a consequence of (3.3.11). After possibly changing the multiplier $m$ in the definition (6.2.24), for (6.2.36) it suffices to prove that

$$|\varphi_k(\xi)T^{kg}_{\mu\nu}[P_{k_1}V^{\psi,\iota_1}, P_{k_2}V^{h,\iota_2}](\xi, s)| \lesssim \varepsilon_1^2 2^{-110\delta' m} 2^{k_2^-}, \tag{6.2.37}$$

$$2^m|\varphi_k(\xi)T^{kg}_{\mu\nu}[\partial_s(P_{k_1}V^{\psi,\iota_1}), P_{k_2}V^{h,\iota_2}](\xi, s)| \lesssim \varepsilon_1^2 2^{-110\delta' m} 2^{k_2^-}, \tag{6.2.38}$$

$$2^m|\varphi_k(\xi)T^{kg}_{\mu\nu}[P_{k_1}V^{\psi,\iota_1}, \partial_s(P_{k_2}V^{h,\iota_2})](\xi, s)| \lesssim \varepsilon_1^2 2^{-110\delta' m} 2^{k_2^-}, \tag{6.2.39}$$

for any $s \in [2^{m-1}, 2^{m+1}]$, $k_1, k_2 \in [-p_0 m - 10, m/100]$, $\mu = (kg, \iota_1)$, $\nu = (wa, \iota_2)$, $\iota_1, \iota_2 \in \{+, -\}$.

**Step 1: proof of** (6.2.37). If $k_1 \leq -4\kappa m - 10$ (so $k_2 \geq -\kappa m - 20$) then we can just use the $L^2$ bounds (3.3.3) and (3.3.5) on the two inputs, and Lemma 3.4 (i). On the other hand, if $k_1 \geq -4\kappa m$ then we decompose $P_{k_1}V^{\psi,\iota_1} = \sum_{j_1} V^{\psi,\iota_1}_{j_1,k_1}$ and $P_{k_2}V^{h,\iota_2} = \sum_{j_2} V^{h,\iota_2}_{j_2,k_2}$ as in (3.3.1). Let $\overline{k} := \max(k, k_1, k_2)$ and recall that $|\Phi_{(kg,+)\mu\nu}(\xi, \eta)| \gtrsim 2^{k_2} 2^{-2\overline{k}^+}$ in the support of the integrals defining the operators $T^{kg}_{\mu\nu}$ (see (3.1.12)).

The contribution of the pairs $(V^{\psi,\iota_1}_{j_1,k_1}, V^{h,\iota_2}_{j_2,k_2})$ with $2^{\max(j_1,j_2)} \leq 2^{0.99m}2^{-6\overline{k}^+}$ is negligible,

$$|T^{kg}_{\mu\nu}[V^{\psi,\iota_1}_{j_1,k_1}, V^{h,\iota_2}_{j_2,k_2}](\xi, s)| \lesssim \varepsilon_1^2 2^{-2m} \quad \text{if} \quad 2^{\max(j_1,j_2)} \leq 2^{0.99m}2^{-6\overline{k}^+}. \tag{6.2.40}$$

Indeed, this follows by integration by parts in $\eta$ (using Lemma 3.1), the bounds (3.2.4), and the observation that the gradient of the phase admits a suitable lower bound $|\nabla_\eta\{s\Lambda_{kg,\iota_1}(\xi - \eta) + s\Lambda_{wa,\iota_2}(\eta)\}| \gtrsim \langle s\rangle 2^{-2k_1^+}$ in the support of the integral. On the other hand, we estimate

$$\begin{aligned}|T^{kg}_{\mu\nu}[V^{\psi,\iota_1}_{j_1,k_1}, V^{h,\iota_2}_{j_2,k_2}](\xi, s)| &\lesssim 2^{-k_2}2^{2\overline{k}^+}\varepsilon_1^2 2^{3k_2/2}\|\widehat{V^{\psi,\iota_1}_{j_1,k_1}}(s)\|_{L^\infty}\|\widehat{V^{h,\iota_2}_{j_2,k_2}}(s)\|_{L^2}\\ &\lesssim \varepsilon_1^2 2^{\delta' m}2^{k_2/2}2^{2k_1^+ + 2k_2^+} \cdot 2^{-k_1}2^{-j_1/2+\delta j_1}2^{-10k_1^+}2^{-j_2}2^{-k_2^-/2-4\delta k_2^-}2^{-10k_2^+}\\ &\lesssim \varepsilon_1^2 2^{2\delta' m}2^{-k_1}2^{-j_1/2+\delta j_1}2^{-j_2}2^{-6\overline{k}^+},\end{aligned}$$

using (3.3.4), (3.3.26), and Lemma 3.4 (i). Since $k_1 \geq -4\kappa m$, this suffices to estimate the contribution of the pairs $(V^{\psi,\iota_1}_{j_1,k_1}, V^{h,\iota_2}_{j_2,k_2})$ for which $2^{\max(j_1,j_2)} \geq 2^{0.99m}2^{-6\overline{k}^+}$, and the bound (6.2.37) follows.

**Step 2: proof of** (6.2.38). Notice that, for any $t \in [0,T]$,

$$\partial_t V^{\psi,\iota_1}(t) = e^{it\Lambda_{kg,\iota_1}}\mathcal{N}^{\psi}(t) = e^{it\Lambda_{kg,\iota_1}}\mathcal{N}^{\psi,2}(t) + e^{it\Lambda_{kg,\iota_1}}\mathcal{N}^{\psi,\geq 3}(t). \tag{6.2.41}$$

The contribution of the nonlinearity $\mathcal{N}^{\psi,\geq 3}$ can be bounded using (6.2.25),

$$\begin{aligned}
&|\varphi_k(\xi)T^{kg}_{\mu\nu}[e^{is\Lambda_{kg,\iota_1}}P_{k_1}\mathcal{N}^{\psi,\geq 3}(s), P_{k_2}V^{h,\iota_2}(s)](\xi)| \\
&\qquad \lesssim 2^{-k_2}2^{2\overline{k}^+}\|\widehat{P_{k_1}\mathcal{N}^{\psi,\geq 3}}(s)\|_{L^\infty}\|P_{k_2}V^h(s)\|_{L^2}2^{3\min(k_1,k_2)/2} \\
&\qquad \lesssim 2^{-1.09m}2^{k_2}2^{-\max(k_1,k_2)}2^{-6\overline{k}^+}.
\end{aligned}$$

This is better than the bounds (6.2.38) since $\max(k_1,k_2) \geq k-10 \geq -\kappa m - 10$.

To bound the contribution of the quadratic components $\mathcal{N}^{\psi,2}$ we recall that $\mathcal{F}\{P_{k_1}\mathcal{N}^{\psi,2}\}(s)$ can be written as a sum of terms of the form

$$\varphi_{k_1}(\gamma)\int_{\mathbb{R}^3}|\rho|^{-1}\langle\gamma-\rho\rangle m_3(\gamma-\rho)\widehat{U^{\psi,\iota_3}}(\gamma-\rho)\widehat{U^{h_4,\iota_4}}(\rho)\,d\rho,$$

where $\iota_3,\iota_4 \in \{+,-\}$, $h_4 \in \{h_{\alpha\beta}\}$, and $m_3$ is a symbol as in (3.2.40) (see (2.1.17)). We combine this with the formula (6.2.24). For (6.2.38) it suffices to prove that, for any $\xi \in \mathbb{R}^3$,

$$\begin{aligned}
&\Big|\varphi_k(\xi)\int_{\mathbb{R}^3\times\mathbb{R}^3}\frac{\varphi_{k_1}(\xi-\eta)m(\xi-\eta,\eta)}{\Lambda_{kg}(\xi)-\Lambda_\mu(\xi-\eta)-\Lambda_\nu(\eta)}e^{-is\Lambda_\nu(\eta)}\widehat{P_{k_2}V^{h,\iota_2}}(\eta,s) \\
&\quad \times m_3(\xi-\eta-\rho)\langle\xi-\eta-\rho\rangle|\rho|^{-1}\widehat{U^{\psi,\iota_3}}(\xi-\eta-\rho,s)\widehat{U^{h_4,\iota_4}}(\rho,s)\,d\eta d\rho\Big| \\
&\lesssim \varepsilon_1^2 2^{k_2^-}2^{-1.005m},
\end{aligned} \tag{6.2.42}$$

provided that $\mu = (kg,\iota_1)$, $\nu = (wa,\iota_2)$, $\iota_1,\iota_2,\iota_3,\iota_4 \in \{+,-\}$, and $k_1,k_2 \in [-p_0m-10, m/10]$.

We decompose the solutions $U^{\psi,\iota_3}$, $U^{h_4,\iota_4}$, and $P_{k_2}V^{h,\iota_2}$ dyadically in frequency and space as in (3.3.1). Then we notice that the contribution when one of the parameters $j_3,k_3,j_4,k_4,j_2$ is large can be bounded using just $L^2$ estimates. For (6.2.42) it suffices to prove that

$$\begin{aligned}
2^{-k_2^-}2^{k_3^+-k_4}|\mathcal{C}_{kg}[e^{-is\Lambda_\theta}V^{\psi,\iota_3}_{j_3,k_3}(s), e^{-is\Lambda_\nu}V^{h,\iota_2}_{j_2,k_2}(s), e^{-is\Lambda_\vartheta}V^{h_4,\iota_4}_{j_4,k_4}(s)](\xi)| \\
\lesssim \varepsilon_1^3 2^{-1.01m}
\end{aligned} \tag{6.2.43}$$

for any $k_2 \in [-p_0m-10, m/10]$, $k_3,k_4 \in [-2m, m/10]$, and $j_2,j_3,j_4 \leq 2m$, where $\theta = (kg, \iota_3)$, $\vartheta = (wa, \iota_4)$, and, with $m \in \mathcal{M}$, $m_3, m_4 \in \mathcal{M}_0$,

$$\begin{aligned}\mathcal{C}_{kg}[f,g,h](\xi) := \int_{\mathbb{R}^3\times\mathbb{R}^3} &\frac{\varphi_k(\xi)\varphi_{k_1}(\xi-\eta)m(\xi-\eta,\eta)}{\Lambda_{kg}(\xi)-\Lambda_\mu(\xi-\eta)-\Lambda_\nu(\eta)} \\ &\times m_3(\xi-\eta-\rho)m_4(\rho)\cdot\widehat{f}(\xi-\eta-\rho)\widehat{g}(\eta)\widehat{h}(\rho)\,d\eta d\rho.\end{aligned} \tag{6.2.44}$$

**Substep 2.1.** Assume first that

$$j_3 \geq 0.99m - 3k_3^+. \tag{6.2.45}$$

Let $k^* := \max\{k_2^+, k_3^+, k_4^+\}$. Let $Y$ denote the left-hand side of (6.2.43). Using Lemmas 3.4 and 3.2 (i) we estimate

$$\begin{aligned}Y &\lesssim 2^{k_3^+-k_4}2^{-2k_2+6\max(k^+,k_2^+)}\|V^{\psi,\iota_3}_{j_3,k_3}(s)\|_{L^2}\|e^{-is\Lambda_\nu}V^{h,\iota_2}_{j_2,k_2}(s)\|_{L^\infty}\|V^{h_4,\iota_4}_{j_4,k_4}(s)\|_{L^2} \\ &\lesssim \varepsilon_1^3 2^{4\delta' m}2^{-j_3}2^{-m}2^{-k_2^-}2^{-k_4^-/2}2^{-dk^*},\end{aligned} \tag{6.2.46}$$

where in the last line we use (3.3.11) and some of the bounds from Lemma 3.15. Since $2^{-k_2^-} \lesssim 2^{0.68m}$ and $j_3 + 3k_3^+ \geq 0.99m$, this suffices to prove (6.2.43) when $k_4^- \geq -0.55m - 10$.

On the other hand, if $k_4 \leq -0.55m - 10$, then we estimate in the Fourier space. Using (3.1.12), (3.3.26), and (3.3.3)–(3.3.4)

$$\begin{aligned}Y &\lesssim 2^{k_3^+-k_4}2^{-2k_2+6\max(k^+,k_2^+)}\|\widehat{V^{\psi,\iota_3}_{j_3,k_3}}(s)\|_{L^\infty} \\ &\quad\times 2^{3k_2/2}\|\widehat{V^{h,\iota_2}_{j_2,k_2}}(s)\|_{L^2}2^{3k_4/2}\|\widehat{V^{h_4,\iota_4}_{j_4,k_4}}(s)\|_{L^2} \\ &\lesssim \varepsilon_1^3 2^{4\delta' m}2^{-j_3/2}2^{k_4}2^{-k_3}2^{-dk^*}.\end{aligned} \tag{6.2.47}$$

Since $k_4 \leq -0.55m - 10$, this suffices to prove (6.2.43) when $k_3 \geq -0.01m - 10$. Finally, if $k_4 \leq -0.55m - 10$ and $k_3 \leq -0.01m - 10$ then $k_2 \geq -\kappa m - 10$ (due to the assumption $k \geq -\kappa m$) and a similar estimate gives

$$\begin{aligned}Y &\lesssim 2^{k_3^+-k_4}2^{-2k_2+6\max(k^+,k_2^+)}2^{3k_3/2}\|\widehat{V^{\psi,\iota_3}_{j_3,k_3}}(s)\|_{L^2} \\ &\quad\times \|\widehat{V^{h,\iota_2}_{j_2,k_2}}(s)\|_{L^\infty}2^{3k_4/2}\|\widehat{V^{h_4,\iota_4}_{j_4,k_4}}(s)\|_{L^2} \\ &\lesssim \varepsilon_1^3 2^{0.01m}2^{-j_3}2^{k_4}2^{-dk^*}.\end{aligned} \tag{6.2.48}$$

This completes the proof of (6.2.43) when $j_3 \geq 0.99m - 3k_3^+$.

**Substep 2.2.** Assume now that

$$j_3 \leq 0.99m - 3k_3^+. \tag{6.2.49}$$

We notice that the $\eta$ gradient of the phase $-s\Lambda_\theta(\xi-\eta-\rho)-s\Lambda_\nu(\eta)$ is $\gtrsim 2^m 2^{-2k_3^+}$

in the support of the integral in (6.2.44). Similarly, the $\rho$ gradient of the phase $-s\Lambda_\theta(\xi-\eta-\rho) - s\Lambda_\vartheta(\rho)$ is $\gtrsim 2^m 2^{-2k_3^+}$ in the support of the integral. Using Lemma 3.1 (integration by parts in $\eta$ or $\rho$), the contribution is negligible unless

$$j_2 \geq 0.99m - 3k_3^+ \quad \text{and} \quad j_4 \geq 0.99m - 3k_3^+. \tag{6.2.50}$$

Given (6.2.50), we estimate first, as in (6.2.47),

$$\begin{aligned} Y &\lesssim 2^{k_3^+ - k_4} 2^{-2k_2 + 6\max(k^+, k_2^+)} \|\widehat{V_{j_3,k_3}^{\psi,\iota_3}}(s)\|_{L^\infty} \\ &\quad \times 2^{3k_2/2} \|\widehat{V_{j_2,k_2}^{h,\iota_2}}(s)\|_{L^2} 2^{3k_4/2} \|\widehat{V_{j_4,k_4}^{h_4,\iota_4}}(s)\|_{L^2} \\ &\lesssim \varepsilon_1^3 2^{4\delta' m} 2^{-j_2 - j_4} 2^{-k_2} 2^{-k_3/2} 2^{-dk^*}. \end{aligned}$$

This suffices if $k_3 \geq -0.2m - 10$. On the other hand, if $k_3 \leq -0.2m - 10$ then we may assume that $\max\{k_2, k_4\} \geq -\kappa m - 10$ (due to the assumption $k \geq -\kappa m$) and estimate as in (6.2.46),

$$\begin{aligned} Y &\lesssim 2^{k_3^+ - k_4} 2^{-2k_2 + 6\max(k^+, k_2^+)} \|e^{-is\Lambda_\theta} V_{j_3,k_3}^{\psi,\iota_3}(s)\|_{L^\infty} \|V_{j_2,k_2}^{h,\iota_2}(s)\|_{L^2} \|V_{j_4,k_4}^{h_4,\iota_4}(s)\|_{L^2} \\ &\lesssim \varepsilon_1^3 2^{4\delta' m} 2^{-m} 2^{-j_2 - j_4} 2^{-5k_2/2} 2^{-3k_4/2} 2^{-dk^*}. \end{aligned}$$

Since $2^{-k_2} \lesssim 2^{p_0 m}$, this suffices to prove (6.2.43) when $k_4 \geq -0.1m - 10$. Finally, if $k_3, k_4 \leq -0.1m - 10$ and $k_2 \geq -\kappa m - 10$ then we estimate as in (6.2.48)

$$\begin{aligned} Y &\lesssim 2^{k_3^+ - k_4} 2^{-2k_2 + 6\max(k^+, k_2^+)} 2^{3k_3/2} \|\widehat{V_{j_3,k_3}^{\psi,\iota_3}}(s)\|_{L^2} \\ &\quad \times \|\widehat{V_{j_2,k_2}^{h,\iota_2}}(s)\|_{L^\infty} 2^{3k_4/2} \|\widehat{V_{j_4,k_4}^{h_4,\iota_4}}(s)\|_{L^2} \\ &\lesssim \varepsilon_1^3 2^{0.01m} 2^{-j_4} 2^{-j_2/2} 2^{-dk^*}, \end{aligned}$$

which suffices. This completes the proof of the the bounds (6.2.38).

**Step 3: proof of** (6.2.39). Notice that, for any $t \in [0, T]$ and $h \in \{h_{\alpha\beta}\}$,

$$\partial_t V^{h,\iota_2}(t) = e^{it\Lambda_{wa,\iota_2}} \mathcal{N}^h(t) = e^{it\Lambda_{wa,\iota_2}} \mathcal{N}^{h,2}(t) + e^{it\Lambda_{wa,\iota_2}} \mathcal{N}^{h,\geq 3}(t). \tag{6.2.51}$$

If $k_1 \leq -0.01m - 10$ then we may assume that $k_2 \geq -\kappa m - 10$ (due to the assumption $k \geq -\kappa m$) and estimate the left-hand side of (6.2.39) using just (3.1.12), (3.3.5), and (4.2.3),

$$C 2^m 2^{-k_2} 2^{6\overline{k}^+} \|\widehat{P_{k_1} V^\psi}\|_{L^2} \|\widehat{P_{k_2} \mathcal{N}^h}\|_{L^2} \lesssim \varepsilon_1^2 2^{k_1^-} 2^{2\kappa m},$$

which suffices. On the other hand, if $k_1 \geq -0.01m - 10$ then we decompose $P_{k_1} V^{\psi,\iota_1} = \sum_{j_1 \geq -k_1^-} V_{j_1,k_1}^{\psi,\iota_1}$, and notice that the contribution of the localized profiles for which $j_1 \geq 0.1m$ can also be bounded in a similar way, using (3.3.26) and estimating $\widehat{P_{k_2} \mathcal{N}^h}$ in $L^1$ to gain a factor of $2^{3k_2/2}$. The contribution of the cubic and higher order nonlinearity can also be bounded in the same way, using the stronger estimates (4.2.43). It remains to prove that

$$2^{-k_2^-}|\varphi_k(\xi)T_{\mu\nu}^{kg}[V_{j_1,k_1}^{\psi,\iota_1}(s), e^{is\Lambda_{wa,\iota_2}}P_{k_2}\mathcal{N}^{h,2}(s)](\xi)| \lesssim \varepsilon_1^2 2^{-1.005m}, \tag{6.2.52}$$

for any $s \in [2^{m-1}, 2^{m+1}]$, $k_2 \in [-p_0 m - 10, m/10]$, $k_1 \geq -0.01m - 10$, and $j_1 \leq 0.1m$.

We examine now the quadratic nonlinearities $\mathcal{N}_{\alpha\beta}^{h,2}$ in (2.1.11). They contain two types of terms: bilinear interactions of the metric components and bilinear interactions of the Klein-Gordon field. We define the trilinear operators

$$\begin{aligned} \mathcal{C}'_{kg}[f,g,h](\xi) := \int_{\mathbb{R}^3\times\mathbb{R}^3} & \frac{\varphi_k(\xi)\varphi_{k_2}(\eta)m(\xi-\eta,\eta)}{\Lambda_{kg}(\xi)-\Lambda_\mu(\xi-\eta)-\Lambda_\nu(\eta)} \\ & \times m_3(\eta-\rho)m_4(\rho)\cdot \widehat{f}(\xi-\eta)\widehat{g}(\eta-\rho)\widehat{h}(\rho)\,d\eta d\rho, \end{aligned} \tag{6.2.53}$$

where $m \in \mathcal{M}$, $m_3, m_4 \in \mathcal{M}_0$. For (6.2.52) it suffices to prove that

$$\begin{aligned} 2^{-k_2^-}2^{|k_3-k_4|}|\mathcal{C}'_{kg}[e^{-is\Lambda_\mu}V_{j_1,k_1}^{\psi,\iota_1}(s), e^{-is\Lambda_{wa,\iota_3}}V_{j_3,k_3}^{h_3,\iota_3}(s), e^{-is\Lambda_{wa,\iota_4}}V_{j_4,k_4}^{h_4,\iota_4}(s)](\xi)| \\ \lesssim \varepsilon_1^2 2^{-1.01m}2^{-\gamma(j_3+|k_3|+j_4+|k_4|)} \end{aligned} \tag{6.2.54}$$

and

$$\begin{aligned} 2^{-k_2^-}|\mathcal{C}'_{kg}[e^{-is\Lambda_\mu}V_{j_1,k_1}^{\psi,\iota_1}(s), e^{-is\Lambda_{kg,\iota_3}}V_{j_3,k_3}^{\psi,\iota_3}(s), e^{-is\Lambda_{kg,\iota_4}}V_{j_4,k_4}^{\psi,\iota_4}(s)](\xi)| \\ \lesssim \varepsilon_1^2 2^{-1.01m}2^{-\gamma(j_3+|k_3|+j_4+|k_4|)}, \end{aligned} \tag{6.2.55}$$

where $s, k_1, k_2, j_1$ are as in (6.2.52), $h_3, h_4 \in \{h_{\alpha\beta}\}$, and $(k_3, j_3), (k_4, j_4) \in \mathcal{J}$.

**Substep 3.1: proof of** (6.2.54). Using just $L^2$ estimates, we may assume that $k_4 \leq k_3 \leq m/10+10$. Notice that the $\eta$ gradient of the phase $-s\Lambda_{kg,\iota_1}(\xi-\eta) - s\Lambda_{wa,\iota_3}(\eta-\rho)$ is $\gtrsim 2^m 2^{-2k_1^+}$ in the support of the integral. Therefore, using integration by parts in $\eta$ (Lemma 3.1), the integral is negligible if $j_3 \leq 0.99m - 3k_1^+$. Similarly, by making the change of variables $\rho \to \eta-\rho$, the integral is negligible if $j_4 \leq 0.99m - 3k_1^+$. Finally, if $\min\{j_3, j_4\} \geq 0.99m - 3k_1^+$ then we use (3.1.12) to estimate the left-hand side of (6.2.54) by

$$\begin{aligned} C2^{-k_2}2^{k_3-k_4}2^{-k_2+6\max(k^+,k_1^+)}\|\widehat{V_{j_1,k_1}^{\psi,\iota_1}}(s)\|_{L^\infty} \\ \times 2^{3k_2/2}\|\widehat{V_{j_3,k_3}^{h_3,\iota_3}}(s)\|_{L^2}2^{3k_4/2}\|\widehat{V_{j_4,k_4}^{h_4,\iota_4}}(s)\|_{L^2} \\ \lesssim \varepsilon_1^3 2^{0.01m}2^{-k_2/2}2^{-k_1^-}2^{-j_3}2^{-j_4(1-\delta')}2^{-d\max\{k_1^+,k_3^+\}}. \end{aligned}$$

This suffices to prove (6.2.54).

**Substep 3.2: proof of** (6.2.55). Using Lemma 3.4 (ii) and (3.3.15), we estimate the left-hand side of (6.2.55) by

$$C2^{-2k_2+6\max(k^+,k_1^+)}\|e^{-is\Lambda_\mu}V^{\psi,\iota_1}_{j_1,k_1}(s)\|_{L^\infty}\|V^{\psi,\iota_3}_{j_3,k_3}(s)\|_{L^2}\|V^{\psi,\iota_4}_{j_4,k_4}(s)\|_{L^2}$$
$$\lesssim \varepsilon_1^3 2^{-1.49m}2^{-j_3}2^{-j_4}2^{-2k_2}2^{-k_3^+-k_4^+}.$$

This suffices if $2^{-0.48m}2^{-j_3}2^{-j_4}2^{-2k_2} \lesssim 2^{-\gamma(j_3+j_4)}$. Otherwise, if $(1-\gamma)(j_3+j_4)+0.48m \leq -2k_2-120$ then we may assume that $j_3 \leq j_4$ (so $j_3 \leq 0.45m-40$ since $k_2 \geq -0.68m-10$) and use (3.3.15) again to estimate the left-hand side of (6.2.55) by

$$C2^{-2k_2+6\max(k^+,k_1^+)}\|\widehat{V^{\psi,\iota_1}_{j_1,k_1}}(s)\|_{L^\infty}2^{3k_2/2}\|e^{-is\Lambda_{kg,\iota_3}}V^{\psi,\iota_3}_{j_3,k_3}(s)\|_{L^\infty}\|V^{\psi,\iota_4}_{j_4,k_4}(s)\|_{L^2}$$
$$\lesssim \varepsilon_1^3 2^{-1.49m}2^{-k_2^-/2}2^{-k_3^-/2}2^{-j_4}.$$

The bounds (6.2.55) follow since $2^{-j_4}2^{-k_3^-/2} \lesssim 1$. This completes the proof of the lemma. □

## 6.3 $Z$-NORM CONTROL OF THE METRIC COMPONENTS

In this section we prove the bounds (2.1.52) for the metric components.

**Proposition 6.9.** *With the hypothesis of Proposition 2.3, for any $t \in [0,T]$, $a,b \in \{1,2,3\}$ and $\alpha,\beta \in \{0,1,2,3\}$ we have*

$$\|V^F(t)\|_{Z_{wa}} + \|V^{\omega_a}(t)\|_{Z_{wa}} + \|V^{\vartheta_{ab}}(t)\|_{Z_{wa}} + \langle t\rangle^{-\delta}\|V^{h_{\alpha\beta}}(t)\|_{Z_{wa}} \lesssim \varepsilon_0. \quad (6.3.1)$$

The rest of the section is concerned with the proof of this proposition. As in the previous section, we need to first renormalize the profiles $V^{h_{\alpha\beta}}$. The nonlinear phase correction is determined only by the quadratic quasilinear components of the nonlinearity

$$\mathcal{Q}^2_{\alpha\beta} := \{-h_{00}\Delta + 2h_{0j}\partial_0\partial_j - h_{jk}\partial_j\partial_k\}h_{\alpha\beta}; \quad (6.3.2)$$

see (2.1.13). The point of the renormalization is to weaken some of the resonant bilinear interactions corresponding to very low frequencies of the metric components.

We define our nonlinear phase correction and the nonlinear profiles associated to the metric components as in subsection 6.2.1. As in (6.2.5), for any $h \in \{h_{\mu\nu}\}$ we define the low frequency component $h^{low}$ by $\widehat{h^{low}}(\rho,s) =$

$\varphi_{\leq 0}(\langle s\rangle^{p_0}\rho)\widehat{h}(\rho,s)$, $p_0 = 0.68$. Then we define the correction

$$\Theta_{wa}(\xi,t) := \int_0^t \Big\{ h_{00}^{low}(s\xi/\Lambda_{wa}(\xi),s)\frac{\Lambda_{wa}(\xi)}{2} + h_{0j}^{low}(s\xi/\Lambda_{wa}(\xi),s)\xi_j + h_{jk}^{low}(s\xi/\Lambda_{wa}(\xi),s)\frac{\xi_j\xi_k}{2\Lambda_{wa}(\xi)}\Big\}\,ds. \tag{6.3.3}$$

Finally, we define the nonlinear (modified) profiles of the metric components by

$$\widehat{V_*^G}(\xi,t) := e^{-i\Theta_{wa}(\xi,t)}\widehat{V^G}(\xi,t), \qquad G \in \{h_{\alpha\beta}, F, \omega_a, \vartheta_{ab}\}. \tag{6.3.4}$$

We notice that the functions $h^{low}$ are real-valued, thus $\Theta_{wa}$ is real-valued. Let $h_{\alpha\beta}^{high} := h_{\alpha\beta} - h_{\alpha\beta}^{low}$. For $X \in \{low,\, high\}$ let

$$\mathfrak{q}_{wa,\pm}^X(\rho,\eta,t) := \pm\widehat{h_{00}^X}(\eta,t)\frac{\Lambda_{wa}(\rho)}{2} + \widehat{h_{0j}^X}(\eta,t)\rho_j \pm \widehat{h_{jk}^X}(\eta,t)\frac{\rho_j\rho_k}{2\Lambda_{wa}(\rho)}. \tag{6.3.5}$$

To derive our main transport equations we start from the formulas

$$\partial_t V^{h_{\alpha\beta}} = e^{it\Lambda_{wa}}\mathcal{N}_{\alpha\beta}^h = e^{it\Lambda_{wa}}\mathcal{Q}_{\alpha\beta}^2 + e^{it\Lambda_{wa}}\mathcal{KG}_{\alpha\beta}^2 + e^{it\Lambda_{wa}}\mathcal{S}_{\alpha\beta}^2 + e^{it\Lambda_{wa}}\mathcal{N}_{\alpha\beta}^{h,\geq 3}; \tag{6.3.6}$$

see (2.1.11). The formulas in the first line of (2.1.36) show that, for $h \in \{h_{\alpha\beta}\}$,

$$-\widehat{\partial_j\partial_k h}(\rho) = \rho_j\rho_k\frac{i\widehat{U^{h,+}}(\rho) - i\widehat{U^{h,-}}(\rho)}{2\Lambda_{wa}(\rho)}, \qquad \widehat{\partial_0\partial_j h}(\rho) = i\rho_j\frac{\widehat{U^{h,+}}(\rho) + \widehat{U^{h,-}}(\rho)}{2}.$$

Thus, using (6.3.2), (6.3.5), and the identities $\widehat{U^{h,\pm}}(\rho,t) = e^{\mp it\Lambda_{wa}(\rho)}\widehat{V^{h,\pm}}(\rho,t)$, we have

$$\widehat{\mathcal{Q}_{\alpha\beta}^2}(\xi,t) = \frac{1}{(2\pi)^3}\sum_{\pm}\int_{\mathbb{R}^3} ie^{\mp it\Lambda_{wa}(\xi-\eta)}\widehat{V^{h_{\alpha\beta},\pm}}(\xi-\eta,t)\mathfrak{q}_{wa,\pm}(\xi-\eta,\eta,t)\,d\eta, \tag{6.3.7}$$

where $\mathfrak{q}_{wa,\pm} = \mathfrak{q}_{wa,\pm}^{low} + \mathfrak{q}_{wa,\pm}^{high}$. Finally, we notice that

$$\dot{\Theta}_{wa}(\xi,t) = \frac{1}{(2\pi)^3}\int_{\mathbb{R}^3}\mathfrak{q}_{wa,+}^{low}(\xi,\eta,t)e^{i\eta\cdot t\xi/\Lambda_{wa}(\xi)}\,d\eta. \tag{6.3.8}$$

Combining (6.3.6)–(6.3.8) we derive our main equations for the modified profiles $V_*^{h_{\alpha\beta}}$,

$$\partial_t\widehat{V_*^{h_{\alpha\beta}}}(\xi,t) = e^{-i\Theta_{wa}(\xi,t)}\{\partial_t\widehat{V^{h_{\alpha\beta}}}(\xi,t) - i\widehat{V^{h_{\alpha\beta}}}(\xi,t)\dot{\Theta}_{wa}(\xi,t)\} = \sum_{a=1}^{6}\mathcal{R}_a^{h_{\alpha\beta}}(\xi,t), \tag{6.3.9}$$

where

$$\mathcal{R}_1^{h_{\alpha\beta}}(\xi,t) := \frac{e^{-i\Theta_{wa}(\xi,t)}}{(2\pi)^3}\int_{\mathbb{R}^3} ie^{it\Lambda_{wa}(\xi)}e^{it\Lambda_{wa}(\xi-\eta)} \times \widehat{V^{h_{\alpha\beta},-}}(\xi-\eta,t)\mathfrak{q}^{low}_{wa,-}(\xi-\eta,\eta,t)\,d\eta, \tag{6.3.10}$$

$$\mathcal{R}_2^{h_{\alpha\beta}}(\xi,t) := \frac{e^{-i\Theta_{wa}(\xi,t)}}{(2\pi)^3}\int_{\mathbb{R}^3} i\{e^{it(\Lambda_{wa}(\xi)-\Lambda_{wa}(\xi-\eta))}\widehat{V^{h_{\alpha\beta}}}(\xi-\eta,t) \times \mathfrak{q}^{low}_{wa,+}(\xi-\eta,\eta,t) - e^{it(\xi\cdot\eta)/\Lambda_{wa}(\xi)}\widehat{V^{h_{\alpha\beta}}}(\xi,t)\mathfrak{q}^{low}_{wa,+}(\xi,\eta,t)\}\,d\eta, \tag{6.3.11}$$

$$\mathcal{R}_3^{h_{\alpha\beta}}(\xi,t) := \frac{e^{-i\Theta_{wa}(\xi,t)}}{(2\pi)^3}\sum_{\iota\in\{+,-\}}\int_{\mathbb{R}^3} ie^{it\Lambda_{wa}(\xi)}e^{-it\Lambda_{wa,\iota}(\xi-\eta)} \times \widehat{V^{h_{\alpha\beta},\iota}}(\xi-\eta,t)\mathfrak{q}^{high}_{wa,\iota}(\xi-\eta,\eta,t)\,d\eta, \tag{6.3.12}$$

$$\mathcal{R}_4^{h_{\alpha\beta}}(\xi,t) := e^{-i\Theta_{wa}(\xi,t)}e^{it\Lambda_{wa}(\xi)}\widehat{\mathcal{KG}^2_{\alpha\beta}}(\xi,t), \tag{6.3.13}$$

$$\mathcal{R}_5^{h_{\alpha\beta}}(\xi,t) := e^{-i\Theta_{wa}(\xi,t)}e^{it\Lambda_{wa}(\xi)}\widehat{\mathcal{S}^2_{\alpha\beta}}(\xi,t), \tag{6.3.14}$$

and

$$\mathcal{R}_6^{h_{\alpha\beta}}(\xi,t) := e^{-i\Theta_{wa}(\xi,t)}e^{it\Lambda_{wa}(\xi)}\widehat{\mathcal{N}^{h,\geq 3}_{\alpha\beta}}(\xi,t). \tag{6.3.15}$$

### 6.3.1 The First Reduction

We return now to the proof of Proposition 6.9. Since $|\widehat{V^G}(\xi,t)| = |\widehat{V^G_*}(\xi,t)|$, in view of the definition (6.3.4) it suffices to prove that

$$\begin{aligned} \|\varphi_k(\xi)\{\widehat{V_*^{h_{\alpha\beta}}}(\xi,t_2) - \widehat{V_*^{h_{\alpha\beta}}}(\xi,t_1)\}\|_{L^\infty_\xi} &\lesssim \varepsilon_0 2^{\delta m}2^{-k^- - \kappa k^-}2^{-N_0k^+},\\ \|\varphi_k(\xi)\{\widehat{V_*^{H}}(\xi,t_2) - \widehat{V_*^{H}}(\xi,t_1)\}\|_{L^\infty_\xi} &\lesssim \varepsilon_0 2^{-\delta m/2}2^{-k^- - \kappa k^-}2^{-N_0k^+}, \end{aligned} \tag{6.3.16}$$

for any $H \in \{F, \omega_a, \vartheta_{ab}\}$, $k \in \mathbb{Z}$, $m \geq 1$, and $t_1, t_2 \in [2^m - 2, 2^{m+1}] \cap [0, T]$. As before, we show first that the bounds (6.3.16) hold if $k$ is too small or if $k$ is too large (relative to $m$).

**Lemma 6.10.** *The bounds* (6.3.16) *hold if* $k \leq -(\delta'/\kappa)m$ *or if* $k \geq \delta' m - 10$.

*Proof.* As in Lemma 6.4, we use Propositions 5.1 and 6.2, and the inequalities

(3.2.62). Thus

$$\begin{aligned}\|\widehat{P_kV^{h_{\alpha\beta}}}(t)\|_{L^\infty} &\lesssim \varepsilon_0 2^{-k-\delta k^-/2}\langle t\rangle^{\delta'/2}2^{-N(0)k^+(1-\delta)/4}2^{-N(2)k^+(3+\delta)/4}\\ &\lesssim \varepsilon_0 2^{-k^- -\delta k^-/2}\langle t\rangle^{\delta'/2}2^{-N_0k^+-dk^+}.\end{aligned}$$

This suffices to prove (6.3.16) if $k\leq -(\delta'/\kappa)m$ or if $k\geq \delta' m-10$, as claimed. □

In the remaining range $k\in[-(\delta'/\kappa)m,\delta' m-10]$, we use the identities (6.3.9)–(6.3.15), so

$$\widehat{V_*^{h_{\alpha\beta}}}(\xi,t_2)-\widehat{V_*^{h_{\alpha\beta}}}(\xi,t_1)=\sum_{a=1}^{6}\int_{t_1}^{t_2}\mathcal{R}_a^{h_{\alpha\beta}}(\xi,s)\,ds.$$

We analyze the contributions of the nonlinear terms $\mathcal{R}_a^{h_{\alpha\beta}}$ separately, and prove the bounds (6.3.16). In fact, in all cases except for the semilinear wave interactions in the terms $\mathcal{R}_5^{h_{\alpha\beta}}$ we can prove stronger bounds, namely

$$\Big|\varphi_k(\xi)\int_{t_1}^{t_2}\mathcal{R}_a^{h_{\alpha\beta}}(\xi,s)\,ds\Big|\lesssim \varepsilon_1^2 2^{-\delta m/2}2^{-k^-}2^{-N_0k^+}, \tag{6.3.17}$$

with a gain of a factor of $2^{-\delta m/2}$ instead of a loss. See Lemmas 6.11, 6.12, 6.13, 6.14, and 6.19. In the case of semilinear wave interactions ($a=5$) we prove slightly weaker bounds, which still suffice for (6.3.16). See Lemma 6.20.

### 6.3.2 The Nonlinear Terms $\mathcal{R}_a^{h_{\alpha\beta}}$, $a\in\{1,2,4,6\}$

In this subsection we consider some of the easier cases, when we can prove the stronger bounds (6.3.17). As in (6.2.24), for $\mu,\nu\in\{(kg,+),(kg,-)\}$ or $\mu,\nu\in\{(wa,+),(wa,-)\}$, and $s\in[0,T]$ we define the operators $T_{\mu\nu}^{wa}$ by

$$T_{\mu\nu}^{wa}[f,g](\xi,s):=\int_{\mathbb{R}^3}\frac{e^{is\Phi_{(wa,+)\mu\nu}(\xi,\eta)}}{\Phi_{(wa,+)\mu\nu}(\xi,\eta)}m(\xi-\eta,\eta)\widehat{f}(\xi-\eta,s)\widehat{g}(\eta,s)\,d\eta, \tag{6.3.18}$$

where $\Phi_{(wa,+)\mu\nu}(\xi,\eta)=\Lambda_{wa}(\xi)-\Lambda_\mu(\xi-\eta)-\Lambda_\nu(\eta)$ (see (2.1.41)) and $m\in\mathcal{M}$ (see (3.2.41)).

**Lemma 6.11.** *The bounds* (6.3.17) *hold for* $m\geq 100$, $k\in[-\kappa m,\delta' m]$, *and* $a=6$.

*Proof.* We use (3.2.63), combined with either (4.2.3)–(4.2.5) or (4.2.43)–(4.2.44). Thus

$$\begin{aligned}\|\widehat{P_k\mathcal{N}_{\alpha\beta}^{h,2}}\|_{L^\infty}&\lesssim\varepsilon_1^2\langle t\rangle^{-1/2+\delta'}2^{-k^-/2}2^{-N(2)k^+},\\ \|\widehat{P_k\mathcal{N}_{\alpha\beta}^{h,\geq 3}}\|_{L^\infty}&\lesssim\varepsilon_1^2\langle t\rangle^{-5/4+6\delta'}2^{-3k^-/4}2^{-N(2)k^+}.\end{aligned} \tag{6.3.19}$$

The bounds in the second line suffice to prove (6.3.17) for $a = 6$. □

**Lemma 6.12.** *The bounds* (6.3.17) *hold for* $m \geq 100$, $k \in [-\kappa m, \delta' m]$, *and* $a = 1$.

*Proof.* This is similar to the proof of Lemma 6.6 and the main point is that the interaction is nonresonant so we can integrate by parts in time. We use the formula (6.3.5), substitute $h = i\Lambda_{wa}^{-1}(U^{h,+} - U^{h,-})/2$, $h \in \{h_{\alpha\beta}\}$, and decompose all the input functions dyadically in frequency. With $U_{low}^{h,\pm}$ and $V_{low}^{h,\pm}$ defined as in (6.2.26), it suffices to prove that

$$\sum_{(k_1,k_2)\in\mathcal{X}_k} 2^{k_1-k_2}\Big\|\varphi_k(\xi)\int_{t_1}^{t_2} e^{is\Lambda_{wa}(\xi)-i\Theta_{wa}(\xi,s)} \times \mathcal{F}\{I[P_{k_1}U^{h_1,-}, P_{k_2}U_{low}^{h_2,\iota_2}]\}(\xi,s)\,ds\Big\|_{L^\infty_\xi} \lesssim \varepsilon_1^2 2^{-\kappa m}, \tag{6.3.20}$$

for $h_1, h_2 \in \{h_{\alpha\beta}\}$ and $\iota_2 \in \{+,-\}$.

We estimate first, using just (3.3.7) and (3.3.3),

$$\begin{aligned}\big\|\mathcal{F}\{I[P_{k_1}U^{h_1,-}, P_{k_2}U_{low}^{h_2,\iota_2}]\}(\xi,s)\big\|_{L^\infty_\xi} &\lesssim \|\widehat{P_{k_1}U^{h_1,-}}\|_{L^\infty}\|\widehat{P_{k_2}U_{low}^{h_2,\iota_2}}\|_{L^1}\\ &\lesssim \varepsilon_1^2 2^{-k_1^- -\kappa k_1^-}2^{2k_2-\delta k_2}2^{2\delta m}2^{-4k_1^+}.\end{aligned} \tag{6.3.21}$$

This suffices to control the contribution of the pairs $(k_1, k_2)$ for which $k_2 \leq -1.01m$. After this reduction it remains to prove that

$$\Big|\int_{t_1}^{t_2}\int_{\mathbb{R}^3} e^{is\Lambda_{wa}(\xi)-i\Theta_{wa}(\xi,s)} m(\xi-\eta,\eta) e^{is\Lambda_{wa}(\xi-\eta)}\widehat{P_{k_1}V^{h_1,-}}(\xi-\eta,s) \times e^{-is\Lambda_{wa,\iota_2}(\eta)}\widehat{P_{k_2}V_{low}^{h_2,\iota_2}}(\eta,s)\,d\eta ds\Big| \lesssim \varepsilon_1^2 2^{-0.01m}2^{k_2}, \tag{6.3.22}$$

for any $\xi$ with $|\xi| \in [2^{k_1-4}, 2^{k_1+4}]$, provided that

$$k_2 \in [-1.01m, -p_0 m + 10], \qquad k_1 \in [-\kappa m - 10, \delta' m + 10]. \tag{6.3.23}$$

To prove (6.3.22) we integrate by parts in time. Notice that $\Lambda_{wa}(\xi) + \Lambda_{wa}(\xi - \eta) - \Lambda_{wa,\iota_2}(\eta) \gtrsim 2^{k_1^-}$ in the support of the integral. The left-hand side of (6.3.22) is dominated by $C(I_{wa} + II_{wa} + III_{wa})(\xi)$, where, with $\mu = (wa, -)$

and $\nu = (wa, \iota_2)$ and $T_{\mu\nu}^{wa}$ defined as in (6.3.18),

$$\begin{aligned}
I_{wa}(\xi) &:= \Big(1 + \int_{t_1}^{t_2} |\dot{\Theta}_{wa}(\xi, s)|\, ds\Big) \sup_{s\in[t_1,t_2]} |T_{\mu\nu}^{wa}[P_{k_1}V^{h_1,-}, P_{k_2}V_{low}^{h_2,\iota_2}](\xi, s)|,\\
II_{wa}(\xi) &:= \int_{t_1}^{t_2} |T_{\mu\nu}^{wa}[\partial_s(P_{k_1}V^{h_1,-}), P_{k_2}V_{low}^{h_2,\iota_2}](\xi, s)|\, ds,\\
III_{wa}(\xi) &:= \int_{t_1}^{t_2} |T_{\mu\nu}^{wa}[P_{k_1}V^{h_1,-}, \partial_s(P_{k_2}V_{low}^{h_2,\iota_2})](\xi, s)|\, ds.
\end{aligned} \tag{6.3.24}$$

As in (6.3.21), assuming $k_1, k_2$ as in (6.3.23), we estimate for any $s \in [t_1, t_2]$

$$|T_{\mu\nu}^{wa}[P_{k_1}V^{h_1,-}, P_{k_2}V_{low}^{h_2,\iota_2}](\xi, s)| \lesssim \varepsilon_1^2 2^{-2k_1^- - \kappa k_1^-} 2^{2k_2} 2^{4\delta m} 2^{-4k_1^+}.$$

It follows from (6.3.3) that $|\dot{\Theta}_{wa}(\xi, s)| \lesssim 2^{k_1^+} \sup_{\alpha,\beta} \|h_{\alpha\beta}(s)\|_{L^\infty} \lesssim 2^{-m+2\delta' m}$. Therefore

$$I_{wa}(\xi) \lesssim \varepsilon_1^2 2^{2k_2} 2^{0.01m}. \tag{6.3.25}$$

Similarly using (6.3.19) and the bounds $\|\widehat{P_{k_2}V_{low}^{h_2,\iota_2}}(s)\|_{L^1} \lesssim \varepsilon_1 2^{2k_2} 2^{4\delta m}$ we have

$$II_{wa}(\xi) \lesssim \varepsilon_1^2 2^{m/2} 2^{2k_2} 2^{2\kappa m}.$$

Finally, using (6.2.33), we see that $III_{wa}(\xi) \lesssim \varepsilon_1^2 2^{2k_2} 2^{0.01m}$, as in (6.3.25). The desired conclusion (6.3.22) follows. □

**Lemma 6.13.** *The bounds* (6.3.17) *hold for* $m \geq 100$, $k \in [-\kappa m, \delta' m]$, *and* $a = 2$.

*Proof.* This is similar to the proof of Lemma 6.7. With $h_1 = h_{\alpha\beta}$ we decompose $V^{h_1} = \sum_{(k_1,j_1)\in\mathcal{J}} V_{j_1,k_1}^{h_1,+}$ as in (3.3.1) and examine the definition (6.3.11). It suffices to prove that

$$\begin{aligned}
\varphi_k(\xi)\Big| \int_{\mathbb{R}^3} \{ & e^{it(\Lambda_{wa}(\xi)-\Lambda_{wa}(\xi-\eta))} \widehat{V_{j_1,k_1}^{h_1,+}}(\xi-\eta, t)\mathfrak{q}_{wa,+}^{low}(\xi-\eta, \eta, t)\\
& - e^{it(\xi\cdot\eta)/\Lambda_{wa}(\xi)} \widehat{V_{j_1,k_1}^{h_1,+}}(\xi, t)\mathfrak{q}_{wa,+}^{low}(\xi, \eta, t)\}\, d\eta\Big| \lesssim \varepsilon_1^2 2^{-m-\kappa m} 2^{-\delta j_1},
\end{aligned}$$

provided that $|k_1 - k| \leq 10$ and $t \in [2^{m-1}, 2^{m+1}]$. Using also the definitions (6.3.5) it suffices to prove that for any multiplier $m \in \mathcal{M}_0$ (see (3.2.40)) and

$\alpha, \beta \in \{0,1,2,3\}$ we have

$$\varphi_k(\xi)\Big|\int_{\mathbb{R}^3} \widehat{h^{low}_{\alpha\beta}}(\eta,t)\{e^{it(\Lambda_{wa}(\xi)-\Lambda_{wa}(\xi-\eta))}\widehat{V^{h_1,+}_{j_1,k_1}}(\xi-\eta,t)m(\xi-\eta)|\xi-\eta| \\ - e^{it(\xi\cdot\eta)/\Lambda_{wa}(\xi)}\widehat{V^{h_1,+}_{j_1,k_1}}(\xi,t)m(\xi)|\xi|\}\,d\eta\Big| \lesssim \varepsilon_1^2 2^{-m-\kappa m}2^{-\delta j_1}. \tag{6.3.26}$$

Recall that $\|\widehat{V^{h_1,+}_{j_1,k_1}}(t)\|_{L^\infty} \lesssim \varepsilon_1 2^{-3k_1/2}2^{-j_1/2+\delta j_1}2^{\delta' m}2^{-4k_1^+}$; see (3.3.26). Thus, without using the cancellation of the two terms in the integral, the left-hand side of (6.3.26) is bounded by

$$C\varepsilon_1 2^{-j_1/2+\delta j_1}2^{2\kappa m}\|\widehat{h^{low}_{\alpha\beta}}(t)\|_{L^1} \lesssim \varepsilon_1^2 2^{-j_1/2+\delta j_1}2^{3\kappa m}2^{-p_0 m}.$$

This suffices to prove (6.2.34) when $j_1$ is large, i.e., $2^{j_1/2} \gtrsim 2^{(1.01-p_0)m}$.

On the other hand, if $2^m \gg 1$ and $j_1/2 \le (1.01-p_0)m = 0.33m$ then we estimate

$$\big|e^{it(\Lambda_{wa}(\xi)-\Lambda_{wa}(\xi-\eta))} - e^{it(\xi\cdot\eta)/\Lambda_{wa}(\xi)}\big| \lesssim 2^{-2p_0 m+m}2^{2\kappa m},$$
$$\big|\widehat{V^{h_1,+}_{j_1,k_1}}(\xi-\eta,t)m(\xi-\eta)|\xi-\eta| - \widehat{V^{h_1,+}_{j_1,k_1}}(\xi,t)m(\xi)|\xi|\big| \lesssim \varepsilon_1 2^{j_1/2}2^{3\kappa m}2^{-p_0 m}, \tag{6.3.27}$$

provided that $|\xi| \approx 2^k$ and $|\eta| \lesssim 2^{-p_0 m}$. Indeed, the first bound follows from the observation that $\nabla\Lambda_{wa}(\xi) = \xi/\Lambda_{wa}(\xi)$. The second bound follows from (3.3.26) once we notice that taking $\partial_\xi$ derivative of the localized profiles $\widehat{V^{h_1,+}_{j_1,k_1}}$ corresponds essentially to multiplication by a factor of $2^{j_1}$. If $j_1/2 \le 0.33m$ it follows that the left-hand side of (6.3.26) is bounded by

$$C\varepsilon_1 2^{-0.34m}\|\widehat{h^{low}_{\alpha\beta}}(t)\|_{L^1} \lesssim \varepsilon_1^2 2^{-p_0 m-0.34m+\kappa m}.$$

This suffices to prove (6.3.26) when $j_1/2 \le 0.33m$, which completes the proof of the lemma. □

**Lemma 6.14.** *The bounds* (6.3.17) *hold for* $m \ge 100$, $k \in [-\kappa m, \delta' m]$, *and* $a = 4$.

*Proof.* Here we analyze bilinear interactions of the Klein-Gordon field with itself. The important observation is that these interactions are still nonresonant, due to Lemma 3.4. Recall that $\mathcal{KG}^2_{\alpha\beta} = 2\partial_\alpha\psi\partial_\beta\psi + \psi^2 m_{\alpha\beta}$. We express $\psi, \partial_\alpha\psi$ in terms of the normalized profiles $U^{\psi,\pm}$ and decompose dyadically in frequency.

It suffices to prove that

$$\Big\|\varphi_k(\xi)\int_{t_1}^{t_2} e^{is\Lambda_{wa}(\xi)-i\Theta_{wa}(\xi,s)}\mathcal{F}\{I[P_{k_1}U^{\psi,\iota_1},P_{k_2}U^{\psi,\iota_2}]\}(\xi,s)\,ds\Big\|_{L^\infty_\xi} \lesssim \varepsilon_1^2 2^{-50\delta' m}, \tag{6.3.28}$$

for any $(k_1,k_2)\in\mathcal{X}_k$ and $\iota_1,\iota_2\in\{+,-\}$.

The estimates (6.3.28) follow easily using the bounds (3.3.5) and (3.3.7) when $\min(k_1,k_2)\leq -3m/5$ or $\max(k_1,k_2)\geq m/20$. In the remaining range we integrate by parts in time. As before, notice that $|\dot{\Theta}_{wa}(\xi,s)|\lesssim 2^{-m+2\delta' m}$. With $T^{wa}_{\mu\nu}$ as in (6.3.18), it suffices to prove that

$$\big|\varphi_k(\xi)T^{wa}_{\mu\nu}[P_{k_1}V^{\psi,\iota_1},P_{k_2}V^{\psi,\iota_2}](\xi,s)\big|\lesssim \varepsilon_1^2 2^{-60\delta' m} \tag{6.3.29}$$

and

$$2^m\big|\varphi_k(\xi)T^{wa}_{\mu\nu}[\partial_s(P_{k_1}V^{\psi,\iota_1}),P_{k_2}V^{\psi,\iota_2}](\xi,s)\big|\lesssim \varepsilon_1^2 2^{-60\delta' m}, \tag{6.3.30}$$

where $\mu=(kg,\iota_1)$, $\nu=(kg,\iota_2)$, $s\in[2^{m-1},2^{m+1}]$, and $k_1,k_2\in[-3m/5,m/20]$.

**Step 1: proof of** (6.3.29). We decompose $P_{k_1}V^{\psi,\iota_1}=\sum_{j_1\geq -k_1^-}V^{\psi,\iota_1}_{j_1,k_1}$ and $P_{k_2}V^{\psi,\iota_2}=\sum_{j_2\geq -k_2^-}V^{\psi,\iota_2}_{j_2,k_2}$ as in (3.3.1). The contribution of the pairs $(V^{\psi,\iota_1}_{j_1,k_1},V^{\psi,\iota_2}_{j_2,k_2})$ with $\max(j_1,j_2)\geq 0.01m$ can be estimated easily, using the observation that $|\Phi_{(wa,+)\mu\nu}(\xi,\eta)|\gtrsim 2^{k^-}2^{-2\max(k_1^+,k_2^+)}$ in the support of the integral (see Lemma 3.4 (i)) and the $L^2$ bounds (3.3.4).

On the other hand, if $\max(j_1,j_2)\leq 0.1m$ then we have to show that

$$\Big|\varphi_k(\xi)\int_{\mathbb{R}^3}\frac{e^{-is[\Lambda_\mu(\xi-\eta)+\Lambda_\nu(\eta)]}}{\Phi_{(wa,+)\mu\nu}(\xi,\eta)}m(\xi-\eta,\eta) \times \widehat{V^{\psi,\iota_1}_{j_1,k_1}}(\xi-\eta,s)\widehat{V^{\psi,\iota_2}_{j_2,k_2}}(\eta,s)\,d\eta\Big|\lesssim \varepsilon_1^2 2^{-\kappa m}. \tag{6.3.31}$$

We observe that

$$|\nabla\Lambda_{kg}(x)-\nabla\Lambda_{kg}(y)|\gtrsim |x-y|/(1+|x|^4+|y|^4)\qquad \text{for any } x,y\in\mathbb{R}^3. \tag{6.3.32}$$

Therefore, using integration by parts in $\eta$ (Lemma 3.1), the left-hand side of (6.3.31) is $\lesssim \varepsilon_1^2 2^{-2m}$ if $\mu=-\nu$. On the other hand, if $\mu=\nu$ then the only space-resonant point (where the gradient of the phase vanishes) is $\eta=\xi/2$ and we insert cutoff functions of the form $\varphi_{\leq 0}(2^{0.4m}(\eta-\xi/2))$ and $\varphi_{>1}(2^{0.4m}(\eta-\xi/2))$. Then we estimate the integral corresponding to $|\eta-\xi/2|\lesssim 2^{-0.4m}$ by $C\varepsilon_1^2 2^{-1.1m}$, by placing the profiles $\mathcal{F}V^{\psi,\iota_1}_{j_1,k_1}$ and $\mathcal{F}V^{\psi,\iota_2}_{j_2,k_2}$ in $L^\infty$. Finally, we estimate the integral corresponding to $|\eta-\xi/2|\gtrsim 2^{-0.4m}$ by $C\varepsilon_1^2 2^{-2m}$, using integration by parts in $\eta$ and (6.3.32). This completes the proof of (6.3.31).

**Step 2: proof of** (6.3.30). Recall that $\partial_t V^{\psi,\iota_1}(t)=e^{it\Lambda_{kg,\iota_1}}\mathcal{N}^{\psi,2}(t)+$

$e^{it\Lambda_{kg,\iota_1}}\mathcal{N}^{\psi,\geq 3}(t)$; see (6.2.41). The contribution of the cubic and higher order nonlinearity $\mathcal{N}^{\psi,\geq 3}$ is easy to estimate, using just the $L^2$ bounds (4.2.52) and the lower bounds $|\Phi_{(wa,+)\mu\nu}(\xi,\eta)| \gtrsim 2^{k^-}2^{-2\max(k_1^+,k_2^+)}$, which hold in the support of the operator.

To bound the contribution of $\mathcal{N}^{\psi,2}$ we define the trilinear operators

$$\begin{aligned}\mathcal{C}_{wa}[f,g,h](\xi) := \int_{\mathbb{R}^3\times\mathbb{R}^3} &\frac{\varphi_k(\xi)\varphi_{k_1}(\xi-\eta)m(\xi-\eta,\eta)}{\Lambda_{wa}(\xi)-\Lambda_\mu(\xi-\eta)-\Lambda_\nu(\eta)} \\ &\times m_3(\xi-\eta-\rho)m_4(\rho)\cdot\widehat{f}(\xi-\eta-\rho)\widehat{g}(\eta)\widehat{h}(\rho)\,d\eta d\rho,\end{aligned} \tag{6.3.33}$$

where $m\in\mathcal{M}$, $m_3,m_4\in\mathcal{M}_0$. For (6.3.30) it remains to prove that

$$2^m\sum_{k_3,k_4\in\mathbb{Z}} 2^{k_3^+-k_4}\big|\mathcal{C}_{wa}[P_{k_3}U^{\psi,\iota_3},P_{k_2}U^{\psi,\iota_2},P_{k_4}U^{h,\iota_4}](\xi,s)\big| \lesssim \varepsilon_1^2 2^{-60\delta' m}, \tag{6.3.34}$$

where $\mu=(kg,\iota_1)$, $\nu=(kg,\iota_2)$, $s\in[2^{m-1},2^{m+1}]$, $h\in\{h_{\alpha\beta}\}$, and $k_1,k_2\in[-3m/5,m/20]$.

Using Lemma 3.4 (ii) and the bounds (3.3.5) and (3.3.11) we estimate

$$\begin{aligned}&\big|\mathcal{C}_{wa}[P_{k_3}U^{\psi,\iota_3},P_{k_2}U^{\psi,\iota_2},P_{k_4}U^{h,\iota_4}](\xi,s)\big| \\ &\quad\lesssim 2^{-k+6\max(k^+,k_2^+)}\|P_{k_3}U^{\psi}\|_{L^2}\|P_{k_2}U^{\psi}\|_{L^2}\|P_{k_4}U^{h}\|_{L^\infty} \\ &\quad\lesssim \varepsilon_1^3 2^{2\delta' m}2^{k_2^-}2^{k_3^-}2^{k_4^-(1-\delta)}\min(2^{-m},2^{k_4^-})2^{-20\max(k_2^+,k_3^+,k_4^+)}.\end{aligned} \tag{6.3.35}$$

This suffices to bound the contribution of the triplets $(k_2,k_3,k_4)$ for which $k_4\leq -1.01m$, or $k_3\leq -0.01m$, or $\max(k_2,k_3,k_4)\geq 10\delta' m$. In the remaining range we further decompose $P_{k_3}U^{\psi,\iota_3}=\sum_{j_3}e^{-it\Lambda_{kg,\iota_3}}V^{\psi,\iota_3}_{j_3,k_3}$ and $P_{k_2}U^{\psi,\iota_2}=\sum_{j_2}e^{-it\Lambda_{kg,\iota_2}}V^{\psi,\iota_2}_{j_2,k_2}$. Notice that the contribution of the pairs $(j_2,j_3)$ for which $\max(j_2,j_3)\geq 0.1m$ can be suitably bounded, using an estimate similar to (6.3.35). For (6.3.34) it remains to prove that

$$\big|\mathcal{C}_{wa}[e^{-it\Lambda_{kg,\iota_3}}V^{\psi,\iota_3}_{j_3,k_3},e^{-it\Lambda_{kg,\iota_2}}V^{\psi,\iota_2}_{j_2,k_2},P_{k_4}U^{h,\iota_4}](\xi,s)\big| \lesssim \varepsilon_1^3 2^{-1.01m}2^{k_4}, \tag{6.3.36}$$

provided that $k_2,k_3\in[-0.1m,10\delta' m]$, $k_4\in[-1.01m,10\delta' m]$, and $j_2,j_3\leq 0.1m$.

To prove this, we insert cutoff functions of the form $\varphi_{\leq 0}(2^{0.35m}(\rho-\xi))$ and $\varphi_{>1}(2^{0.35m}(\rho-\xi))$ in the integral in (6.3.33). The contribution of the integral when $|\rho-\xi|\lesssim 2^{-0.35m}$ (which is nontrivial only if $2^{k_4}\gtrsim 2^{-\kappa m}$) is bounded as claimed by estimating in the Fourier space, with $\widehat{V^{\psi,\iota_3}_{j_3,k_3}}$ and $\widehat{V^{\psi,\iota_2}_{j_2,k_2}}$ estimated in $L^2$ and $\widehat{P_{k_4}U^{h,\iota_4}}$ estimated in $L^\infty$.

On the other hand, the integral when $|\rho-\xi|\gtrsim 2^{-0.35m}$ can be estimated as in the proof of (6.3.31). Using (6.3.32), the $\eta$ integral is bounded by $C\varepsilon_1^2 2^{-1.1m}$, for any $\xi,\rho\in\mathbb{R}^3$. Then we notice that $\|\widehat{P_{k_4}U^{h,\iota_4}}(\rho,s)\|_{L^1_\rho}\lesssim \varepsilon_1 2^{2k_4}2^{\delta' m}$, and the desired conclusion (6.3.36) follows. This completes the proof of the lemma. □

### 6.3.3 Localized Bilinear Wave Interactions

In this subsection we start analyzing the remaining cases, where we have bilinear interactions of the metric components. These cases are more difficult because of the presence of time-resonant frequencies (parallel bilinear interactions), which prevent direct integration by parts in time.

For $b \in \mathbb{Z}$, $\xi \in \mathbb{R}^3$, and multipliers $m \in \mathcal{M}$ we define the bilinear operators

$$J_b[f,g](\xi) = J_{b;\iota_1\iota_2}[f,g](\xi) := \int_{\mathbb{R}^3} m(\xi-\eta,\eta)\widehat{f}(\xi-\eta)\widehat{g}(\eta)\varphi_b\big(\Xi_{\iota_1\iota_2}(\xi-\eta,\eta)\big)\,d\eta, \tag{6.3.37}$$

for $\iota_1, \iota_2 \in \{+,-\}$ (see the definition (3.1.23)). As in the proof of Lemma 3.6, we remark that an expression of the form $\varphi_b\big(\Xi_{\iota_1\iota_2}(\xi-\eta,\eta)\big)\cdot\varphi_k(\xi)\varphi_{k_1}(\xi-\eta)\varphi_{k_2}(\eta)$ can be nontrivial only if either $b \geq -20$ or $2^b \lesssim 2^{k-\max(k_1,k_2)}$.

We start with a lemma.

**Lemma 6.15.** *Assume* $m \geq 10$, $t \in [2^{m-1}, 2^{m+1}] \cap [0,T]$, $l, l_1, l_2 \in \mathbb{Z} \cap [-m, m/5+10]$.

*(i) If* $b \in [-m, 2]$, $(l_1, j_1), (l_2, j_2) \in \mathcal{J}$, *and* $n \in L^\infty(\mathbb{R}^3 \times \mathbb{R}^3)$ *then*

$$\begin{aligned}\Big|\varphi_l(\xi)\int_{\mathbb{R}^3} n(\xi,\eta)\widehat{V^{h_1,\iota_1}_{j_1,l_1}}(\xi-\eta,t)\widehat{V^{h_2,\iota_2}_{j_2,l_2}}(\eta,t)\varphi_{\leq b}(\Xi_{\iota_1\iota_2}(\xi-\eta,\eta))\,d\eta\Big|&\\ \lesssim \varepsilon_1^2\|n\|_{L^\infty}\cdot 2^{\delta' m}\min(2^{-2\max(l_1,l_2)}, 2^{2b-2l})2^{l_1+l_2}2^{-\max(j_1,j_2)}2^{-20(l_1^+ + l_2^+)},&\end{aligned} \tag{6.3.38}$$

*for any* $\xi \in \mathbb{R}^3$, $h_1, h_2 \in \{h_{\alpha\beta}\}$, *and* $\iota_1, \iota_2 \in \{+,-\}$.

*(ii) If* $b \geq (-m+l-l_1-l_2)/2 + \delta m/8$ *then, for any* $\xi \in \mathbb{R}^3$, $h_1, h_2 \in \{h_{\alpha\beta}\}$, *and* $\iota_1, \iota_2 \in \{+,-\}$,

$$\begin{aligned}\sum_{j_1 \geq -l_1^-,\, j_2 \geq -l_2^-} &\big|\varphi_l(\xi)J_b[U^{h_1,\iota_1}_{j_1,l_1}(t), U^{h_2,\iota_2}_{j_2,l_2}(t)]\big|\\ &\lesssim \varepsilon_1^2 2^{2\delta' m - m}\min(2^{-2\max(l_1,l_2)}, 2^{2b-2l})2^{-b+l_1+l_2}2^{-18(l_1^+ + l_2^+)}.\end{aligned} \tag{6.3.39}$$

*As a consequence, for any* $\xi \in \mathbb{R}^3$,

$$\begin{aligned}\sum_{b\leq 2}\sum_{j_1 \geq -l_1^-,\, j_2 \geq -l_2^-} &\big|\varphi_l(\xi)J_b[U^{h_1,\iota_1}_{j_1,l_1}(t), U^{h_2,\iota_2}_{j_2,l_2}(t)]\big|\\ &\lesssim \varepsilon_1^2 2^{2\delta' m - m}2^{\min(l_1,l_2)-l}2^{-18(l_1^+ + l_2^+)}.\end{aligned} \tag{6.3.40}$$

*Proof.* (i) We may assume that $\|n\|_{L^\infty} = 1$. Without loss of generality, by rotation, we may also assume that $j_2 \geq j_1$ and $\xi = (\xi_1, 0, 0)$, $\xi_1 \in [2^{l-1}, 2^{l+1}]$. We notice that the $\eta$ integral in the left-hand side is supported in the set $\mathcal{R}_{\leq b;\xi} := \{|\eta| \approx 2^{l_2},\ |\xi-\eta| \approx 2^{l_1},\ \sqrt{\eta_2^2+\eta_3^2} \lesssim X := \min(2^{l_1}, 2^{l_2}, 2^{b+l_1+l_2-l})\}$ (the last bound on $\sqrt{\eta_2^2+\eta_3^2}$ holds when $b \leq -20$; see (3.1.50)).

With $\mathbf{e}_1 := (1,0,0)$, we estimate the left-hand side of (6.3.38) by

$$C\|\widehat{V_{j_1,l_1}^{h_1,\iota_1}}(t)\|_{L^\infty}\int_{\mathbb{R}^3}|\widehat{V_{j_2,l_2}^{h_2,\iota_2}}(\eta,t)|\mathbf{1}_{\mathcal{R}_{\leq b;\xi}}(\eta)\,d\eta$$
$$\lesssim \|\widehat{V_{j_1,l_1}^{h_1,\iota_1}}(t)\|_{L^\infty}\int_{[0,\infty)\times\mathbb{S}^2,\,|\theta-\mathbf{e}_1|\lesssim 2^{-l_2}X}|\widehat{V_{j_2,l_2}^{h_2,\iota_2}}(r\theta,t)|\,r^2drd\theta.$$

Using (3.3.25), (3.2.7) (with $p=1/\delta$ large) and (3.3.4), we can further estimate the right-hand side of the expression above by

$$C\varepsilon_1 2^{\delta' m/2}2^{-22l_1^+}2^{-l_1^-}\|\widehat{V_{j_2,l_2}^{h_2,\iota_2}}(r\theta,s)\|_{L^2(r^2dr)L^p_\theta}\cdot(2^{-l_2}X)^{2/p'}2^{3l_2/2}$$
$$\lesssim \varepsilon_1^2 2^{\delta' m}2^{-20(l_1^++l_2^+)}2^{-j_2}2^{-l_1-l_2}\min\{2^{\min(l_1,l_2)},2^{b+l_1+l_2-l}\}^2.$$

The bound (6.3.38) follows.

(ii) Notice that (6.3.40) follows from (6.3.38)–(6.3.39) by summation over $b$. To prove (6.3.39) we notice first that the contribution of the pairs $(j_1,j_2)$ for which $\max(j_1,j_2)\geq m+b-\delta m$ is bounded as claimed, due to (6.3.38).

We claim that the contribution of the remaining pairs $(j_1,j_2)$ is negligible,

$$\big|\varphi_l(\xi)J_b[e^{-it\Lambda_{wa,\iota_1}}V_{j_1,l_1}^{h_1,\iota_1}(t),e^{-it\Lambda_{wa,\iota_2}}V_{j_2,l_2}^{h_2,\iota_2}(t)]\big|\lesssim\varepsilon_1^2 2^{-4m}2^{-20(l_1^++l_2^+)},\tag{6.3.41}$$

if $\max(j_1,j_2)\leq m+b-\delta m$. For this we would like to use Lemma 3.1. Notice that, in the support of the integral, we always have the lower bounds

$$\big|\nabla_\eta[\Lambda_{wa,\iota_1}(\xi-\eta)+\Lambda_{wa,\iota_2}(\eta)]\big|=\big|\Xi_{\iota_1\iota_2}(\xi-\eta,\eta)\big|\gtrsim 2^b.\tag{6.3.42}$$

We would like to use Lemma 3.1 with $K\approx 2^{m+b}$ and $\epsilon\approx 2^{\delta m/8}/K$. As in the proof of Lemma 3.6 let $H_{b;\xi}(\eta)=2^{-2b}|\Xi_{\iota_1\iota_2}(\xi-\eta,\eta)|^2$ such that $\varphi_b(\Xi_{\iota_1\iota_2}(\xi-\eta,\eta))=\varphi_0''(H_{b;\xi}(\eta))$, where $\varphi_0''(x):=\mathbf{1}_{[0,\infty)}(x)\varphi_0(\sqrt{x})$. If $2^b\gtrsim 1$ then the function $H_{b;\xi}$ satisfies bounds of the form $|D^\alpha_\eta H_{b;\xi}(\eta)|\lesssim_{|\alpha|}2^{|\alpha|\max(-l_1,-l_2)}$ in the support of the integral, for all multi-indices $\alpha\in\mathbb{Z}^3_+$, and the desired conclusion (6.3.41) follows from Lemma 3.1.

Assume now that $b\leq -20$ (so $2^b\lesssim 2^{l-\max(l_1,l_2)}$) and, as before, $\xi=(\xi_1,0,0)$, $\xi_1>0$. The formula (3.1.50) shows that

$$|D^\alpha_\eta H_{b;\xi}(\eta)|\lesssim_{|\alpha|}2^{-|\alpha|(b+l_1+l_2-l)},\qquad \alpha\in\mathbb{Z}^3_+,\tag{6.3.43}$$

for $\eta\in\mathcal{R}_{b;\xi}=\{|\eta|\approx 2^{l_2},\,|\xi-\eta|\approx 2^{l_1},\,\sqrt{\eta_2^2+\eta_3^2}\approx 2^{b+l_1+l_2-l}\}$. Notice that $\epsilon 2^{\max(j_1,j_2)}+\epsilon 2^{-(b+l_1+l_2-l)}\lesssim 2^{-\delta m/8}$, due to the assumptions $2b+m+l_1+l_2-l\geq\delta m/4$ and $2^{\max(j_1,j_2)}\lesssim K2^{-\delta m}$. The desired bounds (6.3.41) would follow from Lemma 3.1 if we could verify the second bound in (3.1.2). With $f:=K^{-1}s[\Lambda_{wa,\iota_1}(\xi-\eta)+\Lambda_{wa,\iota_2}(\eta)]$, we always have

$$|D^\alpha f(\eta)|\lesssim_{|\alpha|}K^{-1}2^m 2^{-(|\alpha|-1)\min(l_1,l_2)}\tag{6.3.44}$$

in the support of the integral. Since $2^{-\min(l_1,l_2)} \lesssim K2^{-\delta m/8}$, it suffices to verify the bounds (3.1.2) when $|\alpha| = 2$, i.e., $K^{-1}2^m 2^{-\min(l_1,l_2)} \lesssim K2^{-\delta m/8}$. In view of the assumption $2b \geq -m - l_1 - l_2 + l + \delta m$, this holds when $2^{l-\max(l_1,l_2)} \gtrsim 1$, and the desired conclusion (6.3.41) follows in this case.

If $l \leq \max(l_1, l_2) - 40$ then we need to be slightly more careful with the estimates (6.3.44). Since $b \leq -20$, we may assume that $b \leq l - \max(l_1, l_2) + 10$. We may also assume that $\iota_1 = -\iota_2$, since otherwise $\varphi_b(\Xi_{\iota_1\iota_2}(\xi - \eta, \eta)) \equiv 0$. We define $K, \epsilon, f$ as before and notice that the bounds (6.3.44) can be improved to

$$|D^\alpha f(\eta)| \lesssim_{|\alpha|} K^{-1}2^m 2^{l-\max(l_1,l_2)} 2^{-(|\alpha|-1)\max(l_1,l_2)},$$

in the support of the integral. This suffices to verify the bounds (3.1.2) in Lemma 3.1 in the remaining case, and completes the proof of (6.3.41). □

Our main result in this subsection is the following lemma, in which we show that the contribution of non-parallel wave interactions is suitably small.

**Lemma 6.16.** *Assume that $m \geq 100$, $t_1, t_2 \in [2^{m-1}, 2^{m+1}] \cap [0, T]$, $k, k_1, k_2 \in \mathbb{Z}$, $k \in [-\kappa m/4, \delta' m]$, $-p_0 m - 10 \leq k_2 \leq k_1 \leq m/10$, and $q \geq (-m + k - k_1 - k_2)/2 + \delta m/8$. Then*

$$\begin{aligned}\Big\| \varphi_k(\xi) \int_{t_1}^{t_2} e^{is\Lambda_{wa}(\xi) - i\Theta_{wa}(\xi,s)} J_q[P_{k_1}U^{h_1,\iota_1}, P_{k_2}U^{h_2,\iota_2}](\xi, s)\, ds \Big\|_{L^\infty_\xi} & \qquad (6.3.45)\\ \lesssim \varepsilon_1^2 2^{-0.001m} 2^{k_2^-},&\end{aligned}$$

*for any $h_1, h_2 \in \{h_{\alpha\beta}\}$ and $\iota_1, \iota_2 \in \{+, -\}$.*

*Proof.* We notice that the desired bounds follow directly from (6.3.39) if $q \leq -0.002m + 10$ (recall that $k, k_1 \geq -\kappa m/4 - 10$). On the other hand, if

$$q \in [-0.002m + 10, 2] \qquad (6.3.46)$$

then we integrate by parts in time. Notice that

$$\big|\Lambda_{wa}(\xi) - \Lambda_{wa,\iota_1}(\xi - \eta) - \Lambda_{wa,\iota_2}(\eta)\big|^{-1} \lesssim 2^{-2q-k_2} \qquad (6.3.47)$$

in the support of the integral, as a consequence of (3.1.31). We define the operators $T^{wa}_{\mu\nu;q}$ by

$$\begin{aligned}T^{wa}_{\mu\nu;q}[f, g](\xi, s) := \int_{\mathbb{R}^3} & \frac{e^{is\Phi_{(wa,+)\mu\nu}(\xi,\eta)}}{\Phi_{(wa,+)\mu\nu}(\xi, \eta)} m(\xi - \eta, \eta) \qquad (6.3.48)\\ & \times \widehat{f}(\xi - \eta, s)\widehat{g}(\eta, s)\varphi_q\big(\Xi_{\iota_1\iota_2}(\xi - \eta, \eta)\big)\, d\eta,\end{aligned}$$

where $\mu = (wa, \iota_1)$, $\nu = (wa, \iota_2)$, and $\Phi_{(wa,+)\mu\nu}(\xi, \eta) = \Lambda_{wa}(\xi) - \Lambda_\mu(\xi - \eta) - \Lambda_\nu(\eta)$. As in Lemma 6.12, we integrate by parts in time and recall that $|\dot{\Theta}_{wa}(\xi, s)| \lesssim 2^{-m+2\delta' m}$. For (6.3.45) it suffices to prove that, for any

$s \in [2^{m-1}, 2^{m+1}]$,

$$|\varphi_k(\xi) T^{wa}_{\mu\nu;q}[P_{k_1} V^{h_1,\iota_1}, P_{k_2} V^{h_2,\iota_2}](\xi, s)| \lesssim \varepsilon_1^2 2^{-0.001m - 10\delta' m} 2^{k_2^-}, \tag{6.3.49}$$

$$2^m |\varphi_k(\xi) T^{wa}_{\mu\nu;q}[P_{k_1} V^{h_1,\iota_1}, \partial_s(P_{k_2} V^{h_2,\iota_2})](\xi, s)| \lesssim \varepsilon_1^2 2^{-0.001m - 10\delta' m} 2^{k_2^-}, \tag{6.3.50}$$

$$2^m |\varphi_k(\xi) T^{wa}_{\mu\nu;q}[\partial_s(P_{k_1} V^{h_1,\iota_1}), P_{k_2} V^{h_2,\iota_2}](\xi, s)| \lesssim \varepsilon_1^2 2^{-0.001m - 10\delta' m} 2^{k_2^-}. \tag{6.3.51}$$

**Step 1: proof of** (6.3.49). We decompose $P_{k_1} V^{h_1,\iota_1} = \sum_{j_1} V^{h_1,\iota_1}_{j_1,k_1}$ and $P_{k_2} V^{h_2,\iota_2} = \sum_{j_2} V^{h_2,\iota_2}_{j_2,k_2}$. The contribution of $(j_1, j_2)$ with $\max(j_1, j_2) \leq 0.99m$ is negligible, due to Lemma 3.1, the assumptions $q \geq -0.002m - 10$, $k_2 \geq -p_0 m - 10$, and the bounds (6.3.42), (6.3.47). We estimate also

$$\begin{aligned}|\varphi_k(\xi) T^{wa}_{\mu\nu;q}[V^{h_1,\iota_1}_{j_1,k_1}, V^{h_2,\iota_2}_{j_2,k_2}](\xi, s)| &\lesssim 2^{-2q-k_2} \|\widehat{V^{h_1,\iota_1}_{j_1,k_1}}(s)\|_{L^\infty} 2^{3k_2} \|\widehat{V^{h_2,\iota_2}_{j_2,k_2}}(s)\|_{L^\infty} \\ &\lesssim \varepsilon_1^2 2^{5\delta' m} 2^{-4k_1^+} 2^{2k_2 - 2q} (2^{-3k_2/2} 2^{-j_2/2 + \delta j_2})(2^{-3k_1/2} 2^{-j_1/2 + \delta j_1}) \\ &\lesssim \varepsilon_1^2 2^{0.01m} 2^{-2q} 2^{k_2^-/2} 2^{-(j_1+j_2)/2 + \delta(j_1+j_2)}\end{aligned}$$

using (3.3.26). Since $2^{-2q} \lesssim 2^{0.01m}$ (due to (6.3.46)) and $2^{-k_2^-} \lesssim 2^{p_0 m}$, this suffices to control the contribution of the pairs $(j_1, j_2)$ with $j_1 + j_2 \geq 0.99m$. The bounds (6.3.49) follow.

**Step 2: proof of** (6.3.50). Recall (6.2.51). The contribution of the cubic terms $\mathcal{N}^{h,\geq 3}(s)$ can be estimated easily, using the bounds in the second line of (6.3.19). The quadratic nonlinearities $\mathcal{N}^{h,2}$ contain two main types of terms: bilinear interactions of the metric components and bilinear interactions of the Klein-Gordon field (see (2.1.11)). The desired bounds follow from (6.3.53)–(6.3.54) in Lemma 6.17 below.

**Step 3: proof of** (6.3.51). As before, the contribution of the cubic and higher order nonlinearities $\mathcal{N}^{h,\geq 3}(s)$ can be estimated using the bounds in the second line of (6.3.19). The quadratic nonlinearities $\mathcal{N}^{h,2}$ can be estimated using the change of variables $\eta \to \xi - \eta$ and the bounds (6.3.55)–(6.3.56) in Lemma 6.17 below. □

We estimate now the trilinear operators arising in the proof of the previous lemma:

**Lemma 6.17.** *For $m_3, m_4 \in \mathcal{M}_0$, $m \in \mathcal{M}$ (see (3.2.41)) we define the trilinear* *operators*

$$\mathcal{C}^{q,l}_{wa}[f,g,h](\xi) := \int_{\mathbb{R}^3\times\mathbb{R}^3} \frac{\varphi_k(\xi)m(\xi-\eta,\eta)}{\Lambda_{wa}(\xi)-\Lambda_\mu(\xi-\eta)-\Lambda_\nu(\eta)}\varphi_q(\Xi_{\iota_1\iota_2}(\xi-\eta,\eta))\varphi_l(\eta)$$
$$\times m_3(\eta-\rho)m_4(\rho)\cdot\widehat{f}(\xi-\eta)\widehat{g}(\eta-\rho)\widehat{h}(\rho)\,d\eta d\rho, \tag{6.3.52}$$

*where* $q,l\in\mathbb{Z}$, *and* $\mu=(wa,\iota_1),\nu=(wa,\iota_2)$. *Assume that* $m\geq 100$, $s\in[2^{m-1},2^{m+1}]\cap[0,T]$, $k,k_1,k_2\in\mathbb{Z}$, $k\in[-\kappa m/4,\delta' m]$, $-p_0m-10\leq k_2\leq k_1\leq m/10$, *and* $q\geq -0.002m+10$. *Then*

$$2^{-k_2^-}2^{|k_3-k_4|}\big|\mathcal{C}^{q,k_2}_{wa}[U^{h_1,\iota_1}_{j_1,k_1}(s),U^{h_3,\iota_3}_{j_3,k_3}(s),U^{h_4,\iota_4}_{j_4,k_4}(s)](\xi)\big|\lesssim\varepsilon_1^3 2^{-1.002m}, \tag{6.3.53}$$

$$2^{-k_2^-}\big|\mathcal{C}^{q,k_2}_{wa}[U^{h_1,\iota_1}_{j_1,k_1}(s),U^{\psi,\iota_3}_{j_3,k_3}(s),U^{\psi,\iota_4}_{j_4,k_4}(s)](\xi)\big|\lesssim\varepsilon_1^3 2^{-1.002m}, \tag{6.3.54}$$

$$2^{-k_2^-}2^{|k_3-k_4|}\big|\mathcal{C}^{q,k_1}_{wa}[U^{h_2,\iota_2}_{j_2,k_2}(s),U^{h_3,\iota_3}_{j_3,k_3}(s),U^{h_4,\iota_4}_{j_4,k_4}(s)](\xi)\big|\lesssim\varepsilon_1^3 2^{-1.002m}, \tag{6.3.55}$$

*and*

$$2^{-k_2^-}\big|\mathcal{C}^{q,k_1}_{wa}[U^{h_2,\iota_2}_{j_2,k_2}(s),U^{\psi,\iota_3}_{j_3,k_3}(s),U^{\psi,\iota_4}_{j_4,k_4}(s)](\xi)\big|\lesssim\varepsilon_1^3 2^{-1.002m}, \tag{6.3.56}$$

*for any* $(k_1,j_1),(k_2,j_2),(k_3,j_3),(k_4,j_4)\in\mathcal{J}$ *and* $h_1,h_2,h_3,h_4\in\{h_{\alpha\beta}\}$.

*Proof.* Let $Y_1,Y_2,Y_3,Y_4$ denote the expressions in the left-hand sides of (6.3.53), (6.3.54), (6.3.55), and (6.3.56) respectively. We may assume that $\xi=(\xi_1,0,0)$, $\xi_1>0$. We remark that the bounds (6.3.53) and (6.3.55) are different because $k_2$ can be very small, $k_2\geq -p_0m-10$, but $k_1$ cannot be so small, $k_1\geq-\kappa m/4-4$. The same remark applies to the bounds (6.3.54) and (6.3.56).

**Step 1: proof of** (6.3.53)**.** We may assume that $k,k_1\geq -\kappa m/4-10$. We estimate first, using (6.3.47), (6.3.40) (or (3.3.3) if $\min(k_3,k_4)\leq -m+\delta m$), and (3.3.26)

$$\begin{aligned}Y_1&\lesssim 2^{-k_2^-}2^{|k_3-k_4|}2^{-2q-k_2}\|\widehat{U^{h_1,\iota_1}_{j_1,k_1}}(s)\|_{L^\infty}\\&\quad\times\varepsilon_1^2 2^{3k_2}2^{2\delta' m-m}2^{-k_2+\min\{k_3,k_4\}}2^{-10\max\{k_3^+,k_4^+\}}\\&\lesssim\varepsilon_1^3 2^{-0.995m}2^{-j_1/2+\delta j_1}2^{\max\{k_3^-,k_4^-\}}2^{-8\max\{k_3^+,k_4^+\}}2^{-8k_1^+}.\end{aligned}\tag{6.3.57}$$

This suffices to prove the desired bounds unless

$$j_1\leq 0.02m,\quad k_1\in[-\kappa m/4-10,0.01m],\quad \max\{k_3,k_4\}\in[-0.01m,0.01m]. \tag{6.3.58}$$

On the other hand, if these inequalities hold, then we analyze several subcases.

**Substep 1.1.** Assume first that the inequalities in (6.3.58) hold and, in addition,

$$\min\{k_3, k_4\} \le -0.03m - 30. \tag{6.3.59}$$

By symmetry we may assume that $k_4 \le k_3$, therefore $|k_2 - k_3| \le 4$. We fix $\rho$ with $|\rho| \le 2^{k_4+2}$ and estimate the $\eta$ integral by $C2^{-2q-k_2}2^{-0.99m}2^{-4k_1^+}2^{-4k_3^+}$ using Lemma 6.15 with $2^l \approx 2^k$, $2^{l_1} \approx 2^{k_1}$, $2^{l_2} \approx 2^{k_3}$. Thus

$$Y_1 \lesssim 2^{-k_2^-} 2^{k_3-k_4} 2^{-2q-k_2} 2^{-0.99m} 2^{-4k_1^+} 2^{-4k_3^+} 2^{(2-\delta)k_4}.$$

The conclusion follows in this case.

**Substep 1.2.** Assume now that the inequalities in (6.3.58) hold and, in addition,

$$\min\{k_3, k_4\} \ge -0.03m - 30 \quad \text{and} \quad \max\{j_3, j_4\} \ge 0.97m + k_2^- - 100. \tag{6.3.60}$$

We estimate first in the physical space, using (3.1.3), (3.1.34), and (3.3.11),

$$\begin{aligned} Y_1 &\lesssim 2^{-k_2^-} 2^{|k_3-k_4|} 2^{-2q-k_2} \|U^{h_1,\iota_1}_{j_1,k_1}(s)\|_{L^\infty} \|U^{h_3,\iota_3}_{j_3,k_3}(s)\|_{L^2} \|U^{h_4,\iota_4}_{j_4,k_4}(s)\|_{L^2} \\ &\lesssim \varepsilon_1^3 2^{-0.95m} 2^{-2k_2^-} 2^{-(1-\delta)(j_3+j_4)} 2^{-10k_1^+}, \end{aligned} \tag{6.3.61}$$

which suffices if $k_2 \ge -0.3m - 30$.

On the other hand, if $k_2 \in [-p_0 m - 10, -0.3m - 30]$ then we may assume that $|k_3 - k_4| \le 4$ and, by symmetry, $j_3 \le j_4$. Using just $L^2$ estimates on the two components, we notice that the $\rho$ integral is bounded by $C2^{2\delta' m}2^{-j_3}2^{-k_3/2}2^{-j_4}2^{-k_4/2}2^{-4k_3^+}$. This would suffice if $j_3, j_4$ satisfied slightly stronger bounds, such as $j_4 \ge 1.03m + k_2 - 100$ or $j_3 \ge 0.1m$. In the remaining case, when

$$\big|j_4 - m - k_2 + 100\big| \le 0.03m \qquad \text{and} \qquad j_3 \le 0.1m$$

we need to gain by integration by parts in $\eta$.

Recall that $\xi = (\xi_1, 0, 0)$ and insert cutoff functions of the form $\varphi_{\le n}(\rho_2, \rho_3)$ and $\varphi_{>n}(\rho_2, \rho_3)$, where $n := -0.15m + 100$. More precisely, for $* \in \{\le n, > n\}$ we define

$$\begin{aligned} G_*(\xi) := 2^{-k_2^-} \int_{\mathbb{R}^3\times\mathbb{R}^3} & \frac{\varphi_k(\xi) m(\xi-\eta,\eta)\varphi_{k_2}(\eta)\varphi_q(\Xi_{\iota_1\iota_2}(\xi-\eta,\eta))}{\Lambda_{wa}(\xi) - \Lambda_\mu(\xi-\eta) - \Lambda_\nu(\eta)} \varphi_*(\rho_2,\rho_3) \\ & \times m_3(\eta-\rho) m_4(\rho) e^{-is[\Lambda_{wa,\iota_1}(\xi-\eta)+\Lambda_{wa,\iota_3}(\eta-\rho)+\Lambda_{wa,\iota_4}(\rho)]} \\ & \times \widehat{V^{h_1,\iota_1}_{j_1,k_1}}(\xi-\eta,s) \widehat{V^{h_3,\iota_3}_{j_3,k_3}}(\eta-\rho,s) \widehat{V^{h_4,\iota_4}_{j_4,k_4}}(\rho,s)\, d\eta d\rho. \end{aligned}$$

Notice that the $\eta$ derivative of the phase is bounded from below

$$\begin{aligned} \big|\nabla_\eta[\Lambda_{wa,\iota_1}(\xi-\eta) + \Lambda_{wa,\iota_3}(\eta-\rho)]\big| &\ge \frac{|(\rho_2,\rho_3)|}{|\eta-\rho|} - |\eta|\Big(\frac{1}{|\xi-\eta|} + \frac{1}{|\eta-\rho|}\Big) \\ &\gtrsim 2^{n-k_3}, \end{aligned}$$

in the support of the integral defining $G_{>n}$. Thus $|G_{>n}(\xi)| \lesssim 2^{-2m}$, using integration by parts in $\eta$ with Lemma 3.1 (recall that $j_1, j_3 \leq 0.1m$).

Finally we estimate $|G_{\leq n}(\xi)|$ as in the proof of Lemma 6.15 (i), with $p = 1/\delta$,

$$\begin{aligned} |G_{\leq n}(\xi)| &\lesssim 2^{-k_2} 2^{-2q-k_2} 2^{3k_2} \|\widehat{V^{h_1,\iota_1}_{j_1,k_1}}(s)\|_{L^\infty} \|\widehat{V^{h_3,\iota_3}_{j_3,k_3}}(s)\|_{L^\infty} \\ &\quad \times \int_{\mathbb{R}^3} \varphi_{\leq n}(\rho_2,\rho_3) |\widehat{V^{h_4,\iota_4}_{j_4,k_4}}(\rho,s)|\,d\rho \\ &\lesssim \varepsilon_1^2 2^{k_2} 2^{0.04m} \|\widehat{V^{h_4,\iota_4}_{j_4,k_4}}(r\theta,s)\|_{L^2(r^2dr)L^p_\theta} \cdot 2^{(2n-2k_4)/p'} 2^{3k_4/2} \\ &\lesssim \varepsilon_1^3 2^{0.1m} 2^{2n} 2^{k_2} 2^{-j_4}. \end{aligned}$$

The conclusion follows since $2^{-j_4+k_2} \lesssim 2^{-0.97m}$—see (6.3.60)—and $2^{2n} \lesssim 2^{-0.3m}$.

**Substep 1.3.** Assume now that the inequalities in (6.3.58) hold and, in addition,

$$\min\{k_3, k_4\} \geq -0.03m - 30 \quad \text{and} \quad \max\{j_3, j_4\} \leq 0.97m + k_2^- - 100. \tag{6.3.62}$$

By symmetry, we may assume that $k_4 \leq k_3$. We insert first cutoff functions of the form $\varphi_{q'}(\Xi_{\iota_3\iota_4}(\eta-\rho,\rho))$ in the integral (6.3.52). Using (6.3.41) (with $2^{l_1} \approx 2^{k_3}$, $2^{l_2} \approx 2^{k_4}$, $2^l \approx 2^{k_2}$, $b = q'$) the contribution is negligible if $q' \geq k_2^- - 0.025m$. On the other hand, if

$$q'_0 \leq q' \leq k_2^- - 0.025m, \quad \text{where} \quad q'_0 := -m/2 + k_2/2 + 0.03m, \tag{6.3.63}$$

then the $\rho$ integral is bounded by $C\varepsilon_1^2 2^{2\delta' m - 1.02m} 2^{k_3+k_4-k_2} 2^{-8k_3^+}$ (as a consequence of (6.3.39)) for any $\eta$. The desired bounds then follow, once we notice that the $\eta$ integral gains a factor of $2^{3k_2}$.

To bound the contribution of $\varphi_{\leq q'_0}(\Xi_{\iota_3\iota_4}(\eta-\rho,\rho))$, we further insert cutoff functions of the form $\varphi_{\leq n'}(\rho')$ and $\varphi_{>n'}(\rho')$, where $n' := -0.05m + k_4 + 100$ and $\rho' = (\rho_2, \rho_3)$. More precisely, as before, for $* \in \{\leq n', > n'\}$ we define

$$\begin{aligned} G'_*(\xi) &:= 2^{-k_2^-} \int_{\mathbb{R}^3\times\mathbb{R}^3} \frac{\varphi_k(\xi) m(\xi-\eta,\eta) \varphi_{k_2}(\eta) \varphi_q(\Xi_{\iota_1\iota_2}(\xi-\eta,\eta))}{\Lambda_{wa}(\xi) - \Lambda_\mu(\xi-\eta) - \Lambda_\nu(\eta)} \\ &\times \varphi_{\leq q'_0}(\Xi_{\iota_3\iota_4}(\eta-\rho,\rho)) e^{-is[\Lambda_{wa,\iota_1}(\xi-\eta)+\Lambda_{wa,\iota_3}(\eta-\rho)+\Lambda_{wa,\iota_4}(\rho)]} \\ &\times \varphi_*(\rho') m_3(\eta-\rho) m_4(\rho) \widehat{V^{h_1,\iota_1}_{j_1,k_1}}(\xi-\eta,s) \widehat{V^{h_3,\iota_3}_{j_3,k_3}}(\eta-\rho,s) \widehat{V^{h_4,\iota_4}_{j_4,k_4}}(\rho,s)\,d\eta d\rho. \end{aligned}$$

For (6.3.53) it remains to prove that

$$2^{k_3-k_4}|G'_{\leq n'}(\xi)| + 2^{k_3-k_4}|G'_{>n'}(\xi)| \lesssim \varepsilon_1^3 2^{-1.01m}. \tag{6.3.64}$$

Notice that the integral in the definition of $G'_{\leq n'}$ is supported in the set

$$\{(\eta,\rho) : |\rho'| \lesssim 2^{n'},\ \widetilde{\Xi}(\eta,\rho) \lesssim 2^{k_3-k_2} 2^{q'_0}\},$$

due to (3.1.28) and the assumption $\Xi_{\iota_3\iota_4}(\eta-\rho,\rho)\lesssim 2^{q_0'}$. Therefore, using also Lemma 3.19,

$$|G'_{\leq n'}(\xi)|\lesssim 2^{-k_2}2^{2q-k_2}\cdot 2^{2n'}2^{k_4}2^{k_2}(2^{k_3}2^{q_0'})^2\cdot\varepsilon_1^3 2^{-k_1-k_3-k_4}2^{-8k_1^+-8k_3^+}2^{2\delta' m}\\ \lesssim\varepsilon_1^3 2^{-1.02m}2^{2k_4}2^{-4k_3^+}. \tag{6.3.65}$$

Finally we have to bound the functions $G'_{>n'}$. This can be done as in (6.3.65) if $j_3\geq 0.3m$, since the gain of $2^{2n'}$ can be replaced by a gain of $2^{-j_3/2}$ coming from Lemma 3.19. On the other hand, if $j_3\leq 0.3m$ then we claim that

$$|G'_{>n'}(\xi)|\lesssim\varepsilon_1^3 2^{-2m}. \tag{6.3.66}$$

To see this we use integration by parts in $\eta$. We show that

$$|\nabla_\eta\{\Lambda_{wa,\iota_1}(\xi-\eta)+\Lambda_{wa,\iota_3}(\eta-\rho)\}|\gtrsim 2^{-0.46m}2^{-k_2/2} \tag{6.3.67}$$

in the support of the integral defining $G'_{>n'}(\xi)$. In view of Lemma 3.1 (with $K\approx 2^{0.54m}2^{-k_2/2}$, $\epsilon=K^{-1}2^{\delta m}$), and recalling (6.3.58), this would suffice to prove (6.3.66).

To prove (6.3.67), assume for contradiction that it fails, so $\Xi_{\iota_1\iota_3}(\xi-\eta,\eta-\rho)\leq 2^{-0.46m-k_2/2}$ for some $\eta,\rho$ in the support of the integral defining $G'_{>n'}(\xi)$. We may assume also that $m\geq 1/\delta$. Since $\Xi_{\iota_3\iota_4}(\eta-\rho,\rho)\leq 2^{-0.47m+k_2/2}$, it follows from (3.1.28) that $\widetilde{\Xi}(\eta,\rho)\leq 2^{-0.47m-k_2/2+k_3+4}$ and it follows from (3.1.26)–(3.1.27) that $\widetilde{\Xi}(\xi-\eta,\rho)\leq 2^{-0.46m-k_2/2+4}$. Using again (3.1.28) we have $\widetilde{\Xi}(\xi,\rho)\leq 2^{-0.44m-k_2/2}$, in contradiction with the assumption $|\rho'|\geq 2^{n'-4}\geq 2^{-0.05m+k_4+90}$ (recall that $-k_2/2\leq p_0m/2+10\leq 0.34m+10$). This completes the proof of (6.3.66).

**Step 2: proof of** (6.3.54)**.** An estimate similar to (6.3.57) still holds, using (4.1.34)–(4.1.35) instead of (6.3.40). This proves the desired conclusion when $j_1\geq 0.1m$ or when $\max\{k_3,k_4\}\geq 0.01m$. On the other hand, if $j_1\leq 0.1m$ and $\max\{k_3,k_4\}\leq 0.01m$ then we notice that $|\nabla_\eta[s\Lambda_{wa,\iota_1}(\xi-\eta)+s\Lambda_{kg,\iota_3}(\eta-\rho)]|\gtrsim 2^m 2^{-2k_3^+}$ in the support of the integral. Therefore, using integration by parts in $\eta$ with Lemma 3.1, $Y_2$ is negligible if $j_3\leq 0.9m$. Similarly, after making the change of variables $\rho\to\eta-\rho$, $Y_2$ is negligible if $j_4\leq 0.9m$. Finally, if $j_1\leq 0.1m$ and $j_3,j_4\geq 0.9m$ then we estimate

$$Y_2\lesssim 2^{-k_2^-}2^{-2q-k_2}\|\widehat{V_{j_1,k_1}^{h_1,\iota_1}}(s)\|_{L^\infty}2^{3k_2}\|V_{j_3,k_3}^{\psi,\iota_3}(s)\|_{L^2}\|V_{j_4,k_4}^{\psi,\iota_4}(s)\|_{L^2}\\ \lesssim\varepsilon_1^3 2^{0.1m}2^{-j_3}2^{-j_4}2^{-6k_3^+-6k_4^+}.$$

The bounds (6.3.54) follow.

**Step 3: proof of** (6.3.55)**.** We may assume that $k_4\leq k_3$, thus $k,k_1,k_3\geq -\kappa m/4-10$ and $k\leq k_1+6$ and $k_1\leq k_3+6$. The main frequency parameters in the proof are $k_2$ and $k_4$. We estimate first, using (6.3.47) and (6.3.40) (or (3.3.3) if $k_4 \leq -m + \delta m$),

$$\begin{aligned} Y_3 &\lesssim 2^{-k_2^-} 2^{k_3-k_4} 2^{-2q-k_2} 2^{3k_2/2} \|\widehat{U^{h_2,\iota_2}_{j_2,k_2}}(s)\|_2 \cdot \varepsilon_1^2 2^{2\delta' m - m} 2^{k_4-k_1} 2^{-10k_3^+} \\ &\lesssim \varepsilon_1^3 2^{-0.995m} 2^{-j_2-k_2^-} 2^{-8k_3^+}. \end{aligned} \tag{6.3.68}$$

This suffices unless $j_2 + k_2^- \leq 0.01m$. In this case we analyze several subcases.

**Substep 3.1.** Assume first that

$$j_2 + k_2^- \leq 0.01m \qquad \text{and} \qquad k_4 \leq k_2 - 0.03m - 30. \tag{6.3.69}$$

This is similar to the case analyzed in (6.3.59). We fix $\rho$ with $|\rho| \leq 2^{k_4+2}$ and estimate the $\eta$ integral by $C2^{-2q-k_2}2^{-0.99m}2^{k_2-k}2^{-4k_3^+}$ using Lemma 6.15 with $2^l \approx 2^k$, $2^{l_1} \approx 2^{k_2}$, $2^{l_2} \approx 2^{k_3}$. Thus

$$Y_3 \lesssim 2^{-k_2^-} 2^{k_3-k_4} 2^{-2q-k_2} 2^{-0.98m} 2^{k_2-k} 2^{-4k_3^+} 2^{2k_4}.$$

The desired conclusion follows since $2^{k_4-k_2} \lesssim 2^{-0.03m}$; see (6.3.69).

**Substep 3.2.** Assume now that

$$j_2 + k_2^- \leq 0.01m, \qquad k_4 \geq k_2 - 0.03m - 30, \qquad j_3 \geq 0.97m - 100. \tag{6.3.70}$$

As in (6.3.61), we estimate in the physical space, using the bounds (3.1.3), (3.1.34), and (3.3.11),

$$\begin{aligned} Y_3 &\lesssim 2^{-k_2^-} 2^{k_3-k_4} 2^{-2q-k_2} \|U^{h_2,\iota_2}_{j_2,k_2}(s)\|_{L^\infty} \|U^{h_3,\iota_3}_{j_3,k_3}(s)\|_{L^2} \|U^{h_4,\iota_4}_{j_4,k_4}(s)\|_{L^2} \\ &\lesssim \varepsilon_1^3 2^{-0.99m} 2^{-k_2} 2^{-3k_4/2} 2^{-j_3} 2^{-j_4} 2^{-4k_3^+}. \end{aligned} \tag{6.3.71}$$

Since $2^{-k_2} \lesssim 2^{p_0 m} = 2^{0.68m}$ and $2^{-k_4-j_4} \lesssim 1$, this suffices to prove (6.3.55) when $-k_4 \leq 0.55m$. On the other hand, if $k_4 \leq -0.55m$ then we can bound simply, using (3.3.26) and (3.3.3),

$$\begin{aligned} Y_3 &\lesssim 2^{-k_2^-} 2^{k_3-k_4} 2^{-2q-k_2} 2^{3k_2/2} \|\widehat{U^{h_2,\iota_2}_{j_2,k_2}}(s)\|_{L^2} \|\widehat{U^{h_3,\iota_3}_{j_3,k_3}}(s)\|_{L^\infty} 2^{3k_4/2} \|\widehat{U^{h_4,\iota_4}_{j_4,k_4}}(s)\|_{L^2} \\ &\lesssim \varepsilon_1^3 2^{0.01m} 2^{k_4} 2^{-j_3/2+\delta j_3} 2^{-4k_3^+}. \end{aligned}$$

Given (6.3.70), this suffices to prove (6.3.55) if $k_4 \leq -0.55m$.

**Substep 3.3.** Assume now that

$$j_2 + k_2^- \leq 0.01m, \qquad k_4 \geq k_2 - 0.03m - 30, \qquad j_4 \geq 0.97m - 100. \tag{6.3.72}$$

The bounds (6.3.71) still hold, but they only suffice to prove (6.3.55) when $2^{j_3+k_2+3k_4/2} \gtrsim 2^{-0.95m}$. This holds if $2^{j_3+k_2} \gtrsim 2^{0.12m}$ (because $2^{k_4} \gtrsim 2^{-0.71m}$) or when $2^{k_4} \gtrsim 2^{-0.18m}$ (because $2^{j_3+k_2} \gtrsim 2^{-0.68m}$). It remains to prove (6.3.55) in the case when

$$j_3 + k_2 \leq 0.12m \qquad \text{and} \qquad k_4 \leq -0.18m - 50. \tag{6.3.73}$$

Assuming that both (6.3.72) and (6.3.73) hold, we consider the operator defined by $\eta$ integration first. More precisely, with $n := -0.05m + 50$ and $* \in \{\leq n, > n\}$ we define

$$\begin{aligned} H_*(\xi,\rho) := 2^{-k_2^-} \int_{\mathbb{R}^3} & \frac{\varphi_k(\xi) m(\theta, \xi - \theta) \varphi_{k_1}(\xi - \theta) \varphi_q(\Xi_{\iota_1\iota_2}(\theta, \xi - \theta))}{\Lambda_{wa}(\xi) - \Lambda_\mu(\theta) - \Lambda_\nu(\xi - \theta)} \\ & \times \varphi_*(\Xi_{\iota_2\iota_3}(\theta, \xi - \rho - \theta)) m_3(\xi - \rho - \theta) \\ & \times e^{-is[\Lambda_{wa,\iota_2}(\theta) + \Lambda_{wa,\iota_3}(\xi - \rho - \theta)]} \widehat{V^{h_2,\iota_2}_{j_2,k_2}}(\theta, s) \widehat{V^{h_3,\iota_3}_{j_3,k_3}}(\xi - \rho - \theta, s)\, d\theta. \end{aligned} \tag{6.3.74}$$

This corresponds to the $\eta$ integral in the operator in (6.3.52), after making the change of variables $\eta \to \xi - \theta$ and inserting angular cutoff functions of the form $\varphi_*(\Xi_{\iota_2\iota_3}(\theta, \xi - \rho - \theta))$. For (6.3.55) it suffices to prove that, for $* \in \{\leq n, > n\}$ and $\xi \in \mathbb{R}^3$,

$$2^{k_3 - k_4} \int_{\mathbb{R}^3} |H_*(\xi, \rho)| \big|\widehat{U^{h_4,\iota_4}_{j_4,k_4}}(\rho, s)\big|\, d\rho \lesssim \varepsilon_1^3 2^{-1.01m}. \tag{6.3.75}$$

Using just (6.3.38) (with $\xi \to \xi - \rho$, $2^l \approx 2^k$, $2^{l_1} \approx 2^{k_2}$, $2^{l_2} \approx 2^{k_3}$), we bound

$$|H_{\leq n}(\xi, \rho)| \lesssim 2^{-k_2^-} \cdot \varepsilon_1^2 2^{-2q - k_2} 2^{2n - 2k} 2^{2k_2^-} 2^{-8k_3^+} \lesssim \varepsilon_1^2 2^{-0.09m} 2^{-4k_3^+}.$$

The bound (6.3.75) for $H_{\leq n}$ follows since $\big\|\widehat{U^{h_4,\iota_4}_{j_4,k_4}}(\rho, s)\big\|_{L^1_\rho} \lesssim \varepsilon_1 2^{-j_4} 2^{k_4} 2^{\delta' m}$ and $2^{-j_4} \lesssim 2^{-0.97m}$.

On the other hand, we claim that $H_{>n}$ is negligible, i.e., $|H_{>n}(\xi, \rho)| \lesssim \varepsilon_1^2 2^{-4m} 2^{-4k_3^+}$. This follows by integration by parts in $\theta$ using Lemma 3.1 (as in the proof of (6.3.41)), once we notice that the $\theta$ gradient of the phase is bounded from below by $c2^n 2^m \gtrsim 2^{0.9m}$ in the support of the integral, and recall that $2^{\max(j_2, j_3)} \lesssim 2^{0.8m}$ (due to (6.3.72)–(6.3.73)). This completes the proof of (6.3.55) when (6.3.72) holds.

**Substep 3.4.** Finally, assume that

$$j_2 + k_2^- \leq 0.01m, \quad k_4 \geq k_2 - 0.03m - 30, \quad \max(j_3, j_4) \leq 0.97m - 100. \tag{6.3.76}$$

This is similar to the case analyzed in (6.3.62). We insert the cutoff functions $\varphi_{q'}(\Xi_{\iota_3\iota_4}(\eta - \rho, \rho))$ in the integral (6.3.52). Using (6.3.41) (with $l_1 = k_3$, $l_2 = k_4$, $l = k_1$, $b = q'$) the contribution is negligible if $q' \geq -0.025m$. Moreover, if

$$q_1' \leq q' \leq -0.025m, \qquad \text{where} \qquad q_1' := -m/2 - k_4/2 + 0.01m \tag{6.3.77}$$

then we use (6.3.39) to see that the contribution of the $\rho$ integral is bounded by

$C\varepsilon_1^2 2^{-0.02m-m}2^{-2k_1}2^{k_3+k_4}2^{-4k_3^+}$ for any $\eta$. The desired conclusion then follows, once we notice that the $\eta$ integral gains a factor of $2^{-2q}2^{k_2}2^{\delta' m}$.

To bound the contribution of $\varphi_{\leq q_1'}(\Xi_{\iota_3\iota_4}(\eta-\rho,\rho))$, we make the change of variables $\eta\to\xi-\theta$ and insert angular cutoff functions of the form $\varphi_*(\Xi_{\iota_2\iota_3}(\theta,\xi-\rho-\theta))$. More precisely we define

$$\begin{aligned}H_*'(\xi,\rho) := 2^{-k_2^-}\int_{\mathbb{R}^3}&\frac{\varphi_k(\xi)m(\theta,\xi-\theta)\varphi_{k_1}(\xi-\theta)}{\Lambda_{wa}(\xi)-\Lambda_\mu(\theta)-\Lambda_\nu(\xi-\theta)}\varphi_q(\Xi_{\iota_1\iota_2}(\theta,\xi-\theta))\\ &\times\varphi_{\leq q_1'}(\Xi_{\iota_3\iota_4}(\xi-\theta-\rho,\rho))\varphi_*(\Xi_{\iota_2\iota_3}(\theta,\xi-\rho-\theta))m_3(\xi-\rho-\theta)\\ &\times e^{-is[\Lambda_{wa,\iota_2}(\theta)+\Lambda_{wa,\iota_3}(\xi-\rho-\theta)]}\widehat{V_{j_2,k_2}^{h_2,\iota_2}}(\theta,s)\widehat{V_{j_3,k_3}^{h_3,\iota_3}}(\xi-\rho-\theta,s)\,d\theta,\end{aligned}$$

where $n:=-0.05m+50$ as in (6.3.74) and $*\in\{\leq n,>n\}$. For (6.3.53) it remains to prove that

$$2^{k_3-k_4}\int_{\mathbb{R}^3}|H_*'(\xi,\rho)||\widehat{U_{j_4,k_4}^{h_4,\iota_4}}(\rho,s)|\,d\rho\lesssim\varepsilon_1^3 2^{-1.01m}. \tag{6.3.78}$$

We consider first the contribution of $|H_{\leq n}'(\xi,\rho)|$. We use the restrictions

$$\widetilde{\Xi}(\xi-\theta-\rho,\rho)\lesssim 2^{q_1'}\qquad\text{and}\qquad\widetilde{\Xi}(\xi-\theta-\rho,\theta)\lesssim 2^n, \tag{6.3.79}$$

which hold in the support of the defining integrals. Therefore $\widetilde{\Xi}(\xi-\theta,\rho)\lesssim 2^{q_1'}2^{k_3-k_1}$, using (3.1.28). Moreover, since $2^{q_1'}\lesssim 2^n$, we have $\widetilde{\Xi}(\rho,\theta)\lesssim 2^n$ (using (3.1.27)). In addition $\widetilde{\Xi}(\xi-\theta,\theta)\lesssim 2^n 2^{k_3-k_1}$ and then $\widetilde{\Xi}(\xi,\theta)\lesssim 2^n2^{k_3-k}$ (using (3.1.28)). Therefore the support of the $(\theta,\rho)$ integral is included in the set $\{(\theta,\rho):\widetilde{\Xi}(\xi,\theta)\lesssim 2^n2^{k_3-k},\ \widetilde{\Xi}(\xi-\theta,\rho)\lesssim 2^{q_1'}2^{k_3-k_1}\}$. Thus

$$\begin{aligned}2^{k_3-k_4}\int_{\mathbb{R}^3}|H_{\leq n}'(\xi,\rho)||\widehat{U_{j_4,k_4}^{h_4,\iota_4}}(\rho,s)|\,d\rho &\lesssim 2^{k_3-k_4}2^{-2q-2k_2}2^{2k_3^+}\|\widehat{V_{j_2,k_2}^{h_2,\iota_2}}(s)\|_{L^\infty}\|\\ &\quad\times\widehat{V_{j_3,k_3}^{h_3,\iota_3}}(s)\|_{L^\infty}\|\widehat{U_{j_4,k_4}^{h_4,\iota_4}}(s)\|_{L^\infty}\cdot 2^{3k_4}(2^{q_1'}2^{k_3-k_1})^2 2^{3k_2}(2^n2^{k_3-k})^2\\ &\lesssim\varepsilon_1^3 2^{2n}2^{-0.95m}2^{-8k_3^+},\end{aligned}$$

using Lemma 3.19 and (6.3.77) in the last inequality. This gives (6.3.78) when $*=\leq n$.

The same argument also gives the desired bounds (6.3.78) when $*=>n$ if $j_3\geq 0.2n$ or if $k_3\geq 0.01m$. In the remaining case, when $j_3\leq 0.2m$ and $k_3\in[-\kappa m/4-10,0.01m]$ we can integrate by parts in $\theta$, using Lemma 3.1, to see that the contribution is negligible, $|H_{>n}'(\xi,\rho)|\lesssim\varepsilon_1^2 2^{-4m}$. This completes the proof of (6.3.78).

**Step 4: proof of** (6.3.56)**.** As in the proof of (6.3.55) we may assume that $k_4\leq k_3$, thus $k,k_1,k_3\geq-\kappa m/4-10$ and $k\leq k_1+6$ and $k_1\leq k_3+6$. Using

(3.3.4) we have

$$\begin{aligned} Y_4 &\lesssim 2^{-k_2^-}2^{-2q-k_2}\|U^{h_2,\iota_2}_{j_2,k_2}(s)\|_{L^2}2^{3k_2/2}\|U^{\psi,\iota_3}_{j_3,k_3}(s)\|_{L^2}\|U^{\psi,\iota_4}_{j_4,k_4}(s)\|_{L^2} \\ &\lesssim \varepsilon_1^3 2^{0.01m}2^{-k_2^-}2^{-j_2}2^{-j_3}2^{-j_4}2^{-20k_3^+}. \end{aligned} \tag{6.3.80}$$

This suffices if $k_3 \geq 0.06m$ or if $j_3 + j_4 \geq 1.02m$. On the other hand, if $k_3 \leq 0.06m$, $j_2 \leq m - \delta m - 2k_3^+$, and $j_3 \leq m - \delta m - 2k_3^+$ then we notice that $|\nabla_\eta[s\Lambda_{wa,\iota_2}(\xi-\eta) + s\Lambda_{kg,\iota_3}(\eta-\rho)]| \gtrsim 2^{m-2k_3^+}$ in the support of the integral, so $Y_4$ is negligible using integration by parts in $\eta$ and Lemma 3.1. Similarly, $Y_4$ is negligible if $k_3 \leq 0.06m$, $j_2 \leq m - \delta m - 2k_3^+$, and $j_4 \leq m - \delta m - 2k_3^+$.

Finally assume that

$$k_3 \leq 0.06m \qquad \text{and} \qquad j_2 \geq m - \delta m - 2k_3^+. \tag{6.3.81}$$

The bounds (6.3.80) are sufficient to prove (6.3.56) if $j^* := \max\{j_3, j_4\} \geq 0.7m - 10k_3^+$. On the other hand, if $j^* \leq 0.7m - 10k_3^+$ then we examine the definition (6.3.52) and insert cutoff functions of the form $\varphi_{\leq n''}(\rho-\eta/2)$ and $\varphi_{>n''}(\rho-\eta/2)$, where $n'' := -0.29m + 5k_3^+ - 10$. More precisely, for $* \in \{\leq n'', > n''\}$ we define

$$\begin{aligned} I_*(\xi) := 2^{-k_2^-}\int_{\mathbb{R}^3\times\mathbb{R}^3} &\frac{\varphi_k(\xi)m(\xi-\eta,\eta)\varphi_{k_1}(\eta)\varphi_q(\Xi_{\iota_1\iota_2}(\xi-\eta,\eta))}{\Lambda_{wa}(\xi)-\Lambda_\mu(\xi-\eta)-\Lambda_\nu(\eta)} \\ &\times \varphi_*(\rho-\eta/2)m_3(\eta-\rho)m_4(\rho)e^{-is[\Lambda_{kg,\iota_3}(\eta-\rho)+\Lambda_{kg,\iota_4}(\rho)]} \\ &\times \widehat{U^{h_2,\iota_2}_{j_2,k_2}}(\xi-\eta,s)\widehat{V^{\psi,\iota_3}_{j_3,k_3}}(\eta-\rho,s)\widehat{V^{\psi,\iota_4}_{j_4,k_4}}(\rho,s)\,d\eta d\rho. \end{aligned}$$

The contribution of $I_{>n''}(\xi)$ is negligible, using integration by parts in $\rho$ (Lemma 3.1) and (6.3.32). Notice also that $I_{\leq n''}(\xi) = 0$ if $k_4 \leq k_3 - 8$. If $k_4 \geq k_3 - 8$ then we estimate

$$\begin{aligned} |I_{\leq n''}(\xi)| &\lesssim 2^{-k_2^-}2^{-2q-k_2}\|U^{h_2,\iota_2}_{j_2,k_2}(s)\|_{L^2}2^{3k_2/2}2^{3n''}\|\widehat{U^{\psi,\iota_3}_{j_3,k_3}}(s)\|_{L^\infty}\|\widehat{U^{\psi,\iota_4}_{j_4,k_4}}(s)\|_{L^\infty} \\ &\lesssim \varepsilon_1^3 2^{0.02m}2^{-k_2^-}2^{-j_2}\cdot 2^{-0.87m}2^{-4k_3^+}, \end{aligned}$$

using (3.3.4) and (3.3.26). The desired conclusion follows since $2^{-k_2^-}2^{-j_2} \lesssim 2^{-0.3m+2k_3^+}$ (due to (6.3.81) and the assumption $2^{-k_2^-} \lesssim 2^{p_0 m}$). This completes the proof of the lemma. □

We conclude this subsection with an estimate on certain cubic expressions.

**Lemma 6.18.** *Assume that* $m \geq 100$, $t \in [2^{m-1}, 2^{m+1}] \cap [0, T]$, $k, k_1, k_2 \in \mathbb{Z}$, $k \in [-\kappa m, \delta' m]$, $k_1, k_2 \in [-p_0 m - 10, m/10 + 10]$, $h_1 \in \{h_{\alpha\beta}\}$, $\iota_1, \iota_2 \in \{+, -\}$, *and* $q \in [-m, 2]$. *Then*

$$\varphi_k(\xi) \int_{\mathbb{R}^3} |\widehat{P_{k_1} U^{h_1,\iota_1}}(\xi-\eta,t)||\widehat{P_{k_2} H}(\eta,t)|\varphi_{\le q}(\Xi_{\iota_1\iota_2}(\xi-\eta,\eta))\,d\eta \lesssim \varepsilon_1^3 2^{2q} 2^{-m+3\kappa m} 2^{k_1^-+k_2^-} 2^{-10(k_1^++k_2^+)} \tag{6.3.82}$$

*for any* $\xi \in \mathbb{R}^3$, *and* $H = |\nabla|^{-1}\mathcal{N}$, $\mathcal{N} \in \mathcal{QU}$ *(see* (4.3.12)*).*

*Proof.* It follows from Lemma 4.18 and (3.2.7) that

$$\|\widehat{P_{k_2} H}(r\theta,t)\|_{L^2(r^2dr)L^p_\theta} \lesssim \varepsilon_1^2 2^{2\delta' m-m} 2^{-k_2/2} 2^{-15k_2^+}, \tag{6.3.83}$$

for $p = 1/\delta$. The proof of (6.3.82) is similar to the proof of Lemma 6.15 (i). We may assume that $\xi = (\xi_1, 0, 0)$, $\xi_1 \in [2^{k-1}, 2^{k+1}]$. The $\eta$ integral in the left-hand side is supported in the set $\{|\eta| \approx 2^{k_2},\ |\xi-\eta| \approx 2^{k_1},\ \sqrt{\eta_2^2+\eta_3^2} \lesssim 2^{q+k_1+k_2-k}\}$. With $\mathbf{e}_1 := (1,0,0)$, we estimate the left-hand side of (6.3.82) by

$$\begin{aligned}
&C\|\widehat{P_{k_1} U^{h_1,\iota_1}}(t)\|_{L^\infty} \int_{[0,\infty)\times\mathbb{S}^2,\,|\theta-\mathbf{e}_1|\lesssim 2^{q+k_1-k}} |\widehat{P_{k_2} H}(r\theta,t)|\, r^2 dr d\theta \\
&\lesssim \|\widehat{P_{k_1} U^{h_1,\iota_1}}(t)\|_{L^\infty} \|\widehat{P_{k_2} H}(r\theta,t)\|_{L^2(r^2dr)L^p_\theta} \cdot 2^{2(q+k_1-k)/p'} 2^{3k_2/2} \\
&\lesssim \varepsilon_1 2^{-k_1} 2^{\delta' m} 2^{-20k_1^+} \cdot \varepsilon_1^2 2^{3\delta' m-m} 2^{-k_2/2} 2^{-15k_2^+} 2^{2q+2k_1-2k} 2^{3k_2/2}.
\end{aligned}$$

The desired conclusion follows once we recall that $k \ge -\kappa m$. □

### 6.3.4 The Nonlinear Terms $\mathcal{R}_3^{h_{\alpha\beta}}$

We are now ready to bound the quasilinear contributions defined in (6.3.12), for which we can still prove strong bounds.

**Lemma 6.19.** *The bounds* (6.3.17) *hold for* $m \ge 100$, $k \in [-\kappa m/4, \delta' m]$, *and* $a = 3$.

*Proof.* The formulas (6.3.5) and (2.1.29) (with $H = h$) show that

$$\begin{aligned}
\mathfrak{q}_{wa,\iota_1}(\theta,\eta) &= \iota_1 \widehat{h_{00}}(\eta)\frac{\Lambda_{wa}(\theta)}{2} + \widehat{h_{0j}}(\eta)\theta_j + \iota_1\widehat{h_{jk}}(\eta)\frac{\theta_j\theta_k}{2\Lambda_{wa}(\theta)} \\
&= \widehat{F}(\eta) m_{F,\iota_1}(\theta,\eta) + \widehat{\underline{F}}(\eta) m_{\underline{F},\iota_1}(\theta,\eta) + \widehat{\rho}(\eta) m_{\rho,\iota_1}(\theta,\eta) + \widehat{\omega_l}(\eta) m_{\omega_l,\iota_1}(\theta,\eta) \\
&\quad + \widehat{\Omega_a}(\eta) m_{\Omega_a,\iota_1}(\theta,\eta) + \widehat{\vartheta_{ab}}(\eta) m_{\vartheta_{ab},\iota_1}(\theta,\eta),
\end{aligned}$$

where

$$\begin{aligned} m_{F,\iota_1}(\theta,\eta) &:= \iota_1 \frac{|\eta|^2|\theta|^2 - \eta_j\eta_k\theta_j\theta_k}{2|\eta|^2|\theta|}, \\ m_{\underline{F},\iota_1}(\theta,\eta) &:= \iota_1 \frac{|\eta|^2|\theta|^2 + \eta_j\eta_k\theta_j\theta_k}{2|\eta|^2|\theta|}, \\ m_{\rho,\iota_1}(\theta,\eta) &:= -i\frac{\eta_j\theta_j}{|\eta|}, \\ m_{\omega_l,\iota_1}(\theta,\eta) &:= i\frac{\in_{jkl}\,\eta_k\theta_j}{|\eta|}, \\ m_{\Omega_a,\iota_1}(\theta,\eta) &:= \iota_1 \frac{(\in_{kla}\,\eta_j\eta_l + \in_{jla}\,\eta_k\eta_l)\theta_j\theta_k}{2|\eta|^2|\theta|}, \\ m_{\vartheta_{ab},\iota_1}(\theta,\eta) &:= -\iota_1 \frac{\in_{jpa}\in_{kqb}\,\eta_p\eta_q\theta_j\theta_k}{2|\eta|^2|\theta|}. \end{aligned} \tag{6.3.84}$$

We rewrite $\mathfrak{q}_{wa,\iota_1}$ in terms of the normalized solutions $U^{G,\iota_2}$, using (2.1.36),

$$\mathfrak{q}_{wa,\iota_1}(\theta,\eta) = \sum_{\iota_2\in\{+,-\}} \sum_{G\in\{F,\underline{F},\rho,\omega_a,\Omega_a,\vartheta_{ab}\}} \iota_2 \frac{i\widehat{U^{G,\iota_2}}(\eta)}{2|\eta|} m_{G,\iota_1}(\theta,\eta). \tag{6.3.85}$$

We substitute this formula into the definition (6.3.12) and decompose dyadically. Let

$$\begin{aligned} &\mathcal{R}^{h_{\alpha\beta};\iota_1,\iota_2}_{3;k_1,k_2,q}(\xi,t) \\ &:= \frac{-e^{-i\Theta_{wa}(\xi,t)}}{(2\pi)^3} \int_{\mathbb{R}^3} e^{it\Lambda_{wa}(\xi)} \varphi_q(\Xi_{\iota_1\iota_2}(\xi-\eta,\eta)) P_{k_1}\widehat{U^{h_{\alpha\beta},\iota_1}}(\xi-\eta,t) \\ &\times \sum_{G\in\{F,\underline{F},\rho,\omega_a,\Omega_a,\vartheta_{ab}\}} \iota_2 P_{k_2}\widehat{U^{G,\iota_2}}(\eta,t) \frac{m_{G,\iota_1}(\xi-\eta,\eta)}{2|\eta|} \varphi_{\geq 1}(\langle t\rangle^{p_0}\eta)\,d\eta, \end{aligned} \tag{6.3.86}$$

and notice that

$$\varphi_k(\xi)\mathcal{R}^{h_{\alpha\beta}}_3(\xi,t) = \sum_{\iota_1,\iota_2\in\{+,-\}} \sum_{(k_1,k_2)\in\mathcal{X}_k} \sum_{q\leq 4} \varphi_k(\xi)\mathcal{R}^{h_{\alpha\beta};\iota_1,\iota_2}_{3;k_1,k_2,q}(\xi,t).$$

For (6.3.17) we have to prove that if $m \geq 100$, $k \in [-\kappa m/4, \delta' m]$, $\xi \in \mathbb{R}^3$, then

$$\Big| \sum_{\iota_1,\iota_2\in\{+,-\}} \sum_{(k_1,k_2)\in\mathcal{X}_k} \sum_{q\leq 4} \varphi_k(\xi) \int_{t_1}^{t_2} \mathcal{R}^{h_{\alpha\beta};\iota_1,\iota_2}_{3;k_1,k_2,q}(\xi,s)\,ds \Big| \lesssim \varepsilon_1^2 2^{-\delta m/2} 2^{-N_0 k^+}. \tag{6.3.87}$$

The multipiers $m_{G,\iota_1}$ are all sums of symbols of the form $|\theta| m_1(\theta) m_2(\eta)$, with $m_1, m_2 \in \mathcal{M}_0$ (see definition (3.2.40)). We may assume that $k_2 \geq -p_0 m - 10$ in the sum in (6.3.87). Using the bounds $\|P_{k_1}\widehat{U^{h_{\alpha\beta},\iota_1}}\|_{L^1} \lesssim \varepsilon_1 2^{2k_1 - \delta k_1} 2^{\delta' m}$, it is easy to see that the contribution in the sum corresponding to the pairs $(k_1,k_2)$ for which $k_1\leq -p_0m-10$ (thus $|k_2-k|\leq 4$) is bounded by $C2^{-0.1m}$. Also, the contribution of the pairs $(k_1,k_2)$ for which $\max(k_1,k_2)\geq m/10$ is bounded by $C2^{-0.1m}$, using $L^2$ estimates on the profiles. Finally, the contribution of the indices $q$ with $\geq q_0:=(-m+k-k_1-k_2)/2+\delta m/8$ is bounded as claimed, due to Lemma 6.16 (the time dependence of the multiplier $\varphi_{\geq 1}(\langle s\rangle^{p_0}\eta)$ generates an additional term after integration by parts in time, which can be controlled using (6.3.49)). After these reductions, it suffices to prove that

$$\Big|\varphi_k(\xi)\sum_{G\in\{F,\underline{F},\rho,\omega_a,\Omega_a,\vartheta_{ab}\}}\int_{\mathbb{R}^3}\varphi_{\leq q_0}(\Xi_{\iota_1\iota_2}(\xi-\eta,\eta))P_{k_1}\widehat{U^{h_{\alpha\beta},\iota_1}}(\xi-\eta,t)\\ \times P_{k_2}\widehat{U^{G,\iota_2}}(\eta,t)\frac{m_{G,\iota_1}(\xi-\eta,\eta)}{2|\eta|}\varphi_{\geq 1}(\langle t\rangle^{p_0}\eta)\,d\eta\Big|\lesssim \varepsilon_1^2 2^{-\delta m-m}2^{-N_0k^+}, \tag{6.3.88}$$

for any $t\in[2^{m-1},2^{m+1}]$, $\iota_1,\iota_2\in\{+,-\}$, and $k_1,k_2\in[-p_0m-10,m/10]$.

To prove (6.3.88) we use the fact that the multipliers $m_{G,\iota_1}$ satisfy certain null structure conditions. Indeed, notice that

$$\begin{aligned}|v_aw_b-v_bw_a|^2+\big(|v|^2|w|^2-|(v\cdot w)|^2\big)&\lesssim |v||w|\big(|v||w|-|(v\cdot w)|\big)\\ &\lesssim |v|^2|w|^2\Xi_\iota(v,w)^2\end{aligned} \tag{6.3.89}$$

for any $\iota\in\{+,-\}$, $a,b\in\{1,2,3\}$, and $v,w\in\mathbb{R}^3$. Therefore, using the definitions (6.3.84),

$$|m_{G,\iota_1}(\xi-\eta,\eta)|\lesssim 2^{k_1}2^{q_0} \tag{6.3.90}$$

in the support of the integral in (6.3.88), for $G\in\{F,\omega_a,\Omega_a,\vartheta_{ab}\}$. Using (6.3.38), it follows that the integrals in (6.3.88) corresponding to the functions $F,\omega_a,\Omega_a,\vartheta_{ab}$ are bounded by

$$\begin{aligned}C\varepsilon_1^2 2^{\delta' m}2^{k_1+k_2}\cdot 2^{\min(k_1,k_2)}2^{2q_0-2k}\cdot 2^{k_1-k_2}2^{q_0}2^{-4(k_1^++k_2^+)}\\ \lesssim \varepsilon_1^2 2^{2\kappa m}2^{-3m/2}2^{-\min(k_1,k_2)/2},\end{aligned}$$

which suffices.

Clearly, the symbols $m_{\rho,\pm}$ and $m_{\underline{F},\pm}$ in (6.3.84) do not satisfy favorable null structure bounds like (6.3.90). However, we can use the identities (4.3.4)–(4.3.5) to combine the $\rho$ and $\underline{F}$ terms and extract a cancellation. Indeed, notice that

$$\begin{aligned}U^{\rho,+}&=\partial_t\rho-i|\nabla|\rho\\ &=\partial_t(R_0\underline{F}+R_0\tau+|\nabla|^{-1}E_0^{\geq 2})-i|\nabla|(R_0\underline{F}+R_0\tau+|\nabla|^{-1}E_0^{\geq 2})\\ &=-|\nabla|\underline{F}+|\nabla|^{-1}\mathcal{N}^{\underline{F}}-|\nabla|\tau+|\nabla|^{-1}\mathcal{N}^{\tau}+|\nabla|^{-1}\partial_0E_0^{\geq 2}-i\partial_t\underline{F}-i\partial_t\tau-iE_0^{\geq 2}\\ &=-iU^{\underline{F},+}+\big\{-iU^{\tau,+}+|\nabla|^{-1}\mathcal{N}^{\underline{F}}+|\nabla|^{-1}\mathcal{N}^{\tau}+|\nabla|^{-1}\partial_0E_0^{\geq 2}-iE_0^{\geq 2}\big\}.\end{aligned} \tag{6.3.91}$$

Therefore $U^{\rho,\pm} = \mp i U^{\underline{F},\pm} + H^{\pm}$, where the functions $H^{+}$ and $H^{-} = \overline{H^{+}}$ satisfy the bounds (6.3.82) of Lemma 6.18.

We combine the contributions of $U^{\rho,\pm}$ and $U^{\underline{F},\pm}$ in the left-hand side of (6.3.88). Notice that

$$\begin{aligned}\widehat{U^{\rho,\iota_2}}(\eta)m_{\rho,\iota_1}(\xi-\eta,\eta) &+ \widehat{U^{\underline{F},\iota_2}}(\eta)m_{\underline{F},\iota_1}(\xi-\eta,\eta)\\ &= \widehat{U^{\underline{F},\iota_2}}(\eta)\widetilde{m}(\xi-\eta,\eta) + \widehat{H^{\iota_2}}(\eta)m_{\rho,\iota_1}(\xi-\eta,\eta),\end{aligned}$$

as a consequence of (6.3.84) and (6.3.91), where

$$\widetilde{m}(\theta,\eta) := -\iota_2\frac{\eta_j\theta_j}{|\eta|} + \iota_1\frac{|\eta|^2|\theta|^2 + (\eta_j\theta_j)^2}{2|\eta|^2|\theta|} = \frac{\iota_1|\theta|}{8}|\Xi_{\iota_1\iota_2}(\theta,\eta)|^4.$$

The main point is that the combined symbols $\widetilde{m}$ satisfy favorable null structure bounds of the form $|\widetilde{m}(\xi-\eta,\eta)| \lesssim 2^{k_1}2^{q_0}$, similar to (6.3.90) (see also (3.1.24)), in the support of the integral in (6.3.88). As before, this suffices to bound the corresponding contributions in (6.3.88). Finally, the contributions of the functions $H^{\pm}$ to the left-hand side of (6.3.88) are bounded as claimed, as a consequence of (6.3.82). This completes the proof of the lemma. □

### 6.3.5 The Nonlinear Terms $\mathcal{R}_5^{h_{\alpha\beta}}$

Finally, for the contribution of the semilinear quadratic terms we prove weaker bounds, but still sufficient to conclude the proof of (6.3.16).

**Lemma 6.20.** *Assume that $m \geq 100$, $t_1, t_2 \in [2^{m-1}, 2^{m+1}] \cap [0,T]$, and $k \in [-\kappa m/4, \delta' m]$. Then, for any $\xi \in \mathbb{R}^3$,*

$$\Big|\varphi_k(\xi)\int_{t_1}^{t_2}\mathcal{R}_5^{h_{\alpha\beta}}(\xi,s)\,ds\Big| \lesssim \varepsilon_1^2 2^{\delta m/2}2^{-k^-}2^{-N_0k^+}. \tag{6.3.92}$$

*Moreover, for $G \in \{F, \Omega_a, \vartheta_{ab}\}$ we have*

$$\Big|\varphi_k(\xi)\int_{t_1}^{t_2}\mathcal{R}_5^{G}(\xi,s)\,ds\Big| \lesssim \varepsilon_1^2 2^{-\delta m/2}2^{-k^-}2^{-N_0k^+}, \tag{6.3.93}$$

*where (compare with the definitions (2.1.26) with $H = h$)*

$$\begin{aligned}\mathcal{R}_5^{F} &:= \frac{1}{2}[\mathcal{R}_5^{h_{00}} + R_jR_k\mathcal{R}_5^{h_{jk}}],\\ \mathcal{R}_5^{\Omega_a} &:= \in_{akl} R_kR_m\mathcal{R}_5^{h_{lm}},\\ \mathcal{R}_5^{\vartheta_{ab}} &:= \in_{amp}\in_{bnq} R_mR_n\mathcal{R}_5^{h_{pq}}.\end{aligned} \tag{6.3.94}$$

*Proof.* Recall that $\mathcal{R}_5^{h_{\alpha\beta}}(\xi,t) := e^{-i\Theta_{wa}(\xi,t)}e^{it\Lambda_{wa}(\xi)}\widehat{\mathcal{S}^2_{\alpha\beta}}(\xi,t)$ and the decomposition $\mathcal{S}^2_{\alpha\beta} = -(Q^2_{\alpha\beta} + P^2_{\alpha\beta})$, where the nonlinearities $Q^2_{\alpha\beta}$ and $P^2_{\alpha\beta}$ are explicitly given in (2.1.14)–(2.1.15). We decompose $Q^2_{\alpha\beta} + P^2_{\alpha\beta} = \Pi^I_{\alpha\beta} + \mathcal{C}^{I,\geq 3}_{\alpha\beta} - \mathcal{S}^{I,2}_{\alpha\beta}$ as in Lemma 4.25, where $\Pi^I_{\alpha\beta}$ are semilinear null forms and $\mathcal{C}^{I,\geq 3}_{\alpha\beta}$ are semilinear cubic remainders (there are no commutator remainders when $\mathcal{L} = I$). The desired conclusions follow from Lemmas 6.21–6.23 below. □

We prove strong bounds for the cubic terms and the semilinear null forms.

**Lemma 6.21.** *Assume that $m, k, t_1, t_2, \xi$ are as in Lemma 6.20 and $\mathcal{C}^{I,\geq 3}$ is a semilinear cubic remainder of order $(0,0)$ of one of the first two types as defined in (4.3.13). Then*

$$\Big|\varphi_k(\xi)\int_{t_1}^{t_2} e^{-i\Theta_{wa}(\xi,s)}e^{is\Lambda_{wa}(\xi)}\widehat{\mathcal{C}^{I,\geq 3}}(\xi,s)\,ds\Big| \lesssim \varepsilon_1^2 2^{-\delta m/2}2^{-N_0k^+}. \tag{6.3.95}$$

*Proof.* Bilinear interactions of the quadratic and higher order expression are easy to bound, using just (4.3.14). On the other hand, if $\mathcal{C}^{I,\geq 3}$ is a semilinear cubic remainder of the first type as defined in (4.3.13), then we decompose dyadically in frequency and use Lemma 6.18. It remains to bound the contribution of the middle frequencies, so we have to prove that

$$\big|\varphi_k(\xi)\mathcal{F}\{I[P_{k_1}U^{h_1,\iota_1}, P_{k_2}\mathcal{N}]\}(\xi,s)\big| \lesssim \varepsilon_1^2 2^{-1.01m}, \tag{6.3.96}$$

for any $\xi \in \mathbb{R}^3$, $s \in [2^{m-1}, 2^{m+1}]$, $k_1, k_2 \in [-4\kappa m, 10\kappa m]$, $I = I_m$, $m \in \mathcal{M}$, $h_1 \in \{h_{\alpha\beta}\}$, $\iota_1 \in \{+,-\}$, and $\mathcal{N} \in \mathcal{QU}$. Using (3.3.11) and Lemma 4.18 we estimate

$$\|P_k\mathcal{L}I[P_{k_1}U^{h_1,\iota_1}, P_{k_2}\mathcal{N}](s)\|_{L^2} \lesssim 2^{-2m+0.01m},$$
$$\sum_{l\in\{1,2,3\}}\|P_k\{x_l\mathcal{L}I[P_{k_1}U^{h_1,\iota_1}, P_{k_2}\mathcal{N}]\}(s)\|_{L^2} \lesssim 2^{-m+0.01m},$$

for any $\mathcal{L} = \Omega^\gamma$, $|\gamma| \leq 1$. The desired bounds (6.3.96) follow by interpolation; see (3.2.63). □

**Lemma 6.22.** *Assume that $m, k, t_1, t_2, \xi$ are as in Lemma 6.20 and $\Pi$ is a semilinear null form (see Definition 4.21). Then*

$$\Big|\varphi_k(\xi)\int_{t_1}^{t_2} e^{-i\Theta_{wa}(\xi,s)}e^{is\Lambda_{wa}(\xi)}\widehat{\Pi}(\xi,s)\,ds\Big| \lesssim \varepsilon_1^2 2^{-\delta m/2}2^{-N_0k^+}. \tag{6.3.97}$$

*Proof.* After decomposing dyadically in frequency, it remains to prove that

$$\sum_{(k_1,k_2)\in\mathcal{X}_k} Y_{k_1,k_2} \lesssim \varepsilon_1^2 2^{-\delta m/2}2^{-N_0k^+}, \tag{6.3.98}$$

where, with $n \in \mathcal{M}^{null}_{\iota_1\iota_2}$ being a null multiplier as in Definition 4.21,

$$Y_{k_1,k_2} := \Big|\varphi_k(\xi)\int_{t_1}^{t_2} e^{is\Lambda_{wa}(\xi)-i\Theta_{wa}(\xi,s)} \times \int_{\mathbb{R}^3} P_{k_1}\widehat{U^{h_1,\iota_1}}(\xi-\eta,s)P_{k_2}\widehat{U^{h_2,\iota_2}}(\eta,s)n(\xi-\eta,\eta)\,d\eta ds\Big|. \tag{6.3.99}$$

The sum over the pairs $(k_1,k_2)$ in (6.3.98) for which $\min(k_1,k_2)\le -0.01m$ or $\max(k_1,k_2)\ge 0.01m$ can be bounded as claimed using (6.3.40) and disregarding the null structure of the multiplier $n$.

On the other hand, if $|k_1|,|k_2|\le 0.01m$ then we insert cutoff functions of the form $\varphi_{\le q_0}(\Xi_{\iota_1\iota_2}(\xi-\eta,\eta))$ and $\varphi_{>q_0}(\Xi_{\iota_1\iota_2}(\xi-\eta,\eta))$, $q_0 := (-m+k-k_1-k_2)/2+\delta m/8$, in the integral in (6.3.99). The contribution corresponding to $\varphi_{>q_0}(\Xi_{\iota_1\iota_2}(\xi-\eta,\eta))$ can be bounded using Lemma 6.16. The contribution corresponding to $\varphi_{\le q_0}(\Xi_{\iota_1\iota_2}(\xi-\eta,\eta))$ can be bounded using (6.3.38) and the null structure bound $|n(\xi-\eta,\eta)|\lesssim 2^{q_0}$ if $\Xi_{\iota_1\iota_2}(\xi-\eta,\eta)\lesssim 2^{q_0}$. □

Finally, we bound the contributions of the terms $\mathcal{H}_{\alpha\beta} := \mathcal{S}^{I,2}_{\alpha\beta}$ (see the definition (4.3.53)).

**Lemma 6.23.** *Assume that $m,k,t_1,t_2,\xi$ are as in Lemma 6.20. Then*

$$\Big|\varphi_k(\xi)\int_{t_1}^{t_2} e^{-i\Theta_{wa}(\xi,s)}e^{is\Lambda_{wa}(\xi)}\widehat{\mathcal{H}_{\alpha\beta}}(\xi,s)\,ds\Big| \lesssim \varepsilon_1^2 2^{\delta m/2}2^{-k^-}2^{-N_0k^+}. \tag{6.3.100}$$

*Moreover, for $G\in\{F,\Omega_a,\vartheta_{ab}\}$ we have*

$$\Big|\varphi_k(\xi)\int_{t_1}^{t_2} e^{-i\Theta_{wa}(\xi,s)}e^{is\Lambda_{wa}(\xi)}\widehat{\mathcal{H}^G}(\xi,s)\,ds\Big| \lesssim \varepsilon_1^2 2^{-\delta m/2}2^{-N_0k^+}, \tag{6.3.101}$$

*where*

$$\begin{aligned} \mathcal{H}^F &:= \frac{1}{2}[\mathcal{H}_{00}+R_jR_k\mathcal{H}_{jk}],\\ \mathcal{H}^{\Omega_a} &:= \in_{akl} R_kR_m\mathcal{H}_{lm},\\ \mathcal{H}^{\vartheta_{ab}} &:= \in_{amp}\in_{bnq} R_mR_n\mathcal{H}_{pq}. \end{aligned} \tag{6.3.102}$$

*Proof.* Using the definitions and decomposing dyadically in frequency, to prove (6.3.100) it suffices to show that

$$\sum_{(k_1,k_2)\in\mathcal{X}_k} Y'_{k_1,k_2} \lesssim \varepsilon_1^2 2^{\delta m/2}2^{-k^-}2^{-N_0k^+}, \tag{6.3.103}$$

where, with $\vartheta_1, \vartheta_2 \in \{\vartheta_{ab}\}$, $\iota_1, \iota_2 \in \{+,-\}$, and $m_1, m_2 \in \mathcal{M}$,

$$Y'_{k_1,k_2} := \Big|\varphi_k(\xi)\int_{t_1}^{t_2} e^{is\Lambda_{wa}(\xi)-i\Theta_{wa}(\xi,s)}\int_{\mathbb{R}^3} m_1(\xi-\eta)m_2(\eta) \\ \times P_{k_1}\widehat{U^{\vartheta_1,\iota_1}}(\xi-\eta,s)P_{k_2}\widehat{U^{\vartheta_2,\iota_2}}(\eta,s)\,d\eta ds\Big|. \tag{6.3.104}$$

As in Lemma 6.22, the sum over the pairs $(k_1, k_2)$ in (6.3.103) for which $\min(k_1, k_2) \leq -0.01m$ or $\max(k_1, k_2) \geq 0.01m$ can be bounded as claimed using (6.3.40). On the other hand, if $|k_1|, |k_2| \leq 0.01m$ then we insert cut-off functions of the form $\varphi_{\leq q_0}(\Xi_{\iota_1\iota_2}(\xi-\eta,\eta))$ and $\varphi_{>q_0}(\Xi_{\iota_1\iota_2}(\xi-\eta,\eta))$, $q_0 := (-m+k-k_1-k_2)/2+\delta m/8$, in the integral in (6.3.104). The contribution corresponding to $\varphi_{>q_0}(\Xi_{\iota_1\iota_2}(\xi-\eta,\eta))$ can be bounded using Lemma 6.16. Finally, the contribution corresponding to $\varphi_{\leq q_0}(\Xi_{\iota_1\iota_2}(\xi-\eta,\eta))$ can be bounded, using the estimates in the first line of (3.3.7) and the fact that the support of the $\eta$-integral has volume $\lesssim 2^{\min(k_1,k_2)}2^{2q_0+2k_1+2k_2-2k}$ (see the sets $\mathcal{R}_{\leq b;\xi}$ defined in Lemma 6.15), by

$$C2^m 2^{\min(k_1,k_2)}2^{2q_0+2k_1+2k_2-2k}\cdot\varepsilon_1^2 2^{-k_1^- -\kappa k_1^-}2^{-N_0k_1^+}2^{-k_2^- -\kappa k_2^-}2^{-N_0k_2^+} \\ \lesssim \varepsilon_1^2 2^{\delta m/4}2^{-k^-}(2^{\min(k_1,k_2)}2^{-\kappa k_1^-}2^{-\kappa k_2^-})2^{-k^+ +k_1^+ +k_2^+}2^{-N_0k_1^+}2^{-N_0k_2^+}.$$

This suffices to bound the remaining contributions, as claimed in (6.3.103). Notice that this last estimate is tighter than before, and it relies on the strong bounds in (3.3.7), without $2^{C\delta m}$ losses, and on the weak null structure of the nonlinearity.

To prove (6.3.101) we need to notice that $\mathcal{H}^F, \mathcal{H}^{\Omega_a}, \mathcal{H}^{\vartheta_{ab}}$ are defined by multipliers that satisfy suitable null structure bounds. Indeed, we decompose dyadically in frequency, as in (6.3.103)–(6.3.104), and notice that we only need to focus on the pairs $(k_1, k_2)$ for which $|k_1|, |k_2| \leq 0.01m$. The identities (2.1.36) show that the functions $\mathcal{H}^F$, $\mathcal{H}^{\Omega_a}$, $\mathcal{H}^{\vartheta_{ab}}$ are defined by the multipliers

$$n^F_{\iota_1,\iota_2}(\theta,\eta) = C_{\iota_1\iota_2}\frac{(\theta\cdot\eta)^2}{|\theta|^2|\eta|^2}\Big(1-\iota_1\iota_2\frac{(\theta+\eta)\cdot\theta}{|\theta+\eta||\theta|}\frac{(\theta+\eta)\cdot\eta}{|\theta+\eta||\eta|}\Big),$$

$$n^{\Omega_a}_{\iota_1,\iota_2}(\theta,\eta) = C_{\iota_1\iota_2}\frac{(\theta\cdot\eta)^2}{|\theta|^2|\eta|^2}\frac{(\theta+\eta)\cdot\eta}{|\theta+\eta||\eta|}\in_{akl}\frac{(\theta+\eta)_k\theta_l}{|\theta+\eta||\theta|},$$

$$n^{\vartheta_{ab}}_{\iota_1,\iota_2}(\theta,\eta) = C_{\iota_1\iota_2}\frac{(\theta\cdot\eta)^2}{|\theta|^2|\eta|^2}\in_{amp}\in_{bnq}\frac{(\theta+\eta)_m\theta_p}{|\theta+\eta||\theta|}\frac{(\theta+\eta)_n\eta_q}{|\theta+\eta||\eta|},$$

where $\iota_1, \iota_2 \in \{+,-\}$. These multipliers satisfy suitable null structure bounds, namely $|n^*_{\iota_1\iota_2}(\xi-\eta,\eta)| \lesssim 2^{q_0}2^{0.05m}$ if $|\xi|, |\eta|, |\xi-\eta| \in [2^{-0.01m-10}, 2^{0.01m+10}]$ and $\Xi_{\iota_1\iota_2}(\xi-\eta,\eta) \lesssim 2^{q_0}$ (see also (6.3.89)). The desired bounds (6.3.101) follow as in Lemma 6.22. □

# Chapter Seven

# The Main Theorems

## 7.1 GLOBAL REGULARITY AND ASYMPTOTICS

In this section we prove our first set of main theorems. All of the theorems below rely on Proposition 2.3, and some of the ingredients in its proof.

### 7.1.1 Global Regularity

We start with a quantitative global regularity result:

**Theorem 7.1.** *Assume that* $(\overline{g}_{ij}, k_{ij}, \psi_0, \psi_1)$ *is an initial data set on* $\Sigma_0 = \{(x,t) \in \mathbb{R}^4 : t = 0\}$ *that satisfies the smallness conditions* (1.2.5) *and the constraint equations* (1.2.4).

*(i) Then the reduced Einstein-Klein-Gordon system*

$$\begin{aligned} \widetilde{\Box}_{\mathbf{g}} \mathbf{g}_{\alpha\beta} + 2\partial_\alpha\psi\partial_\beta\psi + \psi^2 \mathbf{g}_{\alpha\beta} - F^{\geq 2}_{\alpha\beta}(\mathbf{g}, \partial\mathbf{g}) = 0, \\ \widetilde{\Box}_{\mathbf{g}}\psi - \psi = 0, \end{aligned} \tag{7.1.1}$$

*admits a unique global solution* $(\mathbf{g}, \psi)$ *in* $M = \{(x,t) \in \mathbb{R}^4 : t \geq 0\}$, *with initial data* $(\overline{g}_{ij}, k_{ij}, \psi_0, \psi_1)$ *on* $\Sigma_0$ *(as described in* (1.2.3)*) and satisfying* $\|\mathbf{g} - m\|_{C^4(M)} + \|\psi\|_{C^4(M)} \lesssim \varepsilon_0$. *The solution also satisfies the harmonic gauge conditions in* $M$

$$0 = \mathbf{\Gamma}_\mu = \mathbf{g}^{\alpha\beta}\partial_\alpha \mathbf{g}_{\beta\mu} - (1/2)\mathbf{g}^{\alpha\beta}\partial_\mu \mathbf{g}_{\alpha\beta}, \qquad \mu \in \{0,1,2,3\}. \tag{7.1.2}$$

*(ii) Let* $h_{\alpha\beta} = \mathbf{g}_{\alpha\beta} - m_{\alpha\beta}$ *as in* (2.1.1) *and define the functions* $U^*, V^*$ *as in* (2.1.32)–(2.1.34). *For any* $t \in [0,\infty)$, $\alpha, \beta \in \{0,1,2,3\}$, *and* $a, b \in \{1,2,3\}$ *we have*

$$\begin{aligned} \sup_{n\leq 3,\, \mathcal{L}\in\mathcal{V}^q_n} \langle t\rangle^{-H(q,n)\delta} \big\{ \|(\langle t\rangle|\nabla|_{\leq 1})^\gamma |\nabla|^{-1/2} U^{\mathcal{L}h_{\alpha\beta}}(t)\|_{H^{N(n)}} \\ + \|U^{\mathcal{L}\psi}(t)\|_{H^{N(n)}} \big\} \lesssim \varepsilon_0, \end{aligned} \tag{7.1.3}$$

$$\begin{aligned} \sup_{n\leq 2,\, \mathcal{L}\in\mathcal{V}^q_n} \; \sup_{k\in\mathbb{Z},\, l\in\{1,2,3\}} 2^{N(n+1)k^+} \langle t\rangle^{-H(q+1,n+1)\delta} \\ \big\{ 2^{k/2}(2^{k^-}\langle t\rangle)^\gamma \|P_k(x_l V^{\mathcal{L}h_{\alpha\beta}})(t)\|_{L^2} + 2^{k^+}\|P_k(x_l V^{\mathcal{L}\psi})(t)\|_{L^2} \big\} \lesssim \varepsilon_0, \end{aligned} \tag{7.1.4}$$

*and*

$$\|V^F(t)\|_{Z_{wa}} + \|V^{\omega_a}(t)\|_{Z_{wa}} + \|V^{\vartheta_{ab}}(t)\|_{Z_{wa}} \\ + \langle t\rangle^{-\delta}\|V^{h_{\alpha\beta}}(t)\|_{Z_{wa}} + \|V^{\psi}(t)\|_{Z_{kg}} \lesssim \varepsilon_0, \tag{7.1.5}$$

*where* $H(q,n)$ *is defined as in* (2.1.49).

*(iii) For any* $k \in \mathbb{Z}$, $t \in [0,\infty)$, *and* $\mathcal{L} \in \mathcal{V}_n^q$, $n \le 2$, *the functions* $\mathcal{L}h_{\alpha\beta}$ *and* $\mathcal{L}\psi$ *satisfy the pointwise decay bounds*

$$\|P_k U^{\mathcal{L}h_{\alpha\beta}}(t)\|_{L^\infty} \lesssim \varepsilon_0 \langle t\rangle^{-1+\delta'/2} 2^{k^-} 2^{-N(n+1)k^+ + 2k^+} \min\{1, \langle t\rangle 2^{k^-}\}^{1/2} \tag{7.1.6}$$

*and*

$$\|P_k U^{\mathcal{L}\psi}(t)\|_{L^\infty} \lesssim \varepsilon_0 \langle t\rangle^{-1+\delta'/2} 2^{k^-/2} 2^{-N(n+1)k^+ + 2k^+} \min\{1, \langle t\rangle 2^{2k^-}\}^{1/2}. \tag{7.1.7}$$

*Proof.* **Step 1.** We prove first suitable bounds on the functions $h_{\alpha\beta}$ and $\psi$ for $t \in [0,2]$. Indeed, notice first that, at time $t = 0$,

$$\sum_{|\beta'|\le|\beta|+\beta_0-1\le n} \||\nabla|^{-1/2}|\nabla|^{\gamma}_{\le 1}(x^{\beta'}\partial^\beta\partial_0^{\beta_0}h_{\alpha\beta})(0)\|_{H^{N(n)}} \lesssim \varepsilon_0, \\ \sum_{|\beta'|,|\beta|+\beta_0-1\le n} \|(x^{\beta'}\partial^\beta\partial_0^{\beta_0}\psi)(0)\|_{H^{N(n)}} \lesssim \varepsilon_0, \tag{7.1.8}$$

for any $n \in [0,3]$ and $\beta_0 \in \{0,1\}$, where $x^{\beta'} = x_1^{\beta_1'}x_2^{\beta_2'}x_3^{\beta_3'}$ and $\partial^\beta = \partial_1^{\beta_1}\partial_2^{\beta_2}\partial_3^{\beta_3}$. These bounds follow directly from (1.2.3) and (1.2.5), by passing to the Fourier space and using Lemma 3.3.

We can construct the functions $h_{\alpha\beta}$ and $\psi$ by solving the coupled system

$$(\partial_0^2 - \Delta)h_{\alpha\beta} = \mathcal{N}^h_{\alpha\beta}, \qquad (\partial_0^2 - \Delta + 1)\psi = \mathcal{N}^\psi; \tag{7.1.9}$$

see Proposition 2.1. Using standard energy estimates, similar to (5.1.2)-(5.1.3), the solutions $h_{\alpha\beta}, \psi$ are well defined $C^2$ functions on $\mathbb{R}^3 \times [0,2]$ and satisfy bounds similar to (7.1.8),

$$\sum_{|\beta'|\le|\beta|+\beta_0-1\le n} \||\nabla|^{-1/2}|\nabla|^{\gamma}_{\le 1}(x^{\beta'}\partial^\beta\partial_0^{\beta_0}h_{\alpha\beta})(t)\|_{H^{N(n)}} \lesssim \varepsilon_0, \\ \sum_{|\beta'|,|\beta|+\beta_0-1\le n} \|(x^{\beta'}\partial^\beta\partial_0^{\beta_0}\psi)(t)\|_{H^{N(n)}} \lesssim \varepsilon_0, \tag{7.1.10}$$

for any $t \in [0,2]$, $n \in [0,3]$, and $\beta_0 \in \{0,1\}$.

We would like to show now that the estimates (7.1.10) hold for all $\beta_0 \in [0,3]$. Indeed, using the equations (7.1.9), we can replace $\partial_0^2 h_{\alpha\beta}$ with $\Delta h_{\alpha\beta} + \mathcal{N}^h_{\alpha\beta}$ and $\partial_0^2\psi$ with $\Delta\psi - \psi + \mathcal{N}^\psi$. Simple product estimates using just (7.1.10) and the formulas in Proposition 2.1 show that

$$\sum_{|\beta'|\leq|\beta|+\beta_0-1\leq n,\,\beta_0\in\{2,3\}} \||\nabla|^{-1/2}|\nabla|^{\gamma}_{\leq 1}(x^{\beta'}\partial^{\beta}\partial_0^{\beta_0-2}\mathcal{N}^h_{\alpha\beta})(t)\|_{H^{N(n)}} \lesssim \varepsilon_0,$$
$$\sum_{|\beta'|,|\beta|+\beta_0-1\leq n,\,\beta_0\in\{2,3\}} \|(x^{\beta'}\partial^{\beta}\partial_0^{\beta_0-2}\mathcal{N}^{\psi})(t)\|_{H^{N(n)}} \lesssim \varepsilon_0.$$

Therefore the bounds (7.1.10) hold for all $t \in [0,2]$ and any $n, \beta_0 \in [0,3]$.

**Step 2.** We prove now the bounds (7.1.3)–(7.1.5) for $t \in [0,2]$. Using (7.1.10) we have

$$\begin{aligned} &\||\nabla|^{-1/2}|\nabla|^{\gamma}_{\leq 1}(\nabla_{x,t}\mathcal{L}h_{\alpha\beta})(t)\|_{H^{N(n)}} \lesssim \varepsilon_0,\\ &\|\mathcal{L}\psi(t)\|_{H^{N(n)}} + \|\nabla_{x,t}\mathcal{L}\psi(t)\|_{H^{N(n)}} \lesssim \varepsilon_0, \end{aligned} \tag{7.1.11}$$

for any $t \in [0,2]$, $n \in [0,3]$, and $\mathcal{L} \in \mathcal{V}^q_n$. The energy bounds (7.1.3) follow. The weighted bounds (7.1.4) then follow using Lemma 6.1 and (7.1.10) (for the nonlinear estimates), as in the proof of Proposition 6.2.

We prove now the $Z$-norm bounds (7.1.5). Using (7.1.4) and (3.1.7) we have

$$2^{N(2)k^+}2^{k/2}2^{\gamma k^-}2^j\|Q_{j,k}V^{\Omega h_{\alpha\beta}}(t)\|_{L^2} \lesssim \varepsilon_0, \tag{7.1.12}$$

for any $t, \in [0,2]$, $k \in \mathbb{Z}$, and $\Omega \in \{\Omega_{23}, \Omega_{31}, \Omega_{12}\}$. Moreover,

$$2^{N(0)k^+}2^{-k/2}2^{\gamma k^-}\|Q_{j,k}V^{h_{\alpha\beta}}(t)\|_{L^2} \lesssim \varepsilon_0,$$

due to (7.1.3). Using (3.2.62) it follows that $\|\widehat{P_kV^{h_{\alpha\beta}}}\|_{L^\infty} \lesssim \varepsilon_0 2^{-k-\gamma k^-}2^{-N_0k^+}$ for any $k \in \mathbb{Z}$. Thus $\|V^{h_{\alpha\beta}}(t)\|_{Z_{wa}} \lesssim \varepsilon_0$ as desired.

Similarly, using (7.1.10) for any $t \in [0,2]$ we have

$$\|U^{\psi}(t)\|_{H^{N(0)}} + \|\langle x\rangle^2 U^{\psi}(t)\|_{H^{N(2)}} \lesssim \varepsilon_0. \tag{7.1.13}$$

In particular $\|P_kU^{\psi}(t)\|_{L^1} \lesssim \varepsilon_0$, which gives $\|\widehat{P_kU^{\psi}}(t)\|_{L^\infty} \lesssim \varepsilon_0$ for any $k \in \mathbb{Z}$. This suffices for $k \leq 0$. On the other hand, if $k \geq 0$ then it follows from (7.1.13) that

$$\|P_kU^{\psi}(t)\|_{L^2} \lesssim \varepsilon_0 2^{-N(0)k^+}, \qquad \|\,|x|^2P_kU^{\psi}(t)\|_{L^2} \lesssim \varepsilon_0 2^{-N(2)k^+}.$$

Thus $\|P_kU^{\psi}(t)\|_{L^1} \lesssim \varepsilon_0 2^{-(N(0)+3N(2))/4k^+}$, which gives the desired control $\|V^{\psi}(t)\|_{Z_{kg}} \lesssim \varepsilon_0$.

**Step 3.** To summarize, given suitable initial data we construct the solution $(h_{\alpha\beta}, \psi)$ of the system (7.1.9) on the time interval $[0,2]$ satisfying the bounds (7.1.10). Letting $\mathbf{g}_{\alpha\beta} = m_{\alpha\beta}+h_{\alpha\beta}$, the metric $\mathbf{g}$ (which is close to the Minkowski metric $m$) and the field $\psi$ satisfy the reduced Einstein-Klein-Gordon system (7.1.1) in $\mathbb{R}^3 \times [0,2]$. The harmonic gauge condition (7.1.2) holds at time $t = 0$, due to the constraint equations (1.2.4). Therefore it holds in $\mathbb{R}^3 \times [0,2]$ due to the reduced wave equations (1.1.21).

We apply now Proposition 2.3. A standard continuity argument shows that the solution $(h_{\alpha\beta}, \psi)$ can be extended globally in time, and satisfies the bootstrap bounds (7.1.3)–(7.1.5) and the pointwise smallness bounds $\|h_{\alpha\beta}(t)\|_{L^\infty} \lesssim \varepsilon_0$ for all $t \geq 0$. These pointwise bounds follow from (3.2.9) and (7.1.4) (compare with the proof of (3.3.11)) and are needed to justify that the metric $\mathbf{g}_{\alpha\beta}$ is Lorentzian, so $\mathbf{g}^{\alpha\beta}$ are well defined.

Finally, the bounds (7.1.6)–(7.1.7) are similar to the bounds (3.3.11) and (3.3.13). The profile bounds (7.1.4) and the estimates (3.1.7) show that

$$\begin{aligned} 2^{k/2}(2^{k^-}\langle t\rangle)^\gamma 2^j \|Q_{j,k}V^{\mathcal{L}h}(t)\|_{L^2} + 2^{k^+}2^j\|Q_{j,k}V^{\mathcal{L}\psi}(t)\|_{L^2} \\ \lesssim \varepsilon_0 2^{-N(n+1)k^+}\langle t\rangle^{H(q+1,n+1)\delta}, \end{aligned} \tag{7.1.14}$$

for any $t \in [0,\infty)$, $(k,j) \in \mathcal{J}$, and $\mathcal{L} \in \mathcal{V}_n^q$, $n \leq 2$. The desired bounds (7.1.6)–(7.1.7) follow from the linear estimates in Lemma 3.9. □

### 7.1.2 Decay of the Metric and the Klein-Gordon Field

We prove now several estimates in the physical space. We introduce the tensor-fields

$$L := \partial_t + \partial_r, \quad \underline{L} := \partial_t - \partial_r, \quad \Pi^{\alpha\beta} := r^{-2}\big[\Omega_{12}^\alpha\Omega_{12}^\beta + \Omega_{23}^\alpha\Omega_{23}^\beta + \Omega_{31}^\alpha\Omega_{31}^\beta\big], \tag{7.1.15}$$

where $r := |x|$ and $\partial_r := |x|^{-1}x^j\partial_j$. Notice that

$$m^{\alpha\beta} = -\frac{1}{2}\{L^\alpha\underline{L}^\beta + \underline{L}^\alpha L^\beta\} + \Pi^{\alpha\beta}. \tag{7.1.16}$$

Given a vector-field $V$ we define the (Minkowski) derivative operator $\partial_V := V^\alpha\partial_\alpha$. Let

$$\mathcal{T} := \{L, r^{-1}\Omega_{12}, r^{-1}\Omega_{23}, r^{-1}\Omega_{31}\} \tag{7.1.17}$$

denote the set of "good" vector-fields, tangential to the Minkowski light cones. For $n \in \{0,1,2\}$ and $p \leq 6$ we define also the sets of differentiated metric components

$$\mathcal{H}_{n,p} := \{\partial_1^{a_1}\partial_2^{a_2}\partial_3^{a_3}\mathcal{L}h_{\alpha\beta} : a_1 + a_2 + a_3 \leq p,\, \mathcal{L} \in \mathcal{V}_n^n,\, \alpha,\beta \in \{0,1,2,3\}\}. \tag{7.1.18}$$

We show first that the metric components and their derivatives have suitable decay,

$$|h(x,t)| + \langle t+r\rangle|\partial_V h(x,t)| + \langle t-r\rangle|\nabla_{x,t}h(x,t)| \lesssim \varepsilon_0\langle t+r\rangle^{2\delta'-1}, \tag{7.1.19}$$

where $V \in \mathcal{T}$ is a good vector-field and $h \in \{h_{\alpha\beta}\}$. More precisely:

**Theorem 7.2.** *Assume that* $(\mathbf{g}, \psi)$ *is a global solution of the Einstein-Klein-*

*Gordon system as given by Theorem 7.1.*

*(i) For any $H \in \mathcal{H}_{2,6}$ we have*

$$|H(x,t)| + |\partial_0 H(x,t)| \lesssim \varepsilon_0 \langle t+r\rangle^{\delta'/2-1}. \tag{7.1.20}$$

*(ii) If $H' \in \mathcal{H}_{1,4}$, $H'' \in \mathcal{H}_{0,3}$, $a \in \{1,2,3\}$, and $\Omega \in \{\Omega_{12}, \Omega_{23}, \Omega_{31}\}$ then*

$$\begin{aligned} \langle t+r\rangle|\partial_{r^{-1}\Omega} H'(x,t)| + \langle r\rangle|\partial_L H'(x,t)|| &\lesssim \varepsilon_0\langle t+r\rangle^{2\delta'-1}, \\ \langle t-r\rangle|\partial_a H'(x,t)| + \min(\langle r\rangle, \langle t-r\rangle)|\partial_0 H'(x,t)| &\lesssim \varepsilon_0\langle t+r\rangle^{2\delta'-1}, \\ \langle t+r\rangle|\partial_L H''(x,t)| + \langle t-r\rangle|\partial_0 H''(x,t)| &\lesssim \varepsilon_0\langle t+r\rangle^{2\delta'-1}. \end{aligned} \tag{7.1.21}$$

*(iii) The scalar field decays faster but with limited improvement: for $\Psi = \partial_1^{a_1}\partial_2^{a_2}\partial_3^{a_3}\mathcal{L}_1\psi$ for some $\mathcal{L}_1 \in \mathcal{V}_1^1$ and $a_1+a_2+a_3 \leq 4$ we have*

$$\begin{aligned} &|\Psi(x,t)| + |\partial_0\Psi(x,t) + |\langle\nabla\rangle\Psi(x,t)| \lesssim \varepsilon_0\langle t+r\rangle^{\delta'/2-1}\langle r\rangle^{-1/2}, \\ &|\partial_b\Psi(x,t)| \lesssim \varepsilon_0\langle t+r\rangle^{\delta'/2-3/2} \quad \text{for } b \in \{1,2,3\}. \end{aligned} \tag{7.1.22}$$

*Proof.* **Step 1.** We prove first the bounds (7.1.20), using the profile bounds (7.1.14) and linear estimates. With $\mathcal{L}_2 \in \mathcal{V}_2^2$ and $H = \partial^a\mathcal{L}_2 h_{\alpha\beta} \in \mathcal{H}_{2,6}$, $\partial^a = \partial_1^{a_1}\partial_2^{a_2}\partial_3^{a_3}$, we have

$$\begin{aligned} &|P_k H(x,t)| + |P_k\partial_0 H(x,t)| \\ &\lesssim \sum_{R\in\{|\nabla|^{-1},\mathrm{Id}\}}\sum_{j\geq -k^-} |(\partial^a R P'_k(e^{-it\Lambda_{wa}}Q_{j,k}V^{\mathcal{L}_2 h_{\alpha\beta}}))(x,t)| \end{aligned} \tag{7.1.23}$$

for any $k \in \mathbb{Z}$ and $(x,t) \in M$. Using (3.2.9) and (7.1.14) we thus estimate

$$\begin{aligned} &\|P_k H(t)\|_{L^\infty} + \|P_k\partial_0 H(t)\|_{L^\infty} \\ &\qquad\lesssim 2^{7k^+}2^{k/2}\sum_{j\geq -k^-}\min(1, 2^j\langle t\rangle^{-1})\|Q_{j,k}V^{\mathcal{L}_2 h_{\alpha\beta}}(t)\|_{L^2} \\ &\qquad\lesssim \varepsilon_0\langle t\rangle^{-1+H(3,3)\delta+\delta}2^{-2k^+}\min(1, 2^{k^-}\langle t\rangle)^{1-\delta}. \end{aligned} \tag{7.1.24}$$

The bounds (7.1.20) follow from (7.1.24) if $r \lesssim \langle t\rangle$. On the other hand, if $r = |x| \geq 4\langle t\rangle$ then we still use (7.1.23) first and notice that the contribution of the pairs $(k,j)$ with $2^k \geq 2^{-10}r^{1/2}$ or $2^j \geq 2^{-10}r^{1-\delta}$ can still be bounded as in (7.1.24). On the other hand, if $2^k, 2^j \leq 2^{-10}|x|^{1-\delta}$ and $m \in \mathcal{M}_0$ then we have rapid decay,

$$\left|\int_{\mathbb{R}^3} e^{-it|\xi|}e^{ix\cdot\xi}m(\xi)\widehat{V_{j,k}^{\mathcal{L}_2 h_{\alpha\beta}}}(\xi,t)\,d\xi\right| \lesssim \varepsilon_0|x|^{-10}2^{-10k^+}, \tag{7.1.25}$$

using integration by parts in $\xi$ (Lemma 3.1). The desired bounds (7.1.20) follow.

**Step 2.** We consider now the bounds (7.1.21), which we prove in several stages (more precisely, the bounds (7.1.21) follow from (7.1.26), (7.1.28), (7.1.30) (7.1.33), and (7.1.35)).

We prove first that if $H' \in \mathcal{H}_{1,4}$, $r = |x| \geq 2\langle t\rangle$, and $\mu \in \{0,1,2,3\}$ then

$$|\partial_\mu H'(x,t)| \lesssim \varepsilon_0 r^{\delta'-2}. \tag{7.1.26}$$

Indeed, as in (7.1.23) we estimate

$$|\partial_\mu H'(x,t)| \leq \sum_{(k,j)\in\mathcal{J}} \sum_{R\in\{R_a,\mathrm{Id}\}} \big|\big(R\partial_1^{a_1}\partial_2^{a_2}\partial_3^{a_3} P'_k(e^{-it\Lambda_{wa}} Q_{j,k} V^{\mathcal{L}_1 h_{\alpha\beta}})\big)(x,t)\big|. \tag{7.1.27}$$

Using now (7.1.60), for any $k \in \mathbb{Z}$ we have

$$|x|\big|P_k(\partial_\mu H')(x,t)\big| \lesssim (1+2^k|x|)^{\delta} 2^{k/2} 2^{4k^+} \sum_{j\geq -k^-} \|Q_{j,k} V^{\mathcal{L}_1 h_{\alpha\beta}}(t)\|_{H_\Omega^{0,1}}.$$

The sum over the pairs $(k,j) \in \mathcal{J}$ with $2^j \gtrsim r^{1-\delta}$ is controlled as claimed using (7.1.14). On the other hand, since $r \geq 2\langle t\rangle$, the sum over the pairs $(k,j) \in \mathcal{J}$ with $2^j \leq 2^{-10} r^{1-\delta}$ is negligible, as in (7.1.25). The bounds (7.1.26) follow.

We show now that if $H' \in \mathcal{H}_{1,4}$ and $r \leq 4\langle t\rangle$ then

$$|r^{-1}\Omega_{ab} H'(x,t)| \lesssim \varepsilon_0 \langle t\rangle^{\delta'-2}. \tag{7.1.28}$$

Indeed, this follows using the identities

$$t\Omega_{ab} = x_a\Gamma_b - x_b\Gamma_a, \tag{7.1.29}$$

and the bounds (7.1.20) applied to the functions $\Gamma H'$.

We prove now that if $H' \in \mathcal{H}_{1,4}$ and $r \leq 4\langle t\rangle$ then

$$|\partial_L H'(x,t)| \lesssim \varepsilon_0 \langle r\rangle^{-1} \langle t\rangle^{-1+\delta'}. \tag{7.1.30}$$

This follows from (7.1.20) if $r \lesssim 1$. On the other hand, if $2^4 \leq r \leq 4\langle t\rangle$ then

$$\begin{aligned}|x|\partial_L H'(x,t) = \frac{1}{2}\big\{&|x|\partial^a(U^{\mathcal{L}_1 h_{\alpha\beta}} + \overline{U^{\mathcal{L}_1 h_{\alpha\beta}}})(x,t)\\ &+ ix^b R_b \partial^a (U^{\mathcal{L}_1 h_{\alpha\beta}} - \overline{U^{\mathcal{L}_1 h_{\alpha\beta}}})(x,t)\big\},\end{aligned} \tag{7.1.31}$$

where $\partial^a = \partial_1^{a_1}\partial_2^{a_2}\partial_3^{a_3}$, $\mathcal{L}_1 \in \mathcal{V}_1^1$, $H' = \partial^a \mathcal{L}_1 h_{\alpha\beta}$. Therefore, using the bounds

(7.1.61) below,

$$\begin{aligned}\big||x|\partial_L H'(x,t)\big| &\leq \big|(|x|+ix^bR_b)\partial^a U^{\mathcal{L}_1 h_{\alpha\beta}}(x,t)\big|\\ &\lesssim \sum_{2^k\leq t^{-1}} t2^{3k/2}\|P_kU^{\mathcal{L}_1 h_{\alpha\beta}}(t)\|_{L^2}\\ &+ \sum_{2^k\geq t^{-1},\,(k,j)\in\mathcal{J}} 2^{4k^+}(2^kt)^{2\delta}2^{k/2}\min(1,2^jt^{-1})\|Q_{j,k}V^{\mathcal{L}_1 h_{\alpha\beta}}(t)\|_{H^{0,1}_\Omega}.\end{aligned} \tag{7.1.32}$$

The desired bounds (7.1.30) now follow from (7.1.3) and (7.1.14).

We prove now that if $H'\in\mathcal{H}_{1,4}$, $r\leq 4\langle t\rangle$, and $a\in\{1,2,3\}$ then

$$|\partial_a H'(x,t)|\lesssim \varepsilon_0\langle t\rangle^{\delta'-1}\langle t-r\rangle^{-1}. \tag{7.1.33}$$

Indeed, for this we use the identity

$$(r-t)\partial_r = rL - r^{-1}x^b\Gamma_b. \tag{7.1.34}$$

Using now (7.1.30) and (7.1.20) it follows that $|t-r||\partial_r H'(x,t)|\lesssim\varepsilon_0\langle t\rangle^{\delta'-1}$. Using also (7.1.28) it follows that $|t-r||\partial_a H'(x,t)|\lesssim\varepsilon_0\langle t\rangle^{\delta'-1}$, $a\in\{1,2,3\}$. The bounds (7.1.33) follow using also (7.1.20) in the case $|t-r|\lesssim 1$.

**Step 3.** Finally, we prove that if $H''\in\mathcal{H}_{0,3}$ and $r\leq 4\langle t\rangle$ then

$$|\partial_L H''(x,t)|\lesssim\varepsilon_0\langle t\rangle^{-2+2\delta'}. \tag{7.1.35}$$

This follows from (7.1.30) $t\lesssim 1$ or if $r\gtrsim t$. On the other hand, to prove the bounds when $t\geq 2^8$ and $r\leq t/8$ we notice that

$$\partial_{\underline{L}}\partial_L = \partial_0^2-\partial_r^2 = \partial_0^2-\Delta+(2/r)\partial_r + r^{-2}(\Omega_{12}^2+\Omega_{23}^2+\Omega_{31}^2), \tag{7.1.36}$$

where $\underline{L}=\partial_t-\partial_r$. We apply $\partial_{\underline{L}}$ to $\partial_L H''$ and use the wave equations (2.1.2). Thus

$$\partial_{\underline{L}}\partial_L H'' = W'' := \partial^a\mathcal{N}^h_{\alpha\beta}+(2/r)\partial_r(\partial^a h_{\alpha\beta})+r^{-2}(\Omega_{12}^2+\Omega_{23}^2+\Omega_{31}^2)(\partial^a h_{\alpha\beta}). \tag{7.1.37}$$

We prove now that if $s\geq 2^6$ and $|y|\leq s/2$ then

$$|W''(y,s)|\lesssim\varepsilon_0|y|^{-1}\langle s\rangle^{-2+3\delta'/2}. \tag{7.1.38}$$

Indeed, for the nonlinear terms in $\partial^a\mathcal{N}^h_{\alpha\beta}$ this follows from the formula (2.1.9) and the bounds (7.1.20), (7.1.22) (which is proved below), (7.1.30), and (7.1.33). For the term $(2/r)\partial_r(\partial^a h_{\alpha\beta})$ these bounds follow directly from (7.1.33), while for the term $r^{-2}(\Omega_{12}^2+\Omega_{23}^2+\Omega_{31}^2)(\partial^a h_{\alpha\beta})$ the bounds (7.1.38) follow using (7.1.29) and (7.1.20).

We can now use the identity $\partial_{\underline{L}}\partial_L H''=W''$ and integrate along the vector-field $\underline{L}$ to complete the proof of (7.1.35). Indeed, for any function $F$ and $\lambda\in$

$[0, t/2]$ we have

$$F(x,t) - F\big(x + \lambda\frac{x}{|x|}, t-\lambda\big) = \int_0^\lambda -\frac{d}{ds}F\big(x + s\frac{x}{|x|}, t-s\big)\,ds$$
$$= \int_0^\lambda (\partial_{\underline{L}}F)\big(x + s\frac{x}{|x|}, t-s\big)\,ds.$$

We apply this with $F = \partial_L H''$ and $\lambda = t/8$. Since $|F(x+\lambda x/|x|, t-\lambda)| \lesssim \varepsilon_0 t^{-2+\delta'}$ (due to (7.1.30)) and $|(\partial_{\underline{L}}F)(x+sx/|x|, t-s)| \lesssim \varepsilon_0(s+|x|)^{-1}t^{-2+3\delta'/2}$ for any $s \in [0,\lambda]$ (due to (7.1.38)), it follows that

$$|\partial_L H''(x,t)| \lesssim \varepsilon_0 t^{-2+3\delta'/2}\ln(t/|x|) \qquad \text{for } t \geq 2^8 \text{ and } |x| \leq t/8. \tag{7.1.39}$$

The bounds (7.1.35) follow if $|x| \geq t^{-4}$. Moreover, if $|x'| \leq 2t^{-4}$ then $|\partial_r H''(x',t)| \lesssim \varepsilon_0 t^{-2+\delta'}$ (due to (7.1.33)) and $|\partial_a\partial_0 H''(x',t)| \lesssim \varepsilon_0 t^{-1+\delta'}$, $a \in \{1,2,3\}$ (due to (7.1.20)). In view of (7.1.39) it follows that $|\partial_0 H''(x,t)| \lesssim \varepsilon_0 t^{-2+2\delta'}$ if $|x| \leq 2t^{-4}$, and the desired bounds (7.1.35) follow if $|x| \leq t/8$.

**Step 4.** We prove now the bounds (7.1.22) on the scalar field. These bounds follow in the same way as (7.1.26) if $r \geq 2\langle t\rangle$ or if $r + t \lesssim 1$.

It remains to consider the case $t \geq 8$ and $r \leq 4t$. Notice that $t\partial_b\Psi = \Gamma_b\Psi - x_b\partial_0\Psi$, so the bounds $|\partial_b\Psi(x,t)| \lesssim \varepsilon_0 t^{-3/2+\delta'/2}$ in (7.1.22) follow from the bounds in the first line (for $\partial_0\psi$), and the $L^\infty$ estimates (3.3.13) (for $\Gamma_b\psi$). To summarize, it remains to prove that if $t \geq 8$ and $r \leq 4t$ then

$$|\Psi(x,t)| + |\partial_0\Psi(x,t)| + |\langle\nabla\rangle\Psi(x,t)| \lesssim \varepsilon_0 t^{\delta'/2-1}\langle r\rangle^{-1/2}. \tag{7.1.40}$$

These bounds follow from (3.3.13) if $|r| \lesssim 1$. On the other hand, if $r \geq 1$ then we estimate

$$\begin{aligned}|P_k\Psi(x,t)| &+ |P_k\partial_0\Psi(x,t)| + |P_k\langle\nabla\rangle\Psi(x,t)| \\ &\lesssim \sum_{R\in\{\Lambda_{kg}^{-1},\mathrm{Id}\}}\sum_{j\geq -k^-}\big|(\partial^a RP'_k(e^{-it\Lambda_{kg}}Q_{j,k}V^{\mathcal{L}_1\psi}))(x,t)\big|,\end{aligned} \tag{7.1.41}$$

where $\Psi = \partial^a\mathcal{L}_1\psi$, $\mathcal{L}_1 \in \mathcal{V}_1^1$. Using (7.1.14) we estimate

$$\begin{aligned}\big|(\partial^a RP'_k(e^{-it\Lambda_{kg}}Q_{j,k}V^{\mathcal{L}_1\psi}))(x,t)\big| &\lesssim 2^{4k^+}2^{3k/2}\|Q_{j,k}V^{\mathcal{L}_1\psi}(t)\|_{L^2} \\ &\lesssim \varepsilon_0 2^{-5k^+}2^{3k/2}2^{-j}t^{\delta'/4}.\end{aligned} \tag{7.1.42}$$

If $2^k \lesssim t^{-1/2}$ then we use (7.1.42) to estimate the contribution of the pairs $(k,j)$ with $2^j \geq r^{1-\delta}2^{-10}$. On the other hand, if $2^j \leq r^{1-\delta}2^{-10}$ then we have rapid decay,

$$\big|(\partial^a RP'_k(e^{-it\Lambda_{kg}}Q_{j,k}V^{\mathcal{L}_1\psi}))(x,t)\big| \lesssim \varepsilon_0 2^{3k/2}r^{-4}, \tag{7.1.43}$$

using integration by parts in the Fourier space (Lemma 3.1). Therefore, using (7.1.41),

$$|P_k\Psi(x,t)| + |P_k\partial_0\Psi(x,t)| \lesssim \varepsilon_0(2^{2k}t)r^{-1/2}t^{\delta'/3-1}. \tag{7.1.44}$$

Assume now that $2^{k-40} \geq t^{-1/2}$. Using (3.2.14) and (7.1.14) we have

$$\begin{aligned}\big|(\partial^a RP'_k(e^{-it\Lambda_{kg}}Q_{j,k}V^{\mathcal{L}_1\psi}))(x,t)\big| &\lesssim 2^{5k^+}t^{-3/2}2^{j/2-k^-}t^{\delta}\|Q_{j,k}V^{\mathcal{L}_1\psi}(t)\|_{H^{0,1}_\Omega}\\ &\lesssim \varepsilon_0 2^{-2k^+}t^{-3/2+\delta'/3}2^{-k/2}\end{aligned} \tag{7.1.45}$$

provided that $2^j \leq 2^{k^- -20}t$. Moreover, if $r^{1-\delta} \geq t2^{k+20}$ and $2^j \leq 2^{k^- -20}t$ then

$$\big|(\partial^a RP'_k(e^{-it\Lambda_{kg}}Q_{j,k}V^{\mathcal{L}_1\psi}))(x,t)\big| \lesssim \varepsilon_0 2^{-2k^+}r^{-6},$$

using integration by parts in the Fourier space (Lemma 3.1). Therefore, using (7.1.45),

$$\sum_{2^j\in[2^{-k^-},2^{k^- -20}t]} \big|(\partial^a RP'_k(e^{-it\Lambda_{kg}}Q_{j,k}V^{\mathcal{L}_1\psi}))(x,t)\big| \lesssim \varepsilon_0 2^{-2k^+}t^{-1+2\delta'/5}r^{-1/2}. \tag{7.1.46}$$

To bound the contribution of the sum over $j$ large with $2^j \geq 2^{k^- -20}t$ we notice that

$$\begin{aligned}\big|(\partial^a RP'_k(e^{-it\Lambda_{kg}}Q_{j,k}V^{\mathcal{L}_1\psi}))(x,t)\big| &\lesssim r^{-1}(1+2^k r)^{\delta}2^{k/2}2^{4k^+}\|Q_{j,k}V^{\mathcal{L}_1\psi}\|_{H^{0,1}_\Omega}\\ &\lesssim \varepsilon_0 r^{-1}2^{-4k^+}2^{k/2}2^{-j}t^{\delta'/2-4\delta},\end{aligned} \tag{7.1.47}$$

as a consequence of (7.1.60) and (7.1.14). We use now (7.1.42) if $r \leq 2^{-k}$ and (7.1.47) if $r \geq 2^{-k}$ to conclude that

$$\sum_{2^j\geq 2^{k^- -20}t} \big|(\partial^a RP'_k(e^{-it\Lambda_{kg}}Q_{j,k}V^{\mathcal{L}_1\psi}))(x,t)\big| \lesssim \varepsilon_0 2^{-2k^+}t^{-1+\delta'/2-4\delta}r^{-1/2}.$$

Using also (7.1.46) and (7.1.41), if $2^{k-40} \geq t^{-1/2}$ we have

$$|P_k\Psi(x,t)| + |P_k\partial_0\Psi(x,t)| + |P_k\langle\nabla\rangle\Psi(x,t)| \lesssim \varepsilon_0 2^{-2k^+}t^{-1+\delta'/2-4\delta}r^{-1/2}.$$

The bounds (7.1.40) follow if $r \geq 1$ by summation over $k$, using also (7.1.44). □

Using also the harmonic gauge condition (7.1.2) we can prove some additional bounds on the derivatives of the metric $h_{\alpha\beta}$. More precisely:

**Lemma 7.3.** *With* $(\mathbf{g},\psi)$ *as in Theorem 7.1 and* $\Pi^{\alpha\beta}$ *defined as in* (7.1.15), *we have the additional bounds*

$$\left|V^\alpha L^\beta \partial_{\underline{L}}(\partial^a h_{\alpha\beta})(x,t)\right| + \left|\Pi^{\alpha\beta}\partial_{\underline{L}}(\partial^a h_{\alpha\beta})(x,t)\right| \lesssim \varepsilon_0 \langle t+r\rangle^{-2+3\delta'} \tag{7.1.48}$$

*for any* $(x,t) \in M$, $V \in \mathcal{T}$, *and* $\partial^a = \partial_1^{a_1}\partial_2^{a_2}\partial_3^{a_3}$, $a_1 + a_2 + a_3 \leq 3$.

*Proof.* We write the harmonic gauge condition in the form (4.3.1),

$$m^{\alpha\beta}\partial_\alpha h_{\beta\mu} - (1/2)m^{\alpha\beta}\partial_\mu h_{\alpha\beta} = -g_{\geq 1}^{\alpha\beta}\partial_\alpha h_{\beta\mu} + (1/2)g_{\geq 1}^{\alpha\beta}\partial_\mu h_{\alpha\beta}.$$

Using (7.1.20)–(7.1.21), it follows that, for $\mu \in \{0,1,2,3\}$,

$$m^{\alpha\beta}\{\partial_\alpha(\partial^a h_{\beta\mu}) - (1/2)\partial_\mu(\partial^a h_{\alpha\beta})\} = O(\varepsilon_0\langle t+r\rangle^{-2+3\delta'}), \tag{7.1.49}$$

where $f = O(g)$ means $|f(x,t)| \lesssim g(x,t)$ for all $(x,t) \in M$. We use the formula (7.1.16), and eliminate some of the terms using (7.1.21), to conclude that

$$\begin{aligned} L^\beta \underline{L}^\alpha \partial_\alpha(\partial^a h_{\beta\mu}) - \frac{1}{2}\{L^\alpha \underline{L}^\beta + \underline{L}^\alpha L^\beta\}\partial_\mu(\partial^a h_{\alpha\beta}) + \Pi^{\alpha\beta}\partial_\mu(\partial^a h_{\alpha\beta})& \\ = O(\varepsilon_0\langle t+r\rangle^{-2+3\delta'})&, \end{aligned} \tag{7.1.50}$$

for $\mu \in \{0,1,2,3\}$. The desired conclusions in (7.1.48) follow by multiplying with either $V^\mu$, $V \in \mathcal{T}$, or $\underline{L}^\mu$. □

We prove now almost sharp bounds on two derivatives of the metric tensor, in the region $\{|x| \gtrsim t \gtrsim 1\}$. These bounds are used later to prove weak peeling estimates.

**Lemma 7.4.** *Assume that* $(\mathbf{g},\psi)$ *is a global solution of the Einstein-Klein-Gordon system as given by Theorem 7.1. If* $V_1, V_2 \in \mathcal{T}$ *and* $H \in \mathcal{H}_{0,3}$ *(see (7.1.17)–(7.1.18)) then*

$$\begin{aligned} \langle t-r\rangle^2|\partial_{\underline{L}}^2 H(x,t)| + \langle t-r\rangle\langle r\rangle|\partial_{\underline{L}}\partial_{V_1}H(x,t)| + \langle r\rangle^2|\partial_{V_2}\partial_{V_1}H(x,t)|& \\ \lesssim \varepsilon_0\langle r\rangle^{-1+3\delta'}&, \end{aligned} \tag{7.1.51}$$

*for any* $(x,t) \in M' := \{(x,t) \in M : t \geq 1 \text{ and } |x| \geq 2^{-8}t\}$.

*Proof.* **Step 1.** The bounds are easy when either $V_1$ or $V_2$ is a rotation vector-field. Indeed, if $V_1 = r^{-1}\Omega_{ab}$ then $\partial_{V_1}H$ is a sum of functions of the form $r^{-1}H'$, $H' \in \mathcal{H}_{1,3}$, so

$$\langle t-r\rangle\langle r\rangle|\partial_{\underline{L}}\partial_{V_1}H(x,t)| + \langle r\rangle^2|\partial_{V_2}\partial_{V_1}H(x,t)| \lesssim \varepsilon_0\langle r\rangle^{-1+2\delta'}, \tag{7.1.52}$$

as a consequence of (7.1.21). Moreover, if $V_2 = r^{-1}\Omega_{ab}$ then the commutators $[V_2, \underline{L}]$ and $[V_2, V_1]$ are sums of vector-fields of the form $r^{-1}W$, $W \in \mathcal{T}$.

Therefore, using also (7.1.52),

$$\langle t-r\rangle\langle r\rangle|\partial_{V_2}\partial_{\underline{L}}H(x,t)|+\langle r\rangle^2|\partial_{V_2}\partial_{V_1}H(x,t)|\lesssim\varepsilon_0\langle r\rangle^{-1+2\delta'}. \tag{7.1.53}$$

We show now that

$$\langle t-r\rangle\langle r\rangle|\partial_{\underline{L}}\partial_L H(x,t)|\lesssim\varepsilon_0\langle r\rangle^{-1+2\delta'}. \tag{7.1.54}$$

Recalling that $H=\partial^a h_{\alpha\beta}$ and using the formula (7.1.37), we have

$$\partial_{\underline{L}}\partial_L H=\partial^a\mathcal{N}^h_{\alpha\beta}+(2/r)\partial_r(\partial^a h_{\alpha\beta})+r^{-2}(\Omega_{12}^2+\Omega_{23}^2+\Omega_{31}^2)(\partial^a h_{\alpha\beta}). \tag{7.1.55}$$

As in the proof of (7.1.38), it is easy to see that

$$\begin{aligned}(2/r)|\partial_r(\partial^a h_{\alpha\beta})(x,t)|&\lesssim\langle t-r\rangle^{-1}\langle r\rangle^{-2+2\delta'},\\ r^{-2}|(\Omega_{12}^2+\Omega_{23}^2+\Omega_{31}^2)(\partial^a h_{\alpha\beta})(x,t)|&\lesssim\langle r\rangle^{-3+2\delta'},\end{aligned}$$

using (7.1.21) and (7.1.20). The nonlinearity $\partial^a\mathcal{N}^h_{\alpha\beta}(x,t)$ can be estimated using the formula (2.1.9) and the bounds (7.1.20)–(7.1.22). The desired bounds (7.1.54) follow.

**Step 2.** For (7.1.51) it remains to prove that

$$\langle t-r\rangle^2|\partial_{\underline{L}}^2 H(x,t)|+\langle r\rangle^2|\partial_L^2 H(x,t)|\lesssim\varepsilon_0\langle r\rangle^{-1+3\delta'}, \tag{7.1.56}$$

for any $(x,t)\in M'$. We define the vector-field

$$\widetilde{\Gamma}:=r^{-1}x^b\Gamma_b=r\partial_t+t\partial_r \tag{7.1.57}$$

in $M'$, and notice that

$$\widetilde{\Gamma}=(1/2)[(r+t)L+(r-t)\underline{L}]. \tag{7.1.58}$$

Moreover, using (7.1.21),

$$\begin{aligned}|\partial_L(\widetilde{\Gamma}H)(x,t)|&\lesssim\varepsilon_0\langle r\rangle^{-2+2\delta'},\\ |\partial_{\underline{L}}(\widetilde{\Gamma}H)(x,t)|&\lesssim\varepsilon_0\langle r\rangle^{-1+2\delta'}\langle t-r\rangle^{-1}.\end{aligned} \tag{7.1.59}$$

Using (7.1.58) we have $|\partial_L[(r+t)\partial_L+(r-t)\partial_{\underline{L}}]H(x,t)|\lesssim\varepsilon_0\langle r\rangle^{-2+2\delta'}$, thus

$$|(r+t)\partial_L^2 H(x,t)|\lesssim\varepsilon_0\langle r\rangle^{-2+2\delta'}+|\partial_L H(x,t)|+\langle t-r\rangle|\partial_L\partial_{\underline{L}}H(x,t)|.$$

The bound on $|\partial_L^2 H(x,t)|$ in (7.1.56) follows using also (7.1.21) and (7.1.54).

Similarly, using (7.1.58)–(7.1.59) we have

$$|\partial_{\underline{L}}[(r-t)\partial_{\underline{L}}+(r+t)\partial_L]H(x,t)|\lesssim\varepsilon_0\langle r\rangle^{-1+2\delta'}\langle t-r\rangle^{-1},$$

therefore

$$|(r-t)\partial_{\underline{L}}^2 H(x,t)| \lesssim \varepsilon_0 \langle r\rangle^{-1+2\delta'}\langle t-r\rangle^{-1} + |\partial_{\underline{L}} H(x,t)| + \langle r\rangle |\partial_{\underline{L}}\partial_L H(x,t)|.$$

The bound on $|\partial_{\underline{L}}^2 H(x,t)|$ in (7.1.56) follows using also (7.1.21) and (7.1.54) when $|t-r| \geq 1$, or using (7.1.20) when $|t-r| \leq 1$. This completes the proof of the lemma. □

We prove now the additional linear estimates we used in Theorem 7.2.

**Lemma 7.5.** *(i) For any $f$ in $L^2(\mathbb{R}^3)$, $x \in \mathbb{R}^3$, and $k \in \mathbb{Z}$ we have*

$$\big||x| P_k f(x)\big| \lesssim (1+2^k|x|)^{\delta} 2^{k/2} \|P_k f\|_{H_\Omega^{0,1}}. \tag{7.1.60}$$

*(ii) In addition, if $t \geq 1$, $|x| \leq 8t$, $(k,j) \in \mathcal{J}$ then*

$$\big|(|x| + i x^a R_a) T_m(e^{-it\Lambda_{wa}} f_{j,k})(x)\big| \lesssim (1+2^k t)^{2\delta} 2^{k/2} \min(1, 2^j t^{-1}) \|f_{j,k}^*\|_{H_\Omega^{0,1}}. \tag{7.1.61}$$

*Here $f_{j,k} = P_k' Q_{j,k} f$, $f_{j,k}^* = Q_{j,k} f$ are as in (3.2.2)–(3.2.3), $R_a = |\nabla|^{-1}\partial_a$ are the Riesz transforms, and the linear operators $T_m$ are defined by $T_m g = \mathcal{F}^{-1}(m \cdot \widehat{g})$, where $m \in \mathcal{M}_0$.*

*Finally, if $|x| \in [2^{-20}t, 8t]$ and $2^j \leq t(1+2^k t)^{-\delta}$ then*

$$\big|(|x| + i x^a R_a) T_m(e^{-it\Lambda_{wa}} f_{j,k})(x)\big| \lesssim (1+2^k t)^{2\delta} 2^{-k/2} t^{-1} \|f_{j,k}\|_{H_\Omega^{0,2}}. \tag{7.1.62}$$

*Proof.* (i) Clearly $\|P_k f\|_{L^\infty} \lesssim 2^{3k/2}\|P_k f\|_{L^2}$, so the bounds (7.1.60) follow if $|x| \lesssim 2^{-k}$. On the other hand, if $|x| \geq 2^{-k+20}$ then we estimate

$$|P_k f(x)| \lesssim \int_{\mathbb{R}^3} |P_k f(y)| \cdot 2^{3k}(1+2^k|x-y|)^{-8}\, dy. \tag{7.1.63}$$

Using the Sobolev embedding along the spheres $\mathbb{S}^2$, for any $g \in H_\Omega^{0,1}$ and $p \in [2,\infty)$ we have

$$\big\|g(r\theta)\big\|_{L^2(r^2dr)L^p_\theta} \lesssim_p \sum_{m_1+m_2+m_3 \leq 1} \|\Omega_{23}^{m_1}\Omega_{31}^{m_2}\Omega_{12}^{m_3} g\|_{L^2} \lesssim_p \|g\|_{H_\Omega^{0,1}}. \tag{7.1.64}$$

Therefore, for $x \in \mathbb{R}^3$ with $|x| \geq 2^{-k+20}$ we can estimate the right-hand side of (7.1.63) by

$$\begin{aligned} &C\|P_k f(r\theta)\|_{L^2(r^2dr)L^p_\theta} \|2^{3k}(1+2^k|x-r\theta|)^{-8}\|_{L^2(r^2dr)L^{p'}_\theta} \\ &\lesssim_p \|P_k f\|_{H_\Omega^{0,1}} \cdot 2^{3k} 2^{-k/2} |x| (2^k|x|)^{-2/p'}. \end{aligned}$$

The desired bounds (7.1.60) follow in this case as well, by taking $p$ sufficiently large such that $1 - 1/p' \leq \delta/4$.

(ii) We prove now the bounds (7.1.61). Notice that we may assume $2^j \le t(1+2^k t)^{-\delta}$, since otherwise the bounds follow from (7.1.60). We write

$$(|x| + ix^a R_a)T_m(e^{-it\Lambda_{wa}} f_{j,k})(x) = C \int_{\mathbb{R}^3} (|x| - x\cdot\xi/|\xi|)m(\xi)e^{ix\cdot\xi}e^{-it|\xi|}\widehat{f_{j,k}}(\xi)\, d\xi. \tag{7.1.65}$$

We may also assume $|x| \in [t/2, 2t]$ (otherwise we have rapid decay using integration by parts). The conclusion follows from (7.1.62) and the observation that $\|f_{j,k}\|_{H^{0,2}_\Omega} \lesssim 2^{j+k}\|f^*_{j,k}\|_{H^{0,1}_\Omega}$.

It remains to prove the bounds (7.1.62). We still use the formula (7.1.65), and notice that the desired bounds follow easily if $2^k t \lesssim 1$, so we may assume that $2^k t \ge 2^{50}$. By rotation invariance, we may assume $x = (x_1, 0, 0)$, $x_1 \in [2^{-20}t, 8t]$. Then we bound the right-hand side of (7.1.65) by $Ct\sum_{b,c\in[0,2^{-20}(2^k t)^{1/2}]} |J_{b,c}| + R$, where

$$J_{b,c} := \int_{\mathbb{R}^3} \widehat{f_{j,k}}(\xi)\varphi_{[k-4,k+4]}(\xi)m(\xi)(1-\xi_1/|\xi|)\mathbf{1}_+(\xi_1)e^{ix_1\xi_1 - it|\xi|}\psi_{b,c}(\xi)\, d\xi,$$
$$\psi_{b,c}(\xi) := \varphi_b^{[0,\infty)}(\xi_2/2^\lambda)\varphi_c^{[0,\infty)}(\xi_3/2^\lambda), \qquad 2^\lambda := t^{-1/2}2^{k/2}, \tag{7.1.66}$$

and $R$ is an acceptable remainder that can be estimated using Lemma 3.1.

To prove (7.1.62) it suffices to show that for any $b, c \in [0, 2^{-20}(2^k t)^{1/2}]$ we have

$$|J_{b,c}| \lesssim t^{-2}2^{-k/2}(t2^k)^\delta \|f_{j,k}\|_{H^{0,2}_\Omega}. \tag{7.1.67}$$

Notice that

$$m(\xi)(1-\xi_1/|\xi|)\mathbf{1}_+(\xi_1) = \frac{m(\xi)}{1+\xi_1/|\xi|}\frac{\xi_2^2+\xi_3^2}{|\xi|^2}\mathbf{1}_+(\xi_1) = m'(\xi)\frac{\xi_2^2+\xi_3^2}{|\xi|^2}\mathbf{1}_+(\xi_1), \tag{7.1.68}$$

in the support of the integrals defining $J_{b,c}$, for some suitable symbol $m' \in \mathcal{M}_0$.

We estimate first $|J_{0,0}|$. For any $p \in [2,\infty)$, using also (7.1.64) we have

$$\begin{aligned} |J_{0,0}| &\lesssim \|\widehat{f_{j,k}}(r\theta)\|_{L^2(r^2dr)L^p_\theta}(2^{\lambda-k})^{2/p'}2^{3k/2}2^{2\lambda-2k} \\ &\lesssim_p \|f_{j,k}\|_{H^{0,1}_\Omega}\cdot t^{-2}2^{k/2}(t2^k)^{1/p}, \end{aligned} \tag{7.1.69}$$

where the factor $(2^{\lambda-k})^{2/p'}2^{3k/2}$ is due to the $L^2(r^2dr)L^{p'}_\theta$ norm of the support of the integral, and the factor $2^{2\lambda-2k}$ is due to the null factor $(\xi_2^2+\xi_3^2)/|\xi|^2$ in (7.1.68). This is consistent with the bound (7.1.67), by taking $p$ large enough.

To prove (7.1.67) when $(b,c) \ne (0,0)$ we may assume without loss of generality that $b \ge c$, so $b \ge \max(c,1)$. We integrate by parts in the integral in (7.1.66), up to eight times, using the rotation vector-field $\Omega_{12} = \xi_1\partial_{\xi_2} - \xi_2\partial_{\xi_1}$. Since $\Omega_{12}\{x_1\xi_1 - t|\xi|\} = -\xi_2 x_1$, every integration by parts gains a factor of $t2^{\lambda+b} \approx t^{1/2}2^{k/2+b}$ and loses a factor $t^{1/2}2^{k/2}$. If $\Omega_{12}$ hits the function $\widehat{f_{j,k}}$ twice then we stop integrating by parts and bound the integral by estimating $\Omega_{12}^2\widehat{g_{j,k}}$ in $L^2$. As in (7.1.69) it follows that

$$|J_{b,c}| \lesssim \{\|\widehat{f_{j,k}}(r\theta)\|_{L^2(r^2dr)L^p_\theta} + \|\widehat{\Omega_{12}f_{j,k}}(r\theta)\|_{L^2(r^2dr)L^p_\theta}\} \\ \times (2^{\lambda-k})^{2/p'}2^{3k/2}2^{2\lambda-2k}2^{-b} + \|\Omega_{12}^2\widehat{f_{j,k}}\|_{L^2}(2^{\lambda+b}2^{k/2})(t2^{\lambda+b})^{-2}2^{2\lambda+2b-2k},$$

which gives the bound (7.1.67). This completes the proof of the lemma. □

### 7.1.3 Null and Timelike Geodesics

We consider now the future-directed causal geodesics in our spacetime $M$, and prove that they extend forever (in the affine parametrization) and become asymptotically parallel to the geodesics of the Minkowski space. More precisely:

**Theorem 7.6.** *With* $(\mathbf{g}, \psi)$ *as in Theorem 7.1, assume* $p = (p^0, p^1, p^2, p^3)$ *is a point in* $M$ *and* $v = v^\alpha \partial_\alpha$ *is a null or timelike vector at* $p$*, normalized with* $v^0 = 1$*. Then there is a unique affinely parametrized global geodesic curve* $\gamma : [0, \infty) \to M$ *with*

$$\gamma(0) = p = (p^0, p^1, p^2, p^3), \qquad \dot{\gamma}(0) = v = (v^0, v^1, v^2, v^3). \tag{7.1.70}$$

*Moreover, there is a vector* $v_\infty = (v^0_\infty, v^1_\infty, v^2_\infty, v^3_\infty)$ *such that, for any* $s \in [0, \infty)$,

$$|\dot{\gamma}(s) - v_\infty| \lesssim \varepsilon_0(1+s)^{-1+6\delta'} \quad \text{and} \quad m_{\alpha\beta}v^\alpha_\infty v^\beta_\infty = \mathbf{g}_{\alpha\beta}(p)v^\alpha v^\beta. \tag{7.1.71}$$

*The implicit constant in (7.1.71) is independent of p. As a consequence*

$$|\gamma(s) - v_\infty s - p| \lesssim \varepsilon_0(1+s)^{6\delta'} \qquad \text{for any } s \in [0, \infty). \tag{7.1.72}$$

*Proof.* The proof uses only Theorem 7.2, Lemma 7.3, and the definitions.

**Step 1.** Assume that $T > 0$ and $\gamma = (\gamma^0, \gamma^1, \gamma^2, \gamma^3) : [0, T) \to M$ is a $C^4$ curve satisfying the geodesic equation

$$\ddot{\gamma}^\mu + \dot{\gamma}^\alpha\dot{\gamma}^\beta\mathbf{\Gamma}^\mu{}_{\alpha\beta} = 0, \qquad \mu \in \{0, 1, 2, 3\}, \tag{7.1.73}$$

with initial data (7.1.70). Let $V(s) = (V^0, V^1, V^2, V^3)(s) := \dot{\gamma}(s)$. The identity (7.1.73) implies the norm conservation identity

$$V^\alpha(s)V^\beta(s)\mathbf{g}_{\alpha\beta}(\gamma(s)) = \text{constant} \qquad \text{for } s \in [0, T), \tag{7.1.74}$$

as well as the general identity

$$\frac{d}{ds}\{V^\beta\mathbf{g}_{\alpha\beta}\} = (1/2)V^\mu V^\nu \partial_\alpha h_{\mu\nu}, \qquad \alpha \in \{0, 1, 2, 3\}. \tag{7.1.75}$$

Using first (7.1.74) it follows that $V^\alpha(s)V^\beta(s)\mathbf{g}_{\alpha\beta} \le 0$ for all $s \in [0, T)$. In particular,

$$|V'(s)| \leq 1.1|V^0(s)| \qquad \text{for any } s \in [0,T), \tag{7.1.76}$$

where $V'(s) := (V^1(s), V^2(s), V^3(s))$. Since $V^0(0) = 1$ and $V(s) \neq 0$ (by uniqueness of solutions to the ODE (7.1.73)), we have $V^0(s) > 0$ for all $s \in [0,T)$.

**Step 2.** The idea is to prove that, for any $s \in [0,T)$ and $\alpha \in \{0,1,2,3\}$,

$$V^0(s) \in [3/4, 4/3] \quad \text{and} \quad \int_0^s |\partial_\alpha h_{\mu\nu}(\gamma(u))V^\mu(u)V^\nu(u)|\, du \lesssim \varepsilon_0. \tag{7.1.77}$$

We use again a bootstrap argument. Assume that, for some $T' < T$, the weaker inequalities

$$V^0(s) \in [2/3, 3/2] \qquad \text{and} \qquad \int_0^s |\partial_\alpha h_{\mu\nu}(\gamma(u))V^\mu(u)V^\nu(u)|\, du \leq \varepsilon_1, \tag{7.1.78}$$

hold for any $s \in [0,T')$, where $\varepsilon_1 = \varepsilon_0^{2/3}$ as before. It suffices to show that the stronger inequalities (7.1.77) hold for any $s \in [0,T')$, under the bootstrap assumption (7.1.78).

It follows from (7.1.78) and (7.1.76) that, for any $s \in [0,T')$ and $\mu \in \{0,1,2,3\}$,

$$|V^\mu(s)| \leq 2, \qquad \gamma^0(s) - p^0 \in [2s/3, 3s/2]. \tag{7.1.79}$$

We apply now (7.1.75) with $\alpha = 0$. Let

$$A(s) := -V^\beta(s)\mathbf{g}_{0\beta}(\gamma(s)) = V^0(s) - V^\beta(s)h_{0\beta}(\gamma(s)). \tag{7.1.80}$$

Using (7.1.75) we calculate

$$\partial_s A = -(1/2)V^\mu V^\nu \partial_0 h_{\mu\nu}.$$

Using the bootstrap assumption (7.1.78) it follows that $|A(s) - A(0)| \leq \varepsilon_1$ for any $s \in [0,T')$. Thus $|V^0(s) - 1| \lesssim \varepsilon_1$ (see (7.1.80)), and the bounds $V^0(s) \in [3/4, 4/3]$ in (7.1.77) follow.

To prove the second bounds in (7.1.77) we would like to use (7.1.48), but for this we need to link the vectors $V(s)$ and $L(\gamma(s))$. We define the function $B : [0,T') \to \mathbb{R}$ by

$$B(s) := 1+\gamma^0(s)+(1+\gamma^0(s))^{2\delta'} - \{1+(\gamma^1(s))^2+(\gamma^2(s))^2+(\gamma^3(s))^2\}^{1/2}. \tag{7.1.81}$$

One should think of $B$ as a slight modification (for the purpose of making it increasing along the geodesic curve $\gamma$) of the function $\gamma^0(s) - |\gamma'(s)|$, where $\gamma'(s) := (\gamma^1(s), \gamma^2(s), \gamma^3(s))$ and $|\gamma'(s)| := [(\gamma^1(s))^2 + (\gamma^2(s))^2 + (\gamma^3(s))^2]^{1/2}$.

We calculate

$$(\partial_s B)(s) = V^0(s) + 2\delta' V^0(s)(1+\gamma^0(s))^{2\delta'-1} - \frac{(\gamma^1 V^1)(s) + (\gamma^2 V^2)(s) + (\gamma^3 V^3)(s)}{\{1+(\gamma^1(s))^2+(\gamma^2(s))^2+(\gamma^3(s))^2\}^{1/2}}. \quad (7.1.82)$$

Notice that

$$\sqrt{1+(B(s))^2} \lesssim \langle \gamma^0(s) - |\gamma'(s)|\rangle (1+\gamma^0(s))^{2\delta'}. \quad (7.1.83)$$

Moreover, since the vector $V(s)$ is timelike or null and using (7.1.20), we have

$$0 \leq V^0(s) - |V'(s)| + C\varepsilon_0 V^0(s)(1+\gamma^0(s)+|\gamma'(s)|)^{-1+\delta'}$$

for some constant $C \geq 1$, with $V'(s)$ as in (7.1.76). Therefore

$$(\partial_s B)(s) \geq \big|V^0(s) - |V'(s)|\big| + \Big\{|V'(s)| - \frac{(\gamma' \cdot V')(s)}{[1+|\gamma'(s)|^2]^{1/2}}\Big\}. \quad (7.1.84)$$

We would like now to express the vector $V(s)$ in terms of the good vectors at the point $\gamma(s)$. More precisely, for any $s \in [0, T')$ we would like to decompose

$$V(s) = |V'(s)|L(\gamma(s)) + H'(\gamma(s)) + E(\gamma(s)), \quad (7.1.85)$$

where $H' = H^1\partial_1 + H^2\partial_2 + H^3\partial_3$ is a horizontal vector tangential to the sphere,

$$|H^a(\gamma(s))| \lesssim \sqrt{(\partial_s B)(s)}, \qquad H'(\gamma(s)) \cdot \gamma'(s) = 0, \quad (7.1.86)$$

and $E = E^0\partial_0 + E^1\partial_1 + E^2\partial_2 + E^3\partial_3$ is an error term,

$$|E^\mu(\gamma(s))| \lesssim (\partial_s B)(s), \qquad \mu \in \{0,1,2,3\}. \quad (7.1.87)$$

Indeed, one can simply take $H' = 0$ if $|\gamma'(s)| \leq 1$, since in this case $|(\partial_s B)(s)| \gtrsim 1$ (see (7.1.84)). On the other hand, if $|\gamma'(s)| \geq 1$ then we use the decomposition

$$V = V^0\partial_0 + (V' \cdot \partial_r)\partial_r + H' = |V'|L + H' + \{(V^0 - |V'|)\partial_0 + [(V' \cdot \partial_r) - |V'|]\partial_r\},$$

where $H' = H^a\partial_a$, $H' \cdot \partial_r = 0$. Since $\partial_r = |\gamma'|^{-1}\gamma'$, $|H'|^2 = |V'|^2 - (V' \cdot \partial_r)^2$ and $\partial_s B \geq |V^0 - |V'|| + (|V'| - V' \cdot \partial_r)$ (see (7.1.84)), the desired conclusions (7.1.85)–(7.1.87) follow.

We show now that, for any $s \in [0, T')$ and $\alpha \in \{0,1,2,3\}$,

$$|\partial_\alpha h_{\mu\nu}(\gamma(s)) V^\mu(s) V^\nu(s)| \lesssim \varepsilon_0 (1+|\gamma(s)|)^{-2+5\delta'} + \varepsilon_0 (1+|\gamma(s)|)^{-1+5\delta'} \frac{(\partial_s B)(s)}{(1+B(s)^2)^{1/2}}. \quad (7.1.88)$$

Indeed, in view of (7.1.85) and (7.1.77) the left-hand side of (7.1.88) is bounded by

$$C\{|[\partial_\alpha h_{\mu\nu}L^\mu L^\nu](\gamma(s))| + |[\partial_\alpha h_{\mu\nu}L^\mu (H')^\nu](\gamma(s))|$$
$$+ |[\partial_\alpha h_{\mu\nu}(H')^\mu (H')^\nu](\gamma(s))| + |[\partial_\alpha h_{\mu\nu}V^\mu E^\nu](\gamma(s))| + |[\partial_\alpha h_{\mu\nu}E^\mu E^\nu](\gamma(s))|\}.$$

The terms in the first line are bounded by $C\varepsilon_0(1+|\gamma(s)|)^{-2+4\delta'}$, due to (7.1.21) and (7.1.48). The terms in the second line are $\lesssim \varepsilon_0(1+|\gamma(s)|)^{-1+2\delta'}\langle\gamma^0(s) - |\gamma'(s)|\rangle^{-1}(\partial_s B)(s)$, due to (7.1.21) and (7.1.86)–(7.1.87). The desired bounds (7.1.88) follow using also (7.1.83).

Finally, we can complete the proof of the second estimate in (7.1.77). Since $1+|\gamma(s)| \approx 1+s+p^0+|p'| \approx 1+s+|p|$ (due to (7.1.79)), it follows from (7.1.88) that

$$\begin{aligned}\int_{t_1}^{t_2} |\partial_\alpha h_{\mu\nu}(\gamma(s))V^\mu(s)V^\nu(s)|\,ds &\lesssim \int_{t_1}^{t_2} \varepsilon_0(1+s+|p|)^{-2+5\delta'}\,ds \\ &+ \int_{t_1}^{t_2} \varepsilon_0(1+s+|p|)^{-1+5\delta'}\frac{d}{ds}\log[\sqrt{1+B(s)^2}+B(s)]\,ds \\ &\lesssim \varepsilon_0(1+t_1+|p|)^{-1+6\delta'}\end{aligned} \tag{7.1.89}$$

for any $t_1 \le t_2 \in [0,T')$, where we used integration by parts and the observation that $|B(s)| \lesssim 1+|\gamma(s)| \lesssim 1+s+|p|$. The conclusion (7.1.77) follows by setting $t_1 = 0$.

**Step 3.** We can now prove the conclusions of the theorem. In view of (7.1.77), (7.1.79), and the standard existence theory of solutions of ODEs, the geodesic curve $\gamma$ extends for all values of $s \in [0,\infty)$ as a smooth solution of the equation (7.1.73). Thus one can take $T = \infty$, and the inequalities (7.1.77) are satisfied for all $s \in [0,\infty)$.

We apply (7.1.75) for all $\alpha \in \{0,1,2,3\}$ and integrate between times $s_1 < s_2 \in [0,\infty)$. Using (7.1.89) we have

$$\left|V^\beta(s_2)\mathbf{g}_{\alpha\beta}(\gamma(s_2)) - V^\beta(s_1)\mathbf{g}_{\alpha\beta}(\gamma(s_1))\right| \lesssim \varepsilon_0(1+s_1+|p|)^{-1+6\delta'}.$$

Since $|h_{\alpha\beta}(\gamma(s_2))| + |h_{\alpha\beta}(\gamma(s_1))| \lesssim \varepsilon_0(1+s_1+|p|)^{-1+\delta'}$ it follows that

$$\left|V^\alpha(s_2) - V^\alpha(s_1)\right| \lesssim \varepsilon_0(1+s_1+|p|)^{-1+6\delta'} \tag{7.1.90}$$

for any $s_1 < s_2 \in [0,\infty)$ and $\alpha \in \{0,1,2,3\}$. In particular, $v_\infty := \lim_{s\to\infty} V(s)$ exists. Notice that (7.1.71) follows from (7.1.90) and (7.1.74). The bounds in (7.1.72) then follow by integrating the bounds $|\dot\gamma(u) - v_\infty| \lesssim \varepsilon_0(1+u)^{-1+6\delta'}$ from 0 to $s$. □

## 7.2 WEAK PEELING ESTIMATES AND THE ADM ENERGY

### 7.2.1 Peeling Estimates

In this section we prove weak peeling estimates for the Riemann tensor of our spacetime. The Riemann tensor $\mathbf{R}$ satisfies the symmetry properties

$$\begin{aligned} &\mathbf{R}_{\alpha\beta\mu\nu} = -\mathbf{R}_{\beta\alpha\mu\nu} = -\mathbf{R}_{\alpha\beta\nu\mu} = \mathbf{R}_{\mu\nu\alpha\beta}, \\ &\mathbf{R}_{\alpha\beta\mu\nu} + \mathbf{R}_{\beta\mu\alpha\nu} + \mathbf{R}_{\mu\alpha\beta\nu} = 0. \end{aligned} \tag{7.2.1}$$

In our case, we also have the Einstein-field equations

$$\mathbf{g}^{\mu\nu}\mathbf{R}_{\alpha\mu\beta\nu} = \mathbf{R}_{\alpha\beta} = \mathbf{D}_\alpha\psi\mathbf{D}_\beta\psi + (\psi^2/2)\mathbf{g}_{\alpha\beta}. \tag{7.2.2}$$

The rates of decay of the components of the Riemann tensor are mainly determined by their *signatures*. To define this, we use (Minkowski) frames $(L, \underline{L}, e_a)$, where $L, \underline{L}$ are as in (7.1.15) and $e_a \in \mathcal{T}_h := \{r^{-1}\Omega_{12}, r^{-1}\Omega_{23}, r^{-1}\Omega_{31}\}$. We assign signature $+1$ to the vector-field $L$, $-1$ to the vector-field $\underline{L}$, and 0 to the horizontal vector-fields in $\mathcal{T}_h$. With $e_1, e_2, e_3, e_4 \in \mathcal{T}_h$ we define $\mathrm{Sig}(a)$ as the set of components of the Riemann tensor of total signature $a$, so

$$\begin{aligned} \mathrm{Sig}(-2) &:= \{\mathbf{R}(\underline{L}, e_1, \underline{L}, e_2)\}, \\ \mathrm{Sig}(2) &:= \{\mathbf{R}(L, e_1, L, e_2)\}, \\ \mathrm{Sig}(-1) &:= \{\mathbf{R}(\underline{L}, e_1, e_2, e_3), \mathbf{R}(\underline{L}, L, \underline{L}, e_1)\}, \\ \mathrm{Sig}(1) &:= \{\mathbf{R}(L, e_1, e_2, e_3), \mathbf{R}(L, \underline{L}, L, e_1)\}, \\ \mathrm{Sig}(0) &:= \{\mathbf{R}(e_1, e_2, e_3, e_4), \mathbf{R}(L, \underline{L}, e_1, e_2), \mathbf{R}(L, e_1, \underline{L}, e_2), \mathbf{R}(L, \underline{L}, L, \underline{L})\}. \end{aligned} \tag{7.2.3}$$

These components capture the full curvature tensor, due to (7.2.1).

Notice that we define our decomposition in terms of the Minkowski null pair $(L, \underline{L})$ instead of more canonical null frames (or tetrads) adapted to the metric $\mathbf{g}$ (see, for example, [12], [52], [53]). However, as we explain below, the weak peeling estimates are invariant under natural changes of the frame, and the rate of decay depends only on the signature of the component, except for the components $\mathbf{R}(L, e_1, \underline{L}, e_2)$. More precisely:

**Theorem 7.7.** *Assume that* $(\mathbf{g}, \psi)$ *is as in Theorem 7.1,* $(x,t) \in M' = \{(x,t) \in M : t \geq 1$ *and* $|x| \geq 2^{-8}t\}$, *and* $\Psi_{(a)} \in \mathrm{Sig}(a')$ *for* $a \leq a' \in \{-2,-1,0,1,2\}$. *Then*

$$\begin{aligned} |\Psi_{(-2)}(x,t)| &\lesssim \varepsilon_0 \langle r\rangle^{3\delta'-1}\langle t-r\rangle^{-2}, \\ |\Psi_{(-1)}(x,t)| &\lesssim \varepsilon_0 \langle r\rangle^{5\delta'-2}\langle t-r\rangle^{-1}, \\ |\Psi_{(2)}(x,t)| + |\Psi_{(1)}(x,t)| &\lesssim \varepsilon_0 \langle r\rangle^{7\delta'-3}. \end{aligned} \tag{7.2.4}$$

*Moreover, if* $\Psi^1_{(0)} \in \mathrm{Sig}_1(0) := \{\mathbf{R}(e_1, e_2, e_3, e_4), \mathbf{R}(L, \underline{L}, e_1, e_2), \mathbf{R}(L, \underline{L}, L, \underline{L})\}$

*then*

$$|\Psi^1_{(0)}(x,t)| \lesssim \varepsilon_0 \langle r\rangle^{7\delta'-3}. \tag{7.2.5}$$

*Remark 7.8.* (i) In view of (7.2.2), we notice that the Ricci components decay at most cubically, $\mathbf{R}_{\alpha\beta} = O(\langle r\rangle^{-3+})$ in $M'$. As a result we do not expect uniform estimates of order better than cubic for any components of the Riemann curvature tensor, so the weak peeling estimates in Theorem 7.7 are optimal in this sense, at least up to $r^{C\delta'}$ losses. In fact, the almost cubic decay is also formally consistent with the weak peeling estimates of Klainerman-Nicoló [53, Theorem 1.2 (b)] in the setting of our more general metrics.

(ii) We show in Proposition 7.9 below that our weak peeling estimates are invariant under suitable changes of the frame. Moreover, we also show that the natural $\langle r\rangle^{-3+}$ decay of the signature 0 components $\mathbf{R}(L, e_1, \underline{L}, e_2)$, which is missing in the estimates (7.2.4)–(7.2.5), can be restored if one works with a null vector-field $L$.

*Proof.* The estimates follow from the formulas (7.2.1)–(7.2.3), the bounds on first and second order derivatives of $h_{\alpha\beta}$ in Theorem 7.2 and Lemmas 7.3–7.4, and the general identity

$$\begin{aligned}\mathbf{R}_{\alpha\beta\mu\nu} &= -\partial_\alpha \mathbf{\Gamma}_{\nu\beta\mu} + \partial_\beta \mathbf{\Gamma}_{\nu\alpha\mu} + \mathbf{g}^{\rho\lambda}\mathbf{\Gamma}_{\rho\beta\mu}\mathbf{\Gamma}_{\lambda\alpha\nu} - \mathbf{g}^{\rho\lambda}\mathbf{\Gamma}_{\rho\alpha\mu}\mathbf{\Gamma}_{\lambda\beta\nu}\\ &= \frac{1}{2}\big[\partial_\alpha\partial_\nu h_{\beta\mu} + \partial_\beta\partial_\mu h_{\alpha\nu} - \partial_\beta\partial_\nu h_{\alpha\mu} - \partial_\alpha\partial_\mu h_{\beta\nu}\big]\\ &+ \mathbf{g}^{\rho\lambda}\big[\mathbf{\Gamma}_{\rho\beta\mu}\mathbf{\Gamma}_{\lambda\alpha\nu} - \mathbf{\Gamma}_{\rho\alpha\mu}\mathbf{\Gamma}_{\lambda\beta\nu}\big].\end{aligned} \tag{7.2.6}$$

**Step 1.** In view of (7.1.21)–(7.1.22) and (7.1.51) we have the general bounds

$$\begin{aligned}|\mathbf{\Gamma}_{\alpha\beta\mu}(x,t)| &\lesssim \varepsilon_0 \langle t-r\rangle^{-1}\langle r\rangle^{-1+2\delta'},\\ |\partial_\alpha\partial_\beta h_{\mu\nu}(x,t)| &\lesssim \varepsilon_0 \langle t-r\rangle^{-2}\langle r\rangle^{-1+3\delta'},\\ |\mathbf{R}_{\alpha\beta}(x,t)| &\lesssim \varepsilon_0 \langle r\rangle^{-3+2\delta'},\end{aligned} \tag{7.2.7}$$

for any $(x,t) \in M'$ and $\alpha, \beta, \mu, \nu \in \{0,1,2,3\}$. Also, if $V_1, V_2, V_3 \in \mathcal{T}$ then, using also (7.1.48),

$$\begin{aligned}&|V_1^\alpha V_2^\beta V_3^\mu \mathbf{\Gamma}_{\alpha\beta\mu}(x,t)| + |\underline{L}^\alpha L^\beta V_1^\mu \mathbf{\Gamma}_{\alpha\beta\mu}(x,t)|\\ &\quad + |L^\alpha V_1^\beta \underline{L}^\mu \mathbf{\Gamma}_{\alpha\beta\mu}(x,t)| + |V_1^\alpha \underline{L}^\beta L^\mu \mathbf{\Gamma}_{\alpha\beta\mu}(x,t)| \lesssim \varepsilon_0 \langle r\rangle^{-2+3\delta'}\end{aligned} \tag{7.2.8}$$

and, using (7.1.51),

$$\begin{aligned}|V_1^\alpha V_2^\beta \partial_\alpha\partial_\beta h_{\mu\nu}(x,t)| &\lesssim \varepsilon_0 \langle r\rangle^{-3+3\delta'},\\ |V_1^\alpha \underline{L}^\beta \partial_\alpha\partial_\beta h_{\mu\nu}(x,t)|| &\lesssim \varepsilon_0 \langle t-r\rangle^{-1}\langle r\rangle^{-2+3\delta'}.\end{aligned} \tag{7.2.9}$$

We can now prove the bounds (7.2.4). The bounds on $\Psi_{(-2)}$ and $\Psi_{(2)}$ follow, since all the terms in the right-hand side of (7.2.6) have suitable decay, using (7.2.7) for $\Psi_{(-2)}$ and (7.2.8)–(7.2.9) for $\Psi_{(2)}$.

If $\Psi_{(-1)}$ is one of the curvature components in $\mathrm{Sig}(a')$, $a' \geq -1$, containing at most one vector-field $\underline{L}$, then the desired bounds follow using (7.2.7) and (7.2.9). On the other hand, if $\Psi_{(-1)} = \mathbf{R}(\underline{L}, L, \underline{L}, V)$, $V \in \mathcal{T}$, then we use (7.1.16) and the Einstein equations (7.2.2). Thus

$$\begin{aligned} -\mathbf{R}(\underline{L}, V) &= m^{\alpha\beta}\mathbf{R}(\underline{L}, \partial_\alpha, \partial_\beta, V) + g^{\alpha\beta}_{\geq 1}\mathbf{R}(\underline{L}, \partial_\alpha, \partial_\beta, V) \\ &= -(1/2)\mathbf{R}(\underline{L}, L, \underline{L}, e_1) + \Pi^{\alpha\beta}\mathbf{R}(\underline{L}, \partial_\alpha, \partial_\beta, e_1) + g^{\alpha\beta}_{\geq 1}\mathbf{R}(\underline{L}, \partial_\alpha, \partial_\beta, e_1). \end{aligned} \tag{7.2.10}$$

Using (7.1.20) and (7.2.7) we have $\mathbf{R}(\underline{L}, L, \underline{L}, V) = O(\varepsilon_0 \langle t-r\rangle^{-1}\langle r\rangle^{-2+5\delta'})$, as desired.

If $\Psi_{(1)} = \mathbf{R}(L, e_1, e_2, e_3)$ then the desired bounds in (7.2.4) follow from (7.2.8)–(7.2.9). On the other hand, if $\Psi_1 = \mathbf{R}(L, \underline{L}, L, e_1)$ then we use again the Einstein equations as in (7.2.10), and estimate

$$|\mathbf{R}(L, \underline{L}, L, e_1)| \lesssim |\mathbf{R}(L, e_1)| + |g^{\alpha\beta}_{\geq 1}\mathbf{R}(L, \partial_\alpha, \partial_\beta, e_1)| + \big|\Pi^{\alpha\beta}\mathbf{R}(L, \partial_\alpha, \partial_\beta, e_1)\big|.$$

The desired bounds follow using (7.2.7), (7.2.6), (7.1.20), (7.2.9), and the definition of $\Pi$ in (7.1.15).

**Step 2.** We bound now the components of signature 0. Clearly, using just (7.2.7)–(7.2.9),

$$|\mathbf{R}(e_1, e_2, e_3, e_4)(x,t)| \lesssim \varepsilon_0 \langle r\rangle^{-3+5\delta'}. \tag{7.2.11}$$

We prove now that for any $(x,t) \in M'$

$$|\mathbf{R}(L, \underline{L}, e_1, e_2)(x,t)| \lesssim \varepsilon_0 \langle r\rangle^{-3+5\delta'}. \tag{7.2.12}$$

The quadratic terms involving connection coefficients can be bounded easily,

$$\begin{aligned} &|L^\alpha \underline{L}^\beta e_1^\mu e_2^\nu \mathbf{g}^{\rho\lambda}\mathbf{\Gamma}_{\rho\beta\mu}\mathbf{\Gamma}_{\lambda\alpha\nu}| \\ &\quad\lesssim |L^\alpha \underline{L}^\beta e_1^\mu e_2^\nu g^{\rho\lambda}_{\geq 1}\mathbf{\Gamma}_{\rho\beta\mu}\mathbf{\Gamma}_{\lambda\alpha\nu}| + |L^\alpha \underline{L}^\beta e_1^\mu e_2^\nu m^{\rho\lambda}\mathbf{\Gamma}_{\rho\beta\mu}\mathbf{\Gamma}_{\lambda\alpha\nu}| \\ &\quad\lesssim \varepsilon_0\langle r\rangle^{-3+5\delta'} + |L^\alpha \underline{L}^\beta e_1^\mu e_2^\nu \Pi^{\rho\lambda}\mathbf{\Gamma}_{\rho\beta\mu}\mathbf{\Gamma}_{\lambda\alpha\nu}| \\ &\qquad\qquad + |L^\alpha \underline{L}^\beta e_1^\mu e_2^\nu (L^\rho \underline{L}^\lambda + \underline{L}^\rho L^\lambda)\mathbf{\Gamma}_{\rho\beta\mu}\mathbf{\Gamma}_{\lambda\alpha\nu}| \\ &\quad\lesssim \varepsilon_0\langle r\rangle^{-3+5\delta'}, \end{aligned} \tag{7.2.13}$$

for any $e_1, e_2 \in \mathcal{T}_h$, using (7.1.20), (7.2.7)–(7.2.8), and (7.1.16). Moreover, we also have

$$|\underline{L}^\alpha e_1^\beta L^\nu e_2^\mu \partial_\alpha\partial_\beta h_{\mu\nu}(x,t)| \lesssim \varepsilon_0\langle r\rangle^{-3+5\delta'}. \tag{7.2.14}$$

Indeed, to see this we start with the harmonic gauge condition (7.1.2) and write it in the form

$$m^{\alpha\beta}\partial_\nu\partial_\alpha h_{\beta\mu} - (1/2)m^{\alpha\beta}\partial_\nu\partial_\mu h_{\alpha\beta} = -\partial_\nu(g^{\alpha\beta}_{\geq 1}\partial_\alpha h_{\beta\mu}) + (1/2)\partial_\nu(g^{\alpha\beta}_{\geq 1}\partial_\mu h_{\alpha\beta}).$$

Using now (7.1.16) and multiplying by $2e_1^\nu e_2^\mu$ we have

$$\begin{aligned}
&-(L^\alpha \underline{L}^\beta + \underline{L}^\alpha L^\beta) e_1^\nu e_2^\mu \partial_\nu \partial_\alpha h_{\beta\mu} + 2\Pi^{\alpha\beta} e_1^\nu e_2^\mu \partial_\nu \partial_\alpha h_{\beta\mu} \\
&\qquad = m^{\alpha\beta} e_1^\nu e_2^\mu \partial_\nu \partial_\mu h_{\alpha\beta} - 2e_1^\nu e_2^\mu \partial_\nu (g_{\geq 1}^{\alpha\beta} \partial_\alpha h_{\beta\mu}) + e_1^\nu e_2^\mu \partial_\nu (g_{\geq 1}^{\alpha\beta} \partial_\mu h_{\alpha\beta}).
\end{aligned}$$

Thus

$$\begin{aligned}
|\underline{L}^\alpha L^\beta e_1^\nu e_2^\mu \partial_\nu \partial_\alpha h_{\beta\mu}| &\lesssim |L^\alpha \underline{L}^\beta e_1^\nu e_2^\mu \partial_\nu \partial_\alpha h_{\beta\mu}| + |\Pi^{\alpha\beta} e_1^\nu e_2^\mu \partial_\nu \partial_\alpha h_{\beta\mu}| \\
&\quad + |m^{\alpha\beta} e_1^\nu e_2^\mu \partial_\nu \partial_\mu h_{\alpha\beta}| + |e_1^\nu e_2^\mu \partial_\nu g_{\geq 1}^{\alpha\beta} \partial_\alpha h_{\beta\mu}| + |e_1^\nu e_2^\mu g_{\geq 1}^{\alpha\beta} \partial_\nu \partial_\alpha h_{\beta\mu}| \\
&\quad + |e_1^\nu e_2^\mu \partial_\nu g_{\geq 1}^{\alpha\beta} \partial_\mu h_{\alpha\beta}| + |e_1^\nu e_2^\mu g_{\geq 1}^{\alpha\beta} \partial_\nu \partial_\mu h_{\alpha\beta}|,
\end{aligned}$$

and the bounds (7.2.14) follow using (7.1.51), (7.1.21), and (7.1.20). The desired estimates (7.2.12) now follow from (7.2.13)–(7.2.14), the formula (7.2.6), and the bounds (7.1.51).

Finally, we prove now that for any $(x,t) \in M'$

$$|\mathbf{R}(L, \underline{L}, L, \underline{L})(x,t)| \lesssim \varepsilon_0 \langle r \rangle^{-3+7\delta'}. \tag{7.2.15}$$

Indeed, we start with the Einstein-field equations (7.2.2). Thus

$$\begin{aligned}
-\mathbf{R}(L, \underline{L}) &= m^{\alpha\beta} \mathbf{R}(L, \partial_\alpha, \partial_\beta, \underline{L}) + g_{\geq 1}^{\alpha\beta} \mathbf{R}(L, \partial_\alpha, \partial_\beta, \underline{L}) \\
&= -(1/2)\mathbf{R}(L, \underline{L}, L, \underline{L}) + \Pi^{\alpha\beta} \mathbf{R}(L, \partial_\alpha, \partial_\beta, \underline{L}) + g_{\geq 1}^{\alpha\beta} \mathbf{R}(L, \partial_\alpha, \partial_\beta, \underline{L}).
\end{aligned}$$

We use (7.2.7) and the curvature bounds (7.2.4) proved earlier to estimate the error terms in the identity above. For (7.2.15) it suffices to prove that

$$\left|\Pi^{\mu\nu} \mathbf{R}(\partial_\mu, L, \partial_\nu, \underline{L})(x,t)\right| \lesssim \varepsilon_0 \langle r \rangle^{-3+7\delta'}. \tag{7.2.16}$$

To prove (7.2.16) we apply again the Einstein equations and (7.1.16), so

$$\begin{aligned}
\Pi^{\mu\nu} \mathbf{R}_{\mu\nu} &= \Pi^{\mu\nu} m^{\alpha\beta} \mathbf{R}(\partial_\mu, \partial_\alpha, \partial_\nu, \partial_\beta) + \Pi^{\mu\nu} g_{\geq 1}^{\alpha\beta} \mathbf{R}(\partial_\mu, \partial_\alpha, \partial_\nu, \partial_\beta) \\
&= -\Pi^{\mu\nu} \mathbf{R}(\partial_\mu, L, \partial_\nu, \underline{L}) + \Pi^{\mu\nu} \Pi^{\alpha\beta} \mathbf{R}(\partial_\mu, \partial_\alpha, \partial_\nu, \partial_\beta) \\
&\qquad + \Pi^{\mu\nu} g_{\geq 1}^{\alpha\beta} \mathbf{R}(\partial_\mu, \partial_\alpha, \partial_\nu, \partial_\beta).
\end{aligned}$$

In view of (7.2.7), (7.2.11), and (7.1.20), for (7.2.16) it suffices to prove that

$$\left|\Pi^{\mu\nu} \mathbf{R}(\partial_\mu, W_1, \partial_\nu, W_2)(x,t)\right| \lesssim \varepsilon_0 \langle r \rangle^{-2+6\delta'}, \tag{7.2.17}$$

for any $W_1, W_2 \in \{L, \underline{L}, r^{-1}\Omega_{12}, r^{-1}\Omega_{23}, r^{-1}\Omega_{31}\}$.

The bounds (7.2.17) follow from (7.2.4) and (7.2.11), unless $W_1 = W_2 = \underline{L}$.

In this case we use again the Einstein-field equations and (7.1.16), so

$$\begin{aligned}\mathbf{R}(\underline{L},\underline{L}) &= m^{\alpha\beta}\mathbf{R}(\underline{L},\partial_\alpha,\underline{L},\partial_\beta) + g^{\alpha\beta}_{\geq 1}\mathbf{R}(\underline{L},\partial_\alpha,\underline{L},\partial_\beta)\\ &= \Pi^{\alpha\beta}\mathbf{R}(\underline{L},\partial_\alpha,\underline{L},\partial_\beta) + g^{\alpha\beta}_{\geq 1}\mathbf{R}(\underline{L},\partial_\alpha,\underline{L},\partial_\beta).\end{aligned}$$

The bounds (7.2.17) follow, which completes the proof of the theorem. □

We would like to show now that our weak peeling estimates are invariant under natural changes of frames. Assume $L', \underline{L}', r^{-1}\Omega'_{12}, r^{-1}\Omega'_{23}, r^{-1}\Omega'_{31}$ are vector-fields in $M'$, which are small perturbations of the corresponding Minkowski vector-fields, i.e.,

$$\begin{aligned}&|(L-L')(x,t)| + |(\underline{L}-\underline{L}')(x,t)| + |x|^{-1}|(\Omega_{12}-\Omega'_{12})(x,t)|\\ &\quad + |x|^{-1}|(\Omega_{23}-\Omega'_{23})(x,t)| + |x|^{-1}|(\Omega_{31}-\Omega'_{31})(x,t)| \lesssim \langle r\rangle^{-1+c_0},\end{aligned} \tag{7.2.18}$$

for any $(x,t)\in M'$, where $c_0 \in [2\delta', 1/10]$. We define the associated sets of curvature components, as in (7.2.3),

$$\begin{aligned}\mathrm{Sig}'(-2) &:= \{\mathbf{R}(\underline{L}',e'_1,\underline{L}',e'_2)\},\\ \mathrm{Sig}'(2) &:= \{\mathbf{R}(L',e'_1,L',e'_2)\},\\ \mathrm{Sig}'(-1) &:= \{\mathbf{R}(\underline{L}',e'_1,e'_2,e'_3), \mathbf{R}(\underline{L}',L',\underline{L}',e'_1)\},\\ \mathrm{Sig}'(1) &:= \{\mathbf{R}(L',e'_1,e'_2,e'_3), \mathbf{R}(L',\underline{L}',L',e'_1)\},\\ \mathrm{Sig}'(0) &:= \mathrm{Sig}'_1(0)\cup\mathrm{Sig}'_2(0),\\ \mathrm{Sig}'_1(0) &:= \{\mathbf{R}(e'_1,e'_2,e'_3,e'_4), \mathbf{R}(L',\underline{L}',e'_1,e'_2), \mathbf{R}(L',\underline{L}',L',\underline{L}')\},\\ \mathrm{Sig}'_2(0) &:= \{\mathbf{R}(L',e'_1,\underline{L}',e'_2)\},\end{aligned} \tag{7.2.19}$$

where $e'_1, e'_2, e'_3, e'_4 \in \mathcal{T}'_h := \{r^{-1}\Omega'_{12}, r^{-1}\Omega'_{23}, r^{-1}\Omega'_{31}\}$.

**Proposition 7.9.** *Assume that* $\Psi'_{(a)} \in \mathrm{Sig}'(a')$ *for* $a \leq a' \in \{-2,-1,0,1,2\}$, *and* $\Psi'^1_{(0)} \in \mathrm{Sig}'_1(0)$. *Then* $\Psi'_{(a)}, \Psi'^1_{(0)}$ *satisfy similar bounds as before,*

$$\begin{aligned}|\Psi'_{(-2)}(x,t)| &\lesssim \varepsilon_0\langle r\rangle^{-1+3\delta'}\langle t-r\rangle^{-2},\\ |\Psi'_{(-1)}(x,t)| &\lesssim \varepsilon_0\langle r\rangle^{-2+3\delta'+c_0}\langle t-r\rangle^{-1},\\ |\Psi'_{(2)}(x,t)| + |\Psi'_{(1)}(x,t)| + |\Psi'^1_{(0)}(x,t)| &\lesssim \varepsilon_0\langle r\rangle^{-3+3\delta'+2c_0}.\end{aligned} \tag{7.2.20}$$

*Moreover, if the vector-field* $L'$ *satisfies the almost null bounds*

$$|\mathbf{g}(L',L')(x,t)| \lesssim \langle r\rangle^{-2+2c_0}, \qquad \textit{for any } (x,t)\in M' \tag{7.2.21}$$

*(see Lemma 7.19 below for the construction of such vector-fields associated to almost optical functions), and* $\Psi'^2_{(0)} \in \mathrm{Sig}'_2(0)$, *then we have the additional bounds*

$$|\Psi'^2_{(0)}(x,t)| \lesssim \varepsilon_0\langle r\rangle^{-3+3\delta'+2c_0}. \tag{7.2.22}$$

*Proof.* **Step 1.** We prove first the bounds (7.2.20). Assume that $W_1, \dots, W_4$ are vector-fields in $\{L, \underline{L}, r^{-1}\Omega_{ab}\}$ and let $W_1', \dots, W_4'$ denote their corresponding perturbations. In view of Theorem 7.7 all the components of the curvature tensor in the Minkowski null frame are bounded by $C\varepsilon_0\langle r\rangle^{-1+3\delta'}\langle t-r\rangle^{-2}$. Therefore we can estimate

$$
\begin{aligned}
\big|\mathbf{R}(W_1', W_2', W_3', W_4') - \mathbf{R}(W_1, W_2, W_3, W_4)\big| &\lesssim \big|\mathbf{R}(W_1' - W_1, W_2, W_3, W_4)\big| \\
&+ \big|\mathbf{R}(W_1, W_2' - W_2, W_3, W_4)\big| + \big|\mathbf{R}(W_1, W_2, W_3' - W_3, W_4)\big| \\
&+ \big|\mathbf{R}(W_1, W_2, W_3, W_4' - W_4)\big| + \varepsilon_0\langle r\rangle^{-2+2c_0}\langle r\rangle^{-1+3\delta'}\langle t-r\rangle^{-2}.
\end{aligned}
\tag{7.2.23}
$$

The last remainder term in the right-hand side of (7.2.23) is compatible with all the desired estimates (7.2.20). The other four curvature terms in the right-hand side are all bounded by $C\varepsilon_0\langle r\rangle^{-1+c_0}\langle r\rangle^{-1+3\delta'}\langle t-r\rangle^{-2}$, which still suffices to prove the estimates on $|\Psi'_{(-2)}|$ and $|\Psi'_{(-1)}|$ in the first two lines of (7.2.20).

On the other hand, we can also use (7.2.23) to estimate the terms $|\Psi'_{(2)}|$ and $|\Psi'_{(1)}|$ in the last line of (7.2.20) by

$$
C\varepsilon_0\langle r\rangle^{-1+c_0}\langle r\rangle^{-2+5\delta'}\langle t-r\rangle^{-1}. \tag{7.2.24}
$$

This is because the four curvature terms in the right-hand of (7.2.23) contain only components of signature $\geq -1$, since the change of one vector-field can reduce the signature by at most 2. The remaining term $|\Psi'^{1}_{(0)}|$ is also bounded by the expression (7.2.24), because in the case of the components in $\mathrm{Sig}_1(0)$ the change of one vector-field can only reduce the total signature by 1. The desired bounds (7.2.20) follow.

We remark that this argument fails for the components $\mathbf{R}(L, e_1, \underline{L}, e_2)$ in $\mathrm{Sig}_2(0)$: even if such a component is $O(\langle r\rangle^{-3+})$ for a choice of frame, this bound is not invariant under a small change of the frame (satisfying (7.2.18)) because the replacement of the vector $L$ in the first position would bring in errors of the form $\langle r\rangle^{-1+}\mathbf{R}(\underline{L}, e_1, \underline{L}, e_2)$, which are too big in the wave region.

**Step 2.** We prove now the bounds (7.2.22). Notice first that we may replace $\underline{L}'$, $e_1'$, and $e_2'$ with $\underline{L}$, $e_1$, and $e_2$ respectively, at the expense of components of signature $\geq -1$, thus bounded by the expression in (7.2.24). We may also assume that $L' = L + \rho\underline{L}$, where $|\rho(x,t)| \lesssim \langle r\rangle^{-1+c_0}$. Using (7.2.1) we have

$$
\mathbf{R}(L', e_1, \underline{L}, e_2) - \mathbf{R}(\underline{L}, e_1, L', e_2) = \mathbf{R}(L', \underline{L}, e_1, e_2).
$$

Using the Einstein-field equations (7.2.2) and (7.1.16) we have

$$
\begin{aligned}
\mathbf{R}(e_1, e_2) &= m^{\alpha\beta}\mathbf{R}(\partial_\alpha, e_1, \partial_\beta, e_2) + g_{\geq 1}^{\alpha\beta}\mathbf{R}(\partial_\alpha, e_1, \partial_\beta, e_2) \\
&= -(1/2)[\mathbf{R}(L, e_1, \underline{L}, e_2) + \mathbf{R}(\underline{L}, e_1, L, e_2)] \\
&\quad + \Pi^{\alpha\beta}\mathbf{R}(\partial_\alpha, e_1, \partial_\beta, e_2) + g_{\geq 1}^{\alpha\beta}\mathbf{R}(\partial_\alpha, e_1, \partial_\beta, e_2).
\end{aligned}
$$

In view of (7.2.20) and (7.2.7), we have

$$|\mathbf{R}(L', \underline{L}, e_1, e_2)(x,t)| + |\mathbf{R}(e_1, e_2)(x,t)| + |\Pi^{\alpha\beta}\mathbf{R}(\partial_\alpha, e_1, \partial_\beta, e_2)(x,t)| \lesssim \varepsilon_0 \langle r\rangle^{7\delta'-3}.$$

Combining the last three equations and recalling that $L = L' - \rho\underline{L}$ we have

$$\begin{aligned}
&\mathbf{R}(L', e_1, \underline{L}, e_2) - \mathbf{R}(\underline{L}, e_1, L', e_2) = O(\varepsilon_0\langle r\rangle^{7\delta'-3}),\\
&[\mathbf{R}(L', e_1, \underline{L}, e_2) + \mathbf{R}(\underline{L}, e_1, L', e_2)]\\
&\qquad - 2\rho\mathbf{R}(\underline{L}, e_1, \underline{L}, e_2) - 2g^{\alpha\beta}_{\geq 1}\mathbf{R}(\partial_\alpha, e_1, \partial_\beta, e_2) = O(\varepsilon_0\langle r\rangle^{7\delta'-3}).
\end{aligned}$$

Therefore, to prove (7.2.22) it suffices to show that

$$\rho\mathbf{R}(\underline{L}, e_1, \underline{L}, e_2) + g^{\alpha\beta}_{\geq 1}\mathbf{R}(\partial_\alpha, e_1, \partial_\beta, e_2) = O(\varepsilon_0\langle r\rangle^{-3+3\delta'+2c_0}). \tag{7.2.25}$$

The contributions of the curvature components in the term $\mathbf{R}(\partial_\alpha, e_1, \partial_\beta, e_2)$ in (7.2.25) are all bounded by the expression in (7.2.24), with the exception of the component $\mathbf{R}(\underline{L}, e_1, \underline{L}, e_2)$, which has signature $-2$. Therefore, if we decompose $\partial_\alpha = A_\alpha\underline{L} + B_\alpha L + W$, where $W \cdot \partial_0 = W \cdot \partial_r = 0$, then the left-hand side (7.2.25) is

$$\left[\rho + g^{\alpha\beta}_{\geq 1}A_\alpha A_\beta\right]\mathbf{R}(\underline{L}, e_1, \underline{L}, e_2) + O(\varepsilon_0\langle r\rangle^{-3+3\delta'+2c_0}). \tag{7.2.26}$$

Moreover, $A_\alpha = -(1/2)\partial_\alpha u^0$, where $u^0(x,t) := |x| - t$. Therefore

$$\rho + g^{\alpha\beta}_{\geq 1}A_\alpha A_\beta = \rho + (1/4)g^{\alpha\beta}_{\geq 1}\partial_\alpha u^0\partial_\beta u^0 = \rho - (1/4)L^\alpha L^\beta h_{\alpha\beta} + O(\langle r\rangle^{-2+2\delta'}),$$

where the last identity follows from the explicit formulas (2.1.8). Since $L = L' - \rho\underline{L}$, $\rho = O(\langle r\rangle^{-1+c_0})$, and $\mathbf{g}(L', \underline{L}) = -2 + O(\langle r\rangle^{-1+c_0})$ we have

$$4\rho = \mathbf{g}(L, L) - \mathbf{g}(L', L') + O(\langle r\rangle^{-2+2c_0}).$$

The last two identities and the assumption (7.2.21) show that $\rho + g^{\alpha\beta}_{\geq 1}A_\alpha A_\beta = O(\langle r\rangle^{-2+2c_0})$. The desired conclusion (7.2.25) follows using also (7.2.26). □

### 7.2.2 The ADM Energy

The ADM energy measures the total deviation of our spacetime from the Minkowski solution. In our asymptotically flat case it is calculated by integrating on large spheres on the surfaces $\Sigma_t = \{(x,t) \in M : x \in \mathbb{R}^3\}$, according to the formula

$$E_{ADM}(t) := \frac{1}{16\pi}\lim_{R\to\infty}\int_{S_{R,t}}(\partial_j\mathbf{g}_{nj} - \partial_n\mathbf{g}_{jj})\frac{x^n}{|x|}\,dx, \tag{7.2.27}$$

where the integration is over large (Euclidean) spheres $S_{R,t} \subseteq \Sigma_t$ of radius $R$. Using Stokes theorem and the definitions (2.1.29) and (4.3.2), we can rewrite

$$E_{ADM}(t) = \frac{1}{16\pi} \lim_{R\to\infty} \int_{|x|\leq R} -2\Delta\tau(x,t)\, dx. \tag{7.2.28}$$

We analyze first the density function $-2\Delta\tau$.

**Lemma 7.10.** *We can decompose*

$$-2\Delta\tau = -\delta_{jk}P^2_{jk} + \{(\partial_0\psi)^2 + \psi^2 + \partial_j\psi\partial_j\psi\} + O^1 + \partial_j O^2_j, \tag{7.2.29}$$

*where* $P^2_{jk}$ *are defined as in* (2.1.15),

$$\|O^1(t)\|_{L^1(\mathbb{R}^3)} \lesssim \varepsilon_0^2\langle t\rangle^{-\kappa} \qquad \textit{for any } t \geq 0, \tag{7.2.30}$$

*and*

$$\begin{aligned} O^2_j := &-h_{0j}\partial_0 h_{00} + h_{00}\partial_0 h_{0j} - h_{0n}\partial_0 h_{nj} + h_{nj}\partial_0 h_{0n} - h_{kn}\partial_j h_{kn} \\ &+ h_{0n}\partial_j h_{0n} - h_{0n}\partial_n h_{0j} + h_{kn}\partial_n h_{kj} - h_{00}\partial_j(\tau + \underline{F}) \\ &+ 2h_{0j}\partial_0(\tau + \underline{F}) - h_{nj}\partial_n(\tau + \underline{F}). \end{aligned} \tag{7.2.31}$$

*Proof.* Recall the identity (4.3.5)

$$-2\Delta\tau = \partial_\alpha E^{\geq 2}_\alpha + \underline{F}(\mathcal{N}^h) + \tau(\mathcal{N}^h) = \partial_\alpha E^{\geq 2}_\alpha + \frac{1}{2}\mathcal{N}^h_{00} + \frac{1}{2}\delta_{jk}\mathcal{N}^h_{jk}. \tag{7.2.32}$$

**Step 1.** We use the formulas (2.1.9)–(2.1.15) and (4.3.1) and identify first the $L^1$ errors. Indeed, all the cubic and higher order terms are bounded by $C\varepsilon_0^2\langle t-|x|\rangle^{-1}\langle t+|x|\rangle^{-3+\kappa}$ or by $C\varepsilon_0^2\langle x\rangle^{-1}\langle t+|x|\rangle^{-3+\kappa}$, due to (7.1.20)–(7.1.22), so they are acceptable $L^1$ errors. The semilinear quadratic null forms in $Q^2_{\alpha\beta}$ (see (2.1.14)) are also acceptable $L^1$ errors because

$$\begin{aligned} \big|(m^{\alpha\beta}\partial_\alpha h_1\partial_\beta h_2)(x,t)\big| + \big|(\partial_\mu h_1\partial_\nu h_2 - \partial_\nu h_1\partial_\mu h_2)(x,t)\big| \\ \lesssim \varepsilon_0^2\langle t-|x|\rangle^{-1}\langle t+|x|\rangle^{-3+\kappa}, \end{aligned} \tag{7.2.33}$$

for any $\mu,\nu \in \{0,1,2,3\}$ and $h_1, h_2 \in \{h_{\alpha\beta}\}$. This follows from (7.1.21) and the observation

$$\partial_0 = -\partial_r + L \quad \text{and} \quad \partial_j = (x_j/r)\partial_r + \text{sum of good vector-fields in } \mathcal{T}. \tag{7.2.34}$$

The Klein-Gordon contributions coming from (2.1.12) are the second term in the right-hand side of (7.2.29). It remains to analyze the quadratic semilinear and quasilinear terms involving the metric components, which are

$$\frac{1}{2}(\mathcal{Q}^2_{00} + \mathcal{Q}^2_{jj}) - \frac{1}{2}(P^2_{00} + P^2_{jj}) + \partial_0 E^2_0 + \partial_j E^2_j, \tag{7.2.35}$$

where, using (4.3.1) and (2.1.8),

$$\begin{aligned}E^2_\mu &:= -g_1^{\alpha\beta}\partial_\alpha h_{\beta\mu} + \frac{1}{2}g_1^{\alpha\beta}\partial_\mu h_{\alpha\beta}\\ &= h_{00}\partial_0 h_{0\mu} - h_{0k}(\partial_0 h_{k\mu} + \partial_k h_{0\mu})\\ &+ h_{kn}\partial_k h_{n\mu} - \frac{1}{2}h_{00}\partial_\mu h_{00} + h_{0k}\partial_\mu h_{0k} - \frac{1}{2}h_{kn}\partial_\mu h_{kn}.\end{aligned} \tag{7.2.36}$$

Notice that all the terms in (7.2.35) are of the form $\partial h \cdot \partial h$ or $h \cdot \partial^2 h$. To extract acceptable $L^1$ errors, notice that, for $\mu, \nu \in \{0, 1, 2, 3\}$, $h \in \{h_{\alpha\beta}\}$, and $V \in \mathcal{T}$,

$$\|h(t)\|_{L^p} \lesssim \varepsilon_0 \langle t\rangle^{-\kappa} \quad \text{if } p \geq 3 + 4\kappa, \tag{7.2.37}$$

$$\|\partial_\mu h(t)\|_{L^p} + \|\partial_\mu\partial_\nu h(t)\|_{L^p} \lesssim \varepsilon_0\langle t\rangle^{-\kappa} \quad \text{if } p \geq 2 + 4\kappa, \tag{7.2.38}$$

$$\begin{aligned}&\|(1-\chi_t)\cdot\partial_\mu h(t)\|_{L^p} + \|(1-\chi_t)\cdot\partial_\mu\partial_\nu h(t)\|_{L^p}\\ &\qquad + \|V^\mu\partial_\mu h(t)\|_{L^p} \lesssim \varepsilon_0\langle t\rangle^{-\kappa} \qquad \text{if } p \geq 3/2 + 4\kappa,\end{aligned} \tag{7.2.39}$$

where $\chi_t$ is a smooth characteristic function of the wave region, for example, $\chi_t \equiv 0$ if $t \leq 8$ and $\chi_t(x) = \varphi_{\leq -4}((t-|x|)/t)$ if $t \geq 8$. These bounds follow from Theorem 7.2. Moreover

$$\begin{aligned}&\|(\partial_\alpha h_1 \cdot \partial_\beta h_2)(t)\|_{L^p} + \|(h_1 \cdot \partial_\alpha\partial_\beta h_2)(t)\|_{L^p}\\ &\quad + \|\mathcal{N}^h_{\alpha\beta}(t)\|_{L^p} + \|E^{\geq 2}_\alpha(t)\|_{L^p} + \|\partial_\mu E^{\geq 2}_\alpha(t)\|_{L^p} \lesssim \varepsilon_0\langle t\rangle^{-\kappa},\end{aligned} \tag{7.2.40}$$

for any $p \geq 1 + 2\kappa$. In particular, all the terms in (7.2.35) are barely missing to being acceptable $L^1$ errors. All the semilinear terms that contain a good derivative are acceptable $L^1$ errors.

**Step 2.** We analyze now the quadratic expressions in (7.2.35). Since $(\partial_0^2 - \Delta)h_{\alpha\beta} = \mathcal{N}^h_{\alpha\beta}$, we can use (7.2.40) to write

$$\begin{aligned}\partial_0 E^2_0 \sim \frac{1}{2}\partial_0 h_{00}\partial_0 h_{00} + \frac{1}{2}h_{00}\Delta h_{00} - \partial_0 h_{0k}\partial_k h_{00} - h_{0k}\partial_0\partial_k h_{00}\\ + \partial_0 h_{kn}\partial_k h_{n0} + h_{kn}\partial_0\partial_k h_{n0} - \frac{1}{2}\partial_0 h_{kn}\partial_0 h_{kn} - \frac{1}{2}h_{kn}\Delta h_{kn},\end{aligned}$$

where in this proof $F \sim G$ means $\|F - G\|_{L^1} \lesssim \varepsilon_0^2\langle t\rangle^{-\kappa}$. Using also (7.2.33)

$$\partial_0 E^2_0 \sim \frac{1}{2}\partial_j(h_{00}\partial_j h_{00}) - \partial_k(h_{0k}\partial_0 h_{00}) + \partial_k(h_{nk}\partial_0 h_{n0}) - \frac{1}{2}\partial_j(h_{kn}\partial_j h_{kn}).$$

Therefore, using again (7.2.36), we have $\partial_0 E_0^2 + \partial_j E_j^2 \sim \partial_j O_j^{2,1}$, where

$$\begin{aligned} O_j^{2,1} := & -h_{0j}\partial_0 h_{00} + h_{nj}\partial_0 h_{n0} - h_{kn}\partial_j h_{kn} + h_{00}\partial_0 h_{0j} \\ & - h_{0n}\partial_0 h_{nj} - h_{0n}\partial_n h_{0j} + h_{kn}\partial_k h_{nj} + h_{0n}\partial_j h_{0n}. \end{aligned} \tag{7.2.41}$$

We examine now the terms $\mathcal{Q}_{00}^2$ and $\mathcal{Q}_{jj}^2$. Using the identities (2.1.13)

$$\begin{aligned} \frac{1}{2}(\mathcal{Q}_{00}^2 + \mathcal{Q}_{nn}^2) = & \,\partial_j O_j^{2,2} \\ & + \{\partial_j h_{00}\partial_j(\tau + \underline{F}) - 2\partial_j h_{0j}\partial_0(\tau + \underline{F}) + \partial_n h_{nj}\partial_j(\tau + \underline{F})\}, \end{aligned} \tag{7.2.42}$$

where $(1/2)(h_{00} + \delta_{jk}h_{jk}) = \tau + \underline{F}$ (see (4.3.2)) and

$$O_j^{2,2} := -h_{00}\partial_j(\tau + \underline{F}) + 2h_{0j}\partial_0(\tau + \underline{F}) - h_{nj}\partial_n(\tau + \underline{F}). \tag{7.2.43}$$

The desired formula (7.2.31) follows from (7.2.41) and (7.2.43).

**Step 3.** We identify now the contribution of the semilinear terms. We show first that the semilinear terms in the bracket in the right-hand side of (7.2.42) are acceptable $L^1$ errors. Indeed, using (7.2.34) and (7.2.38)–(7.2.39), we may replace $\partial_j$ with $-(x_j/r)\partial_0$, at the expense of acceptable errors. The semilinear expression in (7.2.42) is

$$\begin{aligned} & \sim \{\partial_0 h_{00}\partial_0(\tau + \underline{F}) + 2(x_j/r)\partial_0 h_{0j}\partial_0(\tau + \underline{F}) + (x_n x_j/r^2)\partial_0 h_{nj}\partial_0(\tau + \underline{F})\} \\ & \sim \partial_0(\tau + \underline{F}) \cdot (L^\alpha L^\beta \partial_0 h_{\alpha\beta}). \end{aligned}$$

This is an acceptable $L^1$ error due to Lemma 7.3 and (7.1.21).

Finally, we examine (2.1.15) and notice that $P_{00} \sim P_{jj}$, due to (7.2.34) and (7.2.38)–(7.2.39). The contribution of $-(1/2)(P_{00}^2 + P_{jj}^2)$ is the first term in the right-hand side of (7.2.29). □

We prove now that the ADM energy is well defined, conserved in time, non-negative, and can be linked to the scattering data of our spacetime. More precisely:

**Proposition 7.11.** *We have $\Delta\tau(t) \in L^1(\mathbb{R}^3)$ and*

$$E_{ADM}(t) = \frac{1}{16\pi}\int_{\mathbb{R}^3} -2\Delta\tau(x,t)\,dx = E_{ADM} \tag{7.2.44}$$

*does not depend on $t \in [0,\infty)$. Moreover, for any $t \geq 0$,*

$$E_{ADM} = \frac{1}{16\pi}\int_{\mathbb{R}^3} \{|U^\psi(t)|^2 + (1/4)\sum_{m,n\in\{1,2,3\}} |U^{\vartheta_{mn}}(t)|^2\}\,dx + O(\varepsilon_0^2\langle t\rangle^{-\kappa}). \tag{7.2.45}$$

*In particular, recalling the scattering profiles $V_\infty^\psi$ and $V_\infty^G$ from (7.3.18), we have*

$$E_{ADM} = \frac{1}{16\pi}\|V_\infty^\psi\|^2 + \frac{1}{64\pi}\sum_{m,n\in\{1,2,3\}}\|V_\infty^{\vartheta_{mn}}\|_{L^2}^2. \tag{7.2.46}$$

*Proof.* **Step 1.** It follows from (7.1.21)–(7.1.22) that

$$\|\partial_\mu E_\nu^{\geq 2}(t)\|_{(L^1\cap L^2)(\mathbb{R}^3)} + \|\mathcal{N}_{\mu\nu}^h\|_{(L^1\cap L^2)(\mathbb{R}^3)} \lesssim \varepsilon_0\langle t\rangle^4, \qquad t\in[0,\infty), \tag{7.2.47}$$

for any $\mu,\nu\in\{0,1,2,3\}$. In particular $\Delta\tau\in L^1(\mathbb{R}^3)$ (due to (4.3.5)), and the identity (7.2.44) holds. To prove that the energy is constant we estimate, for any $t_1\leq t_2\in[0,\infty)$ and $R$ large,

$$\begin{aligned}\Big|\int_{\mathbb{R}^3}\Delta\tau(x,t_2)\varphi_{\leq 0}(|x|/R)\,dx - \int_{\mathbb{R}^3}\Delta\tau(x,t_1)\varphi_{\leq 0}(|x|/R)\,dx\Big|\\ = \Big|\int_{t_1}^{t_2}\int_{\mathbb{R}^3}\partial_0\partial_j\tau(x,s)\cdot\partial_j\varphi_{\leq 0}(|x|/R)\,dx\Big|\\ \lesssim |t_2-t_1|R^{-1/4}\sup_{s\in[t_1,t_2]}\sum_{j\in\{1,2,3\}}\|\partial_0\partial_j\tau(s)\|_{L^{4/3}}.\end{aligned}$$

Using now (4.3.5) and (7.2.47) it follows that

$$\lim_{R\to\infty}\Big|\int_{\mathbb{R}^3}\Delta\tau(x,t_2)\varphi_{\leq 0}(|x|/R)\,dx - \int_{\mathbb{R}^3}\Delta\tau(x,t_1)\varphi_{\leq 0}(|x|/R)\,dx\Big| = 0,$$

for any $t_1\leq t_2\in[0,\infty)$, so the function $E_{ADM}$ is constant in time.

**Step 2.** We prove now the identity (7.2.45). We start from (7.2.29) and notice that the contribution of the Klein-Gordon field $\psi$ is given by the integral of $|U^\psi(t)|^2$, as claimed. The divergence term $\partial_j O_j^2$ does not contribute, because $\|O_j^2(t)\|_{L^{4/3}}\lesssim\varepsilon_0^2$.

Finally, to calculate the contribution of $\delta_{jk}P_{jk}$ we would like to use Lemma 4.25. To apply it, we use the bounds (7.2.49) proved below. In particular, using also (7.2.34) we have

$$\begin{aligned}&\big\|(m^{\alpha\beta}\partial_\alpha Rh\cdot\partial_\beta R'h')(t)\big\|_{L^1}\\ &+\big\|(\partial_\mu Rh\cdot\partial_\nu R'h' - \partial_\nu Rh\cdot\partial_\mu R'h')(t)\big\|_{L^1}\lesssim\varepsilon_0^2\langle t\rangle^{-\kappa},\end{aligned} \tag{7.2.48}$$

for any $h,h'\in\{h_{\alpha\beta}\}$, $\mu,\nu\in\{0,1,2,3\}$, and compounded Riesz transforms $R,R'$.

We can now use the calculations in Lemma 4.25 with $\mathcal{L}=\mathcal{L}_1=\mathcal{L}_2=Id$. The cubic and higher order terms are all acceptable $L^1$ errors, due to (7.2.40). Also, most of the quadratic terms in (4.3.82) can be estimated using (7.2.33), for example,

$$\begin{aligned}R_jR_k\partial_nG_1\cdot R_jR_k\partial_nG_2 &= \partial_jR_kR_nG_1\cdot\partial_nR_jR_kG_2\\ &\sim\partial_nR_kR_nG_1\cdot\partial_jR_jR_kG_2 = \partial_kG_1\cdot\partial_kG_2,\end{aligned}$$

for any $G_1, G_2 \in \{F, \underline{F}, \omega_a, \vartheta_{ab}\}$, where $f \sim g$ means $\|f - g\|_{L^1} \lesssim \varepsilon_0^2 \langle t \rangle^{-\kappa}$ as before. The terms containing $\tau$ are all acceptable $L^1$ errors, due to (4.3.5) and (7.2.49) below. The only remaining terms are $(1/2)\partial_j\vartheta_{mn}\cdot\partial_j\vartheta_{mn} \sim (1/2)\partial_0\vartheta_{mn}\cdot\partial_0\vartheta_{mn} \sim (1/4)(\partial_j\vartheta_{mn}\cdot\partial_j\vartheta_{mn} + \partial_0\vartheta_{mn}\cdot\partial_0\vartheta_{mn})$, coming from $-\delta_{jk}A^{\vartheta\vartheta}_{jk}$, which lead to the identity (7.2.45). □

We collect now some bounds on the Riesz transforms of the metric components, which are used in Proposition 7.11 and Theorem 7.23 below.

**Lemma 7.12.** *Assume that $R = R_1^{a_1}R_2^{a_2}R_3^{a_3}$, $a_1+a_2+a_3 \leq 6$, is a compounded Riesz transform, $h \in \{h_{\alpha\beta}\}$, $\mathcal{N}^h \in \{\mathcal{N}^h_{\alpha\beta}\}$, and $V \in \mathcal{T}$ is a good vector-field. Then*

$$\begin{aligned}
\|\partial_\mu Rh(t)\|_{L^p} &\lesssim \varepsilon_0\langle t\rangle^{-\kappa} && \text{if } p \geq 2+4\kappa,\\
\|V^\mu\partial_\mu Rh(t)\|_{L^q} &\lesssim \varepsilon_0\langle t\rangle^{-\kappa} && \text{if } q \geq 3/2+4\kappa,\\
\big\||\nabla|^{-1}R\mathcal{N}^h(t)\big\|_{L^q} + \big\||\nabla|^{-1}R\partial_\mu E^{\geq 2}_\alpha(t)\big\|_{L^q} &\lesssim \varepsilon_0\langle t\rangle^{-\kappa} && \text{if } q \geq 3/2+4\kappa,
\end{aligned} \tag{7.2.49}$$

*for any $\mu, \alpha \in \{0,1,2,3\}$. In addition, if $t \leq 1$ or if $t \geq 1$ and $|x| \in [2^{-10}t, 2^{10}t]$ then*

$$\begin{aligned}
&|Rh(x,t)| + |\partial_\mu Rh(x,t)| \lesssim \varepsilon_0\langle t\rangle^{-1+\kappa},\\
&|V^\mu\partial_\mu Rh(x,t)| \lesssim \varepsilon_0\langle t\rangle^{-4/3+\kappa},\\
&\big||\nabla|^{-1}R\mathcal{N}^h(x,t)\big| + \big||\nabla|^{-1}RE^{\geq 2}_\alpha(x,t)\big| + \big||\nabla|^{-1}R\partial_\mu E^{\geq 2}_\alpha(x,t)\big| \lesssim \varepsilon_0\langle t\rangle^{-4/3+\kappa}.
\end{aligned} \tag{7.2.50}$$

*Proof.* The bounds in the first line of (7.2.49) follow directly from (7.2.38). The bounds in the third line follow from (7.2.40) and the Hardy-Littlewood-Sobolev inequality. To estimate $\|V^\mu\partial_\mu Rh(t)\|_{L^q}$ we notice that the contribution of $(1-\chi_t)\partial_\mu h(t)$ is bounded easily, due to the estimate on the first term in the left-hand side of (7.2.39). Moreover,

$$\|V^\mu R(\chi_t\partial_\mu h(t))\|_{L^q} \lesssim \|V^\mu\chi_t\partial_\mu h(t)\|_{L^q} + \big\|[V^\mu, R](\chi_t\partial_\mu h(t))\big\|_{L^q}. \tag{7.2.51}$$

The first term is bounded by $C\varepsilon_0\langle t\rangle^{-\kappa}$, due to (7.2.39). The second term is a Calderón commutator that can be estimated using the general bound

$$\begin{aligned}
\big|[V^\mu, R](g)(x)\big| &\lesssim \int_{\mathbb{R}^3} |g(y)||x-y|^{-3}|V^\mu(x) - V^\mu(y)|\,dy\\
&\lesssim \int_{\mathbb{R}^3} |g(y)|t^{-1}|x-y|^{-2}\,dy.
\end{aligned} \tag{7.2.52}$$

Therefore, recalling that $|g(x)| \lesssim \varepsilon_0\chi_t(x)\langle t-|x|\rangle^{-1}t^{2\delta'-1}$ (see (7.1.21)), we have

$$\big\|[V^\mu, R](\chi_t\partial_\mu h(t))\big\|_{L^q} \lesssim t^{-1}\|g\|_{L^{p_1}} \lesssim \varepsilon_0 t^{-1}t^{2\delta'-1}t^{2/p_1},$$

where $1/p_1 = 1/q + 1/3$ (by fractional integration). The desired bounds (7.2.49) follow using also (7.2.51).

We notice now that the pointwise bounds (7.2.50) follow easily from (7.1.3) if $t \lesssim 1$. On the other hand, if $t \geq 8$ then we fix $K_0$ the largest integer such that $2^{K_0} \leq \langle t\rangle^8$ and notice that the contribution of low or high frequencies $P_{\leq -K_0-1}h + P_{\geq K_0+1}h$ is suitably bounded due to (7.1.3). For the medium frequencies we use the general estimate

$$\|RP_{[-K_0,K_0]}f\|_{L^\infty} \lesssim \log(2+t)\|P_{[-K_0,K_0]}f\|_{L^\infty}. \tag{7.2.53}$$

The bounds in the first line of (7.2.50) follow from (7.1.20) and (7.1.21).

To prove the bounds in the second line we estimate

$$\begin{aligned}|V^\mu R(P_{[-K_0,K_0]}\partial_\mu h)(x,t)| &\lesssim \|RP_{[-K_0,K_0]}(V^\mu \partial_\mu h)(t)\|_{L^\infty} \\ &\quad + \big|[V^\mu, RP_{[-K_0,K_0]}](\partial_\mu h)(x,t)\big|.\end{aligned}$$

The first term in the right-hand side is bounded by $\varepsilon_0 t^{-2+4\delta'}$, due to (7.1.21) and (7.2.53). Since $|x| \approx t$ the second term can be bounded as in (7.2.52),

$$\begin{aligned}\big|[V^\mu, RP_{[-K_0,K_0]}](\partial_\mu h)(x,t)\big| &\lesssim \int_{\mathbb{R}^3} |\partial_\mu h(y)| t^{-1}|x-y|^{-2}\, dy \\ &\lesssim \varepsilon_0 \int_{\mathbb{R}^3} \frac{\langle t+|y|\rangle^{4\delta'-1}}{\langle t-|y|\rangle t|x-y|^2}\, dy,\end{aligned}$$

where we used (7.1.21) for the second estimate. The last integral in the inequalities above is bounded by $Ct^{-4/3+5\delta'}$, which suffices to prove the estimates in the second line of (7.2.50).

The estimates in the third line follow from Hölder's inequality once we notice that the functions $\mathcal{N}^h$, $E^{\geq 2}_{\alpha}$, and $\partial_\mu E^{\geq 2}_{\alpha}$ are all bounded by $C\varepsilon_0^2 t^{-4/3+5\delta'}$ in $L^p$ for all $p \geq 3-\delta'$. □

### 7.2.3 The Linear Momentum

With $\Sigma_t$ as before, let $N$ denote its associated future-pointing unit normal vector-field. We define the second fundamental form

$$k_{ab} := -\mathbf{g}(\mathbf{D}_{\partial_a}N, \partial_b) = \mathbf{g}(N, \mathbf{D}_{\partial_a}\partial_b) = N^\alpha \mathbf{\Gamma}_{\alpha ab}, \qquad a,b \in \{1,2,3\}. \tag{7.2.54}$$

Let $\overline{g}_{jk} = \mathbf{g}_{jk}$ denote the induced (Riemannian) metric on $\Sigma_t$. Our main result in this section is the following:

**Proposition 7.13.** *We define the linear momentum*

$$\mathbf{p}_a(t) := \frac{1}{8\pi}\lim_{R\to\infty}\int_{S_{R,t}} \pi_{ab}\frac{x^b}{|x|}\, dx, \qquad \pi_{ab} := k_{ab} - (\mathrm{tr}k)\,\overline{g}_{ab}, \tag{7.2.55}$$

*where $S_{R,t} \subseteq \Sigma_t$ denotes the sphere of radius $R$ as before. Then the functions $\mathbf{p}_a$, $a \in \{1,2,3\}$ are well defined and independent of $t \in [0,\infty)$. Moreover, for any $t \geq 0$,*

$$\mathbf{p}_a = -\frac{1}{16\pi}\int_{\mathbb{R}^3}\{2\partial_0\psi\partial_a\psi + \frac{1}{2}\sum_{m,n\in\{1,2,3\}}\partial_0\vartheta_{mn}\partial_a\vartheta_{mn}\}(t)\,dx + O(\varepsilon_0^2\langle t\rangle^{-\kappa}). \tag{7.2.56}$$

*In particular*

$$\sum_{a\in\{1,2,3\}}\mathbf{p}_a^2 \leq E_{ADM}^2, \tag{7.2.57}$$

*so the ADM mass $M_{ADM} := \big(E_{ADM}^2 - \sum_{a\in\{1,2,3\}}\mathbf{p}_a^2\big)^{1/2} \geq 0$ is well defined.*

*Proof.* This is similar to the proof of Lemma 7.10 and Proposition 7.11.

**Step 1.** Since $t$ is fixed, the quadratic and higher order terms do not contribute to the integral as $R\to\infty$, so we may redefine

$$\mathbf{p}_a(t) = \frac{1}{8\pi}\lim_{R\to\infty}\int_{S_{R,t}}\pi_{ab}^1\frac{x^b}{|x|}\,dx, \qquad \pi_{ab}^1 := k_{ab}^1 - \delta_{ab}\delta_{jk}k_{jk}^1, \tag{7.2.58}$$

where $k_{ab}^1 := \mathbf{\Gamma}_{0ab}$ is the linear part of the second fundamental form $k$,

$$\begin{aligned}2k_{ab}^1 &= \partial_a h_{0b} + \partial_b h_{0a} - \partial_0 h_{ab} = R_aR_b(-2|\nabla|\rho - \partial_0 F + \partial_0\underline{F})\\ &\quad + (\in_{akl} R_bR_k + \in_{bkl} R_aR_k)(|\nabla|\omega_l + \partial_0\Omega_l) - \in_{apm}\in_{bqn}\partial_0R_pR_q\vartheta_{mn},\end{aligned}$$

using (2.1.29). Recall that $R_mR_n\vartheta_{mn} = 0$ and $\delta_{mn}\vartheta_{mn} = -2\tau$ (see (2.1.26) and (4.3.2)). Thus

$$2\partial_b\pi_{ab}^1 = -\in_{akl}|\nabla|R_k(|\nabla|\omega_l + \partial_0\Omega_l) + 2\partial_a\partial_0\tau. \tag{7.2.59}$$

Using (2.1.26) and then (4.3.1) we calculate

$$\begin{aligned}-\in_{akl}|\nabla|R_k(|\nabla|\omega_l + \partial_0\Omega_l) &= (\delta_{km}\delta_{an} - \delta_{kn}\delta_{am})R_kR_m\big(|\nabla|^2h_{0n} + \partial_0\partial_bh_{nb}\big)\\ &= -(\delta_{an} + R_aR_n)\big(|\nabla|^2h_{0n} + \partial_0\partial_bh_{nb}\big)\\ &= -(\delta_{an} + R_aR_n)\big(-\Delta h_{0n} + \partial_0^2h_{0n} + \partial_0E_n^{\geq 2}\big)\\ &= -(\delta_{an} + R_aR_n)\big(\mathcal{N}_{0n}^h + \partial_0E_n^{\geq 2}\big).\end{aligned}$$

Using (4.3.5) and the last two identities we calculate (after cancelling four terms)

$$2\partial_b\pi_{ab}^1 = -\mathcal{N}_{0a}^h - \partial_aE_0^{\geq 2} - \partial_0E_a^{\geq 2}. \tag{7.2.60}$$

**Step 2.** Using Stokes theorem, (7.2.47), and (7.2.60), the limit in (7.2.58) exists and

$$\mathbf{p}_a(t) = \frac{1}{16\pi}\int_{\mathbb{R}^3}2\partial_b\pi_{ab}^1\,dx = -\frac{1}{16\pi}\int_{\mathbb{R}^3}\{\mathcal{N}_{0a}^h + \partial_aE_0^{\geq 2} + \partial_0E_a^{\geq 2}\}\,dx. \tag{7.2.61}$$

Moreover, using (7.2.59),

$$\partial_0(2\partial_b\pi^1_{ab}) = -\in_{akl}\partial_k(|\nabla|\partial_0\omega_l + \partial_0^2\Omega_l) + 2\partial_a\partial_0^2\tau.$$

In view of (4.3.4)–(4.3.5) and (7.2.47), we have

$$\big\|(|\nabla|\partial_0\omega_l + \partial_0^2\Omega_l)(t)\big\|_{L^{4/3}} + \big\|\partial_0^2\tau(t)\|_{L^{4/3}} \lesssim \varepsilon_0\langle t\rangle^4,$$

and the same argument as in the proof of Proposition 7.11 shows that the components $\mathbf{p}_a(t)$ are constant in time.

To prove the identity (7.2.56) we use the formula (7.2.61), extract the time decaying components, and then let $t \to \infty$, just as in the proofs of Lemma 7.10 and Proposition 7.11. Indeed, all the cubic and higher order terms and all the quadratic null terms lead to time decaying contributions. Thus using (2.1.9)–(2.1.15) and (4.3.1),

$$\mathcal{N}^h_{0a} + \partial_a E_0^{\geq 2} + \partial_0 E_a^{\geq 2} \sim 2\partial_0\psi\partial_a\psi + \mathcal{Q}^2_{0a} - P^2_{0a} + \partial_0 E^2_a + \partial_a E^2_0,$$

where $F \sim G$ means $\|F - G\|_{L^1} \lesssim \varepsilon_0^2\langle t\rangle^{-\kappa}$ as before. Also, as we know from the proof of Proposition 7.11, derivative terms of the form $\partial(h\cdot\partial h)$ do not contribute to the integral in the limit $t \to \infty$. As in the proof of Lemma 7.10, the terms $\mathcal{Q}^2_{0a}$, $\partial_0 E^2_a$, and $\partial_a E^2_0$ are sums of derivatives and $L^1$ acceptable errors. The only terms that contribute in the limit are the terms $2\partial_0\psi\partial_a\psi$ and $(1/2)R_pR_q\partial_0\vartheta_{mn}R_pR_q\partial_a\vartheta_{mn}$ coming from $-P^2_{0a}$ after removing the $L^1$ acceptable errors (as in Proposition 7.11). In view of (7.2.61), this leads to the desired formula (7.2.56). Finally, the inequality (7.2.57) follows using also (7.2.45) and letting $t \to \infty$. □

### 7.2.4 Gauge Conditions and Parameterizations

Our main result, Theorem 1.3 works in any parameterization of the initial time slice. It turns out that some gauge choices allow us to simplify the metric up to quadratic $O(\varepsilon_0^2)$ terms. We now explore this and its relation to the Hodge decomposition in (2.1.29). We first observe that $\vartheta$ represents a "minimal" expression of the metric on any time slice.

**Proposition 7.14.** *Let $t \geq 0$ and consider $\Sigma_t$ a fixed time slice in the spacetime constructed in Theorem 1.3; let $\overline{g}$ be the induced Riemannian metric. There exists a choice of* spatially *harmonic coordinates on $(\Sigma_t, \overline{g})$ for which the metric $\overline{g}_{jk}$ coincides linearly with the $\vartheta$ component of $\mathbf{g}$ in the sense that*

$$\|\nabla_x(\overline{g}_{jk} - \in_{jpm}\in_{kqn} R_pR_q\vartheta_{mn})\|_{H^3} \lesssim \varepsilon_0^2\langle t\rangle^{2\delta'-1}. \tag{7.2.62}$$

*In particular, we see that $\tau$ is related to the scalar curvature of $\Sigma_t$,*

$$\|\overline{R} + \Delta\tau\|_{H^2} \lesssim \varepsilon_0^2\langle t\rangle^{2\delta'-1}. \tag{7.2.63}$$

*Remark 7.15.* Conversely, using Lemma 7.16, we see that $\|\nabla_x \widetilde{g}\|^2_{L^2}$ is always at least as large as $\|\nabla_x \vartheta\|^2_{L^2}$.

*Proof of Proposition 7.14.* We start by computing the spatial covariant derivative. Starting from the formula

$$\begin{aligned}\overline{\mathbf{\Gamma}}_j - \{\partial_t h_{0j} - (1/2)\partial_j h_{00}\} &= \mathbf{\Gamma}_j + \{\mathbf{\Gamma}_{j00} - \partial_t h_{0j} + (1/2)\partial_j h_{00}\} \\ &\quad - \big\{(g^{pq}_{\geq 1} - \overline{g}^{pq}_{\geq 1})\mathbf{\Gamma}_{jpq} + g^{00}_{\geq 1}\mathbf{\Gamma}_{j00} + 2g^{0p}_{\geq 1}\mathbf{\Gamma}_{jp0}\big\},\end{aligned}$$

we define the first order corrector $\chi^{1,j}$ by

$$\partial_p \chi^{1,j} := \delta^{qj} R_p \{-R_0 h_{0q} + (1/2) R_q h_{00}\}. \tag{7.2.64}$$

By the bootstrap bounds (2.1.50), we have $\nabla_x \chi^{1,j} \in H^4$, so that $\chi^{1,j}$ is well defined and in $C^3$. In fact, adapting the proof of (7.1.20), we can see that

$$|\chi^{1,j}(x,t)| \lesssim \varepsilon_0 \langle x \rangle / \langle |x| + t \rangle^{1-2\delta'}.$$

Let $x^j$ be the usual coordinates on $\Sigma_t$. We look for (spatially) harmonic coordinates of the form $y^j = x^j + \phi^j = x^j + \chi^{1,j} + \psi^j$, where $\psi^j$ satisfies

$$\begin{aligned}\Delta_{\overline{g}} \psi^j &= \overline{\mathbf{\Gamma}}^j - \Delta_{\overline{g}} \chi^{1,j} \\ &= \big\{\overline{\mathbf{\Gamma}}^j - \delta^{jq}\left[\partial_t h_{0q} - (1/2)\partial_q h_{00}\right]\big\} - \big\{\overline{g}^{pq}_{\geq 1}\partial_p\partial_q \chi^{1,j} + \overline{\mathbf{\Gamma}}^r \partial_r \chi^{1,j}\big\}.\end{aligned}$$

Direct calculations, using (7.1.20), show that the right-hand side is in $H^2$, so that, by elliptic regularity, $\psi^j \in \dot{H}^1 \cap C^3$. In addition,

$$\begin{aligned}&\|\nabla_{x,t}\phi^j\|_{L^\infty} \lesssim \varepsilon_0 \langle t \rangle^{2\delta'-1}, \\ &\|\partial_p \phi^j - \delta^{qj} R_p\{-R_0 h_{0q} + (1/2)R_q h_{00}\}\|_{H^4} \lesssim \varepsilon_0^2 \langle t \rangle^{2\delta'-1},\end{aligned} \tag{7.2.65}$$

so the mapping $x^j \mapsto y^j$ is a global diffeomorphism. Let $\widetilde{g}_{jk} = \delta_{jk} + \widetilde{h}_{jk}$ be the metric in the new coordinates $y^j$, so that

$$\overline{g}_{jk} = \mathbf{g}_{jk} = \widetilde{g}_{pq}\frac{\partial y^p}{\partial x^j}\frac{\partial y^q}{\partial x^k} = \widetilde{g}_{jk} + \widetilde{g}_{pk}\partial_j\phi^p + \widetilde{g}_{jq}\partial_k\phi^q + \widetilde{g}_{pq}\partial_j\phi^p\partial_k\phi^q. \tag{7.2.66}$$

In particular, using (7.2.65), we find that

$$\begin{aligned}&\|\widetilde{h}_{jk}\|_{L^\infty} \lesssim \varepsilon_0, \\ &\|\widetilde{h}_{jk} - h_{jk} - R_0 R_j h_{0k} - R_0 R_k h_{0j} + R_j R_k h_{00}\|_{H^4} \lesssim \varepsilon_0^2 \langle t \rangle^{2\delta'-1}.\end{aligned}$$

Therefore

$$\|\nabla_x(\widetilde{h}_{jk} - \in_{jpm}\in_{kqn} R_pR_q\vartheta_{mn})\|_{H^3} \lesssim \varepsilon_0^2\langle t\rangle^{2\delta'-1}$$
$$+ \|\nabla_x(\underline{F} + R_0\rho)\|_{H^3} + \|\nabla_x(R_0\omega - \Omega)\|_{H^3} \lesssim \varepsilon_0^2\langle t\rangle^{2\delta'-1},$$

where we use Lemmas 4.19 and 4.20 and the identities (see (4.3.4))

$$\underline{F} + R_0\rho + \tau = |\nabla|^{-2}[\underline{F}[\mathcal{N}^h] + \tau[\mathcal{N}^h]] + |\nabla|^{-2}\partial_t E_0^{\geq 2}. \tag{7.2.67}$$

We now turn to (7.2.63). Since $\widetilde{g}_{jk}$ denotes the metric in spatial harmonic coordinates, the same computations as in (1.1.14) give that

$$\big\|\overline{R} + (1/2)\widetilde{g}^{ab}\widetilde{g}^{jk}\partial_{y^j}\partial_{y^k}\widetilde{g}_{ab}\big\|_{H^2} \lesssim \varepsilon_0^2\langle t\rangle^{2\delta'-1}.$$

Since $x$ and $y$ derivatives of $\widetilde{g}_{jk}$ agree up to quadratic errors, we deduce that

$$\big\|\overline{R} + (1/2)\delta^{ab}\delta^{jk}\partial_{x^j}\partial_{x^k}\big(h_{jk} - \partial_j\phi^k - \partial_k\phi^j\big)\big\|_{H^2} \lesssim \varepsilon_0^2\langle t\rangle^{2\delta'-1}.$$

Using (2.1.29) and (7.2.65), we obtain that

$$\|\overline{R} + \Delta\tau\|_{H^2} \lesssim \varepsilon_0^2\langle t\rangle^{2\delta'-1} + \|\Delta(\underline{F} + R_0\rho)\|_{H^2} + \|R_0\omega - \Omega\|_{H^4} \lesssim \varepsilon_0^2\langle t\rangle^{2\delta'-1}.$$

This completes the proof. □

**Lemma 7.16.** *The decomposition* (2.1.29) *is orthogonal in* $L^2$, *i.e,*

$$\sum_{\alpha,\beta=0}^{4} \|\partial g_{\alpha\beta}\|_{L^2}^2 = 2\|\partial F\|_{L^2}^2 + 2\|\partial \underline{F}\|_{L^2}^2 + \|\partial\rho\|_{L^2}^2$$
$$+ \sum_{j=1}^{3}\big[\|\partial\omega_j\|_{L^2}^2 + 2\|\partial\Omega_j\|_{L^2}^2\big] + \sum_{j,k=1}^{3}\|\partial\vartheta_{jk}\|_{L^2}^2.$$

*for any derivative* $\partial = \partial_\nu$.

*Proof of Lemma 7.16.* This follows by direct computation, starting with (2.1.29) and using the identities $\partial_j\Omega_j = \partial_j\omega_j = 0$ and $\partial_j\vartheta_{jk} = \partial_j\vartheta_{kj} = 0$. □

**Proposition 7.17.** *Given initial data* $(\Sigma_0, \overline{g}, k)$ *satisfying the constraint equations* (1.2.4), *we can choose harmonic coordinates such that, for all* $t \geq 0$,

$$\|\underline{F}(t)\|_{L^\infty} + \|\rho(t)\|_{L^\infty} + \|\omega(t)\|_{L^\infty} + \|\Omega(t)\|_{L^\infty} + \|\nabla_{x,t}\tau(t)\|_{H^2} \lesssim \varepsilon_0^2\langle t\rangle^{\delta'-1}$$
$$\|\nabla_{x,t}\omega(t)\|_{H^3} + \|\nabla_{x,t}\Omega(t)\|_{H^3} \lesssim \varepsilon_0^2. \tag{7.2.68}$$

*In addition, if the initial slice* $\Sigma_0 \subset M$ *is maximal (i.e., if* $\overline{g}^{ab}k_{ab} = 0$*), then we* *can also assume that*

$$\langle t\rangle^{1-\delta'}\|F(t)\|_{L^\infty}+\|\nabla_{x,t}F(t)\|_{H^3}\lesssim\varepsilon_0^2.$$

*Thus the metric is purely determined by $\vartheta$ and $\underline{F},\rho$ at the linear level.*

*Remark 7.18.* While the assumption that the initial time slice is maximal is a geometric assumption, Cauchy surfaces can often be deformed to maximal Cauchy surfaces; see [5].

*Proof of Proposition 7.17.* It suffices to verify (7.2.68) for the initial components as the nonlinear evolution will only contribute terms of order $O(\varepsilon_0^2)$; see, e.g., the proofs in section 6.3. We assume that we start with spatial coordinates on $\Sigma_0$ that are *spatially harmonic* as in Proposition 7.14 and are then extended to spacetime coordinates on $M$ that satisfy the harmonic gauge condition (1.1.17). Using Proposition 7.14, we have

$$\|\nabla_x(F-\underline{F})(t=0)\|_{H^3}+\|\nabla_x\Omega(t=0)\|_{H^3}\lesssim\varepsilon_0^2.$$

In addition, given our definition of initial data in (1.2.3), we see that

$$h_{0j}(t=0)=0,$$
$$\partial_t h_{0a}(t=0)=-(\frac{1}{2}\partial_a(F-\underline{F})-\in_{alm}\partial_l\Omega_m-\partial_a\tau)(t=0)+O(\varepsilon_0^2)=O(\varepsilon_0^2),$$

from which we deduce that $h_{0j}$ and hence $\rho$ and $\omega$ remain of size $O(\varepsilon_0^2)$. Using (4.3.4) and (7.2.67), we can then extend these bounds to $\underline{F}$ and $\Omega$.

On the other hand, using (1.2.3), we also see that

$$(F+\underline{F})(t=0)=0,\qquad \partial_t(F+\underline{F})(t=0)=\overline{g}^{ab}k_{ab}.$$

Therefore if the initial slice is maximal then $\partial_t F(0)=O(\varepsilon_0^2)$. □

## 7.3 ASYMPTOTICALLY OPTICAL FUNCTIONS AND THE BONDI ENERGY

Our final application concerns the construction of Bondi energy functions, with good monotonicity properties along null infinity. We would like to thank Yakov Shlapentokh-Rothman for useful discussions on this topic.

### 7.3.1 Almost Optical Functions and the Friedlander Fields

In order to get precise information on the asymptotic behavior of the metric in the physical space we need to understand the bending of the light cones caused by the long-range effect of the nonlinearity (i.e., the modified scattering).

In the Minkowski space, the outgoing light cones correspond to the level sets of the optical function $u^0 = |x| - t$. In our case, the analogous objects we use are what we call *almost (or asymptotically) optical functions* $u$, which are close to $u^0$ but better adapted to the null geometry of our problem. Recall that the metrics we consider here have slow $O(\langle r\rangle^{-1+})$ decay at infinity, and we expect a nontrivial deviation that is not radially isotropic.

We first define and construct a suitable class of almost optical functions.

**Lemma 7.19.** *There exists a $C^4$ almost optical function $u : M' \to \mathbb{R}$ satisfying the properties*

$$u(x,t) = |x| - t + u^{cor}(x,t), \qquad \mathbf{g}^{\alpha\beta}\partial_\alpha u \partial_\beta u = O(\varepsilon_0\langle r\rangle^{-2+6\delta'}) \tag{7.3.1}$$

*and, for any $\mu \in \{0,1,2,3\}$,*

$$u^{cor} = O(\varepsilon_0\langle r\rangle^{3\delta'}), \quad \partial_\mu u^{cor} = O(\varepsilon_0\langle r\rangle^{3\delta'-1}), \quad \partial_\mu(L^\alpha\partial_\alpha u^{cor}) = O(\varepsilon_0\langle r\rangle^{3\delta'-2}). \tag{7.3.2}$$

*In addition, $u^{cor}$ is close to $\Theta_{wa}/|x|$ (see (6.3.3)) in the vicinity of the light cone, i.e., if $(x,t) \in M'$ and $\big|t - |x|\big| \leq t/10$ then*

$$\left|u^{cor}(x,t) - \frac{\Theta_{wa}(x,t)}{|x|}\right| \lesssim \varepsilon_0\langle r\rangle^{-1+3\delta'}(\langle r\rangle^{p_0} + \langle t - |x|\rangle). \tag{7.3.3}$$

*Remark 7.20.* The classical approach—see, for example, [12]—is to construct exact optical functions, satisfying the stronger identity $\mathbf{g}^{\alpha\beta}\partial_\alpha u \partial_\beta u = 0$ instead of the approximate identity in (7.3.1). We could do this too, but we prefer to work here with almost optical functions instead of exact optical functions because they are easier to construct and their properties still suffice for our two main applications (the improved peeling estimates in Proposition 7.9 and the construction of the Bondi energy in Theorem 7.23).

*Proof.* We define the function $H_L : M \to \mathbb{R}$,

$$H_L := \frac{1}{2}L^\alpha L^\beta h_{\alpha\beta} = -\frac{1}{2}g^{\alpha\beta}_{\geq 1}\partial_\alpha u^0 \partial_\beta u^0 + \frac{1}{2}g^{\alpha\beta}_{\geq 2}\partial_\alpha u^0 \partial_\beta u^0, \tag{7.3.4}$$

where the identity holds due to (2.1.8). Notice that

$$H_L = O(\varepsilon_0\langle t + r\rangle^{-1+\delta'}), \qquad \partial_\mu H_L = O(\varepsilon_0 r^{-1}\langle t + r\rangle^{-1+3\delta'}), \tag{7.3.5}$$

in $M$, as a consequence of (7.1.20), (7.1.21), and (7.1.48). We will define $u^{cor}$ such that

$$L^\alpha\partial_\alpha u^{cor} = H_L, \tag{7.3.6}$$

in addition to the bounds in (7.3.2)–(7.3.3).

**Step 1.** For $s, b \in [0, \infty)$ we define the projections $\Pi_b^-$ and $\Pi_b^+$,

$$\begin{aligned}\Pi_b^- H_L(s) &:= L^\alpha L^\beta \mathcal{F}^{-1}\{\varphi_{\le 0}(\langle b\rangle^{p_0}\xi)\widehat{h_{\alpha\beta}}(\xi, s)\},\\ \Pi_b^+ H_L(s) &:= L^\alpha L^\beta \mathcal{F}^{-1}\{\varphi_{\ge 1}(\langle b\rangle^{p_0}\xi)\widehat{h_{\alpha\beta}}(\xi, s)\},\end{aligned} \tag{7.3.7}$$

as in section 6.3, where $p_0 = 0.68$. Then we define the correction $u^{cor}$ by integrating the low frequencies of $H_L$ from 0 to $t$ and the high frequencies from $t$ to $\infty$. More precisely, let

$$\begin{aligned}u_1^{cor}(x,t) := &-\int_t^\infty (\Pi_s^+ H_L)(x + (s-t)x/|x|, s)\, ds + \int_0^t (\Pi_s^- H_L)(sx/|x|, s)\, ds\\ &- \int_t^\infty \big\{(\Pi_s^- H_L)(x + (s-t)x/|x|, s) - (\Pi_s^- H_L)(sx/|x|, s)\big\}\, ds,\end{aligned} \tag{7.3.8}$$

and

$$\begin{aligned}u_2^{cor}(x,t) := &-\int_{|x|}^\infty (\Pi_s^+ H_L)(sx/|x|, s + t - |x|)\, ds + \int_0^{|x|} (\Pi_s^- H_L)(sx/|x|, s)\, ds\\ &- \int_{|x|}^\infty \big\{(\Pi_s^- H_L)(sx/|x|, s + t - |x|) - (\Pi_s^- H_L)(sx/|x|, s)\big\}\, ds.\end{aligned} \tag{7.3.9}$$

We fix a smooth function $\chi_1 : \mathbb{R} \to [0,1]$ supported in $(-\infty, 2]$ and equal to 1 in $(-\infty, 1]$. Let $\chi_2 := 1 - \chi_1$ and define

$$u^{cor}(x,t) := u_1^{cor}(x,t)\chi_1(|x| - t) + u_2^{cor}(x,t)\chi_2(|x| - t). \tag{7.3.10}$$

Notice that, formally, one can rewrite the formula (7.3.8) as

$$u_1^{cor}(x,t) \approx -\int_t^\infty H_L(x + (s-t)x/|x|, s)\, ds + \int_0^\infty (\Pi_s^- H_L)(sx/|x|, s)\, ds, \tag{7.3.11}$$

which is consistent with the desired transport identity (7.3.6). However, the two infinite integrals in (7.3.11) do not converge, and we need to reorganize them as in (7.3.8) to achieve convergence. A similar remark applies to the definition of $u_2^{cor}$ in (7.3.9).

We prove now the bounds (7.3.2). Notice that, for any $b \ge 0$ and $(y,s) \in M$,

$$\begin{aligned}|\Pi_b^- H_L(y,s)| + |y||\nabla_{y,s}(\Pi_b^- H_L)(y,s)| &\lesssim \varepsilon_0 \langle |y| + s\rangle^{3\delta' - 1},\\ \langle y\rangle |(\Pi_b^+ H_L)(y,s)| &\lesssim \varepsilon_0 \langle |y| + s\rangle^{3\delta' - 1}\langle b\rangle^{p_0},\end{aligned} \tag{7.3.12}$$

due to (7.3.5) and Lemma 7.21. Finally, for $(x,t) \in M'$ and $s \ge \max(\rho, 0)$, $\rho \in [\max(t, |x|) - 3, \max(t, |x|) + 3]$, using the bounds in the first line of (7.3.12) we have

$$\begin{aligned}|(\Pi_b^- H_L)(x+(s-\rho)x/|x|,s+t-\rho)-(\Pi_b^- H_L)(sx/|x|,s)|\\ \lesssim \varepsilon_0\langle t-|x|\rangle\langle s\rangle^{3\delta'-2}.\end{aligned} \tag{7.3.13}$$

Using the definitions (7.3.8)–(7.3.10), it follows that the functions $u_1^{cor}$, $u_2^{cor}$ are well defined for any $(x,t)\in M'$, and moreover satisfy the estimates

$$\chi_a(|x|-t)|u_a^{cor}(x,t)|\lesssim \varepsilon_0\langle x\rangle^{3\delta'},\qquad \chi_a(|x|-t)|\partial_\mu u_a^{cor}(x,t)|\lesssim \varepsilon_0\langle x\rangle^{3\delta'-1}, \tag{7.3.14}$$

for $a\in\{1,2\}$ and $\mu\in\{0,1,2,3\}$. Using again the gradient bounds in (7.3.12) and (7.3.5), we also have $|u_1^{cor}(x,t)-u_2^{cor}(x,t)|\lesssim\varepsilon_0\langle x\rangle^{3\delta'-1}$ if $|x|-t\in[1,2]$. Finally,

$$L^\alpha\partial_\alpha u_1^{cor}=H_L \text{ if } |x|-t\leq 2\qquad\text{and}\qquad L^\alpha\partial_\alpha u_2^{cor}=H_L \text{ if } |x|-t\geq 1,$$

and the desired bounds in (7.3.2) follow.

**Step 2.** We calculate now in $M'$

$$\begin{aligned}\mathbf{g}^{\alpha\beta}\partial_\alpha u\partial_\beta u&=(m^{\alpha\beta}+g_{\geq1}^{\alpha\beta})\partial_\alpha(u^0+u^{cor})\partial_\beta(u^0+u^{cor})\\&=2m^{\alpha\beta}\partial_\alpha u^0\partial_\beta u^{cor}+g_{\geq1}^{\alpha\beta}\partial_\alpha u^0\partial_\beta u^0+O(\varepsilon_0\langle r\rangle^{-2+6\delta'})\\&=2L^\beta\partial_\beta u^{cor}-2H_L+O(\varepsilon_0\langle r\rangle^{-2+6\delta'}),\end{aligned}$$

using (7.3.4) and (7.3.2). The bounds in (7.3.1) follow using also (7.3.6).

Finally, to prove (7.3.3) we examine the definition (6.3.3) and notice that

$$\frac{\Theta_{wa}(x,t)}{|x|}=\int_0^t(\Pi_s^- H_L)(sx/|x|,s)\,ds. \tag{7.3.15}$$

Therefore, using (7.3.12), if $(x,t)\in M'$ and $t=|x|$

$$\left|u^{cor}(x,t)-\frac{\Theta_{wa}(x,t)}{|x|}\right|\lesssim\varepsilon_0\langle r\rangle^{-1+3\delta'+p_0}.$$

The estimates (7.3.3) follow using also the bounds $\partial_r u^{cor}=O(\varepsilon_0\langle r\rangle^{3\delta'-1})$. □

**Lemma 7.21.** *We have*

$$|x|\left|[L^\alpha L^\beta,\pi_b^\pm]\nabla_{x,t}h_{\alpha\beta}\right|(x,s)\lesssim\varepsilon_0\langle|x|+s\rangle^{3\delta'-1}, \tag{7.3.16}$$

*where*

$$\mathcal{F}\{\pi_b^- f\}(\xi)=\varphi_{\leq0}(\langle b\rangle^{p_0}\xi)\widehat{f}(\xi),\qquad \pi_b^+=1-\pi_b^-$$

*is the multiplier from (7.3.7).*

*Proof.* Clearly, it suffices to consider $\pi = \pi_b^-$. We rewrite the commutator as

$$T[H] := \langle b\rangle^{-2p_0} \int_{\mathbb{R}^3} \{\langle b\rangle^{-p_0}|x-y| \cdot \check{\varphi}_{\leq 0}(\langle b\rangle^{-p_0}(x-y))\}\, H(y,s)\frac{I(x)-I(y)}{|x-y|}\, dy,$$

where $H$ denotes a generic derivative of a component of $h$, $H = \partial_\alpha h_{\beta\gamma}$, and $I$ denotes a generic tensor in $\{1, x_j/|x|, x_j x_k/|x|^2\}$. The bounds (7.3.16) then follows by direct integration using that

$$|H(y)| \lesssim \varepsilon_0 \langle |y|+s\rangle^{2\delta'-1}\langle |y|-s\rangle^{-1}, \qquad |x|\left|\frac{I(x)-I(y)}{|x-y|}\right| \lesssim 1,$$

as follows from Theorem 7.2. □

We will prove now asymptotic formulas in the physical space for some of the metric components and for the Klein-Gordon field. These formulas will be used in the Bondi energy analysis in subsection 7.3.2 below.

Recall the definitions

$$\begin{aligned} \widehat{V_*^G}(\xi,t) &= \widehat{V^G}(\xi,t)e^{-i\Theta_{wa}(\xi,t)}, \qquad G \in \{F, \omega_a, \vartheta_{ab}\},\\ \widehat{V_*^\psi}(\xi,t) &= \widehat{V^\psi}(\xi,t)e^{-i\Theta_{kg}(\xi,t)}; \end{aligned} \tag{7.3.17}$$

see (6.3.4) and (6.2.6). It follows from (6.3.16) and (6.2.14) that there are functions $V_\infty^G \in Z_{wa}$, $G \in \{F, \omega_a, \vartheta_{ab}\}$, and $V_\infty^\psi \in Z_{kg}$ such that, for any $t \geq 0$,

$$\sum_{G\in\{F,\omega_a,\vartheta_{ab}\}} \|V_*^G(t) - V_\infty^G\|_{Z_{wa}} + \|V_*^\psi(t) - V_\infty^\psi\|_{Z_{kg}} \lesssim \varepsilon_0 \langle t\rangle^{-\delta/2}. \tag{7.3.18}$$

We define the smooth characteristic function $\chi_{wa}$ of the wave region by

$$\chi_{wa} : \{(x,t) \in M : t \geq 8\} \to [0,1], \qquad \chi_{wa}(x,t) := \varphi_{\leq 0}((|x|-t)/t^{0.4}). \tag{7.3.19}$$

We define also the function

$$\nu_{kg} : \{(x,t) \in M : t \geq 8,\ |x| < t\} \to \mathbb{R}^3, \qquad \nu_{kg}(x,t) := \frac{x}{\sqrt{t^2-|x|^2}}, \tag{7.3.20}$$

such that $\nu_{kg}(x,t)$ is the critical point of the function $\xi \to x\cdot\xi - t\sqrt{1+|\xi|^2}$. We are now ready to state our main proposition describing the solutions in the physical space.

**Proposition 7.22.** *For any $G \in \{F, \omega_a, \vartheta_{ab}\}$ and $t \in [8, \infty)$ we have*

$$U^G(x,t) = \frac{-i\chi_{wa}(x,t)}{4\pi^2|x|}\int_0^\infty e^{i\rho u(x,t)}\varphi_{[-k_0,k_0]}(\rho)\widehat{V_*^G}(\rho x/|x|,t)\rho\,d\rho + U^G_{rem}(x,t),$$

$$\begin{aligned}U^\psi(x,t) = \frac{\mathbf{1}_+(t-|x|)}{\sqrt{8\pi^3}e^{i\pi/4}}\frac{e^{-i\sqrt{t^2-|x|^2}}t}{(t^2-|x|^2)^{5/4}}e^{i\Theta_{kg}(\nu_{kg}(x,t),t)}\widehat{P_{[-k_0,k_0]}V_*^\psi}(\nu_{kg}(x,t),t)\\ + U^\psi_{rem}(x,t),\end{aligned} \tag{7.3.21}$$

*where $u$ is an almost optical function satisfying (7.3.1)–(7.3.3) and $k_0$ denotes the smallest integer for which $2^{k_0} \geq t^{\delta'}$. The remainders $U^G_{rem}$ and $U^\psi_{rem}$ satisfy the $L^2$ bounds*

$$\sum_{G\in\{F,\omega_a,\vartheta_{ab}\}} \|U^G_{rem}(t)\|_{L^2} + \|U^\psi_{rem}(t)\|_{L^2} \lesssim \varepsilon_0 t^{-\delta}, \qquad \text{for any } t \geq 8. \tag{7.3.22}$$

*Proof.* **Step 1.** We prove first the conclusions concerning the variables $U^G$. We start from the formula

$$\widehat{U^G}(\xi,t) = e^{-it|\xi|}e^{i\Theta_{wa}(\xi,t)}\widehat{V_*^G}(\xi,t), \tag{7.3.23}$$

and extract acceptable $L^2$ remainders until we reach the desired formula.

We may assume that $t \gg 1$ and let $J_0$ denote the smallest integers for which $2^{J_0} \geq t^{1/3}$. We define

$$\begin{aligned}V^G_{*,1} &:= (I - P_{[-k_0,k_0]})V^G_* + P_{[-k_0-2,k_0+2]}[\varphi_{\geq J_0+1}\cdot P_{[-k_0,k_0]}V^G_*],\\ V^G_{*,2} &:= P_{[-k_0-2,k_0+2]}[\varphi_{\leq J_0}\cdot P_{[-k_0,k_0]}V^G_*],\end{aligned} \tag{7.3.24}$$

and notice that $V^G_* := V^G_{*,1} + V^G_{*,2}$. We show first that

$$\|V^G_{*,1}(t)\|_{L^2} \lesssim \varepsilon_0 t^{-\delta'/4}. \tag{7.3.25}$$

Indeed, notice that $\|(I - P_{[-k_0,k_0]})V^G_*(t)\|_{L^2} \lesssim \varepsilon_0 t^{-\delta'/4}$, due to (7.1.3) and (7.3.17). To bound the remaining term we examine the definition (6.3.3) and notice that

$$\big|D^\alpha_\xi[e^{\pm i\Theta_{wa}(\xi,t)}]\big| \lesssim_\alpha t^{|\alpha|(1-p_0+2\delta')}t^{2\delta'} \qquad \text{if } t^{-\delta'} \lesssim |\xi| \lesssim t^{\delta'}. \tag{7.3.26}$$

Let $A_t$ denote the operator on the Euclidean space $\mathbb{R}^3$ defined by the Fourier multiplier $\xi \to e^{-i\Theta_{wa}(\xi,t)}\varphi_{[-k_0-2,k_0+2]}(\xi)$. Notice that

$$\begin{aligned}P_{[-k_0,k_0]}V^G_* &= A_t[P_{[-k_0,k_0]}V^G]\\ &= A_t[\varphi_{\leq J_0-4}\cdot P_{[-k_0,k_0]}V^G] + A_t[\varphi_{\geq J_0-3}\cdot P_{[-k_0,k_0]}V^G].\end{aligned}$$

In view of (7.3.26), the kernel of the operator $A_t$ decays rapidly if $|x| \gtrsim 2^{J_0}$, so

$$\big\|\varphi_{\geq J_0+1} \cdot A_t[\varphi_{\leq J_0-4} \cdot P_{[-k_0,k_0]}V^G]\big\|_{L^2} \lesssim \varepsilon_0 t^{-1}. \tag{7.3.27}$$

Moreover, using (7.1.14),

$$\big\|A_t[\varphi_{\geq J_0-3} \cdot P_{[-k_0,k_0]}V^G]\big\|_{L^2} \lesssim \varepsilon_0 2^{-J_0} t^{2\delta'} \lesssim \varepsilon_0 t^{-1/3+2\delta'}. \tag{7.3.28}$$

The bounds (7.3.25) follow, using the definition. Therefore

$$\|U^G_{rem,1}(t)\|_{L^2} \lesssim \varepsilon_0 t^{-\delta'/4}, \text{ where } U^G_{rem,1}(t) := \mathcal{F}^{-1}\{e^{-it|\xi|}e^{i\Theta_{wa}(\xi,t)}\widehat{V^G_{*,1}}(\xi,t)\}. \tag{7.3.29}$$

We define now

$$U^G_{rem,2} := (1-\chi_{wa}) \cdot \mathcal{F}^{-1}\{e^{-it|\xi|}e^{i\Theta_{wa}(\xi,t)}\widehat{V^G_{*,2}}(\xi,t)\}. \tag{7.3.30}$$

Using integration by parts in $\xi$ (Lemma 3.1) and the formulas (7.3.24) and (7.3.26) we have rapid decay,

$$\|U^G_{rem,2}(t)\|_{L^2} \lesssim \varepsilon_0 t^{-1}. \tag{7.3.31}$$

To estimate the main term $\chi_{wa} \cdot \mathcal{F}^{-1}\{e^{-it|\xi|}e^{i\Theta_{wa}(\xi,t)}\widehat{V^G_{*,2}}(\xi,t)\}$ we write it in the form

$$U^G_3(x,t) := \frac{\chi_{wa}(x,t)}{8\pi^3}\int_{\mathbb{R}^3} e^{ix\cdot\xi}e^{-it|\xi|}e^{i\Theta_{wa}(\xi,t)}\widehat{V^G_{*,2}}(\xi,t)\,d\xi. \tag{7.3.32}$$

We can extract more remainders by inserting angular cutoffs. Notice that if we insert the factor $\varphi_{\geq 1}(t^{0.49}(x/|x| - \xi/|\xi|))$ in the integral above then the corresponding contribution is a rapidly decreasing $L^2$ remainder. Passing to polar coordinates $x = r\omega$, $\xi = \rho\theta$, $\omega, \theta \in \mathbb{S}^2$, it remains to estimate the integral

$$\begin{aligned} U^G_4(x,t) := \frac{\chi_{wa}(x,t)}{8\pi^3}\int_0^\infty \int_{\mathbb{S}^2} & e^{ir\rho\omega\cdot\theta}e^{-it\rho}e^{i\rho\Theta_{wa}(\theta,t)} \\ & \times \widehat{V^G_{*,2}}(\rho\theta,t)\varphi_{\leq 0}(t^{0.49}(\omega-\theta))\,\rho^2 d\rho d\theta. \end{aligned} \tag{7.3.33}$$

For any $\omega, \theta \in \mathbb{S}^2$ and $\rho \in [2^{-10}t^{-\delta'}, 2^{10}t^{\delta'}]$ we have

$$|e^{i\rho\Theta_{wa}(\theta,t)} - e^{i\rho\Theta_{wa}(\omega,t)}| + |\widehat{V^G_{*,2}}(\rho\theta,t) - \widehat{V^G_{*,2}}(\rho\omega,t)| \lesssim \varepsilon_0 t^{4\delta'}|\theta-\omega|. \tag{7.3.34}$$

Indeed, using the definitions we have $|\Omega_\theta[e^{\pm i\rho\Theta_{wa}(\theta,t)}]| \lesssim t^{2\delta'}$, where $\Omega_\theta$ is any of the rotation vector-fields in the variable $\theta \in \mathbb{S}^2$. The bounds (7.3.34) follow using also (3.3.25). Therefore we can further replace the angular variable $\theta$ with $\omega$ in two places in the integral in (7.3.33), at the expense of acceptable errors.

It remains to estimate the integral

$$U_5^G(x,t) := \frac{\chi_{wa}(x,t)}{8\pi^3} \int_0^\infty e^{-it\rho} e^{i\rho\Theta_{wa}(\omega,t)} \widehat{V_{*,2}^G}(\rho\omega,t) \times \Big\{ \int_{\mathbb{S}^2} e^{ir\rho\omega\cdot\theta} \varphi_{\leq 0}(t^{0.49}(\omega-\theta))\, d\theta \Big\} \rho^2 d\rho. \tag{7.3.35}$$

The integral over $\mathbb{S}^2$ in (7.3.35) does not depend on $\omega$ and can be calculated explicitly. We may assume $\omega = (0,0,1)$ and use spherical coordinates to write this integral in the form

$$2\pi \int_0^\pi e^{ir\rho\cos y} \varphi_{\leq 0}(t^{0.49} 2\sin(y/2)) \sin y\, dy = 8\pi \int_0^1 e^{ir\rho(1-2z^2)} \varphi_{\leq 0}(t^{0.49} 2z) z\, dz$$
$$= \frac{2\pi}{r\rho} e^{ir\rho} \int_0^\infty e^{-i\alpha} \varphi_{\leq 0}\Big(\frac{2t^{0.49}}{\sqrt{2r\rho}}\sqrt{\alpha}\Big)\, d\alpha = \frac{-2\pi i}{r\rho} e^{ir\rho} + O(t^{-2}).$$

We substitute this into (7.3.35), and it remains to estimate the integral

$$U_6^G(x,t) := \frac{-i\chi_{wa}(x,t)}{4\pi^2 r} \int_0^\infty e^{-it\rho} e^{i\rho\Theta_{wa}(\omega,t)} \widehat{V_{*,2}^G}(\rho\omega,t) e^{ir\rho} \rho\, d\rho. \tag{7.3.36}$$

In view of (7.3.27)–(7.3.28) we can replace now the factor $\widehat{V_{*,2}^G}(\rho\omega,t)$ in (7.3.36) with $\mathcal{F}\{P_{[-k_0,k_0]} V_*^G\}(\rho\omega,t)$. Then we replace $\Theta_{wa}(\omega,t) = \Theta_{wa}(x/|x|,t)$ with $u^{cor}(x,t)$, up to acceptable errors (due to (7.3.3)). The desired formula in (7.3.21) follows, since $u(x,t) = |x| - t + u^{cor}(x,t)$.

**Step 2.** The proof for the Klein-Gordon variable $U^\psi$ is similar. We start from the formula

$$\widehat{U^\psi}(\xi,t) = e^{-it\sqrt{1+|\xi|^2}} e^{i\Theta_{kg}(\xi,t)} \widehat{V_*^\psi}(\xi,t), \tag{7.3.37}$$

and extract acceptable $L^2$ remainders until we reach the desired formula in (7.3.21). We may assume $t \gg 1$, set $J_0$ such that $2^{J_0} \approx t^{1/3}$, and define

$$\begin{aligned} V_{*,1}^\psi &:= (I - P_{[-k_0,k_0]}) V_*^\psi + P_{[-k_0-2,k_0+2]} [\varphi_{\geq J_0+1} \cdot P_{[-k_0,k_0]} V_*^\psi], \\ V_{*,2}^\psi &:= P_{[-k_0-2,k_0+2]} [\varphi_{\leq J_0} \cdot P_{[-k_0,k_0]} V_*^\psi]. \end{aligned} \tag{7.3.38}$$

The definition (6.2.4) shows that

$$\big|D_\xi^\alpha [e^{\pm i\Theta_{kg}(\xi,t)}]\big| \lesssim_\alpha t^{|\alpha|(1-p_0+2\delta')} t^{2\delta'} \qquad \text{if } t^{-\delta'} \lesssim |\xi| \lesssim t^{\delta'}. \tag{7.3.39}$$

As in (7.3.25) it follows that $\|V_{*,1}^\psi(t)\|_{L^2} \lesssim \varepsilon_0 t^{-\delta'/4}$, so the contribution of $V_{*,1}^\psi$ is an acceptable remainder. The contribution of $V_{*,2}^\psi$ is also a remainder in the region $\{|x| \geq t\}$, due to Lemma 3.1. It remains to estimate the main contribution,

$$U_2^{\psi}(x,t) := \frac{\mathbf{1}_+(t-|x|)}{8\pi^3} \int_{\mathbb{R}^3} e^{ix\cdot\xi} e^{-it\sqrt{1+|\xi|^2}} e^{i\Theta_{kg}(\xi,t)} \widehat{V_{*,2}^{\psi}}(\xi,t)\, d\xi. \tag{7.3.40}$$

We can insert a cutoff function of the form $\varphi_{\leq 0}(t^{0.49}(\xi - \nu_{kg}(x,t)))$ in the integral above to localize near the critical point, and then replace $e^{i\Theta_{kg}(\xi,t)}$ and $\widehat{V_{*,2}^{\psi}}(\xi,t)$ with $e^{i\Theta_{kg}(\nu_{kg}(x,t),t)}$ and $\widehat{V_{*,2}^{\psi}}(\nu_{kg}(x,t),t)$ respectively, at the expense of acceptable remainders (using (7.3.39) and (3.3.26) respectively).

Therefore, the desired identity in (7.3.21) holds if $\nu_{kg}(x,t) \geq 2^{k_0+4}$ (with the main term vanishing). On the other hand, if $\nu_{kg}(x,t) \leq 2^{k_0+4}$ (thus $t-|x| \gtrsim t^{1-4\delta'}$) then the remaining $\xi$ integral in (7.3.40) can be estimated explicitly,

$$\begin{aligned}\int_{\mathbb{R}^3} & e^{ix\cdot\xi} e^{-it\sqrt{1+|\xi|^2}} \varphi_{\leq 0}(t^{0.49}(\xi - \nu_{kg}(x,t)))\, d\xi \\ &= e^{-i\sqrt{t^2-|x|^2}} \frac{e^{-i\pi/4}(2\pi)^{3/2} t}{(t^2-|x|^2)^{5/4}} + O(t^{-7/4}),\end{aligned}$$

where the approximate identity follows from the standard stationary phase formula. After these reductions, the remaining main term is

$$U_3^{\psi}(x,t) := \frac{\mathbf{1}_+(t-|x|)}{\sqrt{8\pi^3}} e^{i\Theta_{kg}(\nu_{kg}(x,t),t)} \widehat{V_{*,2}^{\psi}}(\nu_{kg}(x,t),t) \frac{e^{-i\sqrt{t^2-|x|^2}} e^{-i\pi/4} t}{(t^2-|x|^2)^{5/4}}.$$

The desired conclusion in (7.3.21) follows using (3.3.26). □

### 7.3.2 The Bondi Energy

We can define now a more refined concept of energy function. For this we fix $t \geq 1$, define the hypersurface $\Sigma_t := \{(x,t) \in M : x \in \mathbb{R}^3\}$, and let $\overline{g}_{jk} = \mathbf{g}_{jk}$ denote the induced (Riemannian) metric on $\Sigma_t$. Let $\overline{g}^{jk}$ denote the inverse of the matrix $\overline{g}_{jk}$, $\overline{g}^{jk}\overline{g}_{jn} = \delta_n^k$, let $\overline{D}$ denote the covariant derivative on $\Sigma_t$ induced by the metric $\overline{g}$. Notice that

$$|\overline{g}^{jk} - \mathbf{g}^{jk}| \lesssim \varepsilon_0^2 \langle t+r\rangle^{-2+2\delta'}, \qquad \overline{\Gamma}_{njk} = \mathbf{\Gamma}_{njk}, \tag{7.3.41}$$

for any $n,j,k \in \{1,2,3\}$. With $u$ an almost optical function as defined in (7.3.1)–(7.3.2) let

$$\mathbf{n}_j := \partial_j u (\overline{g}^{ab}\partial_a u \partial_b u)^{-1/2} \tag{7.3.42}$$

denote the unit vector-field in $\Sigma_t' := \{x \in \Sigma_t : |x| \geq 2^{-8}t\}$, normal to the level sets of the function $u$. In this section we use the metric $\overline{g}$ to raise and lower indices.

We fix a function $u$ as in Lemma 7.19. For $R \in \mathbb{R}$ and $t$ large (say $t \geq 2|R|+10$) we define the modified spheres $S_{R,t}^u := \{x \in \Sigma_t : u(x,t) = R\}$. We would like to define

$$E_{Bondi}(R) := \frac{1}{16\pi} \lim_{t\to\infty} \int_{S^u_{R,t}} W_j \mathbf{n}^j \, d\sigma, \tag{7.3.43}$$

for a suitable vector-field $W$, where $d\sigma = d\sigma(\overline{g})$ is the surface measure induced by the metric $\overline{g}$. The main issue is the convergence as $t \to \infty$ of the integrals in the formula above, for $R$ fixed. For this we need to be careful with both the choice of surfaces of integration $S^u_{R,t}$ and the choice of the vector-field $W$.

Our main theorem in this section is the following:

**Theorem 7.23.** *Let*

$$W_j := \overline{g}^{ab}(\partial_a h_{jb} - \partial_j h_{ab}). \tag{7.3.44}$$

*Then the limit in (7.3.43) exists, and $E_{Bondi} : \mathbb{R} \to \mathbb{R}$ is a well-defined continuous and increasing function. Moreover,*

$$\begin{aligned} \lim_{R\to-\infty} E_{Bondi}(R) &= E_{KG} := \frac{1}{16\pi} \|V^\psi_\infty\|^2_{L^2}, \\ \lim_{R\to\infty} E_{Bondi}(R) &= E_{ADM}. \end{aligned} \tag{7.3.45}$$

*Proof.* **Step 1.** We decompose $W = W^1 + W^{\geq 2}$,

$$W^1_j := \delta^{ab}(\partial_a h_{jb} - \partial_j h_{ab}), \qquad W^{\geq 2}_j := \overline{g}^{ab}_{\geq 1}(\partial_a h_{jb} - \partial_j h_{ab}). \tag{7.3.46}$$

To calculate the linear contribution of $W^1_j$ we use the divergence theorem

$$\int_{S^u_{R,t}} W^1_j \mathbf{n}^j \, d\sigma = \int_{B^u_{R,t}} \overline{D}^j W^1_j \, d\mu, \tag{7.3.47}$$

where $B^u_{R,t} := \{x \in \Sigma_t : u(x,t) \leq R\}$ is the ball of radius $R$. Then we calculate

$$\overline{D}^j W^1_j = \overline{g}^{jk}\{\partial_k W^1_j - \overline{\Gamma}^m{}_{jk} W^1_m\} = \delta^{jk}\partial_k W^1_j + \overline{g}^{jk}_1 \partial_k W^1_j - \delta^{jk}\overline{\Gamma}^m{}_{jk} W^1_m + E_1, \tag{7.3.48}$$

where $E_1$ is a cubic and higher order term and $\overline{g}^{jk}_1 = g^{jk}_1 = -h_{jk}$ is the linear part of the metric $\overline{g}^{jk}$. As in the proof of Lemma 7.10, the cubic term satisfies the bounds $\|E_1(t)\|_{L^1} \lesssim \varepsilon_0^2 t^{-\kappa}$. The point of the identity (7.3.48) is that the linear part $\delta^{jk}\partial_k W^1_j = -2\Delta\tau$ satisfies the equations (7.2.32). We can therefore apply the results of Lemma 7.10, and write

$$\begin{aligned} \overline{D}^j W^1_j &\sim -\delta_{jk} P^2_{jk} + \{(\partial_0\psi)^2 + \psi^2 + \partial_j\psi\partial_j\psi\} + \partial_j O^2_j \\ &\quad - h_{jk}\partial_j W^1_k - \delta^{jk}\overline{\Gamma}^m{}_{jk} W^1_m \\ &\sim -\delta_{jk} P^2_{jk} + \{(\partial_0\psi)^2 + \psi^2 + \partial_j\psi\partial_j\psi\} + \partial_j\{O^2_j - h_{jk} W^1_k\} \\ &\quad + \{\partial_j h_{jk} W^1_k - \overline{\Gamma}_{mjj} W^1_m\}. \end{aligned} \tag{7.3.49}$$

As in the previous section, $F \sim G$ means $\|F-G\|_{L^1} \lesssim \varepsilon_0^2 t^{-\kappa}$, and $O_j^2$ is defined in (7.2.31).

We show now that the last semilinear term in (7.3.49) is an acceptable $L^1$ error. Indeed, using (4.3.1) and the definitions,

$$\begin{aligned}\partial_j h_{jk} W_k^1 - \overline{\Gamma}_{mjj} W_m^1 &= \frac{1}{2}\partial_k h_{jj}(\partial_a h_{ka} - \partial_k h_{aa})\\ &= \frac{1}{2}\partial_k(2\tau - F + \underline{F})\cdot(\in_{klm}\partial_l\Omega_m - 2\partial_k\tau).\end{aligned}$$

Using now (7.2.48) and (4.3.5) with (7.2.49) we have

$$\partial_j h_{jk} W_k^1 - \overline{\Gamma}_{mjj} W_m^1 \sim 0. \tag{7.3.50}$$

Moreover, since $\widetilde{O}_j^2 := O_j^2 - h_{jk}W_k^1$ is quadratic,

$$\partial_j\{O_j^2 - h_{jk}W_k^1\} \sim \overline{g}^{kj}\overline{D}_k\widetilde{O}_j^2.$$

Using the divergence theorem, the formulas (7.3.47), (7.3.49), and the proof of Proposition 7.11, we have

$$\begin{aligned}\int_{S^u_{R,t}} W_j^1\mathbf{n}^j\,d\sigma = \int_{B^u_{R,t}} &\{[(\partial_0\psi)^2 + \psi^2 + \partial_j\psi\partial_j\psi](t)\\ &+\frac{1}{2}\sum_{m,n\in\{1,2,3\}} |\partial_0\vartheta_{mn}(t)|^2\}\,d\mu + \int_{S^u_{R,t}} \widetilde{O}_j^2\mathbf{n}^j\,d\sigma + O(\varepsilon_0^2\langle t\rangle^{-\kappa}).\end{aligned} \tag{7.3.51}$$

**Step 2.** We examine now the term $\widetilde{O}_j^2\mathbf{n}^j$ in the surface integral above. This is a quadratic term, thus generically bounded by $C\varepsilon_0\langle t\rangle^{-2+}$, so its integral does not vanish as $t\to\infty$. However, many of its pieces have additional structure (such as good vector-fields), and therefore satisfy slightly better estimates and do not contribute in the limit as $t\to\infty$.

More precisely, in view of (7.2.50), (4.3.4), and (4.3.5), if $|x|\in[t/8,8t]$ then

$$\begin{aligned}|R(|\nabla|\rho - \partial_0\underline{F})(x,t)| + |R(\partial_0\rho + |\nabla|\underline{F})(x,t)| + |R(|\nabla|\Omega_j - \partial_0\omega_j)(x,t)|\\ + |R(\partial_0\Omega_j + |\nabla|\omega_j)(x,t)| + |R\partial_\mu\tau(x,t)| \lesssim \varepsilon_0 t^{-5/4}\end{aligned} \tag{7.3.52}$$

for any $i\in\{1,2,3\}$, $\mu\in\{0,1,2,3\}$, and any compounded Riesz transform $R = R_1^{a_1}R_2^{a_2}R_3^{a_3}$, $a_1+a_2+a_3\le 6$. Moreover, with $\widehat{x}_j := x^j/|x|$, $j\in\{1,2,3\}$, we have

$$t^{-1}|(\mathbf{n}^j - \widehat{x}_j)(x,t)| + |(\partial_0+\partial_r)RG(x,t)| + |(\partial_j + \widehat{x}_j\partial_0)RG(x,t)| \lesssim \varepsilon_0 t^{-5/4}, \tag{7.3.53}$$

for any $j\in\{1,2,3\}$ and $G\in\{F,\underline{F},\rho,\omega_a,\Omega_a,\vartheta_{ab}\}$, as a consequence of (7.2.34),

(7.2.50), and Lemma 7.19.

We regroup now the terms in the formula (7.2.31), with respect to the non-differentiated metric component. Using (2.1.29) and (7.3.52)–(7.3.53) we write

$$\begin{aligned}\mathbf{n}^j\{h_{00}\partial_0 h_{0j} - h_{00}\partial_j(\tau + \underline{F})\} &\simeq h_{00}\widehat{x}_j(-R_j\partial_0\rho - \partial_j\underline{F} + \in_{jkl}\partial_0 R_k\omega_l)\\ &\simeq h_{00}\widehat{x}_j \in_{jkl} |\nabla| R_k\Omega_l \simeq 0,\end{aligned}$$

where in this section $f \simeq g$ means $|(f-g)(x,t)| \lesssim \varepsilon_0^2 t^{-2-\kappa}$ for all $(x,t)$ with $t \geq 1$ and $|x| \in [t/4, 4t]$. The other terms in $\mathbf{n}^j(O_j^2 - h_{jk}W_k^1)$ can be simplified in a similar way, using also (7.2.34),

$$\mathbf{n}^j\{-h_{0j}\partial_0 h_{00} + 2h_{0j}\partial_0(\tau + \underline{F})\} \simeq h_{0j}\widehat{x}_j\partial_0(-F + \underline{F}) \simeq h_{0j}\partial_j(F - \underline{F}),$$

$$\begin{aligned}&\mathbf{n}^j\{-h_{0n}\partial_0 h_{nj} + h_{0n}\partial_j h_{0n} - h_{0n}\partial_n h_{0j}\}\\ &\quad\simeq h_{0n}\{\partial_j h_{nj} + \widehat{x}_j \in_{nab}\partial_j R_a\omega_b - \widehat{x}_j \in_{jab}\partial_n R_a\omega_b\}\\ &\quad\simeq h_{0n}\{\partial_n(\underline{F} - F) + \in_{nab}\partial_a\Omega_b - \in_{nab}\partial_0 R_a\omega_b\} \simeq h_{0n}\partial_n(\underline{F} - F),\\ &\mathbf{n}^j\{h_{nj}\partial_0 h_{0n} - h_{nj}\partial_n(\tau + \underline{F}) - h_{nj}W_n^1\}\\ &\quad\simeq h_{nj}\{-\partial_j h_{0n} - \widehat{x}_j\partial_n\underline{F} + \widehat{x}_j(\partial_n h_{aa} - \partial_a h_{an})\}\\ &\quad\simeq h_{nj}\{\partial_j R_n\rho - \widehat{x}_j\partial_n\underline{F} - \in_{nab}\partial_j R_a\omega_b - \widehat{x}_j \in_{nab}\partial_a\Omega_b\}\\ &\quad\simeq h_{nj}\{-\widehat{x}_j R_n(\partial_0\rho + |\nabla|\underline{F}) + \widehat{x}_j \in_{nab} R_a(\partial_0\omega_b - |\nabla|\Omega_b)\} \simeq 0.\end{aligned}$$

Summing up these identities and using (2.1.8) we get

$$\mathbf{n}^j\widetilde{O}_j^2 \simeq \mathbf{n}^j h_{kn}(\partial_n h_{kj} - \partial_j h_{kn}) \simeq -\mathbf{n}^j\overline{g}_{\geq 1}^{kn}(\partial_n h_{kj} - \partial_j h_{kn}). \tag{7.3.54}$$

Therefore, $\mathbf{n}^j\widetilde{O}_j^2 + \mathbf{n}^j W_j^{\geq 2} \simeq 0$, and (7.3.51) gives

$$\begin{aligned}\int_{S^u_{R,t}} W_j\mathbf{n}^j\,d\sigma = \int_{B^u_{R,t}} &\{[(\partial_0\psi)^2 + \psi^2 + \partial_j\psi\partial_j\psi](t)\\ &+ (1/2)\sum_{m,n\in\{1,2,3\}} |\partial_0\vartheta_{mn}(t)|^2\}\,d\mu + O(\varepsilon_0^2\langle t\rangle^{-\kappa}).\end{aligned} \tag{7.3.55}$$

**Step 3.** We fix now $R \in \mathbb{R}$ and let $t \to \infty$. At this stage, for the limit to exist it is important that the almost optical function $u$ has the properties stated in Lemma 7.19.

Recall the scattering profiles $V_\infty^\psi$ and $V_\infty^{\vartheta_{ab}}$ defined in (7.3.17)–(7.3.18). We show first that, for any $R \in \mathbb{R}$,

$$\lim_{t\to\infty}\int_{B^u_{R,t}} [(\partial_0\psi)^2 + \psi^2 + \partial_j\psi\partial_j\psi](x,t)\,d\mu = \|V_\infty^\psi\|_{L^2}^2. \tag{7.3.56}$$

Indeed, we notice that

$$\int_{\mathbb{R}^3}[(\partial_0\psi)^2+\psi^2+\partial_j\psi\partial_j\psi](x,t)\,dx=\|U^\psi(t)\|_{L^2}^2=\|V_*^\psi(t)\|_{L^2}^2, \tag{7.3.57}$$

and $d\mu=dx(1+O(\varepsilon_0\langle t\rangle^{-1+2\delta'}))$. Therefore, for (7.3.56) it suffices to prove that

$$\lim_{t\to\infty}\int_{\mathbb{R}^3\setminus B^u_{R,t}}[(\partial_0\psi)^2+\psi^2+\partial_j\psi\partial_j\psi](x,t)\,dx=0. \tag{7.3.58}$$

Recalling that $U^\psi(t)=\partial_0\psi(t)-i\Lambda_{kg}\psi(t)$ and $|u(x,t)-|x|+t|\lesssim\varepsilon_0\langle x\rangle^{3\delta'}$ (see Lemma 7.19), for (7.3.58) it suffices to prove that, for operators $A\in\{I,\langle\nabla\rangle^{-1},\partial_j\langle\nabla\rangle^{-1}\}$ we have

$$\lim_{t\to\infty}\int_{|x|\geq t-t^{1/2}}|AU^\psi(x,t)|^2\,dx=0. \tag{7.3.59}$$

This follows as in the proof of Proposition 7.22. Indeed, with $k_0,J_0$ defined as the smallest integers for which $2^{k_0}\geq t^{\delta'}$ and $2^{J_0}\geq t^{1/3}$, we have

$$\big\|A(I-P_{[-k_0,k_0]})U^\psi(t)\big\|_{L^2}+\big\|AP_{[-k_0-2,k_0+2]}(\varphi_{\geq J_0+1}\cdot P_{[-k_0,k_0]}U^\psi)(t)\big\|_{L^2}\lesssim\varepsilon_0t^{-\delta},$$

due to (7.1.14). Moreover, the remaining component $AP_{[-k_0-2,k_0+2]}(\varphi_{\leq J_0}\cdot P_{[-k_0,k_0]}U^\psi)(t)$ is rapidly decreasing in the region $\{|x|\geq t-t^{1/2}\}$, so the desired limit (7.3.58) follows.

**Step 4.** To calculate the contribution of the metric components $|\partial_0\vartheta_{ab}|^2$ we use Proposition 7.22. For $v\in\mathbb{R}$, $\theta\in\mathbb{S}^2$, and $t\geq 10$ we define

$$L_{ab}(v,\theta,t):=\Re\Big\{\frac{-i}{4\pi^2}\int_0^\infty e^{i\rho v}\varphi_{[-k_0,k_0]}(\rho)\widehat{V_*^{\vartheta_{ab}}}(\rho\theta,t)\rho\,d\rho\Big\}. \tag{7.3.60}$$

We show first that, for any $R\in\mathbb{R}$, $a,b\in\{1,2,3\}$, and $t\geq(2+|R|)^{10}$ we have

$$\int_{B^u_{R,t}}|\partial_0\vartheta_{ab}(x,t)|^2\,d\mu=\int_{[-t^{0.4}/8,R]\times\mathbb{S}^2}|L_{ab}(v,\theta,t)|^2\,dvd\theta+O(\varepsilon_0^2t^{-\delta}). \tag{7.3.61}$$

Indeed, we can first replace the measure $d\mu$ by $dx$, at the expense of an acceptable error. Then, using (7.3.25), (7.3.30), and (7.3.31), we may assume that the integration is over the domain $D^u_{[-t^{0.4}/8,R],t}$ where $D^u_{[A,B],t}:=\{x\in\Sigma_t:u(x,t)\in[A,B]\}$, since the integration in the interior region $\{|x|\leq t-t^{0.4}/20\}$ of $|\partial_0\vartheta_{ab}(t)|^2$ produces an acceptable error. In addition, we may replace $\partial_0\vartheta_{ab}(x,t)$ with $|x|^{-1}L_{ab}(u(x,t),x/|x|,t)$ at the expense of acceptable errors, due to the first identity in (7.3.21). To summarize,

$$\int_{B^u_{R,t}} |\partial_0\vartheta_{ab}(x,t)|^2\,d\mu = \int_{D^u_{[-t^{0.4}/8,R],t}} |x|^{-2}|L_{ab}(u(x,t),x/|x|,t)|^2\,dx + O(\varepsilon_0^2 t^{-\delta}). \tag{7.3.62}$$

We pass now to polar coordinates $x = r\theta$, and then make the change of variables $r \to v := u(r\theta,t)$. The approximate identity (7.3.61) follows using also (7.3.2).

We apply now (7.3.18) to conclude that

$$\lim_{t\to\infty}\int_{B^u_{R,t}} |\partial_0\vartheta_{ab}(x,t)|^2\,d\mu = \int_{(-\infty,R]\times\mathbb{S}^2} |L^\infty_{ab}(v,\theta)|^2\,dvd\theta, \tag{7.3.63}$$

where $L^\infty_{ab}(v,\theta) = \lim_{t\to\infty} L_{ab}(v,\theta,t)$ in $L^2(\mathbb{R}\times\mathbb{S}^2)$ is given by

$$L^\infty_{ab}(v,\theta) := \Re\Big\{\frac{-i}{4\pi^2}\int_0^\infty e^{i\rho v}\widehat{V^{\vartheta_{ab}}_\infty}(\rho\theta)\rho\,d\rho\Big\}. \tag{7.3.64}$$

Combining this with (7.3.55) and (7.3.56), we have

$$\lim_{t\to\infty}\int_{S^u_{R,t}} W_j\mathbf{n}^j\,d\sigma = \|V^\psi_\infty\|^2_{L^2} + \frac{1}{2}\sum_{m,n\in\{1,2,3\}}\int_{(-\infty,R]\times\mathbb{S}^2} |L^\infty_{mn}(v,\theta)|^2\,dvd\theta.$$

Recalling the definitions (7.3.43) we have

$$E_{Bondi}(R) = \frac{1}{16\pi}\|V^\psi_\infty\|^2_{L^2} + \frac{1}{32\pi}\sum_{m,n\in\{1,2,3\}}\int_{(-\infty,R]\times\mathbb{S}^2} |L^\infty_{mn}(v,\theta)|^2\,dvd\theta, \tag{7.3.65}$$

which is clearly well defined, continuous, and increasing on $\mathbb{R}$. The limit as $R \to -\infty$ in (7.3.45) follows since $L^\infty_{mn} \in L^2(\mathbb{R}\times\mathbb{S}^2)$. To prove the limit as $R \to \infty$ we use (7.2.45) and let $t \to \infty$. Clearly $\lim_{t\to\infty}\|U^\psi(t)\|^2_{L^2} = \|V^\psi_\infty\|^2_{L^2}$ and, using (7.2.48),

$$\lim_{t\to\infty}\big\{\|U^{\vartheta_{mn}}(t)\|^2_{L^2} - 2\|\partial_0\vartheta_{mn}(t)\|^2_{L^2}\big\} = 0.$$

The desired limit as $R \to \infty$ in (7.3.45) follows using also (7.3.61). □

### 7.3.3 The interior energy

We see that the total Klein-Gordon energy $E_{KG}$ defined in (7.3.45) is part of the null Bondi energy $E_{Bondi}(R)$, for all $R \in \mathbb{R}$. This is consistent with the geometric intuition, because the matter travels at speeds lower than the speed of light, and accumulates at the future timelike infinity, not at null infinity. We show now that this Klein-Gordon energy can be further radiated by taking limits along suitable timelike cones.

**Proposition 7.24.** *For $\alpha \in (0,1)$ let*

$$E_{i^+}(\alpha) := \frac{1}{16\pi} \lim_{t\to\infty} \int_{S_{\alpha t,t}} (\partial_j h_{nj} - \partial_n h_{jj}) \frac{x^n}{|x|}\, dx, \tag{7.3.66}$$

*where the integration is over the Euclidean spheres $S_{\alpha t,t} \subseteq \Sigma_t$ of radius $\alpha t$. Then the limit in (7.3.66) exists, and $E_{i^+} : (0,1) \to \mathbb{R}$ is a well-defined continuous and increasing function. Moreover, we have*

$$\lim_{\alpha\to 0} E_{i^+}(\alpha) = 0, \qquad \lim_{\alpha\to 1} E_{i^+}(\alpha) = E_{KG}. \tag{7.3.67}$$

*Proof.* We notice that the definition (7.3.66) is similar to the definition of the ADM energy in (7.2.27). We could also use a more "geometric" definition involving the vector-field $W$ (see (7.3.44)), but this would make no difference here as $t \to \infty$, because the integrand $|\partial_j h_{nj} - \partial_n h_{jj}|$ is already bounded by $C_\alpha \langle t\rangle^{-2+4\delta'}$.

**Step 1.** We may assume that $t$ is large, say $t \geq \alpha^{-10} + (1-\alpha)^{-10}$. Using Stokes theorem and the definitions (4.3.2), we can rewrite

$$\int_{S_{\alpha t,t}} (\partial_j h_{nj} - \partial_n h_{jj}) \frac{x^n}{|x|}\, dx = \int_{|x|\leq \alpha t} -2\Delta\tau(x,t)\, dx. \tag{7.3.68}$$

The density function $-2\Delta\tau$ was analyzed in Lemma 7.10. The contributions of the error terms $O^1$ and $\partial_j O_j^2$ decay as $t \to \infty$, and the contribution of the metric components $-\delta_{jk} P_{jk}^2$ also decays because $|\partial_\beta h_{\mu\nu}(x,t)| \lesssim_\alpha \varepsilon_0 \langle t\rangle^{-2+2\delta'}$ in the ball $\{|x| \leq \alpha t\}$ (due to (7.1.21)). Thus

$$\int_{|x|\leq \alpha t} -2\Delta\tau(x,t)\, dx = \int_{|x|\leq\alpha t} \{(\partial_0\psi)^2 + \psi^2 + \partial_j\psi\partial_j\psi\}(x,t)\, dx + O_\alpha(\varepsilon_0^2 t^{-\kappa}). \tag{7.3.69}$$

**Step 2.** To apply Proposition 7.22 we would like to show now that

$$\int_{|x|\leq\alpha t} \{(\partial_0\psi)^2 + \psi^2 + \partial_j\psi\partial_j\psi\}(x,t)\, dx = \int_{|x|\leq\alpha t} |U^\psi(x,t)|^2\, dx + O_\alpha(\varepsilon_0^2 t^{-\kappa}). \tag{7.3.70}$$

Indeed, the real part of $U^\psi$ is $\partial_0\psi$, as needed. The imaginary part of $U^\psi$ is $-\langle\nabla\rangle\psi$, and some care is needed because the functions $\psi$ and $\partial_j\psi$ are connected to $\langle\nabla\rangle\psi$ by nonlocal operators.

In view of (7.1.22), we may replace the integral over the ball $\{|x| \leq \alpha t\}$ with a suitably smooth version, using the function $\chi_1(t^{-0.9}(|x| - \alpha t))$, where $\chi_1 : \mathbb{R} \to [0,1]$ is a smooth function supported in $(-\infty, 2]$ and equal to 1 in $(-\infty, 1]$. For (7.3.70) it suffices to prove that

$$\int_{\mathbb{R}^3} \chi_1(t^{-0.9}(|x| - \alpha t))\{\psi^2 + \partial_j\psi\partial_j\psi - (\langle\nabla\rangle\psi)^2\}(x,t)\, dx = O_\alpha(\varepsilon_0^2 t^{-\kappa}). \tag{7.3.71}$$

To prove this we notice that

$$\int_{\mathbb{R}^3} \chi_1(t^{-0.9}(|x|-\alpha t))G(x)\overline{G(x)}\,dx = C\int_{\mathbb{R}^3\times\mathbb{R}^3} \widehat{G}(\xi)\overline{\widehat{G}(\eta)}K_{1,t}(\xi-\eta)\,d\xi d\eta, \tag{7.3.72}$$

where

$$K_{1,t}(\rho) := \int_{\mathbb{R}^3} \chi_1(t^{-0.9}(|x|-\alpha t))e^{ix\cdot\rho}\,dx. \tag{7.3.73}$$

We apply the identity (7.3.72) for $G \in \{\langle\nabla\rangle\psi, \psi, \partial_j\psi\}$. For (7.3.71) it suffices to prove that

$$\Big|\int_{\mathbb{R}^3\times\mathbb{R}^3} \widehat{\psi}(\xi,t)\overline{\widehat{\psi}(\eta,t)}\big(\langle\xi\rangle\langle\eta\rangle - 1 - \xi_j\eta_j\big)K_{1,t}(\xi-\eta)\,d\xi d\eta\Big| \lesssim_\alpha \varepsilon_0^2 t^{-\kappa}. \tag{7.3.74}$$

Using integration by parts it is easy to see that $|K_{1,t}(\rho)| \lesssim_\alpha t^3(1+t^{0.9}|\rho|)^{-10}$ (the rapid decay here is the main reason for replacing the characteristic function of the ball $\{|x| \le \alpha t\}$ by the smooth approximation $\chi_1(t^{-0.9}(|x|-\alpha t))$). Moreover, $\big|\langle\xi\rangle\langle\eta\rangle - 1 - \xi_j\eta_j\big| \lesssim \langle\xi\rangle\langle\eta\rangle|\xi-\eta|$, thus

$$\big|\langle\xi\rangle\langle\eta\rangle - 1 - \xi_j\eta_j\big|\,|K_{1,t}(\xi-\eta)| \lesssim_\alpha \langle\xi\rangle\langle\eta\rangle t^{2.1}(1+t^{0.9}|\xi-\eta|)^{-9}.$$

Therefore, the left-hand side of (7.3.74) is bounded by

$$C_\alpha\|\widehat{\psi}(\xi,t)\langle\xi\rangle\|_{L^2}^2 t^{-0.5} \lesssim_\alpha \varepsilon_0^2 t^{-0.5},$$

as desired. This completes the proof of (7.3.70).

**Step 3.** We use now (7.3.21), pass to polar coordinates, and change variables to calculate

$$\begin{aligned}
&\int_{|x|\le\alpha t} |U^\psi(x,t)|^2\,dx \\
&= \frac{1}{8\pi^3}\int_{[0,\alpha t]\times\mathbb{S}^2} \frac{t^2r^2}{(t^2-r^2)^{5/2}}\Big|\widehat{P_{[-k_0,k_0]}V_*^\psi}\Big(\frac{r\theta}{\sqrt{t^2-r^2}},t\Big)\Big|^2\,drd\theta + O_\alpha(\varepsilon_0^2t^{-\delta}) \\
&= \frac{1}{8\pi^3}\int_{[0,\alpha/\sqrt{1-\alpha^2}]\times\mathbb{S}^2} \rho^2\big|\widehat{P_{[-k_0,k_0]}V_*^\psi}(\rho\theta,t)\big|^2\,d\rho d\theta + O_\alpha(\varepsilon_0^2t^{-\delta}).
\end{aligned}$$

Therefore, using also (7.3.68)–(7.3.70) and letting $t\to\infty$, we have

$$E_{i^+}(\alpha) = \frac{1}{16\pi}\frac{1}{8\pi^3}\int_{[0,\alpha/\sqrt{1-\alpha^2}]\times\mathbb{S}^2} \rho^2\big|\widehat{V_\infty^\psi}(\rho\theta)\big|^2\,d\rho d\theta, \tag{7.3.75}$$

which is clearly a well-defined continuous and increasing function of $\alpha$ satisfying (7.3.67) (recall that $E_{KG} = 1/(16\pi)\|V_\infty^\psi\|_{L^2}^2 = 1/(16\pi)(8\pi^3)^{-1}\|\widehat{V_\infty^\psi}\|_{L^2}^2$). $\square$

# Bibliography


[1] S. Alinhac, The null condition for quasilinear wave equations in two space dimensions I. Invent. Math. **145** (2001), 597–618.

[2] S. Alinhac, The null condition for quasilinear wave equations in two space dimensions. II. Amer. J. Math. **123** (2001), 1071–1101.

[3] R. Arnowitt, S. Deser, and C. Misner, Republication of: The dynamics of general relativity. Gen. Relativ. Gravit. **40** (2008), 1997–2027.

[4] R. Bartnik, The mass of an asymptotically flat manifold. Comm. Pure Appl. Math. **39** (1986), 661–693.

[5] R. Bartnik, P. T. Chrusciel, and N. O Murchadha, On maximal surfaces in asymptotically flat space-times. Commun. Math. Phys. **130** (1990), 95–109.

[6] L. Bieri and N. Zipser, Extensions of the stability theorem of the Minkowski space in General Relativity. AMS/IP Studies in Advanced Mathematics, 45. American Mathematical Society, Providence, RI; International Press, Cambridge, MA, 2009. xxiv+491 pp.

[7] L. Bigorgne, D. Fajman, J. Joudioux, J. Smulevici, and M. Thaller, Asymptotic stability of Minkowski space-time with non-compactly supported massless Vlasov matter. Preprint (2020), arXiv:2003.03346.

[8] P. Bizon and A. Wasserman, On existence of mini-boson stars. Commun. Math. Phys. **215** (2000), 357–373.

[9] A. Carlotto and R. Schoen, Localizing solutions of the Einstein constraint equations. Invent. Math. **205** (2016), 559–615.

[10] O. Chodosh and Y. Shlapentokh-Rothman, Time-periodic Einstein-Klein-Gordon bifurcations of Kerr. Comm. Math. Phys. **356** (2017), 1155–1250.

[11] D. Christodoulou, Global solutions of nonlinear hyperbolic equations for small initial data. Comm. Pure Appl. Math. **39** (1986), 267–282.

[12] D. Christodoulou and S. Klainerman, The global nonlinear stability of the Minkowski space. Princeton Mathematical Series, 41. Princeton University Press, Princeton, NJ, 1993. x+514 pp.

[13] M. Dafermos, I. Rodnianski, and Y. Shlapentokh-Rothman, Decay for solutions of the wave equation on Kerr exterior spacetimes III: The full subextremal case $|a| < M$. Ann. of Math. (2) **183** (2016), 787–913.

[14] J.-M. Delort, Existence globale et comportement asymptotique pour l'équation de Klein-Gordon quasi-linéaire à données petites en dimension 1. Ann. Sci. École Norm. Sup. **34** (2001), 1–61.

[15] J.-M. Delort, D. Fang, and R. Xue, Global existence of small solutions for quadratic quasilinear Klein-Gordon systems in two space dimensions. J. Funct. Anal. **211** (2004), 288–323.

[16] Y. Deng, A. D. Ionescu, and B. Pausader, The Euler-Maxwell system for electrons: global solutions in 2D. Arch. Ration. Mech. Anal. **225** (2017), 771–871.

[17] Y. Deng, A. D. Ionescu, B. Pausader, and F. Pusateri, Global solutions of the gravity-capillary water-wave system in three dimensions. Acta Math. **219** (2017), 213–402.

[18] Y. Deng and F. Pusateri, On the global behavior of weak null quasilinear wave equations. Preprint (2018), arXiv:1804.05107.

[19] D. Fajman, J. Joudioux, and J. Smulevici, The stability of the Minkowski space for the Einstein-Vlasov system. Preprint (2017), arXiv:1707.06141.

[20] R. Friedberg, T. D. Lee, and Y. Pang, Mini-soliton stars. Phys. Rev. D **35** (1987), 3640–3657.

[21] H. Friedrich, On the existence of n-geodesically complete or future complete solutions of Einstein's field equations with smooth asymptotic structure. Commun. Math. Phys. **107** (1986), 587–609.

[22] V. Georgiev, Global solution of the system of wave and Klein-Gordon equations. Math. Z. **203** (1990), 683–698.

[23] P. Germain and N. Masmoudi, Global existence for the Euler-Maxwell system. Ann. Sci. Éc. Norm. Supér. **47** (2014), 469–503.

[24] P. Germain, N. Masmoudi, and J. Shatah, Global solutions for 3D quadratic Schrödinger equations. Int. Math. Res. Not. (2009), 414–432.

[25] P. Germain, N. Masmoudi, and J. Shatah, Global solutions for the gravity water waves equation in dimension 3. Ann. of Math. (2) **175** (2012), 691–754.

[26] E. Giorgi, S. Klainerman, and J. Szeftel, A general formalism for the stability of Kerr. Preprint (2020), arXiv:2002.02740.

[27] Y. Guo, A. D. Ionescu, and B. Pausader, Global solutions of the Euler-Maxwell two-fluid system in 3D. Ann. of Math. (2) **183** (2016), 377–498.

[28] Y. Guo and B. Pausader, Global smooth ion dynamics in the Euler-Poisson system. Comm. Math. Phys. **303** (2011), 89–125.

[29] S. Gustafson, K. Nakanishi, and T.-P. Tsai, Scattering theory for the Gross-Pitaevskii equation in three dimensions. Commun. Contemp. Math. **11** (2009), 657–707.

[30] D. Häfner, P. Hintz, and A. Vasy, Linear stability of slowly rotating Kerr black holes. Invent. Math. **223** (2021), no. 3, 1227–1406.

[31] C. A. R. Herdeiro and E. Radu, Kerr black holes with scalar hair. Phys. Rev. Lett. **112** (2014) 221101.

[32] P. Hintz and A. Vasy, The global non-linear stability of the Kerr–de Sitter family of black holes. Acta Math. **220** (2018), 1–206.

[33] P. Hintz and A. Vasy, Stability of Minkowski space and polyhomogeneity of the metric. Ann. PDE **6** (2020), no. 1, Paper No. 2, 146 pp.

[34] M. Ifrim and A. Stingo, Almost global well-posedness for quasilinear strongly coupled wave-Klein-Gordon systems in two space dimensions. Preprint (2019), arXiv:1910.12673.

[35] A. D. Ionescu and V. Lie, Long term regularity of the one-fluid Euler-Maxwell system in 3D with vorticity. Adv. Math. **325** (2018), 719–769.

[36] A. D. Ionescu and B. Pausader, The Euler-Poisson system in 2D: global stability of the constant equilibrium solution. Int. Math. Res. Not. (2013), 761–826.

[37] A. D. Ionescu and B. Pausader, Global solutions of quasilinear systems of Klein-Gordon equations in 3D. J. Eur. Math. Soc. (JEMS) **16** (2014), 2355–2431.

[38] A. D. Ionescu and B. Pausader, On the global regularity for a Wave-Klein-Gordon coupled system. Acta Math. Sin. (Engl. Ser.) **35** (Special Issue in honor of Carlos Kenig on his 65th birthday) (2019), 933–986.

[39] A. D. Ionescu and F. Pusateri, Nonlinear fractional Schrödinger equations in one dimension. J. Funct. Anal. **266** (2014) 139–176.

[40] A. D. Ionescu and F. Pusateri, Global solutions for the gravity water waves system in 2d. Invent. Math. **199** (2015), 653–804.

[41] A. D. Ionescu and F. Pusateri, Global regularity for 2D water waves with surface tension. Mem. Amer. Math. Soc. **256** (2018), no. 1227, v+124 pp.

[42] F. John, Blow-up of solutions of nonlinear wave equations in three space dimensions. Manuscripta Math. **28** (1979), 235–268.

[43] F. John and S. Klainerman, Almost global existence to nonlinear wave equations in three space dimensions. Comm. Pure Appl. Math. **37** (1984), 443–455.

[44] S. Katayama, Global existence for coupled systems of nonlinear wave and Klein-Gordon equations in three space dimensions. Math. Z. **270** (2012), 487–513.

[45] T. Kato, The Cauchy problem for quasi-linear symmetric hyperbolic systems. Arch. Rational Mech. Anal. **58** (1975), 181–205.

[46] J. Kato and F. Pusateri, A new proof of long range scattering for critical nonlinear Schrödinger equations. Differential and Integral Equations **24** (2011), 923–940.

[47] D. J. Kaup, Klein-Gordon geon. Phys. Rev. **172** (1968), 1331–1342.

[48] S. Klainerman, Long time behaviour of solutions to nonlinear wave equations. Proceedings of the International Congress of Mathematicians, Vol. 1, 2 (Warsaw, 1983), 1209–1215, PWN, Warsaw, (1984).

[49] S. Klainerman, Uniform decay estimates and the Lorentz invariance of the classical wave equation. Comm. Pure Appl. Math. **38** (1985), 321–332.

[50] S. Klainerman, Global existence of small amplitude solutions to nonlinear Klein-Gordon equations in four space-time dimensions. Comm. Pure Appl. Math. **38** (1985), 631–641.

[51] S. Klainerman, The null condition and global existence to nonlinear wave equations. Nonlinear systems of partial differential equations in applied mathematics, Part 1 (Santa Fe, N.M., 1984), 293–326, Lectures in Appl. Math., 23, Amer. Math. Soc., Providence, RI, 1986.

[52] S. Klainerman and F. Nicolò, The evolution problem in general relativity. Progress in Mathematical Physics, 25. Birkhäuser Boston, Inc., Boston, MA, 2003. xiv+385 pp.

[53] S. Klainerman and F. Nicolò, Peeling properties of asymptotically flat solutions to the Einstein vacuum equations. Class. Quantum Grav. **20** (2003), 3215–3257.

[54] S. Klainerman and J. Szeftel, Global nonlinear stability of Schwarzschild spacetime under polarized perturbations. Preprint (2018), arXiv:1711.07597.

[55] S. Klainerman, Q. Wang, and S. Yang, Global solution for massive Maxwell-

Klein-Gordon equations. Preprint (2018), arXiv:1801.10380.

[56] P. G. LeFloch and Y. Ma, The hyperboloidal foliation method. Series in Applied and Computational Mathematics, 2. World Scientific Publishing Co. Pte. Ltd., Hackensack, NJ, 2014. x+149 pp.

[57] P. G. LeFloch and Y. Ma, The global nonlinear stability of Minkowski space for self-gravitating massive fields. Comm. Math. Phys. **346** (2016), 603–665.

[58] P. G. LeFloch and Y. Ma, The global nonlinear stability of Minkowski space for self-gravitating massive fields. Series in Applied and Computational Mathematics, 3. World Scientific Publishing Co. Pte. Ltd., Hackensack, NJ, 2018. xi+174 pp.

[59] P. G. LeFloch and Y. Ma, The global nonlinear stability of Minkowski space. Einstein equations, f(R)-modified gravity, and Klein-Gordon fields. Preprint (2017), arXiv:1712.10045.

[60] S. L. Liebling and C. Palenzuela, Dynamical boson stars. Living Rev. Relativ. **15** (2012), Article 6.

[61] H. Lindblad, On the asymptotic behavior of solutions to the Einstein vaccum equations in wave coordinates. Commun. Math. Phys. **353** (2017), 135–184.

[62] H. Lindblad and I. Rodnianski, Global existence for the Einstein vacuum equations in wave coordinates. Commun. Math. Phys. **256** (2005), 43–110.

[63] H. Lindblad and I. Rodnianski, The global stability of Minkowski spacetime in harmonic gauge. Ann. of Math. (2) **171** (2010), 1401–1477.

[64] H. Lindblad and M. Taylor, Global stability of Minkowski space for the Einstein-Vlasov system in the harmonic gauge. Preprint (2017), arXiv:1707.06079.

[65] C. W. Misner, K. S. Thorne, and J. A. Wheeler, Gravitation, W. H. Freeman and Co., San Francisco, Calif., 1973. ii+xxvi+1279+iipp.

[66] H. Okawa, V. Cardoso, and P. Pani, Collapse of self-interacting fields in asymptotically flat spacetimes: do self-interactions render Minkowski spacetime unstable? Phys. Rev. D **89** (2014), no. 4, 041502,

[67] E. Poisson, A Relativist's Toolkit: The Mathematics of Black-hole Mechanics. Cambridge University Press, Cambridge, 2004. xvi+233 pp.

[68] R. Schoen and S. T. Yau, On the proof of the positive mass conjecture in general relativity. Comm. Math. Phys. **65** (1979), no. 1, 45–76.

[69] J. Shatah, Normal forms and quadratic nonlinear Klein-Gordon equations. Comm. Pure Appl. Math. **38** (1985), 685–696.

[70] Y. Shlapentokh-Rothman, Exponentially growing finite energy solutions for the Klein-Gordon equation on sub-extremal Kerr spacetimes. Comm. Math. Phys. **329** (2014), 859–891.

[71] J. Simon, A wave operator for a nonlinear Klein–Gordon equation. Lett. Math. Phys. **7** (1983), 387–398.

[72] J. Speck, The global stability of the Minkowski spacetime solution to the Einstein-nonlinear system in wave coordinates. Anal. PDE **7** (2014), 771–901.

[73] R. Wald, General relativity. University of Chicago Press, Chicago, IL, 1984. xiii+491 pp.

[74] Q. Wang, An intrinsic hyperboloid approach for Einstein Klein-Gordon equations. Preprint (2016), arXiv:1607.01466.

# Index

Lightning Source UK Ltd.
Milton Keynes UK
UKHW021233101022
410236UK00008B/1224

9 780691 233048